Polymer Science Overview

Volume I

Polymer Science Overview
Volume I

Edited by **Stacy Spoon**

NY RESEARCH PRESS
New York

Published by NY Research Press,
23 West, 55th Street, Suite 816,
New York, NY 10019, USA
www.nyresearchpress.com

Polymer Science Overview: Volume I
Edited by Stacy Spoon

International Standard Book Number: 978-1-63238-362-4 (Hardback)

Printed in China.

Contents

Preface

It was Jöns Jakob Berzelius in 1833, who coined the term 'polymer'. Also known as macromolecular science, polymer science is a sub-area of materials science. Basically, this field aims to target concepts related to synthetic polymers like plastic and elastomers. Also known as macromolecular science, the origins of the field of polymer science can be traced to the rubber industries in pre-columbian Mexico. However, the first modern example of polymer science was Henri Braconnot's experiment in the 1830s, when he produced derivatives and new synthetic materials such as celluloid and cellulose acetate. It was only in 1922, with the work of Hermann Staudinger that the molecular nature of polymers was understood. By World War 2, a strong commercial polymer industry had emerged.

Polymer Science is a multidisciplinary field, in which researchers refer to disciplines of chemistry, physics, and engineering. Mainly, this area of study concerns three specific sub-disciplines which are:

1. Polymer Chemistry: Also known as macromolecular chemistry, this subject studies the chemical synthesis, chemical characteristics, and chemical compositions of polymers
2. Polymer Physics or macromolecular physics is all about the study of bulk properties of polymers and polymer related engineering applications
3. Polymer Characterization is a discipline, which covers the analysis of structures chemically evolved or developed and morphological studies.

Polymer Science also focuses on physical properties with respect to composition and structural aspects of polymers.

This book is a combined effort of authors from all over the world, who're experts in the field of polymer science. Without their hard work, support and timely submissions, this book would not have been possible. I also wish to extend my gratefulness to the team at the publishing house, for their constant support at each and every step. Lastly, I wish to thank my family for all the support they provided for the completion of this project.

Editor

1

Curcumin-Loaded Chitosan/Gelatin Composite Sponge for Wound Healing Application

Van Cuong Nguyen,[1] Van Boi Nguyen,[1] and Ming-Fa Hsieh[2]

[1] *Department of Chemical Engineering, Industrial University of Ho Chi Minh City, 12 Nguyen Van Bao Street, Go Vap, Ho Chi Minh City 70000, Vietnam*
[2] *Department of Biomedical Engineering, Center for Nanotechnology and Institute of Biomedical Technology, Chung Yuan Christian University, 200 Chung Pei Road, Chungli, 32023 Taoyuan, Taiwan*

Correspondence should be addressed to Van Cuong Nguyen; nvc@hui.edu.vn

Academic Editor: Hai-Yin Yu

Three composite sponges were made with 10% of curcumin and by using polymers, namely, chitosan and gelatin with various ratios. The chemical structure and morphology were evaluated by FTIR and SEM. These sponges were evaluated for water absorption capacity, antibacterial activity, *in vitro* drug release, and *in vivo* wound healing studies by excision wound model using rabbits. The *in vivo* study presented a greater wound closure in wounds treated with curcumin-composite sponge than those with composite sponge without curcumin and untreated group. These obtained results showed that combination of curcumin, chitosan and gelatin could improve the wound healing activity in comparison to chitosan, and gelatin without curcumin.

1. Introduction

Medicines derived from plants play an important role in the healthcare of many cultures, both ancient and modern. Curcumin (diferuloylmethane, a polyphenol) is an active agent of the perennial herb *Curcuma longa* (commonly known as turmeric). Curcumin is a naturally occurring polyphenolic phytoconstituent which presents anticancer, antioxidant, anti-inflammatory, hyperlipidemic, antibacterial, wound healing, and hepatoprotective activities [1, 2]. The therapeutic efficacy of curcumin is limited due to its poor aqueous solubility and extensive first pass metabolism [3–5]. Topical formulation of curcumin (curcumin incorporated collagen matrix) was a feasible and productive approach to support dermal wound healing [1]. Therefore, development of novel curcumin formulation and delivery systems is required. The topical delivery of naturally occurring compounds, such as curcumin or catechins, is able to increase its solubility, stability, and pharmacological activities, resulting in improvement of therapeutic effects [6].

Additionally, chitosan, a polysaccharide biopolymer derived from naturally occurring chitin, displays unique polycationic, chelating, and film-forming properties due to the presence of active amino and hydroxyl functional groups. Chitosan also exhibits a number of interesting biological activities, including antimicrobial activity, induced disease resistance in plants, and diverse stimulating or inhibiting activities toward a number of human cell types [7, 8]. Moreover, chitosan can be used to prevent or treat wound and burn infections due to its intrinsic antimicrobial properties and its ability to deliver extrinsic antimicrobial agents to wounds and burns [9]. Additionally, it can accelerate the functions of inflammatory cells, macrophages, and fibroblasts [10, 11]. Chitosan and its derivatives are also used to increase the stability of the drug in which the drugs are loaded in chitosan film or chitosan nanoparticles, resulting in enhancement of drug accumulation [12] and toxicity to cancer cells [13, 14]. Gelatin is also a natural polymer derived from collagen of animal skin and bones. It is biocompatible, hydrophilic, and biodegradable under normal physiological conditions.

Gelatin is known to prevent fluid loss due to exudation, resulting in enhancement of its wound healing properties [15, 16]. Thus, the present study focused on the preparation and evaluation of composite sponge made with two natural polymers: chitosan and gelatin at different proportions. Moreover, curcumin was loaded into the chitosan and gelatin sponge for wound healing application.

2. Materials and Methods

2.1. Materials. Chitosan is extracted from crab shells which was prepared from our lab. The degree of deacetylation of the chitosan was approximately 85%. Curcumin was purchased from Vietnam Institute of Dietary Supplement (Hanoi, Vietnam) and acetic acid and gelatin were purchased from Sigma-Aldrich Chemical Co (USA). All chemicals used in this study were of reagent grade.

2.2. Preparation of Composite Sponge. Chitosan (1.0 g) was dissolved in 1% aqueous acetic acid (100 mL) to form a 1% chitosan solution (w/v) at room temperature. Gelatin (10%) solution was prepared by dissolving gelatin powder (10 g) in 100 mL of deionized water at room temperature. The composite sponges were prepared by dissolving chitosan solution into the gelatin solution with different portions (3 : 1, 1 : 1, and 1 : 3) of chitosan and gelatin, named as SP3, SP2 and SP1, respectively. The resulting solution was vigorously mixed under magnetic stirrer until the opaque aqueous solution was obtained. The solution was sonicated for 5 min for the removal of entrapped air bubbles. Then the air bubble-free solution was poured into separated petridishes and frozen at −40°C for 24 h followed by lyophilization to obtain composite sponge. For curcumin encapsulated in composite sponge, 1 mL of curcumin solution (0.5% w/v in absolute ethanol) was added to opaque aqueous solution. Then the solution was poured into separated petridishes, frozen at −40°C for 24 h, and followed by lyophilization to obtain curcumin composite sponge.

2.3. Characterization of Composite Sponge. The chemical structure and morphology of chitosan/gelatin composite sponges were evaluated using FT-IR (PerkinElmer) and SEM (JSM-5500).

2.4. Water Absorption Test. The water absorption capacity of the composite sponge was determined by incubating the composite sponge in phosphate-buffered saline (PBS) at pH 7.4 at 37°C for 2 h. The wet weight of the swollen sponge was measured immediately after gently blotting with filter paper to remove absorbed water on surface, followed by lyophilization and reweighing. The water content of the sponge was calculated as follows:

$$E_{\text{ad}} = \left[\frac{W_e - W_0}{W_0}\right] \times 100, \quad (1)$$

where E_{ad} is the percentage water adsorption of chitosan-gelatin sponge at equilibrium. W_e and W_0 represent the weight of the chitosan-gelatin sponge at equilibrium and initial weight of the chitosan-gelatin sponge, respectively.

2.5. Weight Loss of Prepared Sponges. The prepared sponges were immersed in PBS buffer solution at 37°C. After 24 h, the sponges were taken out, washed with deionized water, frozen, and lyophilized. The weights of sponges were measured after lyophilization. Weight loss of sponges was calculated through the following equation: $\%W = (W_0 - We)/W_0 \times 100$, where W_0 and W_e are the initial weight of sponge (g) and weight of sponge after lyophilization (g), respectively.

2.6. In Vitro Drug Release. The composite sponges (3 × 3 cm) were dispersed in 20 mL phosphate buffer solution, pH 7.4 at 37°C. All the supernatants were pipetted out periodically and replaced with equivalent volume of fresh phosphate buffer solution. The concentration of released drug was then determined by UV-Vis spectrometer at 420 nm using a standard curve of curcumin in ethanol. The percentage of curcumin released was determined as curcumin release (%) = curcumin released at time t/total curcumin loaded in sponge × 100.

2.7. Evaluation of Antimicrobial Activity. The measurements of the antimicrobial activities of individual chitosan and composite sponge were conducted by agar diffusion method using *Pseudomonas aeruginosa*. The zone of inhibition was determined by placing a definite size of sponge in inoculated solidified nutrient agar medium in a petriplate. Petriplates were incubated for 24 h at 37°C. This was done in triplicate with each film for each organism, and an average diameter of zone of inhibition was noted.

2.8. Cytotoxicity. The cytotoxicity of curcumin sponge was conducted on mouse fibroblast cells (L929) indirect MTT method [17]. The cells were seeded into 96-well microplates at a density of 10^3 cells/well at 37°C under humidified atmosphere containing 5% CO_2. Triplicates of sterilized sample were taken and then incubated in serum containing media for 24 h at 37°C. 100 μL of the media from each sample was taken and transferred into each well then incubated for another 24 h. After that, MTT solution was added to each well followed by 4 hours of incubation at 37°C. Subsequently, the medium was removed and violet crystals formed were solubilized with DMSO. After shaking slowly twice for 5 seconds, the absorbance of each well was determined using Eliza reader at 570 nm. The cell viability was expressed as a percentage of the control.

2.9. Evaluation of Wound Healing Activity. The wound healing activity was evaluated by excision wound model in albino rabbits (1.7 to 2.0 kg). All *in vivo* experiments were carried out in accordance with the ethical guidelines for Animal Care and Use Committee of Medical and Pharmaceutical University (Ho Chi Minh City). The animals were anesthetized by using lidocaine 25% prior to the test. The dorsal hair of the rabbit was removed. Full thickness wound of 2.5 × 2.5 cm^2 was excised from the back of the rabbit. The wound was covered with an equal size of drug-loaded sponge or blank sponge. The cotton gauge was used for control. The sponge was removed and replaced by fresh sponge every three days. The wound closure was observed at different periods of treatment using a digital camera. Wound area was calculated on the 0th, 3rd, 12nd, 15th and 21st days. On day 21st, histological examinations of skin wound tissue were carried out. The skin wound tissue of animal was excised, fixed with 10% formalin, and stained with hematoxylin-eosin (H&E) reagent for histological observations.

3. Results and Discussion

The composite sponges were prepared in different ratios of chitosan and gelatin (3 : 1, 1 : 1, and 1 : 3) (Figure 1). The results indicated that increasing of gelatin concentration leads to increase of folding endurance. The composite sponge SP1 showed the highest folding endurance among all composite sponges (data not shown). To examine the chemical structures of various sponges prepared under different feeding ratios, the prepared sponges were examined by FTIR spectra. The –CH– stretching vibration of the chitosan and gelatin is at 2936–2969 cm^{-1}, while peaks of chitosan at 1650 and 1560 cm^{-1} correspond to the amide carbonyl group (amide I) and the bending frequency of the amide N–H group (amide II), respectively (Figure 2). The FT-IR of composite sponge exhibited a mixture of characteristic absorptions because of amine groups of chitosan and the carboxylic acid groups of gelatin. The peaks of C = O of amide I for chitosan at 1650 cm^{-1} was shifted to 1669 cm^{-1} in composite sponges. The intensity of C = O peaks increased with the decline ratio of gelatin in composite sponge from 1 : 3 to 3 : 1 (chitosan/gelatin). These results can be attributed to the interaction of $-NH^{3+}$ of chitosan and the COO^- of gelatin in three composites [18, 19].

FIGURE 1: Prepared composite sponge with ratio of chitosan and gelatin (3 : 1) (a) curcumin-sponge and (b) blank sponge.

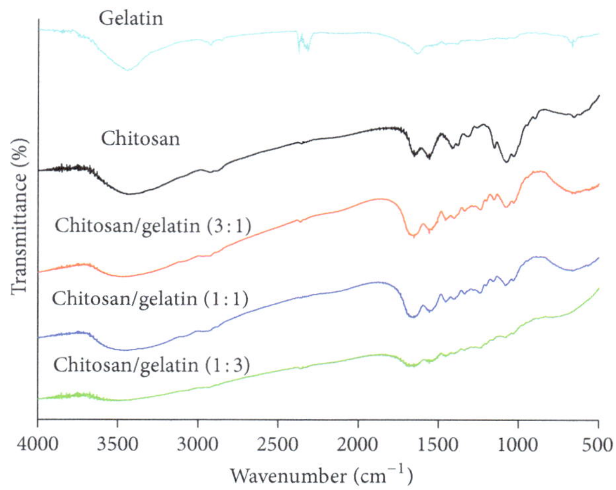

FIGURE 2: FT-IR spectra of prepared composite sponge with various ratios of chitosan and gelatin.

Figure 3 shows the cross-section morphology of chitosan/gelatin composite sponges. According to the figure, the sponges appeared as a very porous structure. Additionally, the roughness of chitosan/gelatin sponge surface was decreased with increasing of gelatin content (Figures 3(a)–3(c)). It was found that the large pores were observed in cross-section morphology of chitosan/gelatin sponges (Figures 3(d)–3(f)). The pore sizes were 43.06, 35.6, and 29.9 μm for SP3, SP2, and SP1, respectively. The sponges with low gelatin level represent regular pore size. On the other hand, the irregular pore size was observed in cross-section morphology of sponge with high gelatin content. The irregular pore size appearing in the sponge with high gelatin level may be deduced from the incompatibility between chitosan and gelatin molecules.

Water absorption capacity is an important parameter for biological applications and wound healing. It presents the capacity of sponge to absorb wound exudates. As the gelatin contents of prepared sponges were increased, the water adsorption capacities were significantly decreased. This is because lower water content led to smaller pore size during the preparation of the sponges. As a result, the water uptake was relatively lower. The water uptake capacity was 227%, 351%, and 2352% for ratio of chitosan/gelatin 1 : 3, 1 : 1, and 3 : 1, respectively. The high water content in prepared sponges was explained by the hydrophilic and swelling properties of gelatin [20]. The SP3 presented maximum water absorption capacity among all prepared sponges; the sponge can retain water in porous pore surrounding by cell walls. Additionally, it decreased about 10-fold for SP1 and SP2 which may attribute to the swelling and dissolving of gelatin that could decrease the pore size leading to less water absorption and therefore could not retain much water within their network structure. Furthermore, the water absorption capacity of drug-loaded sponge was not significantly different from blank sponge. The similar results were also reported in previous study [18]. These results were further confirmed by weight loss of sponges. Table 1 presents that the weight losses were 81.5, 66.5 and 16% for SP1, SP2, and SP3, respectively. The weight loss is reasonable as more gelatin can resist the dissolution in water. The water uptake may be related to the pore. The more gelatin, the smaller pores, and pore volume were observed.

The *in vitro* release profile of curcumin-sponge was conducted in phosphate buffer solution. The average percent release of curcumin was approximately 91%, 55%, and 37% for

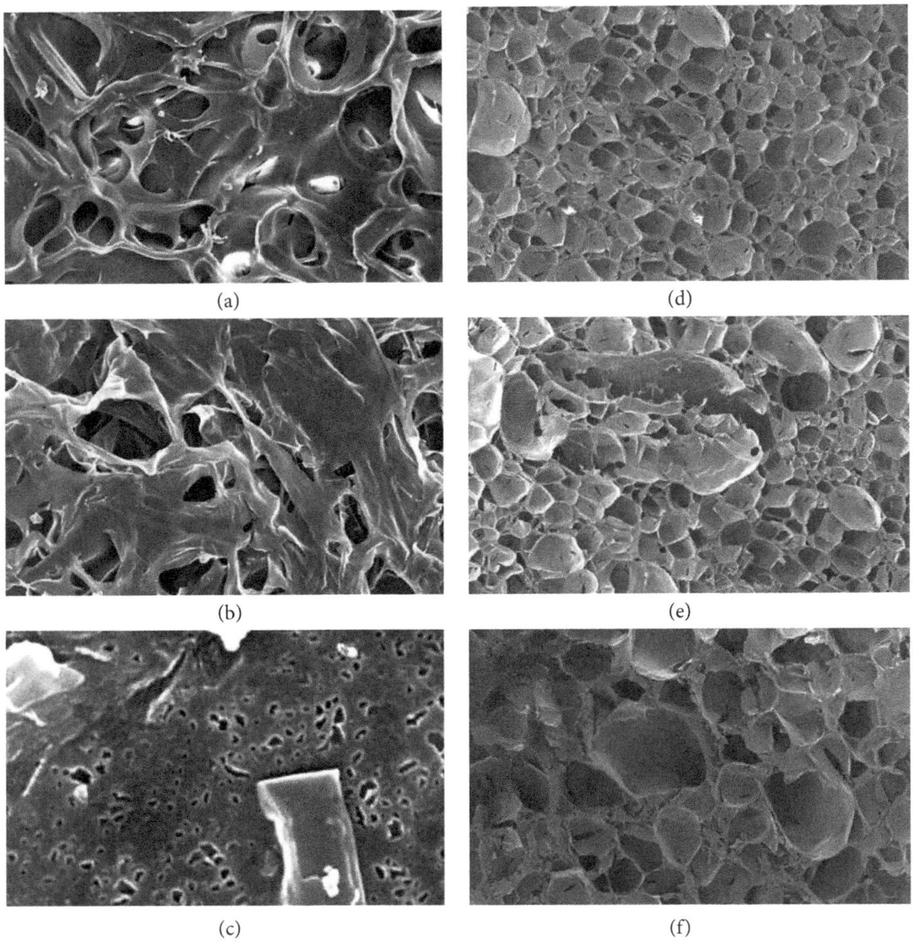

FIGURE 3: SEM images of surface morphology of chitosan/gelatin sponges (a) SP3, (b) SP2, and (c) SP1. Cross-section morphology of chitosan/gelatin sponges (d) SP3, (e) SP2, and (f) SP1 (scale bar at 100 μm).

TABLE 1: Composition and physical parameter of prepared sponges.

Sample no.	Feeding ratio		Average water uptake (%)	Weight loss (%)
	Chitosan	Gelatin		
SP1	1	3	227	81.5
SP2	1	1	351	66.5
SP3	3	1	2352	16

TABLE 2: The curcumin release profile of prepared sponges.

Time (h)	Drug release (%)		
	SP1	SP2	SP3
0	0.00	0.00	0.00
1	16.35	1.38	0.53
6	30.85	4.47	1.62
24	43.53	10.21	7.38
30	49.44	13.59	8.15
48	59.44	29.73	9.21
54	63.41	33.97	14.26
72	78.12	43.26	23.09
96	90.97	54.71	36.91

SP1, SP2, and SP3, respectively, in 96 h. The curcumin release increased with increase of gelatin content in prepared sponge (Table 2). This is because the hydrophilicity of curcumin was relatively lower. According to the partition ratio between sponge phase and water phase, the drug in the sponge having a higher gelatin content is preferred to be released, instead of being dissolved by the sponge.

From the results of antibacterial activity studies of prepared sponges against *P. aeruginosa*, it can be seen that the inhibition zone is larger in curcumin composite sponge than that of chitosan sponge without curcumin. It can be also seen that the SP3 sponge exhibited a higher inhibition zone than those of SP1 and SP2 sponges (Figure 4). The inhibition zones were 21 mm and 17.5 mm for curcumin-loaded SP3 and SP3 without curcumin, respectively.

The cytotoxicity of prepared sponges to L929 cells was evaluated using MTT assay method. The results demonstrate the cell viability after 24 h incubation with medium released

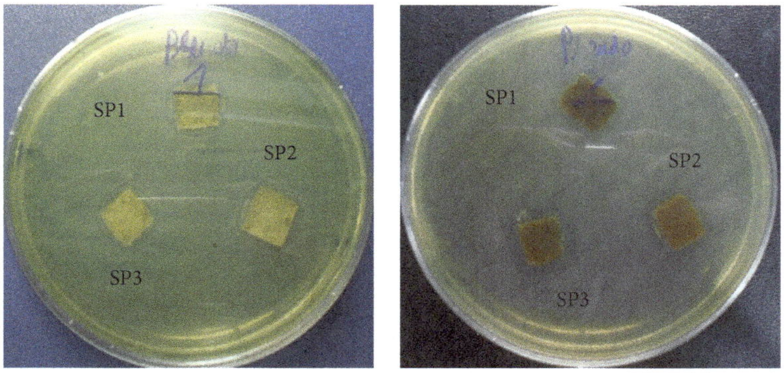

(a) (b)

FIGURE 4: Antibacterial activity of prepared sponges against *Pseudomonas aeruginosa:* (a) sponges without curcumin and (b) curcumin-sponges.

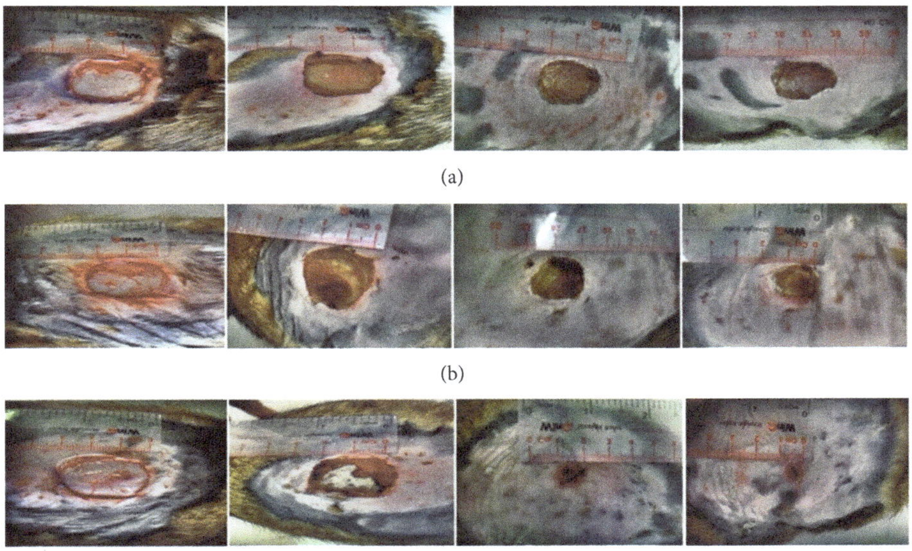

(a)

(b)

(c)

FIGURE 5: Photographs of macroscopic appearance of wound repair covered with (a) control, (b) blank sponge, and (c) curcumin sponge, at day 0, day 3, day 12, and day 15, respectively.

from sponges. The results present that cell viability decreased to approximately 90% in the presence of medium containing sponge without curcumin and curcumin sponge. The cell viability suggests that these sponges have generally low cytotoxicity to the L929 fibroblast cells (data not shown).

The SP3 sponge was selected for wound healing test. The percentage of wound closure in untreated and treated groups was measured on 3rd, 12nd, 15th, and 21st post wound day and the results are presented in Table 3. The wound closure was similar in three groups on 21st post wound day. The wound closure was 99.49%, 97.63%, and 94.30% for curcumin-SP3, blank SP3 and control group, respectively. It was found that the wounds treated with curcumin composite and composite without curcumin have shown faster wound healing than wound untreated group. Moreover, the wound closure on the 12nd and 15th day was significantly increased in curcumin SP3 in comparison with SP3 without curcumin (Figure 5). These results enhance the wound healing activity and could be a good material to be employed as wound dressing.

Histological Examination. The histological examination of three different kinds of wound dressing applied on skin wound of rabbit is presented in Figure 6. As can be seen in Figure 6, the tissue is in the acute stage of wound healing for control (Figure 6(a)). The tissue is under subacute stage of wound healing due to that there are many eosinophil, allergic reactions may be considered (Figure 6(b)). The

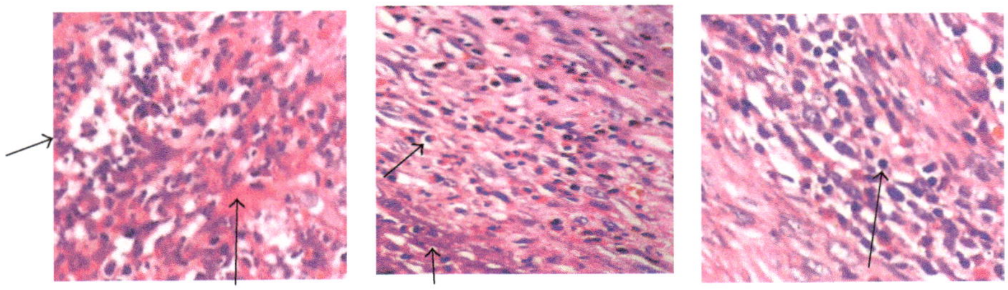

(a) (b) (c)

Figure 6: The H&E stained sections of twelve-day postsurgery of wound treated with (a) control: Short arrow—polymorphonuclear neutrophil (PMN), long arrow—fibroblast. The tissue is in the acute stage of wound healing. (b) Blank sponge (without curcumin): short arrow—eosinophil, long arrow—necrotic tissue. The tissue is under subacute stage of wound healing due to that there are many eosinophil and allergic reaction may be considered, and (c) curcumin sponge: arrow—lymphocyte, the tissue is under subacute stage of wound healing (400x).

Table 3: Wound closure of animal treated with blank sponge and curcumin sponge.

	Wound closure (%) (mean ± SD)			
Time	3rd day	12nd day	15th day	21st day
Control	15.36 ± 0.54	30.00 ± 1.05	48.80 ± 1.71	94.30 ± 1.93
SP3	14.79 ± 0.58	49.41 ± 1.93	77.96 ± 3.91	97.63 ± 4.28
Curcumin-SP3	26.53 ± 1.14	90.82 ± 2.11	95.41 ± 3.62	99.49 ± 3.81

wound is under subacute stage for curcumin-loaded sponge (Figure 6(c)). The granulation tissue and collagen alignment were more advanced in the sponge without curcurmin and curcumin-loaded sponge-treated wound as compared to the untreated wounds (control).

4. Conclusion

In this work, composite sponge of curcumin/chitosan/gelatin was prepared at the various ratios of chitosan and gelatin. The structure and morphology were characterized using FT-IR and SEM. The composite of curcumin, chitosan, and gelatin improved water uptake ability, antibacterial activity, and wound closure. Based on the results of drug release of curcumin, the higher content of gelatin in composite sponge exhibited a faster release behavior up to 240 min. These composite sponges were found to enhance the formation of collagen and wound closure *in vivo* and therefore improved the wound healing activity.

Acknowledgments

The authors would like to thank the Nafosted and Industrial University of Ho Chi Minh City for financial support. The authors thank Miss Tu and Miss Minh Nhat for their help on sample preparations.

References

[1] D. Gopinath, M. R. Ahmed, K. Gomathi, K. Chitra, P. K. Sehgal, and R. Jayakumar, "Dermal wound healing processes with curcumin incorporated collagen films," *Biomaterials*, vol. 25, no. 10, pp. 1911–1917, 2004.

[2] P. Basnet and N. Skalko-Basnet, "Curcumin: an anti-inflammatory molecule from a curry spice on the path to cancer treatment," *Molecules*, vol. 16, no. 6, pp. 4567–4598, 2011.

[3] A. Kunwar, A. Barik, R. Pandey, and K. I. Priyadarsini, "Transport of liposomal and albumin loaded curcumin to living cells: an absorption and fluorescence spectroscopic study," *Biochimica et Biophysica Acta*, vol. 1760, no. 10, pp. 1513–1520, 2006.

[4] A. Kumar, S. Dogra, and A. Prakash, "Protective effect of curcumin (Curcuma longa), against aluminium toxicity: possible behavioral and biochemical alterations in rats," *Behavioural Brain Research*, vol. 205, no. 2, pp. 384–390, 2009.

[5] B. T. Kurien and R. H. Scofield, "Increasing aqueous solubility of curcumin for improving bioavailability," *Trends in Pharmacological Sciences*, vol. 30, no. 7, pp. 334–335, 2009.

[6] C.-H. Chen, M.-F. Hsieh, Y.-N. Ho et al., "Enhancement of catechin skin permeation via a newly fabricated mPEG-PCL-graft-2-hydroxycellulose membrane," *Journal of Membrane Science*, vol. 371, no. 1-2, pp. 134–140, 2011.

[7] B. Wang, K. Chen, S. Jiang et al., "Chitosan-mediated synthesis of gold nanoparticles on patterned poly(dimethylsiloxane) surfaces," *Biomacromolecules*, vol. 7, no. 4, pp. 1203–1209, 2006.

[8] H. Yi, L.-Q. Wu, W. E. Bentley et al., "Biofabrication with chitosan," *Biomacromolecules*, vol. 6, no. 6, pp. 2881–2894, 2005.

[9] T. Dai, M. Tanaka, Y.-Y. Huang, and M. R. Hamblin, "Chitosan preparations for wounds and burns: antimicrobial and wound-healing effects," *Expert Review of Anti-Infective Therapy*, vol. 9, no. 7, pp. 857–879, 2011.

[10] R. Jayakumar, M. Prabaharan, P. T. Sudheesh Kumar, S. V. Nair, and H. Tamura, "Biomaterials based on chitin and chitosan in

wound dressing applications," *Biotechnology Advances*, vol. 29, no. 3, pp. 322–337, 2011.

[11] R. A. A. Muzzarelli, "Chitins and chitosans for the repair of wounded skin, nerve, cartilage and bone," *Carbohydrate Polymers*, vol. 76, no. 2, pp. 167–182, 2009.

[12] S.-J. Yang, F.-H. Lin, K.-C. Tsai et al., "Folic acid-conjugated chitosan nanoparticles enhanced protoporphyrin IX accumulation in colorectal cancer cells," *Bioconjugate Chemistry*, vol. 21, no. 4, pp. 679–689, 2010.

[13] Y.-Q. Ye, F.-Y. Chen, Q.-A. Wu et al., "Enhanced cytotoxicity of core modified chitosan based polymeric micelles for doxorubicin delivery," *Journal of Pharmaceutical Sciences*, vol. 98, no. 2, pp. 704–712, 2009.

[14] H. Zhou, W. Yu, X. Guo et al., "Synthesis and characterization of amphiphilic glycidol-chitosan-deoxycholic acid nanoparticles as a drug carrier for doxorubicin," *Biomacromolecules*, vol. 11, no. 12, pp. 3480–3486, 2010.

[15] A. Tanaka, T. Nagate, and H. Matsuda, "Acceleration of wound healing by gelatin film dressings with epidermal growth factor," *Journal of Veterinary Medical Science*, vol. 67, no. 9, pp. 909–913, 2005.

[16] E. J. Chong, T. T. Phan, I. J. Lim et al., "Evaluation of electrospun PCL/gelatin nanofibrous scaffold for wound healing and layered dermal reconstitution," *Acta Biomaterialia*, vol. 3, no. 3, pp. 321–330, 2007.

[17] K. Madhumathi, P. T. Sudheesh Kumar, S. Abhilash et al., "Development of novel chitin/nanosilver composite scaffolds for wound dressing applications," *Journal of Materials Science*, vol. 21, no. 2, pp. 807–813, 2010.

[18] M. Dai, X. Zheng, X. Xu et al., "Chitosan-alginate sponge: preparation and application in curcumin delivery for dermal wound healing in rat," *Journal of Biomedicine and Biotechnology*, vol. 2009, Article ID 595126, 8 pages, 2009.

[19] M. C. Silva and C. T. Andrade, "Evaluating conditions for the formation of chitosan/gelatin microparticles," *Polimeros*, vol. 19, no. 2, pp. 133–137, 2009.

[20] T. V. L. Hima Bindu, M. Vidyavathi, K. Kavitha, T. P. Sastry, and R. V. Suresh kumar, "Preparation and evaluation of ciprofloxacin loaded chitosan-gelatin composite films for wound healing activity," *International Journal of Drug Delivery*, vol. 2, no. 2, 2011.

2

Phase Analysis of the Cellulose Triacetate-Nitromethane System

Anna B. Shipovskaya,[1] Natalia O. Gegel,[1] Sergei L. Shmakov,[1] and Sergei Yu. Shchyogolev[2]

[1] *Institute of Chemistry, Saratov State University, 83 Ulitsa Astrakhanskaya, Saratov 410012, Russia*
[2] *Institute of Biochemistry and Physiology of Plants and Microorganisms, 13 Prospect Entuziastov, Saratov 410049, Russia*

Correspondence should be addressed to Anna B. Shipovskaya, shipovskayaab@rambler.ru

Academic Editor: Jan-Chan Huang

A comprehensive study was made on the cellulose triacetate-nitromethane system to explore its phase separation within ranges 2–25 wt.% and $-5 \div +80^\circ$C by means of polarization light and electron microscopy, the turbidity spectrum method, differential thermal and X-ray analyses, and rheological techniques. The physical state of the polymer was identified within the phase coexistence boundaries on the phase diagram which included three types of phase separation (amorphous (with a UCST at $T_{cr} = 57^\circ$C and $c_{cr} = 7.3$ wt.%), crystal, and liquid crystal). The boundaries of the regions determining the coexistence of the liquid crystal (LC) and the partly crystal phase were found to be inside the region of amorphous liquid-liquid phase separation. For cellulose ester-solvent systems, this state diagram is the first experimental evidence for the possibility of coexistence of several phases with amorphous, LC, and crystal polymer ordering.

1. Introduction

The structure of polymer-solvent systems and their behavior in particular conditions are reflected/described by the corresponding state diagrams. The information contained therein determines the character of any observed structural transformations, most often connected with the appearance of new disperse phases [1, 2]. Knowledge of the mechanism of phase separation kinetics opens ways to directly influence the system's morphology and, correspondingly, to design materials and products with desired properties.

However, drawing phase diagrams for polymeric systems based on cellulose esters is associated with serious experimental and theoretical difficulties. The structural and molecular heterogeneity of cellulose derivatives, their high molecular masses, diffusion coefficient low values, high viscosities of even moderately concentrated solutions, and specific intermolecular and molecule-solvent interactions hinder obtaining correct (consistent) data. This may lead to ambiguous or even contradictory conclusions on the state diagrams of the same (e.g., nitrocellulose-glycerol) systems, (for a detailed discussion, see [1]).

This work explores phase separation in the cellulose triacetate (CTA) nitromethane system. This system was chosen because of the wide practical application of CTA and the possibility of producing fibers and films with good physicomechanical properties (including low factors of unevenness and variation) from its nitromethane solutions or by means of nitromethane-vapor modification of the polymer [3, 4].

One version of the diagram of state of this system was plotted in [5] by means of the turbidity spectrum method [2, 6]. A sample obtained from Khimvolokno Corp. (Vladimir City, Russian Federation) was used, and the diagram showed phase separation regions of both liquid-liquid (with a UCST) and liquid-crystal type. The boundary curve of the system's absolute instability was named the "quasispinodal", since it was not tangent to the binodal at the apparent UCST (as it should according to the classical thermodynamics of solutions) but was located tens of degrees lower. This circumstance might be caused by long relaxation processes in the system with its high viscosity and, hence, natural restrictions of the thermokinetic approach used in this paper to draw the phase diagram. The methodology described in [2, 6] is for the location of configurative points at the true spinodal only for systems with relatively low viscosities, where the relaxation process rates are comparable with the phase separation ones. Except [5], no other information on the state diagram of the CTA-nitromethane system has been found by us in the literature.

We used a CTA sample that had its physicochemical characteristics close to those of the polymer from [5] but was produced at another industrial enterprise. To assess the type of phase separation (amorphous, crystal, etc.) and the equilibrium degree of the boundary curves obtained, the range of experimental and calculational methods used was widened, and special approaches were applied to allow for some features of the phase behavior of the system.

The aim of this work was to make a comprehensive analysis of phase separation in the CTA-nitromethane system and its properties both outside and inside the phase separation range, with identification of the polymer's phase state within the phase coexistence regions on the state diagram.

2. Materials and Methods

A commercial granulated CTA sample of wood origin (Khimvolokno Corp., Engels City, Russian Federation), used for the production of acetate fiber, was taken served as an object of our research. Its characteristics are as follows: acetylation degree, 62.2%, viscosity-averaged molecular mass, $\overline{M}_\eta$ = 70 kDa; ashness, 0.15. The CTA sample was separated from its high- and low-molecular-weight fractions. According to our fractionation (sequential precipitation) data, the polymolecularity of the sample was rather low, with the range of most probable molecular masses being 60–80 kDa. Nitromethane (analytical grade) with $d = 1.14$ g/ml, b.p. = 101.2°C, and m.p. = −28.5°C was used as a low-molecular-weight component. It is an electron donor, dipolar, and aprotic solvent [7] that preferably solvates hydroxylic groups in cellulose acetates [8, 9].

The CTA-nitromethane system was studied within a concentration range of c = 2–25 wt.% and a temperature range of T = −5–80°C. The research objects were prepared by swelling a weighed polymeric sample at room temperature for 24 h with subsequent dissolution on a water bath at T = 70–80°C. Dissolution was monitored by polarization microscopy and was considered finished when no undissolved particles of the polymer were seen. The prepared systems were kept at room temperature under atmospheric pressure.

In every experiment, the composition of the solution/mixture was constant and the temperature varied. Before optical and rheological measurements, the system had been thermostatted at T = 70–80°C for 1.0–1.5 h with subsequent cooling down to the temperature of the current experiment. The cooling rate was set according to the peculiarities of the technique used. In separate experiments, measurements were made at raising temperature.

Optical measurements were conducted on an MPSU-1 polarization microscope equipped with a UKRT-3 temperature control and regulation tool. Photos were made with a Zenith camera with a 3-4 s exposition. The polarizer and analyzer were located so as to see a dark field for an isotropic (though optically active) solution. Phase separation was observed visually by changes in the system's morphology with a microscope (new-phase particles on a dark field). First, at rapid cooling (≈1°C/min) of a layer of the solution (~1 mm), the phase transition temperature was roughly estimated. Then, it was determined more precisely in the a ±5°C region around the estimated value in the modes of stepwise lowering and increasing the temperature with at a ≈1°C/h rate and a duration of thermostatting of not shorter than 4–7 h at every temperature. This procedure was repeated until the discrepancies between the temperature values obtained in these two modes did not exceed 0.5°C.

Absorbance was measured on an FRK-56 photoelectrocolorimeter within the wavelength range λ = 400–670 nm. The solutions, after preliminary thermostatting at 70–80°C, were quickly (≈3-4°C/min) cooled down to the desired temperature. During the subsequent thermostatting (t = 5–8 h), turbidity spectra were recorded. The phase conversion rate V_τ was estimated from those segments where the turbidity τ (variant 1) or induction period (variant 2) on the $\tau = f(t)$ dependence increased most rapidly at λ = 540 nm, T = const, and c = const.

The stability boundary of the solution of the current concentration was determined by extrapolation of the temperature dependence of the turbidity increase rate $V_\tau(T)$ to zero. The extrapolation $V_\tau^{-1}(T) \rightarrow 0$ allowed finding the temperature of absolute instability of the system. The effective radius of particles $\overline{r}_\lambda$ was calculated according to [2, 6], with the relative refraction index being 1.07. The optical activity of both phases was neglected.

Melting points were determined by differential thermal analysis (DTA) on a Kurnakov pyrometer with an Al_2O_3 reference and a heating rate of 1.5°C/min. Diffraction patterns of CTA (powder or film) were recorded by X-ray diffraction analysis at small angles on a DRON-3 diffractometer with Cu-K$_\alpha$ light at U = 22 kV and J_a = 20 mA.

Rheological studies were performed on a Reotest-1 viscometer within a shear stress range $\lg\sigma$ = 1.7–2.7 (Pa). Deformations at every temperature were applied periodically, in fixed periods of time. Between these deformation cycles, the system was at rest in the working unit (the coaxial cylinders H) of the viscometer under isobaric isothermal conditions.

Electron microscopic studies were conducted on EM-5 and Tesla-242E electron microscopes. The solution was sprayed with a UZDN-1 ultrasound low-frequency disperser.

The spinodal was calculated within the framework of Flory-Huggins' lattice model [10] by the following equation:

$$\chi_1 = \frac{1}{2(1-v_2)} + \frac{1}{2zv_2}, \quad (1)$$

where χ_1 is the Flory-Huggins interaction parameter, z is the average number of elementary units in a macromolecule ($z = n/2$, n being the polymerization degree), and v_2 is the volume fraction of the second component (CTA) in the mixture.

The binodal for biphasic amorphous separation in the polymer-low-molecular-weight liquid system was approximated according to the algorithm described in [11], with the use of the experimentally found coordinates of configurative points.

3. Results and Discussion

Our polarization microscopic observations showed that phase separation occurred in visually homogeneous (within the microscope field) CTA solutions in nitromethane heated up to 70–80°C—there arose light emitting particles (new-phase centers) of primarily spherical shape. In the course of thermostatting at some T the number of these new-phase centers increased up to some limiting value within 1–6 hours (depending on the polymer concentration and temperature). Particle shape did not change. The lower was the T inside the found concentration-temperature range of phase separation, the greater was the maximum size of the particles being formed. Figures 1(b)–1(d) provide photos reflecting the changes in the CTA-nitromethane solution morphology ($c = 8$ wt.%) at lowering temperature and in the course of thermostatting at a constant T.

When the temperature was raised, the disperse phase particles dissolved. The formation-dissolution temperature of the heterogeneous structure fixed in the cooling-heating mode was considered as the coordinate of the boundary configurative points on the state diagram. The set of such coordinates obtained for several c was taken as the boundary curve of phase separation.

It should be noted that for polymer concentrations ~5–11 wt%, the transition of the homogeneous system to its metastable state is preceded by the appearance of anisometric birefringent (in polarized light) fragments (Figure 1(a)). This could be regarded as a manifestation of fluctuation structures (which are not separate phases in their essence) owing to the critical retardation effect (with diffusion coefficient vanishing) ([2, pages 238–240], [12]). The shape of such long-lived heterophasic fluctuations differs from the usual quasispherical one, possibly because of intercompensation of the strong polymer-polymer and polymer-solvent interactions. Nitromethane, solvating a small quantity of the hydroxylic groups remaining in CTA, does not hinder the formation of frequent intermolecular contacts by the free acetate groups. Consequently, the weaker interactions, previously unnoticeable/negligible on the background of the stronger ones, are now given an opportunity to develop. In particular, at lowering temperature, the nitromethane solvation of the residual hydroxylic groups (which the intermolecular interaction is realized by) in CTA weakens. This promotes engagement of the critically frozen polymeric molecules into spongy anisotropic (nonspherical) aggregates-fluctuations, which are observed with a microscope during hours. The low rates of relaxation processes provide the longevity of such fluctuations.

However, another explanation for the appearance of these anisometric anisotropic formations is possible. For example, [13] theoretically justifies the possible formation of a lyotropic LC phase in the cellulose acetate-nitromethane system. Taking account of the relatively great sizes (comparable with those of the disperse-phase particles) of the birefringent formations, one can hypothesize the origination of particles of a highly ordered LC phase. However, as the LC particle formation rate is significantly lower than the amorphous particle formation rate, the kinetics observed with a polarization microscope at decreasing T (Figures 1(b)–1(d)) reflect amorphous phase separation in the CTA-nitromethane system. Other possible evidence for the hypothesized superposition of two types of phase separations will be given below.

Allowing for some specific peculiarities (giant concentration fluctuations) of the system's transfer from its single-phase state to the range of phase separation, the turbidity spectrum method [2, 6] was also used for locating configurative points and for studying the kinetics of changes of the new-phase particle parameters in the metastable range. In this case, the turbidity caused by concentration fluctuations is set almost instantaneously and does not change in the future. Correspondingly only turbidity kinetics is experimentally measured which is related to the size changes of the new-phase particles [2].

Figure 2 presents an example of turbidity changes in time at fixed temperatures for the CTA-nitromethane system with a polymer concentration of 2 wt%. For other compositions, the $\tau = f(t)$ dependences were similar. One can see that in a certain temperature range, τ sharply increased during thermostatting, and the average effective particle size $\bar{r}_\lambda$ increased also. Such a behavior, along with our polarization microscopy data, attests that phase separation occults by the mechanism of nuclei formation and growth of the new-phase particles.

Figure 3 shows typical (for solutions with $c \sim 2$–7 wt%) temperature dependences of both direct and inverse rates of phase separation increase (turbidity), calculated from the segments of both initial (the induction period) and intense changes of τ. As temperature increased, V_τ decreased, which is characteristic of systems with an UCST. The geometrical loci with $V_\tau(T) = 0$ and $V_\tau^{-1}(T) = 0$ were taken as the binodal curve and the absolute instability curve (spinodal), respectively.

Within a certain concentration-temperature range, the growth of τ and $\bar{r}_\lambda$ began after an induction period. It turned out that for $c \sim 8$–10 wt%, the temperature dependences of both direct and inverse rates of turbidity increase, calculated by method 1 in the segment of intense growth of τ and the new-phase particles (Figure 4(a)), were similar to the $V_\tau = f(T)$ curves obtained for lower concentration solutions (Figure 3). However, the temperature dependences of the rate determined from the induction period segment (i.e., at the first stage of phase separation) of the kinetic curve $\tau = f(t)$ (method 2) were of extreme character (Figure 4(b)). Note that the temperatures obtained by extrapolation to zero of both left-hand and right-hand branches of the $V_\tau = f(T)$ curve with a maximum turned out to be almost equal to those obtained by the $V_\tau(T) \rightarrow 0$ and $V_\tau^{-1}(T) \rightarrow 0$ extrapolations for V_τ, calculated by method 1.

Such a different character of the temperature dependence of V_τ at different stages of phase separation gives evidence for the complex type of phase separation in the system. The $V_\tau(T)$ dependence is known to be represented by a smooth curve with no minima or maxima for the liquid-liquid phase separation, where as for the liquid-crystal one, the function $V_\tau(T)$ is of extreme character [2, 12]. In the latter case, the right-hand branch of the $V_\tau(T)$ dependence

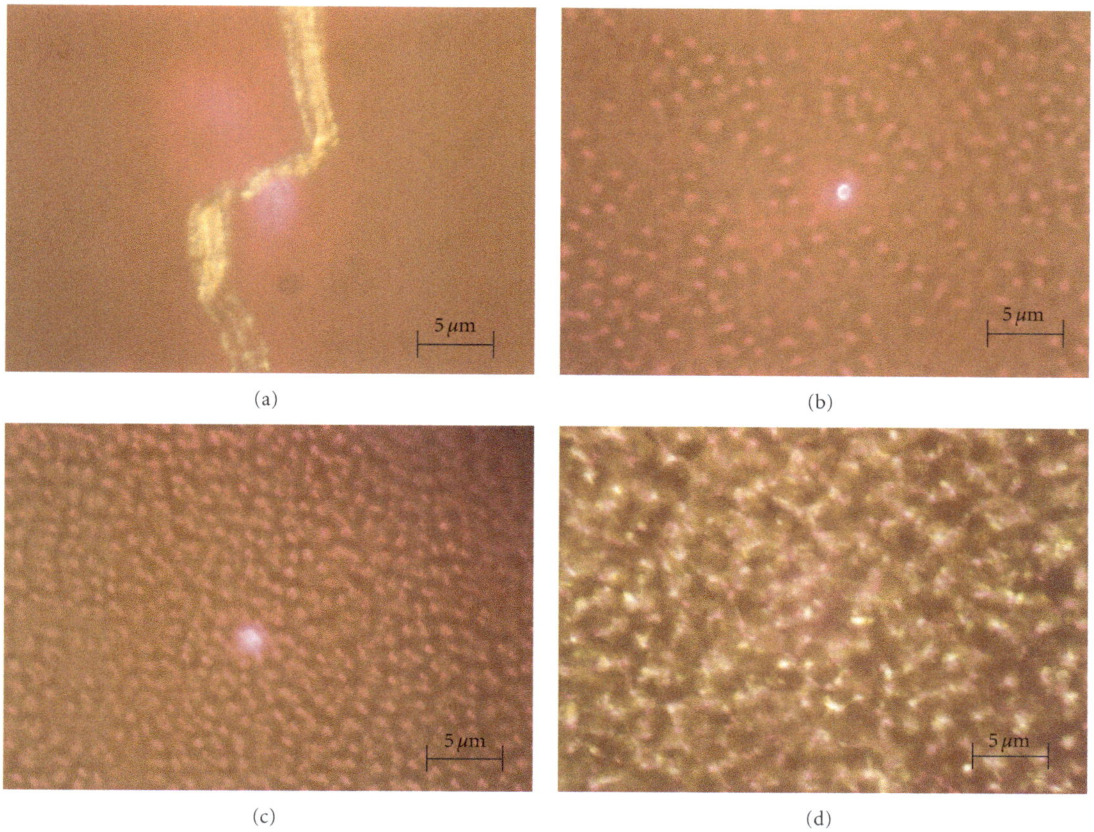

FIGURE 1: Morphology of the CTA-nitromethane system with a polymer concentration of 8 wt% at temperatures of 55° (a), 50° (b), (c), and 25°C (d) after thermostatting for 30 min (a), (b), 1 h (c), and 2 h (d). The microphotos were made in polarized light.

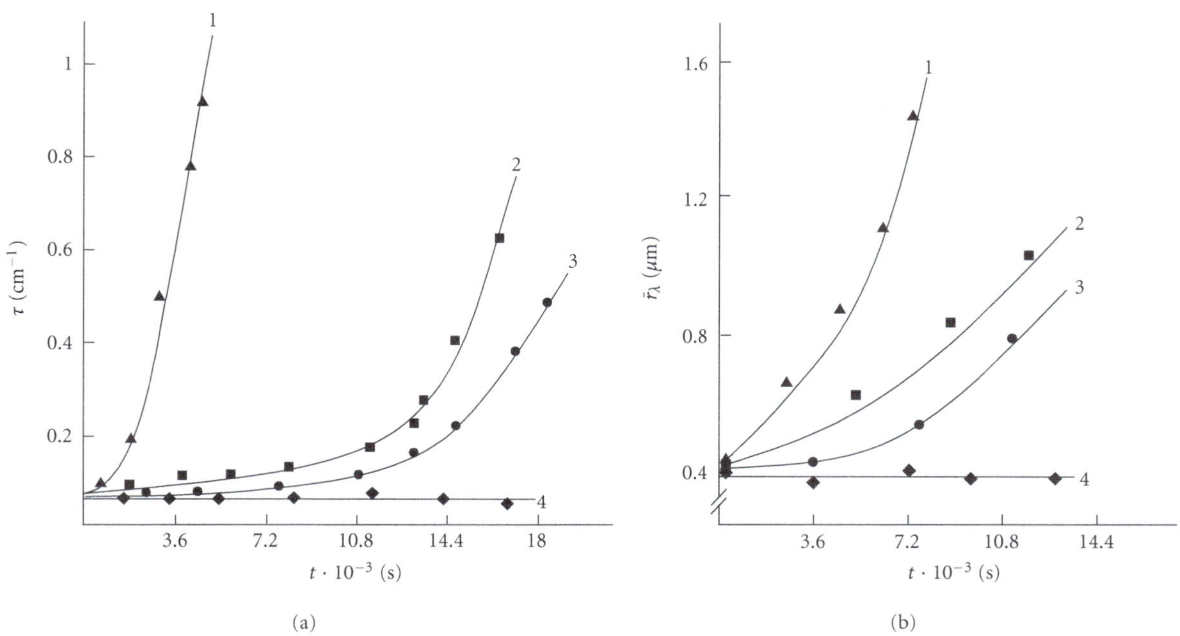

FIGURE 2: Kinetics of the turbidity τ at $\lambda = 540$ nm (a) and the particle size $\bar{r}_\lambda$ (b) for the CTA-nitromethane system with a polymer concentration of 2 wt% at temperatures of –5° (1), 8° (2), 11° (3), and 20°C (4).

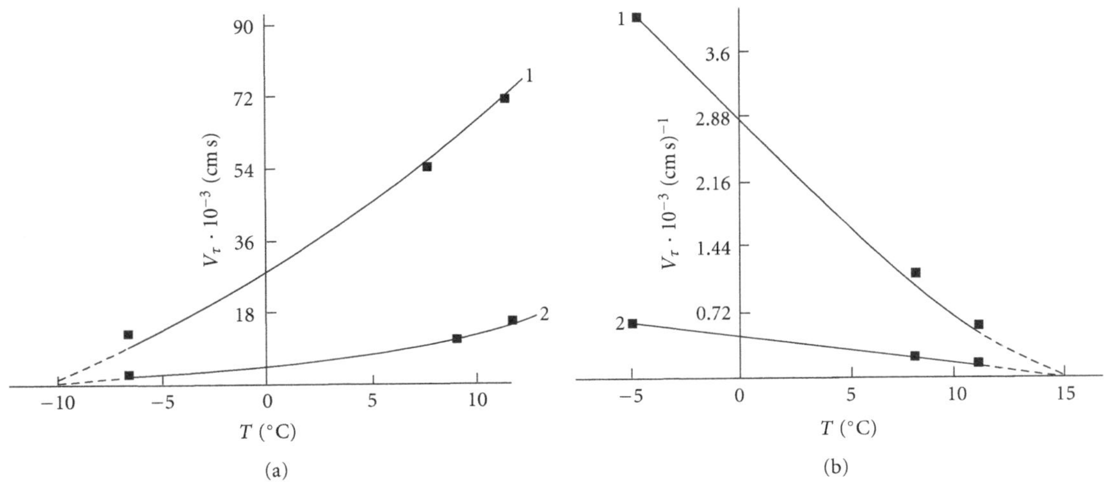

FIGURE 3: Estimation of the boundary temperatures of phase separation for the CTA-nitromethane system with a polymer concentration of 2 wt% by extrapolation of the direct (a) and inverse rates (b) of turbidity growth, calculated at the stage of intense τ changes (1, method 1) and from the initial fragment of the kinetic $\tau = f(t)$ curve (2, method 2).

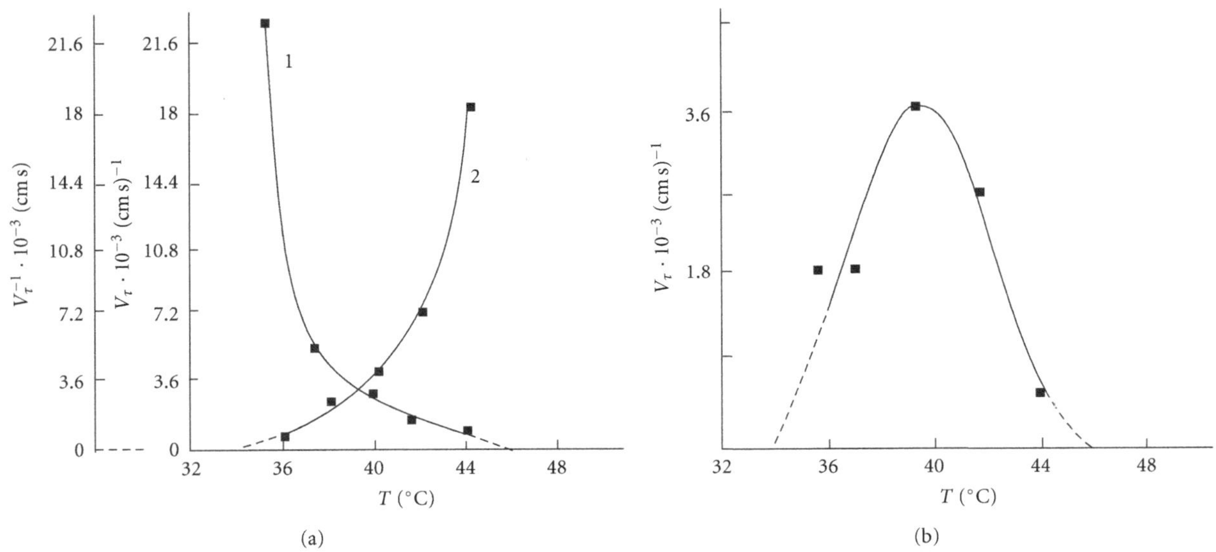

FIGURE 4: Estimation of the boundary temperatures of phase separation for the CTA-nitromethane system with a polymer concentration of 9 wt%: (a) by the extrapolations $V_\tau(T) \rightarrow 0$ (1) and $V_\tau^{-1}(T) \rightarrow 0$ (2) for the rate calculated from the $\tau = f(t)$ fragment of intense turbidity growth (method 1); (b) by the extrapolation $V_\tau(T) \rightarrow 0$ for V_τ determined from the initial fragment of the kinetic $\tau = f(t)$ curve (method 2).

(c = const) at extrapolation intersects the abscissa axis at the temperature of liquidus, whereas the left-hand one gives the vitrification temperature [12]. Probably, in the composition range with an extreme $V_\tau = f(T)$ dependence (method 2), particles with either amorphous or highly oriented structure (e.g., of crystal or LC nature) originate at the first instant of time when the system transfers to its metastable state. As the kinetics of both crystallization and LC ordering are extended in time and the viscosity of the CTA-nitromethane mixture of such compositions is rather high, the system is apparently in the metastable range with respect to these types of phase separation [14]. By virtue of these causes, relatively quick amorphous phase separation is fixed at further thermostatting. In this case, the formation and growth of new-phase centers are facilitated by the fact that the probability of appearance of the corresponding fluctuation ensembles is significantly higher because there is no need for strict geometrical packing of the segments (as is required to form crystallites) or for realization of a certain macromolecular conformation (to form LC structures). The establishment of crystal or LC equilibria (if any) within this concentration-temperature range could be expected in a longer period and at a high degree of supersaturation of the polymer-concentrated phase.

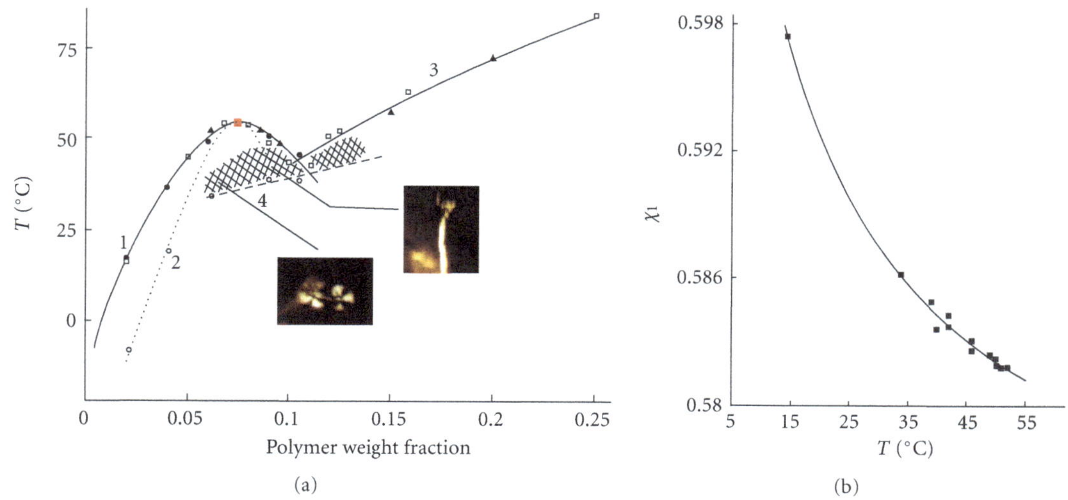

FIGURE 5: (a) Phase diagram of the cellulose triacetate-nitromethane system: squares denote the data of polarization microscopy, the circles are from the turbidity spectrum method (dark ones: extrapolation $V_\tau(T) \to 0$ by method 1 and that of the right-hand branch $V_\tau(T) \to 0$ by method 2; open: extrapolation $V_\tau^{-1}(T) \to 0$ by method 1 and that of the left-hand branch $V_\tau(T) \to 0$ by method 2), and the triangles mean DTA; (1), (2) the binodal (the solid line) and spinodal (the dashed line), approximated and calculated, respectively, from Flory-Huggins' theory with experimentally found compositions of the coexisting phases, (3) the liquidus, (4) the vitrification curve. The dashed area depicts the concentration-temperature range of LC state existence. Insets: the system's morphology in polarized light (see text for explanation). (b) Temperature dependence of the calculated Huggins interaction parameter χ_1.

Having the extreme dependence of the turbidity change rate for moderately concentrated systems ($c \sim$ 8–10 wt%) as the starting point (Figure 4(b)), we can hypothesize as follows. The geometrical locus of points obtained by zero extrapolation of the right-hand branch of the extreme $V_\tau = f(T)$ curve (method 2) or by the condition $V_\tau(T) = 0$ (method 1) reflects the location of the binodal. The Zero extrapolation of the left-hand branch of the $V_\tau = f(T)$ curve apparently gives the temperature coordinates of the vitrification curve. A more detailed interpretation of this hypothesis is given below.

Separate experiments to find configurative points of the state diagram were conducted by means of DTA. The T value corresponding to the minimum point of the endothermic peak on the thermogram was taken as the temperature of phase separation.

The found compositions of the phases coexisting at a certain temperature were used to approximate the binodal (see above) and to calculate the spinodal by Flory-Huggins' theory [10], as well as to estimate the interaction parameter χ_1. The number of elementary units in a macromolecule was taken as $z = 161$, and the fitted temperature dependence of χ_1 was expressed as

$$\chi_1 = 0.5708 + \frac{0.7016}{T - 261}. \tag{2}$$

Figure 5(a) shows our version of the phase diagram of the CTA-nitromethane system plotted as a result of our experiments, approximations, and calculations. The $\chi_1 = f(T)$ dependence is depicted in Figure 5(b). This state diagram is characterized by a UCST with the following coordinates: critical temperature, $T_{cr} = 57°C$; critical concentration, $c_{cr} =$ 7.3 wt%. No LCST was detected, though the heating of the solutions in sealed ampoules was up to 200°C, much higher than the boiling point of the pure solvent. The polarization microscopy, turbidity spectra, and DTA data are in good agreement in respect to the temperatures of phase separation round. The reproducibility and reversibility (noted above) of the temperature-concentration location of the phase transition points enable us to consider the experimental and calculated boundary curves (the binodal, spinodal, liquidus, and vitrification curve) to be equilibrium ones.

The interaction parameter χ_1 near T_{cr} tends to ~0.5, which corresponds to the regularities of the thermodynamics of solutions of semirigid polymers at amorphous phase separation. It tends to infinity at $T \to 261$ K, rather than at $T \to 0$ K (see (2) for χ_1). This temperature lies close to the melting point of nitromethane, where it loses the properties of a solvent. In other words, the energy of one nitromethane molecule is insufficient to transfer it to the medium of polymeric segments. As we see, the introduction of an additional term into the expression for χ_1 in the first approximation of Flory-Huggins' theory seems rather reasonable.

The state diagram shown in Figure 5(a) is sophisticated, with two types of phase separation, namely, amorphous (under the binodal cupola) and crystal (below the liquidus curve 3), and a range of coexistence of various phases. This phase diagram is in fundamental accord (i.e., by phase separation types) with that obtained in [5] by the turbidity spectrum method (with the application of thermokinetic approach) for a polymer sample with close physicochemical characteristics but produced at a different industrial enterprise. The only difference is in the interpretation of the absolute instability curve obtained by the turbidity spectrum method and located (in our case, as well as in [5]) below the phase separation curve of the amorphous type and the initial segment of the liquidus. We divide it into the spinodal (curve 2) and the vitrification curve (curve 4).

The phase state of the system is such: at temperatures above the binodal and liquidus, all the mixtures are homogeneous transparent solutions. If the temperature is lowered and the configurative point moves to the metastable range, the CTA-nitromethane system of various concentrations may have various morphologies, depending on the polymer concentration and the distance (along the temperature axis towards lower T's) from the stability range. The state diagram contains several ranges. For low-concentrated solutions (below ~ 3%), under conditions of a rather high kinetic mobility of the components' molecules, quick processes of amorphous phase separation are observed with lowering temperature. They are fixed visually, for example, by the appearance of opalescence; in some cases, flocky precipitation may appear. For medium-concentrated solutions (c~ −5.53 wt%) with low viscosities, the system splits into two phases in time. A Dense white precipitate is formed in the polymer-concentrated phase, and the other phase is an opalescent solution with suspended particles.

Exploring solutions with higher concentrations is hindered by the slow establishment of equilibrium and the absence of sharp interphases. For example, at amorphous stratification (inside the binodal cupola), incomplete phase separation is observed within $\sim 5.5 < c < \sim 10$ wt% because of the high viscosity, low values of the diffusion coefficient [15], the longevity of relaxation processes, and the critical retardation effect (see above). Structure formation processes are strongly influenced by the time factor, as well as by the choice of the initial temperature point. Cooling of the solutions with such concentrations down to temperatures below the compatibility curve on the state diagram is accompanied by the formation of gels with their characteristic instability. Being owing transparent at first, such gels then get turbid and undergo syneresis in time to ordering and orientation processes [16]. Such a behavior can be explained by allowing for some peculiarities of phase separation in solutions of crystallizable polymers, including CTA [17, 18]. For such objects, a state with an amorphous type of phase separation is often realized at first, with subsequent transition to crystallization or to possible formation of an intermediate LC state.

Solutions with $c > \sim 10$ wt% form no gels. Decreasing the temperature results in intense dimness of the system and in its transformation to a kind of paste. The higher the polymer concentration in the system and the lower the temperature inside the phase separation region, the higher the rate of this process. At room temperatures, crystallization occurs in the CTA-nitromethane system with the mentioned compositions. It is evidenced by the X-ray diffraction patterns of the films cast from such systems.

Figure 6 presents the diffraction patterns of powder CTA isolated from its nitromethane solution with concentrations 12 and 25 wt% (curves 1 and 2, resp.) at 22 ± 2°C. For comparison, we give a diffraction pattern of a CTA film cast from a methylene chloride-ethanol (90 : 10) mixture, which is a volatile liquid (curve 3). Because of the high rate of

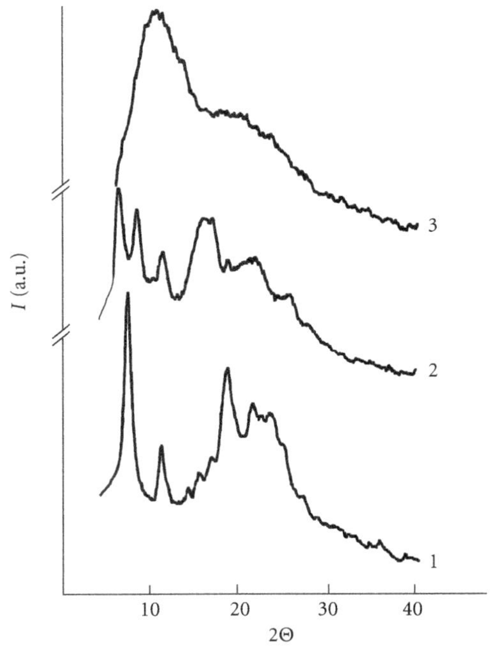

Figure 6: Diffraction patterns of CTA: powder isolated/precipitated from a nitromethane solution with polymer concentrations of 12 (1) and 25 wt% (2); a film obtained from a solution in the process solvent (methylene chloride + ethanol (90 : 10)) with a concentration of 3 wt% (3); $T = 22 \pm 2$°C.

evaporation of this mixture (~3-4 h), the macromolecules are obviously fixed in that conformation which they had in a dilute solution (e.g., coiled). Hence, such films hafe an amorphous structure. The diffractograms of the CTA powder obtained from nitromethane solutions resemble that of crystal CTA [18].

The maximum number of intense and resolved reflexes is observed for the CTA isolated from its nitromethane solution with a lower weight concentration (Figure 6, curve 1). This can be explained by taking into account that the specific polymer-solvent interaction and kinetic factors (associated with the high viscosity, low rate of diffusion and relaxation processes, etc.) promote better contact between the chains along the nonsolvated segments of the polymeric macromolecules. When the solvent is evaporated (the configurative point isothermally goes right along the composition axis), the macrochains become ordered and a rightly oriented phase is accumulated, and then fixed on complete solvent removal. Naturally, the diffraction patterns of the CTA powder isolated from a solution with a higher concentration (i.e., more distant along the concentration axis at the early stage of phase separation) show less intense reflexes (curve 2).

The fact of high-crystalline CTA isolation from its dilute solutions (including nitromethane ones) is well known [17–19]. Highly acetylated CTA is a polymer of regular structure with sufficiently homogeneous distribution of acetate groups along the macromolecular chain. The higher the acetylation degree of a sample, the more crystallizable this sample is [19]. The molar fraction of crystallizing units in this polymer

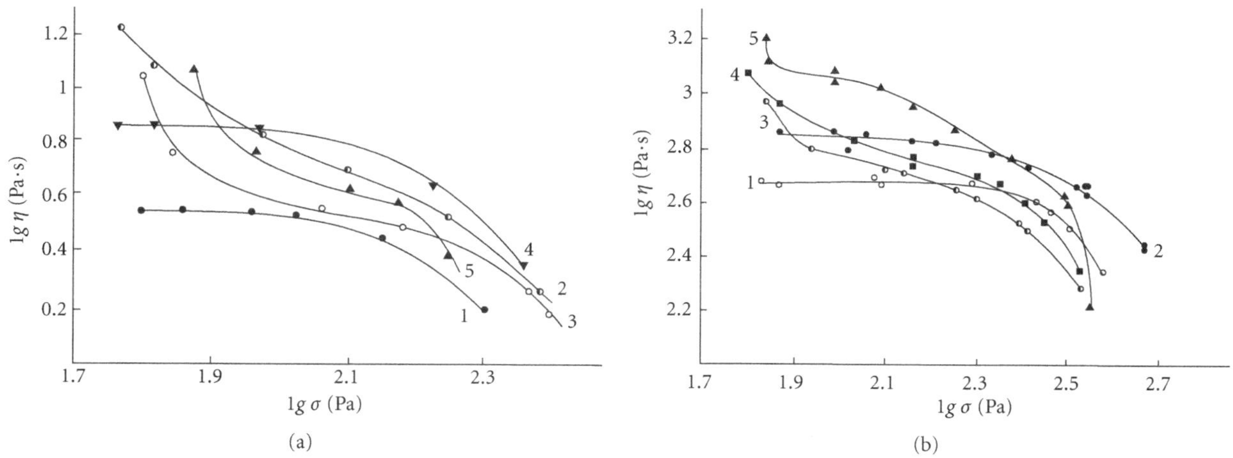

FIGURE 7: Flow curves of the CTA-nitromethane system with polymer concentrations of (a) 6 wt% at $T = 60°$ (1), 43° (2), and 35°C (3), and 8 wt% at 60° (4) and 40°C (5). The duration of thermostatting, 1 h; (b) 13 wt% at 50° (1), 45° (2), 38° (3), 30° (4), and 25°C (5), the duration of thermostatting 1 h (1, 2), 2 h (3–5).

calculated according to Flory [10] was 0.92 for the CTA sample under study.

It is natural to suppose that at lowering temperature, crystal precipitation of CTA from its nitromethane solutions will be the final stage of phase separation for the solutions under study. However, as this process is extended in time, every stage of phase transformations will show a variety of physical states, structural and morphological forms of the polymer.

Our rheological studies of the CTA-nitromethane system have shown that classical flow curves (usual for polymeric solutions) with regions of the maximum Newtonian viscosity (below $\lg\sigma\sim 2$–2.2 (Pa)) and structural viscosity (Figure 7(a), curves 1 and 4; Figure 7(b), curves 1 and 2) are observed in the thermodynamic miscibility range. In the concentration-temperature range under the binodal and the liquidus curve, the flow character and, hence, the shape of viscosity rheograms change. As the rheokinetics of phase transitions in the CTA-nitromethane system is discussed in [20] in detail, we now dwell on some peculiarities of the rheological behavior, reflecting changes in the phase state of the system.

For the range $5.5 < c < 10$ wt% at lowering temperature and of the crossing of the binodal, rheological curves characteristic of structured gellike systems are realized (Figure 7(a), curves 2, 3, and 5). In the region of relatively small shear stresses, a sharp increase in viscosity occurs, which speaks for the appearance of structures with an ultimate strength. This manifests itself as continuous lowering of η in the whole σ range and as the absence of a region of constant viscosity on the flow curves. Besides, in a certain temperature range there occurs such an intense drop in viscosity that it becomes lower than its value corresponding to a higher temperature, that is, the homogeneous solution range (see, e.g., curves 4 and 5). Such a viscosity drop cannot be explained only by gel formation; it testifies to a change in the flow mechanism, where the appearing structures possess lower resistance. It is natural to suppose that the decrease in η is caused by the formation of a highly ordered phase whose viscosity may be much lower than that of the isotropic phase [21, 22]. Similar data have also been obtained for many rigid-chain synthetic polymers at the transition of their solutions from isotropic to LC state [21, Pages160–165].

Cellulose acetates belong to the class of linear polymers with enhanced chain rigidity and a high asymmetry degree [22]. Spontaneous transition to an LC state is possible for them under certain thermodynamic conditions, which has been confirmed in the literature [23–25]. The possibility of transition of the CTA-nitromethane system to an LC phase is also supported by [13, 26]; the effect of spontaneous elongation of CTA-based fibers and films in nitromethane vapor [3, 27–29] speaks for this as well. We note that the selective solvation of the residual OH groups in CTA with nitromethane predetermines the realization of rather rigid, extended helical conformations of macromolecules [27, 29]. According to [30], structure formation in such a polymer could go through a stage of LC ordering.

The rheologically based temperature dependences of viscosity have a maximum peculiar to LC systems (Figure 8, curves 1′-2″, 3). Such character of the temperature dependence of viscosity was also observed in studying the rheokinetic regularities of phase separation in the CTA-nitromethane system with relatively low shear stresses ($\lg\sigma <$ 1(Pa)) on a Heppler viscometer. The lower the temperature inside the found range of mesophase formation, the higher the rate of formation of highly oriented structures. All this gives evidence for a spontaneous process of CTA macromolecular ordering, with the factor of external impacts apparently playing merely a stimulating role in structural self-organization.

The formation of an anisotropic phase was observed with an optical microscope in crossed Polaroids as well (i.e., with no shear deformations applied). Birefringent supramolecular structures appeared in the temperature range of the anomalous drop in η. For example, helicoidal (in the form of twisted ribbons) anisometric structures and even anisotropic

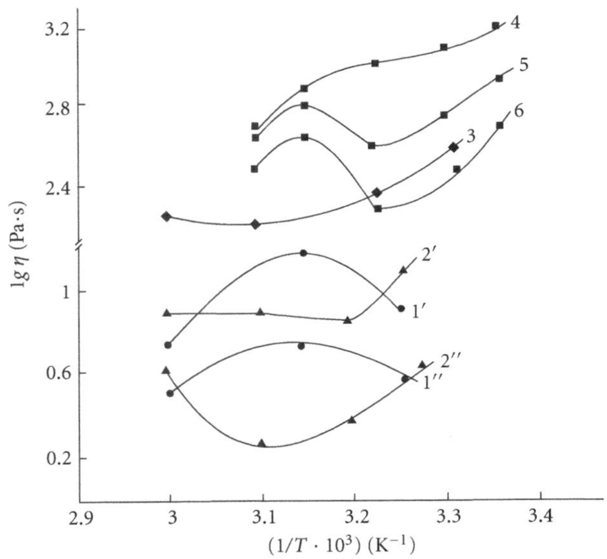

FIGURE 8: Dependence of viscosity on the inverse temperature for a nitromethane solution of CTA with polymer concentrations of 6 (1′, 1″), 8 (2′, 2″), 9 wt% (3), and 13 wt% (4, 5, 6) at the following shear stresses ($\lg \sigma$, (Pa)): 1.8 (1′, 2′, 4), 2.1 (3), 2.2 (1″, 2″, 5), and 2.5 (6).

drops crossed with a Maltese cross (see insets in Figure 5(a)) were observed. All the optically active structures were mobile; turning the object table of the polarization microscope by 40–50° led to the disappearance of their images. The reversibility of the anisotropic phase formation was established in the cooling-heating cycle without shearing.

Precipitation of these LC-related structures with, for example, ethanol results in the formation of fibrous precipitates (Figures 10(a)–10(c)), which also confirms the realization of an LC phase [21, Pages 100–107]. The fresh sediment is optically anisotropic (Figure 10(a)). Electron microscopic images of the CTA precipitate isolated from an anisotropic solution by means of spraying onto mica and shaded with carbon show its morphology as elongated chaotically oriented fibrils (Figure 10(d)) [20].

The observed indications of supramolecular orientation of the structural elements (Figures 9 and 10) give evidence of the helicoidal ordering of the LC phase in the CTA-nitromethane system. On the basis of our results and the literature data on studying the LC state in solutions of cellulose derivatives [23–25, 31, 32], the LC ordering type in the said system can be characterized as cholesteric.

No anomalous rheological curves and no extremum on the $\lg \eta = f(T^{-1})$ curve were obtained when $c < 5.5$ wt%. Within $11 < c < 14$ wt% in a narrow temperature range between the liquidus and the vitrification curve (Figure 5, curves 3 and 4), all the above peculiarities of rheological behavior were observed, namely, the anomalous drop in viscosity (Figure 7(b), curve 3–5) and the nonlinearity of the temperature dependence of η (Figure 8, curves 5 and 6). However, the critical stresses above which plastic flow of the system and a maximum on the $\lg \eta = f(T^{-1})$ curve were observed were higher for the given range. When σ was lower than its critical value, η increased with decreasing T (Figure 8, curve 4). As well as for the low-concentrated systems (curves *1 ′–2 ″*), the higher was σ, the more pronounced was the peak on the $\lg \eta = f(T^{-1})$ (curves 5 and 6). Such behavior is typical of LC solutions of many cellulose derivatives [22, 23, 25] and other synthetic rigid-chain polymers [21]. This can be explained by hypothesizing viscosity anisotropy, that is, the dependence of internal friction on the orientation of the macromolecules' long axes along the direction of the flow [33]. Viscosity rose with no limit in the temperature range below the vitrification curve, up to the system's transition to a nonflowing state.

Obtaining reliable information on the morphological properties of a system with the polymer concentration above 14 wt% and below the liquidus temperatures is rather difficult. When being cooled, such systems quickly get turbid and pass into a pastelike substance. This may be caused either by the formation of LC structures or by the crystallization process.

Our experimental data allowed the concentration-temperature field of LC state formation in the CTA-nitromethane system to be determined (dashed partition on the phase diagram, Figure 5(a)). One can notice its narrowness in both variables (temperature and concentration), which is in accord with Flory's theoretical concepts [34] and experimental studies on other systems of the cellulose ether/ester-solvent type [21–23]. It is interesting that the threshold concentration (~5.5 wt%), that is, the value from which anomalous viscosity rheograms, critical shear stresses, anisotropic effects, and so forth are observed was confirmed by calculations as well. The asymmetry degree calculated by [21] for the most acetylated CTA is ~140, which corresponds (according to [34]) to a weight polymer concentration of 5.7%. Beginning with the latter value, an LC phase can be formed in CTA solutions.

The absence of any LC structures in the top part of the binodal cupola, somewhat below its right-hand branch, attracts attention. This may be caused by the fact that as temperature increases, the chain rigidity diminishes and becomes lower than the critical value necessary for LC phase formation. This effect was confirmed in special experiments to study the hydrodynamic properties of dilute solutions of CTA samples with various $\overline{M}_\eta$ obtained by fractionation of the source polymer.

Our detection and location (in concentration-temperature coordinates) of the boundaries of LC phase formation in the CTA-nitromethane system allow the essentially different character of the temperature dependence of the turbidity kinetics for c ~8–10 wt% (see above and Figure 4) to be explained. Whereas at the first stage (the induction period) the temperature dependence of the τ growth rate goes through a maximum (as for liquid-crystal phase separation), at the subsequent stage this rate increases with temperature monotonically as for liquid-liquid phase separation. In particular, when the temperature is lowered, the rate may vanish during the induction period and may tend to huge values (theoretically, to infinity) after this period.

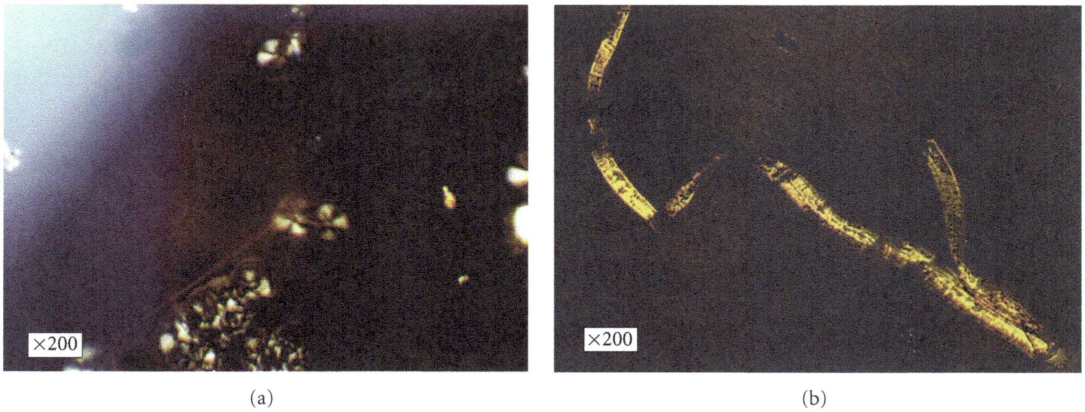

FIGURE 9: Photos of the CTA—nitromethane system in polarized light with no shear deformations applied: (a) denotes the mixture with $C = 6$ wt% at $T = 35°C$; (b) $C = 9$ wt% at $T = 45°C$.

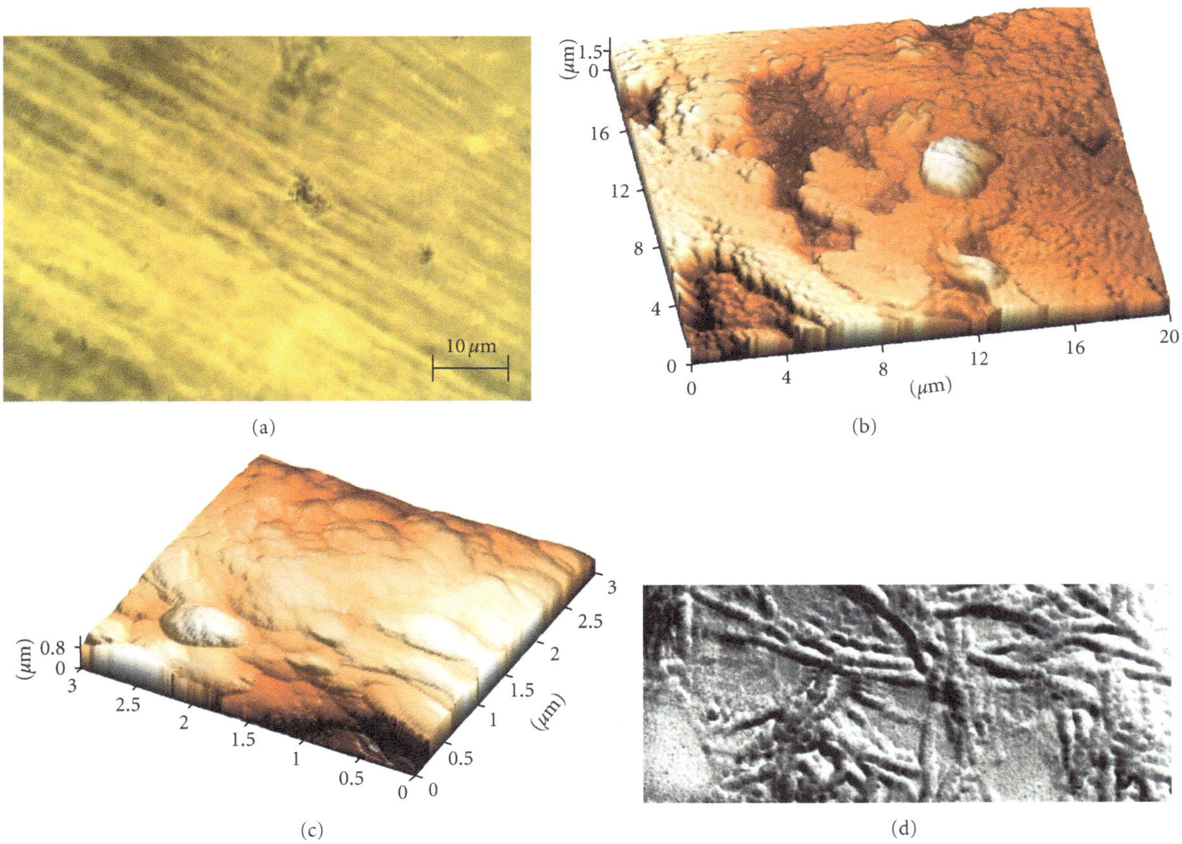

FIGURE 10: Morphology of the sediment obtained from the anisotropic CTA—nitromethane system by precipitation to ethanol (a)–(c) and by spraying onto mica (d): (a) is the structure of the source sediment in polarized light under crossed Polaroids; (b), (c) are the topographical image (scanning microscopy) of the surface (20 × 20 and 3 × 3 μm, resp.) of the same sediment (a) dried at room temperature; (d) is the structure of the sprayed sediment (electronic microscopy).

As is known, the LC state combines liquid and solid properties: there exists long-range order in one direction on the bases of rigid bonds, whereas other direction(s) may provide free sliding of layers, that is, short-range order, as in a liquid. If we hypothesize the appearance of LC particles rather than true crystallites as a disperse phase corresponding to the right-hand branch of the binodal at phase separation, then, apparently, non-full-dimensional centers (e.g., one-dimensional ones for a nematic, two-dimensional ones for a smectic) must appear at first. Their formation obeys all the regularities of crystallization. When the temperature is decreased, the configurative point crosses the LC liquidus. Extrapolation to the zero rate gives the temperature at which the heat movement in the corresponding directions is frozen so as to prevent the macromolecules acquiring the conformations necessary for the appearance of LC centers. If we considered full-dimensional crystalline centers, such a temperature would mean full loss of macromolecular mobility and vitrification of the polymer. However, since free dimension(s) is/are available, the process continues.

After the appearance of LC centers in some way, their further growth proceeds by other free dimensions (two for a nematic, one for a smectic), that is, preferably in one or two possible directions "sideways". The centers serve as a matrix for substance accumulation and easily orient the approaching chains with the need neither in the random appearance of the necessary conformation nor in the strict geometrical packing of segments. In this respect, the sideways growth is like liquid-liquid phase separation (except the induction period, of course). One can see that all the regularities of the corresponding kinetics are observed. As temperature decreases, the turbidity growth rate monotonically increases up to huge values, which enables extrapolation to infinity. Whereas the heat movement in some dimensions of the center is frozen down to stopping its growth in "breadth" (by crystal order), the heat movement in the other dimensions appears to be enough for providing high rates of the "sideways" growth (by liquid order).

In view of all our experiments and calculations, the above discussion, and the results of the works [34–37], one can assume that bringing the configurative point to the dashed area in Figure 5(a), between the binodal cupola and the vitrification curve, results in equilibrium between the isotropic and the LC phases. Bringing it below the liquidus would give LC-crystal equilibrium. In accordance with the principle of phase equilibria independence, the appearance of LC structures under the binodal occurs in the metastable (by theamorphous equilibrium) state. Naturally, the phases corresponding to LC ordering disappear after the final equilibrium is reached.

The unusual location of the LC state region in the concentration-temperature field of the phase diagram (Figure 5(a)) may have been caused by two factors. One is the specificity of the polymer-solvent interaction, which manifests itself in the variety of physical states: homogeneous and anisotropic solutions, gels, visually observed biphasic systems, and so on. The other is superposition of different types of phase separation. To the best of our knowledge, their state diagram is the first experimental confirmation of Papkov's hypothetical phase diagram [35], combining amorphous, LC, and crystal equilibria. The only difference is that the LC and crystal (partly) equilibria are inside the liquid-liquid phase equilibrium. Hence, the relations between the coexisting phases are understandable. At the same temperature (a constant χ_1), the lowest concentration of the equilibrium isotropic phase is peculiar to the liquid (amorphous) equilibrium. Therefore, the equilibrium concentration of the isotropic phase for the LC equilibrium is lower than the concentration of the same phase for the crystal equilibrium. It is quite probable that the narrow corridor of coexistence of the isotropic and anisotropic phases (replacing the wide biphasic range "isotropic solution + LC"), which follows from Flory's calculations for rigid-chain macromolecules [34] (with Frenkel's supplement for semirigid chains to allow for their specific interaction with the solvent [36]) and separates the biphasic region from the single-phase LC region (not specified in this article), is sharply skewed towards higher polymer concentrations. Taking into account the decrease in the CTA macromolecular rigidity with increasing temperature, this corridor may be expected to degenerate into a bend on the binodal.

In spite of the sophisticated character of the state diagram obtained, it obeys Gibbs' phase rule and the corresponding theoretical predictions [35, 38, 39], and it specifies the physical state of the polymer (insufficiently interpreted in [5]), corresponding to the boundaries separating the biphasic fields from the isotropic solution for all the three types of phase separation identified. Note also that the possibility of transition of cellulose esters to a crystal state through an LC phase was suggested in [37]. Though some results of this work were questioned and disputed, the general conclusion therein was supported in [35]. We finally note that with the exception of [35, 37], no papers are known to us in which models quantitatively describing the LC transition in cellulose derivative solutions have been presented.

Acknowledgments

The authors are indebted to Professor V. J. Klenin and the late Dr. G. N. Timofeyeva for the fruitful discussions of their results. They are thankful to Mr. D. N. Tychinin for his aid in the paper preparation.

References

[1] A. E. Chalykh, V. K. Gerasimov, and Y. M. Mikhailov, *Diagrams of the Phase State of Polymeric Systems*, Yanus-K, Moscow, Russia, 1998.

[2] V. J. Klenin, *Thermodynamics of Systems Containing Flexible-Chain Polymers*, Elsevier, Amsterdam, The Netherlands, 1999.

[3] A. B. Shipovskaya and G. N. Timofeeva, "The liquid-crystalline state of cellulose acetates: from past to present. Experience of the Saratov school," *Fibre Chemistry*, vol. 38, no. 1, pp. 17–21, 2006.

[4] G. N. Timofeyeva, E. S. Sveshnikova, and A. B. Shipovskaya, "The orientation phenomena in system acetate of cellulose of different degree of replacement—nitromethane," in *Proceedings of the Composite—2004 International Conference*, pp. 336–341, SGTU Press, Saratov, Russia, 2004.

[5] G. N. Timofeyeva, V. M. Averianova, and V. J. Klenin, "Thermokinetic studying of phase division in system triacetate cellulose-nitrometan," *Vysokomolekulyarnye Soedineniya B*, vol. 19, no. 1, p. 7, 1977.

[6] V. J. Klenin, S. Yu. Shchyogolev, and V. I. Lavrushin, *Characteristic Functions of Light Scattering of Disperse Systems*, Saratov State University, Saratov, Russia, 1977.

[7] C. Reichardt, *Solvents and Solvent Effects in Organic Chemistry*, VCH, Weinheim, Germany, 1988.

[8] W. R. Moor and J. Schuttlewort, "Solvation of cellulose derivatives," *Journal of Polymer Science B*, vol. 132, no. 1, pp. 89–91, 1964.
[9] A. S. Buntjakov and V. M. Averyanova, "Structure of solutions and films of cellulose acetate," *Journal of Polymer Science Part C*, no. 38, pp. 109–120, 1972.
[10] P. J. Flory, *Principles of Polymer Chemistry*, Cornell University Press, Ithaca, NY, USA, 1953.
[11] S. L. Shmakov, "Phase equilibrium in solutions of star-shaped macromolecules: an improved Okada-Numasawa model," *Polymer*, vol. 43, no. 4, pp. 1491–1495, 2001.
[12] V. I. Klenin, I. V. Fedusenko, and Y. I. Klokhtina, "Structure of crystallizable polymers in solutions: effect of dissolution technique," *Vysokomolekularnye Soedineniya*, vol. 45, no. 12, pp. 2054–2062, 2003.
[13] A. E. Chalykh, U. V. Nikulova, and A. S. Nikulov, "Phase balance and phase structure in system polystyrene-polivinil-metilovyj an ether," *Chemistry and Chemical Technology*, vol. 51, no. 12, pp. 28–32, 2008.
[14] S. P. Papkov, *Physicochemical Principles of Processing Polymer Solutions*, Chimiya press, Moscow, Russia, 1971.
[15] A. B. Shipovskaya and G. N. Timofeyeva, "Peculiarities of phase separation in the variacetylated cellulose acetate—mesophasogenic solvent (nitromethane) system," in *Struktura i Dinamika Molekuliarnykh Sistem*, vol. 10, part 2, pp. 222–225, Kazan University Press, Kazan, Russia, 2003.
[16] P. V. Kozlov and E. V. Gert, "Individual crystals of cellulose acetates," in *Chemistry and Technology of Cellulose Derivatives*, pp. 156–160, MITXT press, Moscow, Russia, 1971.
[17] A. Kuppel, H. Bittiger, and E. Husemann, "Das faserdiagramm des Triacetylcellulose-Nitromethankomplexes," *Kolloid-Zeitschrift und Zeitschrift für Polymere*, vol. 250, no. 6, pp. 623–624, 1972.
[18] P. Messien, "Cristallinite des acetates et triacetates de cellulose," *Bulletin de la Société Royale des Sciences de Liège*, vol. 37, no. 9-10, pp. 445–454, 1968.
[19] A. B. Shipovskaya, G. N. Timofeeva, N. O. Gegel, and S. Yu. Shchegolev, "Phase separation and liquid-crystal state of cellulose triacetate in nitromethane," *Journal of Engineering Physics and Thermophysics*, vol. 81, no. 6, pp. 1222–1231, 2008.
[20] S. P. Papkov and V. G. Kulichikhin, *Liquid Crystal State of Polymers*, Khimiya press, Moscow, Russia, 1977.
[21] V. G. Kulichikhin and L. K. Golova, "Liquid crystalline state of cellulose and its derivatives," *Chemistry of Wood*, vol. 5, pp. 9–27, 1985.
[22] V. V. Myasoyedova, S. Y. Belov, and G. A. Krestov, "Relations between rheological properties and phase transitions in non-aqueous solutions of cellulose triacetate and cellulose in the trifluoroacetic acid—1,2-dichloroethane mixture," *Vysokomolekulyarnye Soedineniya A*, vol. 29, no. 6, pp. 1149–1152, 1987.
[23] B. Y. Yunusov, O. A. Khanchich, M. V. Shablygin, O. A. Nikitina, and A. T. Serkov, "Structural transformations at solidifying anisotropic solutions of cellulose acetates in trifluoroacetic acid," *Vysokomolekulyarnye Soedineniya B*, vol. 25, no. 4, pp. 292–294, 1983.
[24] G. H. Meeten and P. Navard, "Gel formation and liquid crystalline order in cellulose triacetate solutions," *Polymer*, vol. 24, no. 7, pp. 815–819, 1983.
[25] S. M. Aharoni, "Rigid backbone polymers—8. Effects of the nature of the solvent on the lyotropic mesomorphicity of cellulose acetate," *Molecular Crystals and Liquid Crystals*, vol. 56, no. 7, pp. 237–241, 1980.
[26] G. N. Timofeeva and V. M. Averianova, "Studying kinetics phase division in student not forming system triacetate cellulose-nitrometan a rheological method," *Colloid Journal*, vol. 42, no. 2, pp. 393–397, 1980.
[27] A. B. Shipovskaya, S. L. Shmakov, and G. N. Timofeeva, "Phase processes and energetics of spontaneous change in the dimensions of acetate fibers in nitromethane vapors," *Polymer Science—Series A*, vol. 48, no. 5, pp. 509–519, 2006.
[28] A. B. Shipovskaya and G. N. Timofeeva, "New information concerning spontaneous elongation of acetate fibers," *Polymer Science—Series A*, vol. 49, no. 11, pp. 1210–1217, 2007.
[29] A. B. Shipovskaya and G. N. Timofeeva, "Self-organizing a macropierkul of cellulose acetates under the influence of steams of mezofazogen solvents," in *Sinergetika v Estestvennykh Naukakh*, N. O. Gegel and S. Yu. Shchyogolev, Eds., pp. 178–181, TverGU Press, Tver Cuty, Russia, 2008.
[30] S. Y. Frenkel, V. G. Baranov, B. M. Ginzburg, and G. K. Eliashevich, "Specific interactions polymer-solvent at crystallization and occurrence of a supramolecular order," *Doklady Akademii Nauk SSSR*, vol. 179, no. 5, pp. 1155–1158, 1968.
[31] H. Song, Y. Niu, Z. Wang, and J. Zhang, "Liquid crystalline phase and gel-sol transitions for concentrated microcrystalline cellulose (MCC)/1-ethyl-3-methylimidazolium acetate (EMIMAc) solutions," *Biomacromolecules*, vol. 12, no. 4, pp. 1087–1096, 2011.
[32] D. D. Grinshpan, T. A. Savitskaya, N. G. Tsygankova et al., "Cellulose acetate sulfate as a lyotropic liquid crystalline polyelectrolyte: synthesis, properties, and application," *International Journal of Polymer Science*, pp. 1–17, 2010.
[33] V. G. Kulichikhin, G. I. Kudryavtsev, and S. P. Papkov, "Rheological properties of liquid-crystalline polymer solutions," *International Journal of Polymeric Materials*, vol. 9, no. 3-4, pp. 239–256, 1982.
[34] P. J. Flory, "Phase equilibria in solutions of rodlike particles," *Proceedings of the Royal Society A*, vol. 234, no. 1, pp. 60–72, 1956.
[35] S. P. Papkov, "About a mutual combination liquid, liquid crystal and crystal phase balance in connection with a problem of morphology of natural polymers," *Vysokomolekulyarnye Soedineniya A*, vol. 26, no. 5, pp. 1083–1089, 1984.
[36] S. Y. Frenkel, "Thermokinetics of formation of ordered structures in polymer solutions and gels," *Pure and Applied Chemistry*, vol. 38, no. 1-2, pp. 117–149, 1974.
[37] A. Ciferri and W. R. Krigbaum, "Self-assembly of natural polymers via liquid crystalline phases," *Molecular Crystals and Liquid Crystals*, vol. 69, no. 3-4, pp. 273–280, 1981.
[38] W. Hu, D. Frenkel, and V. B. F. Mathot, "Lattice-model study of the thermodynamic interplay of polymer crystallization and liquid-liquid demixing," *Journal of Chemical Physics*, vol. 118, no. 22, pp. 10343–10348, 2003.
[39] A. Y. Grosberg and A. R. Khokhlov, *Statistical Physics of Macromolecules*, Nauka press, Moscow, Russia, 1989.

3

Ultrasonic, Molecular and Mechanical Testing Diagnostics in Natural Fibre Reinforced, Polymer-Stabilized Earth Blocks

C. Galán-Marín,[1] C. Rivera-Gómez,[1] and F. Bradley[2]

[1] *Departamento Construcciones Arquitectónicas I, Escuela Técnica Superior de Arquitectura, University of Seville, Avda. Reina Mercedes, 2, 41012 Seville, Spain*
[2] *Department of Architecture, Faculty of Engineering, University of Strathclyde, 16 Richmond Street, Glasgow G11XQ, UK*

Correspondence should be addressed to C. Galán-Marín; cgalan@us.es

Academic Editor: Gonzalo Martínez-Barrera

The aim of this research study was to evaluate the influence of utilising natural polymers as a form of soil stabilization, in order to assess their potential for use in building applications. Mixtures were stabilized with a natural polymer (alginate) and reinforced with wool fibres in order to improve the overall compressive and flexural strength of a series of composite materials. Ultrasonic pulse velocity (UPV) and mechanical strength testing techniques were then used to measure the porous properties of the manufactured natural polymer-soil composites, which were formed into earth blocks. Mechanical tests were carried out for three different clays which showed that the polymer increased the mechanical resistance of the samples to varying degrees, depending on the plasticity index of each soil. Variation in soil grain size distributions and Atterberg limits were assessed and chemical compositions were studied and compared. X-ray diffraction (XRD), X-ray fluorescence spectroscopy (XRF), and energy dispersive X-ray fluorescence (EDXRF) techniques were all used in conjunction with qualitative identification of the aggregates. Ultrasonic wave propagation was found to be a useful technique for assisting in the determination of soil shrinkage characteristics and fibre-soil adherence capacity and UPV results correlated well with the measured mechanical properties.

1. Introduction

The development of building systems has been inextricably linked throughout history with the evolution of construction materials and the technological advancements related to harvesting and exploiting our planet's natural resources [1]. In recent years, the construction sector has been under increasing pressure to reduce its CO^2 emissions and the volume of natural resources which it is responsible for consuming. Environmental concerns relating to the specification of contemporary materials which often involve energy-intensive and oil-dependent processes have become increasingly recognised [2] and with buildings, cities, and their associated infrastructure playing such a significant role in depleting our global resources, it is vital that material utilisation within buildings is specified with care in order to reduce the impact on our planet's resources and delicate ecosystems.

The purpose of this research was to explore the potential for developing a low embodied energy construction materials obtained where possible from natural, renewable resources. The main barrier to the use of natural materials at present, particularly in developed countries, is their perceived poor mechanical properties and durability in comparison with synthetic materials such as steel, concrete, and other ceramics. This study therefore explores the mechanical properties of an innovative, natural, unfired, composite brick designed to reduce both embodied energy values and CO^2 emissions.

Earth construction is not only cost effective, as a result of the inclusion of low-cost raw materials, but it also uses locally sourced, benign materials. As a building system, it is considered to be highly energy efficient due to the excellent thermal properties which earthen materials exhibit and it also possesses a low embodied energy when raw clay is utilised [3]. Adobe blocks, for example, do not undergo any energy-intensive firing processes since they are simply sun-dried and therefore harness solar energy directly. To put this in context, the energy required to produce an adobe block is only 5 (kWh)/cubic meter compared to about 1000 (kWh)/cubic metre for a fired brick and 400–500 (kWh)/cubic metre for concrete [4].

Earth construction is therefore becoming an increasingly valued natural building material and its durability benefits and minimization of pollution and waste characteristics—particularly in industrial countries—are also being progressively recognised [5]. With regards to unfired earth construction, Heath et al. [6] have recently shown that there is structural potential for utilizing commercially manufactured unfired bricks but concluded that additional research needs to be carried out into structural behaviour and methods for minimizing moisture susceptibility. This project therefore examines an innovative, sustainable, natural earth product to assess its initial performance against an extensive series of mechanical and analytical laboratory tests [7].

Chemical soil stabilization involves changing the properties of a soil by adding chemicals or additives. This occurs either by creating a matrix, which binds or coats the grains, or by means of a physiochemical reaction between the grains and the additive materials. Cement is one of the most widely used chemical stabilizers for compressed earth blocks (CEBs) and adding it before compaction improves the characteristics of the material, particularly its resistance to water [8–13]. A proportion of at least 5-6% of cement is generally needed to obtain satisfactory results [14]. When compaction of moist soil is used in combination with cement stabilisation, it not only improves compressive strength and water resistance compared to earth construction techniques such as "adobe," but also improves dimensional stability and tolerances improving construction quality and integrity [15].

Another method of stabilizing soil is with nonhydraulic lime (quicklime or slaked lime). This technique is commonly used for road construction, although it is mainly adopted in temporary roads. The use of this type of stabilizer is not recommended, however, for the manufacture of CEBs as these bricks require a fairly low moisture content and a soil with a relatively high sand content. For stabilization purposes the amounts generally used range from 6 to 12% that is equivalent to the proportion of cement used [16]. The disadvantage of using lime alone is its negative impact on durability as described in [17].

Cementing and waterproofing cohesive soils can be achieved with small amounts of natural or synthetic polymers proportionally less than 2% by dry weight of soil. Typical polymers used in soils comprise cement-resin mixes such as polymer cements or organic resins. These range from epoxy, acrylic, polyacrylate, polyurethane, polymers derived from tomato pulp to alginate, which is an extract from seaweed [18]. There are other recently researched methods relating to the stabilization of clay bricks described in [19] and a variety of techniques and compositions currently under investigation relating to fired and unfired bricks. This research work, however, focuses on 100% natural material ingredients, namely, clay, lignin, wool, and alginate.

2. Materials and Methods

The main objectives of this research were to analyse the effect on the mechanical properties of alginate added to hand-moulded bricks stabilized with natural fibre and to determine an optimal ratio for wool and alginate within three different soil types. The samples and methods that were selected are described in this section. X-ray diffraction (XRD), energy dispersive X-ray fluorescence (EDXRF), ultrasonic pulse velocity (UPV), and compression and bending tests were all conducted to provide a wide spectrum of data for analysis.

Table 1: Physical characteristics, grain size, and Atterberg limits of the three soils.

Physical characteristics	Errol	Ibstock	Raeburn
Sand content	22.50%	27.50%	35.00%
Silt content	45.00%	47.50%	40.00%
Clay content	32.00%	25.00%	25.00%
Classification I.S.S.S.	Silty clay loam	Silt loam	Loam
Liquid limit	34.8%	25.9%	25.9%
Plastic limit	19.1%	16.4%	16.8%
Plasticity index	15.7%	9.5%	9.1%

Figure 1: Photograph of the three soil types used.

2.1. Soil. The materials used in these experiments were three different types of clay soils, alginate, wool, and lignin. The physical properties and Atterberg limits of the three different types of alluvial soils used in this experimental investigation are described in Table 1. All the soils were supplied by Scottish brick manufacturers; Errol (from the East Coast of Scotland) and Ibstock and Raeburn from Glasgow (see Figure 1). All three soils had different colours and textures but importantly their particle-size distributions were all within the maximum limits specified for utilisation within CEBs.

The moisture content (in mass percentage) at which clays and silts pass from semisolid into plastic states and then into a liquid state is defined by the Atterberg limits, which are empirical divisions between the solid, plastic, and liquid limits of a clay. The upper and lower limits of the range of water content over which soils exhibit plastic behaviour are defined by liquid and plastic limits and the water content range between these values is termed the plasticity index [20]. The Errol soil has a much higher liquid limit compared to the other soils as can be seen in Table 1. The clay in each soil sample acts like cement in concrete, binding all the larger particles in the soil whereas the silt and sand particles behave as filters in the soil matrix in a similar manner to aggregates.

Errol soil is described as a silty clay loam and contains a significantly higher proportion of clay compared to either an Ibstock or Raeburn soil. The Ibstock soil is classified as a silt loam and the Raeburn soil is classified as a loam [20]. With regards to their plasticity indexes, it is interesting to note the quite remarkable variation (see Table 1). All soils were additionally analysed and characterized by utilizing X-ray diffraction (XRD) and X-ray fluorescence (EDXRF) tests.

2.2. Alginate. Seaweed is abundant within the coastal waters of countries across the globe and during the last two centuries has been used for a wide variety of products from food and medical products to soda ash production for soap and glass production. In terms of its chemical composition, alginic acid, also called algin or alginate, is a polysaccharide or carbohydrate molecule and it is obtained by extracting alginate salts from the cell walls of brown seaweeds. These alginate salts make up between 20 and 60% of the dry matter of the algae and take the form of different compounds including sodium alginate, calcium alginate, and magnesium alginate. The physical and chemical properties of alginates are nowadays being increasingly investigated and the polymer composition of this natural molecule is increasingly being understood to have a structural function within the cell walls and intercellular mucilage of seaweed [21]. The alginate matrix therefore contributes to the flexibility and mechanical strength of algae [22] in a similar manner to the way that cellulose and pectin components affect land-based plants [23]. Different algal species as well as geographical and environmental conditions influence alginate matrices which gives rise to the variations in properties that different alginates can exhibit.

Alginates are extremely versatile and exhibit important gelling properties as well as high water holding characteristics. Importantly in this research, they have the ability to act as a natural binding matrix within composite systems. Their natural propensity for improving viscosity and stabilizing emulsions has facilitated their widespread use today within the medical, pharmaceutical, and food industries where they are widely used as dental impression materials and gelling agents. Their colours range from white to yellowish brown and they are sold in a variety of forms including filamentous, granular, powdered, or gel forms.

Within the geotechnical engineering sector it, patents have been approved for the use of alginates within in situ stabilization of contaminated and non-contaminated soils [24] and a few previous tests such as those of Friedemann et al. [25] and Galán-Marín et al. [18, 26] have also been carried out incorporating alginate into building materials.

The initial selected proportions of the composite materials was derived from work previously carried out at the Laboratory of the Building Construction Department at the University of Seville and has been the subject of a patent [27]. The alginate used in our research was supplied by FMC Biopolymer, Girvan, Scotland (UK), under the name of seaweed extract and contained sodium alginate, sodium carbonate, and inorganic salt. In these experiments, we used an alginate paste (gluey, brown liquid), which is a product of the first stage of alginate extraction from seaweed.

2.3. Fibre. Traditionally, natural fibres have been used as soil reinforcement where available, to improve certain engineering properties of the soil. Vegetal fibres, derived from plants such as coir, jute, sisal, bamboo, wood, palm leaf, coconut leaf truck, cotton, hemp, and grass, have been tested as reinforcing materials not only for soils, but also within various polymer matrix composites [28] for utilisation within various industries.

Vegetal fibres such as coir can come in different varieties and the individual fibre cells are narrow and hollow, with thick walls made of cellulose. Coir is a relatively waterproof fibre and is one of the few natural fibers resistant to damage by salt water. Jute, in contrast, is a long, soft, shiny vegetable fibre similar to industrial hemp and flax (linen) and can be spun into coarse, strong threads and when woven is called hessian or burlap [29]. Another natural fibre that has recently been utilised in CEB research has been produced from cassava peels [17] and sugarcane bagasse ash [30] so there are a wide variety of vegetal fibres currently being examined in clay composites with regards to strength and flexural properties. Organic products containing cellulose fibres do however have several drawbacks such as an incompatibility with hydrophobic polymer matrices [31] and a propensity to show little resistance to prolonged moisture. For this reason this project has examined the behaviour of animal fibres which to date have tended to be overlooked as a constituent in unfired brick reinforcement.

Wool fibre is composed of a protein known as keratin. Generally, wool fibres measure 40–127 mm in length and 14–40 μ in width. Their cross-sectional shape is oval in form and the fibre grows in the form of a wave with a certain amount of twist. Its mechanical properties include a tensile strength 120–174 MPa, an elasticity component of 25–35% elongation at break and Young's modulus of 2.3–3.4 MPa [32]. Alternative figures listed in CES Edupak (2012) [29] give a tensile strength of between 40 Mpa and 200 Mpa and Young's modulus values between 3.9 MPa and 5.2 Mpa. Different species of sheep produce quite different types of wool with varied fibre length, diameter, and other differing physical characteristics. The molecular structure of wool fibres is interesting in that they comprise two different types of cell. Internal cells are referred to as the cortex and then outside these cells are external cuticle cells (or scales) that form a sheath around the fibre, overlapping like roof tiles. This structure gives wool its uniqueness compared to the variety of other fibres being used within natural composites today.

For the tests carried out in this project, wool fibre was added as the natural reinforcement within earth blocks, totally untreated and taken straight from the animal fleece so that no artificial additives were introduced. In addition, the wool was hand-cut, by trimming the top 10 mm strand of fibre, as longer fibres would have been too long to create a homogenous mix. All the specimens for this study were prepared and manufactured with the addition randomly oriented of a small amount (0.5–0.25%) of this raw, unprocessed wool according to recommendations from previous

Table 2: Mixes used (by weight).

Soil mix no.	Proportions	Soil	Alginate	Lignum	Wool	Water
1	Unstabilized soil	80.0%	—	0.5%	—	19.5%
2	Soil + alginate	79.5%	19.75%	0.5%	—	0.25%
3	Soil + 0.25% wool	79.5%	—	0.5%	0.25%	19.75%
4	Soil + alginate + 0.50% wool	79.0%	19.5%	0.5%	0.50%	0.50%
5	Soil + alginate + 0.25% wool	79.5%	19.5%	0.5%	0.25%	0.25%

Table 3: Mass percentages of each mineral and clay proportions.

Soil	Calcite	Quartz	Phyllosilicates	Feldspars	Illite	Kaolinite	Chlorite
Errol	<5	41	52	Traces	50	38	12
Ibstock	—	39	59	Traces	36	64	Traces
Raeburn	—	34	62	<5	27	69	4

research which looked at the impact of various proportions [18].

2.4. Lignin. Lignin is a treacle-like resin extracted from wood during the production of cellulose. Lignosulfonate and lignin products are therefore derived from a natural raw material. In all three mixes 0.5% of lignin sulfonate (under name of Additive A, Traffaid 45 by Borregaard LignoTech) was added to improve the workability of the soil mixture because it greatly facilitates the mixing of clay with low proportions of water. This additive is commonly used in the manufacturing processes for unfired bricks.

3. Sample Manufacturing Process

Material preparation was carried out manually in the laboratory of the University of Seville. Machine mixing was carried out using hand compaction and no extra compression was added. The mixtures adopted a proportion of close to 80 : 20 soil: (water + stabilizer) ratio after making necessary adjustments to previously carried out corrective dosages [26]. It was decided to choose a water/soil ratio of 19.5/80% (adding 0.5% of lignum) to get a normal consistency and low total shrinkage for mixes without the addition of alginate. Ratios of 0.25–0.5% of wool were added to mixes, where alginate was added to the mix and 0.25% only where no alginate was present (see Table 2). Soil Mixes numbered 1, 2, and 3 were used as contrast dosages to compare the effect of either the fibre or the polymer alone. Soil Mixes numbered 4 and 5 were manufactured to detect the appropriate quantity of fibre reinforcement. Reinforcing wool fibres were cut to the required length, soaked in water for 24 hours, to improve mixing, and then added randomly, but in a homogeneous way, to the moist soil using a 5 litre mixer until a completely homogeneous composite was achieved. All the brick samples used in this study were prismatic specimens (160 × 40 × 40 mm) in accordance with the European standards format for the mechanical testing of mortar tests for masonry [33].

In this research study three different soils were tested and five different mix combinations. For each batch, seven specimens were tested. All specimens were first placed in an oven at 50°C to dry for 24 hours and subsequently dried at room temperature for 48 hours before unmolding. Specimens were cured in the uncontrolled laboratory environment at 20–25°C and about 65% RH. A different consistency and workability was observed during the manufacturing process, for the Errol mixes. This was due to the higher percentages of illite within the Errol soil compared to Raeburn or Ibstock, which allowed more water to be absorbed within the crystal matrix.

3.1. X-Ray Diffraction (XRD). In order to identify and quantify the presence of the different mineral components within the clays and the phyllosilicates in each soil, the researchers followed a standard protocol, set out in a standard preparation protocol procedure, entitled PNT07LRX0044 [34]. The experiments were carried out in the laboratory within the University of Seville and the protocol determined the percentage composition of the small illite, kaolinite, and chlorite grains within each sample, using the oriented aggregates method which led to the proportions described in Table 3.

The oriented aggregates study included dissolving the samples in ethylene glycol and then subjecting them to a heat treatment between 350 degrees and 550 degrees. Figures 2, 3, and 4 show the phyllosilicate and quartz fractions over the majority of each soil type and the percentage and proportion of each mineral is shown in Table 2. Testing and analysis was performed in this way relating to all the different groups of clays prevailing within each soil.

3.2. X-Ray Fluorescence Spectroscopy (XRF). XRF is mostly a quantitative technique—the peak-height for any element being directly related to the concentration of that element within the sampling volume. However, extreme care must be taken because two or more elements can interact with each other, resulting in contamination and thus skewing results. XRF tests showed (see Table 4) the chemical composition of three soils (samples dried at 110°C).

3.3. Energy Dispersive X-Ray Fluorescence (EDXRF). Energy dispersive X-ray fluorescence is one of two general types of X-ray fluorescence techniques used for elemental analysis

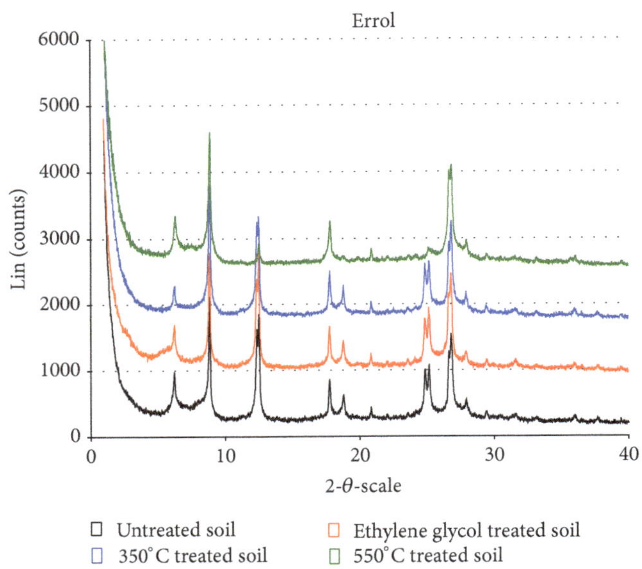

FIGURE 2: XRD patterns of the Errol soil oriented aggregates.

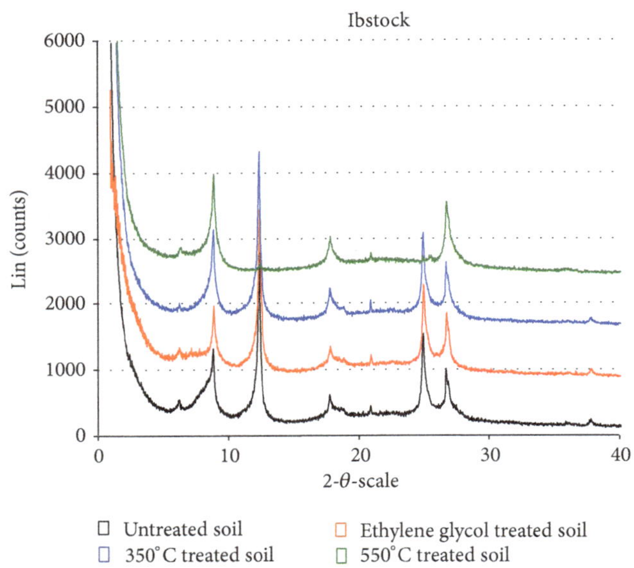

FIGURE 3: XRD patterns of the Ibstock soil oriented aggregates.

applications. These tests show the chemical composition of the three soils (samples dried at 110°C), their main elements, and traces (see Table 5).

3.4. Ultrasonic Pulse Velocity Testing (UPV). Application of ultrasonic methods for the testing of materials including polymer composites has a long-lasting tradition. In this context, ultrasound's physical nature as a mechanical wave is used and knowledge of sound wave propagation characteristics in a tested medium allows for a theoretical analysis of a phenomenon.

On the basis of wave parameters on the boundary of an area, conclusions can be drawn concerning geometric

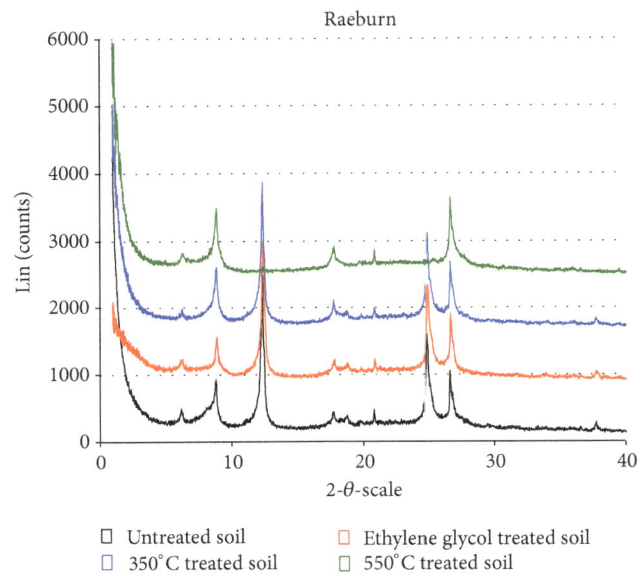

FIGURE 4: XRD patterns of the Raeburn soil oriented aggregates.

TABLE 4: Chemical composition of three soils.

		Errol	Ibstock	Raeburn
SiO_2	%	56,53	56,93	51,72
Al_2O_3	%	17,02	19,57	20,19
Fe_2O_3	%	7,09	4,81	6,51
MnO	%	0,10	0,06	0,10
MgO	%	2,71	1,21	1,50
CaO	%	2,07	0,60	0,74
Na_2O	%	1,53	0,23	0,28
K_2O	%	3,18	2,31	2,22
TiO_2	%	0,95	1,02	0,95
P_2O_5	%	0,15	0,12	0,18
SO_3	%	0,03	0,02	0,03
PC	%	6,02	10,59	13,34
TOTAL	%	97,39	97,47	97,77

properties and the distribution of physical properties within a medium—in this case unfired natural bricks.

Ultrasonic tests were carried out with the ultrasonic model brand BPV Krautkramer. This equipment gives the delay time from when a transmitted wave leaves the probe until the wave is received back by the probe. It is measured by a liquid crystal five digit display which measures the reading time in microseconds. The accuracy is ±0.1 μs. Measurements were recorded by cylindrical transducers and the emission frequency of the probes was 50 KHz. A single ultrasonic head was used and the time of a sound wave transition through tested samples (τ), was expressed in microseconds. The sound wave velocity (V) through a sample was then calculated using the following formula:

$$V = \frac{h}{\tau}, \quad (1)$$

TABLE 5: Chemical composition of three soils (main elements and traces).

	Cl ppm	Sc ppm	V ppm	Cr ppm	Co ppm	Ni ppm	Cu ppm	Zn ppm	Ga ppm	Ge ppm	As ppm	Se ppm	Br ppm
Errol	32	15	106	111	16	53	38	86	22	N.D.	14	N.D.	N.D.
Ibstock	35	12	83	102	15	51	28	61	23	N.D.	4	0	N.D.
Raeburn	34	12	83	101	17	57	29	68	24	N.D.	6	0	N.D.
	Rb ppm	Sr ppm	Y ppm	Zr ppm	Nb ppm	Mo ppm	Ag ppm	Cd ppm	Sn ppm	Sb ppm	Te ppm	I ppm	Cs ppm
Errol	106	152	24	178	15	1	N.D.	N.D.	4	2	N.D.	N.D.	3
Ibstock	111	184	28	188	17	1	N.D.	N.D.	4	3	N.D.	N.D.	5
Raeburn	102	217	28	167	15	1	N.D.	N.D.	4	4	N.D.	N.D.	5
	Ba ppm	La ppm	Ce ppm	Nd ppm	Sm ppm	Yb ppm	Hf ppm	Ta ppm	Tl ppm	Pb ppm	Bi ppm	Th ppm	U ppm
Errol	582	40	72	33	3	1	4	N.D.	0	22	N.D.	11	1
Ibstock	400	45	78	35	6	3	4	N.D.	0	23	0	13	2
Raeburn	402	41	81	37	7	2	4	0	N.D.	24	N.D.	14	2

FIGURE 5: Ultrasonic tests.

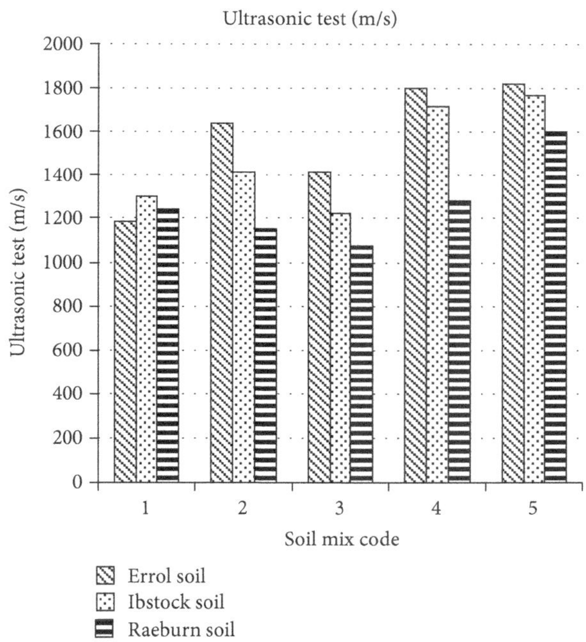

FIGURE 6: Graphical comparison of the UPV results on the five different mixes of the three types of soils.

where h was the sample thickness. All the UPV tests were carried out on the prism specimens as per the guidelines of UNE-EN 583-1/A1 [32]. Test pieces with dimensions 160 mm × 40 mm × 40 mm were tested perpendicularly to the 40 mm × 40 mm plane as shown in Figure 5 and the results are shown in Figure 6. A comparison between a series of mechanical and UPV test results for all soils and mixes is shown in Table 6.

4. Mechanical Test Results

4.1. Compression Tests. After breaking samples roughly in half, three-point bending strength tests were used to determine compressive strength and a total of 210 compressive strength tests were carried out (see Table 6 and Figure 7).

4.2. Flexural Tests. Bending strengths were determined by carrying out a three-point bending test on the specimens, in agreement with the specifications of UNE-EN 1015-11:2000 European standards. This standard is for the determination of bending strength of mortars used for rough castings and mortar linings, but it was decided to adopt this standard in the absence of other specific regulations. Several papers including the study carried out by Raut et al. [19] have illustrated the wide range of testing and curing regimes currently being carried out in different laboratories as well as the range of sizes of earth brick samples. Furthemore and Heath et al. [6] have additionally discussed the current limitations in Eurocode 6: Design of masonry structures (BS EN 1996-2:2006) and the fact that it does not currently have a section referring to earth masonry. Within this context, it was felt that UNE-EN1015-11:2000 was an appropriate standard to adopt.

All the flexural tests were conducted at room temperature (20°C) on a Codein S.L., MCO-30/139 machine (maximum

TABLE 6: Mechanical and UPV tests results of three soils.

Mix code	01 Unstabilized soil	02 Soil + alginate	03 Soil + 0.25% wool	04 Soil + alginate + 0.50% wool	05 Soil + alginate + 0.25% wool
Errol soil					
Compressive strength (MPa)	2,23	3,77	3,05	4,37	4,44
Flexural strength (MPa)	1,12	1,06	1,1	1,08	1,45
Ultrasonic testing (m/s)	1182	1637	1416	1798	1818
Ibstock soil					
Compressive strength (MPa)	2,06	2,49	1,89	3,43	3,59
Flexural strength (MPa)	0,97	0,98	0,96	1,28	1,61
Ultrasonic testing (m/s)	1298	1413	1222	1720	1769
Raeburn soil					
Compressive strength (MPa)	2,44	2,24	1,88	2,69	3,75
Flexural strength (MPa)	1,12	1,1	0,93	1,11	1,24
Ultrasonic testing (m/s)	1240	1153	1075	1280	1604

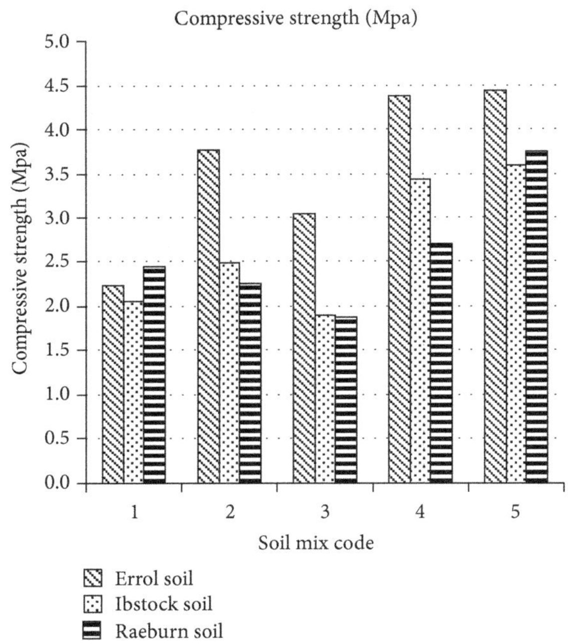

FIGURE 7: Graphical comparison of the compressive strength results on the five different mixes of the three types of soils.

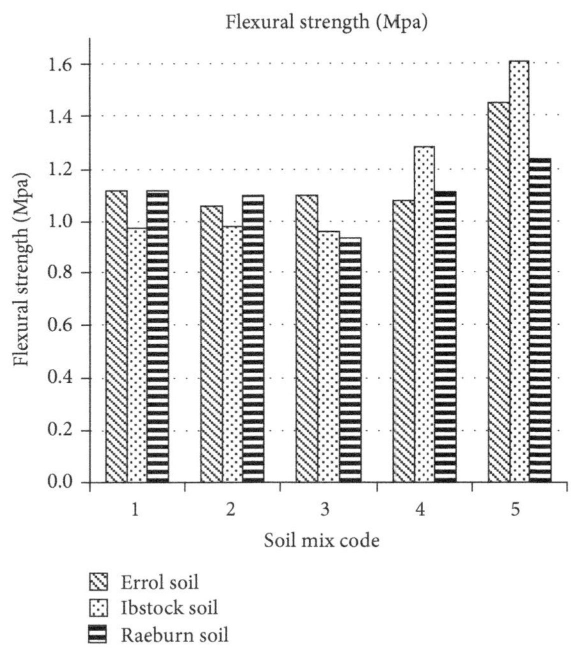

FIGURE 8: Graphical comparison of the flexural strength results on the five different mixes of the three types of soils.

load 10 kN) in a three-point bending configuration. The support length was adapted to the size of sample, and at least seven specimens were tested for each mix under study. A total of 105 tensile tests were therefore conducted both with the fibrous and non-fibrous samples and the mechanical properties determined from these tests included the flexural modulus, the ultimate stress, and the ultimate strain (see Table 6 and Figure 8).

5. Results and Discussion

It is generally accepted that incorporation of fibres increases flexural strength. Our tests have not shown the expected improved flexural strength, indeed sometimes just the opposite, as is shown in Figure 8 for the Ibstock soil. In fact the addition of just fibre to soils, without the presence of alginate (as indicated in the comparison between mixes 03 and 01), does not increase flexural strength. Additionally, compressive strength in samples with no alginate and only fibres is increased only in the soil with the high plasticity index, namely, Errol. Indeed, both the flexural and compressive strengths of the Raeburn and Ibstock soils decreased in the mixture only containing wool fibres without the polymer.

In contrast, adding wool into the soil mixed with alginate always increased flexural resistance, especially if a proportion of 0.25% of wool was added. The incremental strength improvement was particularly significant (65%), for the Ibstock soil specimens, where flexural resistance changed from 0.97 MPa (plain soil specimens) to 1.60 MPa (0.25% wool + soil + alginate mixes).

As the XRD analytical results show, higher percentages of illite can be found in the Errol soil (compared to Raeburn or Ibstock). This phyllosilicate appears to allow more water to be absorbed within the crystal matrix giving rise to a drier mix and this finding is consistent with the plasticity indexes obtained for the different soils showed in Table 1. The consequences of this drier and therefore stiffer consistency can be observed in the mechanical results.

UPV measurements (see Figure 6) demonstrated that Errol specimens, in most of the mixes tested, provided much higher compactness and with regards to mechanical tests, Errol specimens reached higher resistance values in the compression tests compared with the other soils. Flexural tests, however were less conclusive, with Errol providing higher flexural strengths in mix 3, Raeburn in mix 2, and Ibstock in mixes 4 and 5. Flexural strength values were equal in the Raeburn and Errol soils in mix 1 and the UPV results showed a pulse velocity increase for mixes 4 and 5 compared to mixtures 1, 2, and 3. Specimens of any type of soil tested for mix 5 (the mix with the lower quantities of wool reinforcement) showed higher UPV values than mix 4. This was particularly the case with the Raeburn soil which had the lowest plasticity index. This compactness decrease could be caused by the fact that the shrinkage values for the fibre were much higher than the soil shrinkage.

Lower compressive and flexural resistance values were obtained when larger quantities of fibre were used in mix 4. This could be explained as follows: the development of strength properties in the fibre/soil mixes mostly depends on the formation of fibre-matrix bonds as has been shown in previous studies [35]. Bonding is affected by fibre dimensions, surface textures, and the number of fibres present in a given volume of material. Increasing wool fibre quantities gives rise to fibre agglomeration and folding of fibres (balling) and this can result in a decrease in the bond strength within the specimens, which in turn leads to lower compressive strength values. In addition to the XRD, XRF, and EDXRF results, the ultrasonic results were incorporated into this study's testing regime in order to indicate the prevalence of voids and compare the relative material densities. The results of these tests within each sample confirm that mix 5 had the best overall engineering properties and it is suggested that this was due to the quality of bonding within the composite matrix and the overall homogeneity of the mixture.

The highest compressive strength, in all three soil types, was obtained with the composite specimens including both alginate and wool reinforcement and better results were obtained with the reduced quantity of wool, that is, the 0.25% mix 04. By adding wool to the mix, the Ibstock soil specimens stabilized with alginate improved their compressive strengths by 74% whilst the Raeburn soil specimens improved their compressive strengths by 54%.

Various research papers have shown that hygrometric shrinkage and its associated cracking of earth-based materials can be greatly reduced by introducing fibres into the mixture and this study confirmed these findings by demonstrating that shrinkage due to the drying process was significantly reduced with the inclusion of natural fibres into the soil mix. As would be expected, specimens of plain soil had a very quick (and almost without warning), brittle failure mode. In contrast, fibre-reinforced mixes deformed, after the ultimate load was reached and fine cracks could be seen on the surface giving warning before failure.

6. Conclusions

Soil characterization through XRD, XRF, and EDXRF tests has proved to be very important in order to understand the different mechanical behaviour of the different stabilised soils (with the same fibre and stabilizer content) and therefore the effect of the stabilization itself.

Mixes 2, 4, and 5 (those that included the alginate polymer as a stabilizer) showed better results in UPV tests in every type of soils, especially though the Errol soil and these results are consistent with the compression test results. As would be expected, flexural resistance values generally increased in mixes 4 and 5, where alginate and fibre were used simultaneously.

The use of UPV, in this study, has added an interesting additional data set with results which closely align with the mechanical compressive strength results. Errol soil, due to its higher content of illite, contains a crystalline structure that facilitates a higher level of water absorption compared to the Raeburn and Ibstock kaolinite crystalline structures. Improved results were therefore obtained with the Errol soil due to the higher plasticity index related to a higher proportion of illite within the clay fraction.

The addition of short wool fibres (10 mm long) randomly oriented to the mixes leads to a decrease in bulk density which correspondingly decreases the compressive strength of the specimens. Therefore UPV measurements were useful in determining the resultant final porosity of the dried mix, after the shrinkage process.

Fibre water adsorption and soil-fibre surface friction, due to the drying shrinkage characteristics of a fibre, depend on the available water and this in turn depends on the characteristics of the soil plasticity for different types of clay. The higher plasticity index obtained for the Errol soil seems to be responsible for its different behaviour observed in the various tests compared with the Ibstock and Raeburn soils, in mixtures with similar water proportions.

As a result of the range of data presented within this paper, it has therefore been shown that it is possible to prepare 100% green composites from natural fibres and natural polymers with mechanical performance results within ranges compatible with producing unfired bricks. Further research is currently under development to investigate methods for improving composite bonding and interaction and ultimately durability.

Acknowledgments

The authors wish to acknowledge the IUACC "Instituto Universitario de Arquitectura y Ciencias de la Construcción" for the necessary support to develop this research. They would also wish to acknowledge the CITIUS "Centro de Investigación, Tecnología e Innovación" of the University of Seville, X-Ray Laboratory, where these tests were carried out.

References

[1] J. M. Saiter, L. Dobircau, and N. Leblanc, "Are 100% green composites and green thermoplastics the new materials for the future?" *International Journal of Polymer Science*, vol. 2012, Article ID 280181, 7 pages, 2012.

[2] M. F. Ashby, *Materials and the Environment: Eco-Informed Material Choice*, Elsevier, Oxford, UK, 2009.

[3] P. McHenry, *Adobe and Rammed Earth Construction. Adobe, Build It Yourself*, 1994.

[4] A. Debnath, S. V. Singh, and Y. P. Singh, "Comparative assessment of energy requirements for different types of residential buildings in India," *Energy and Buildings*, vol. 23, no. 2, pp. 141–146, 1995.

[5] J. O. Arumala and T. Gondal, *Compressed Earth Building Blocks for Affordable Housing*, RICS, COBRA, London, UK, 2007.

[6] A. Heath, D. Maskell, P. Walker, M. Lawrence, and C. Fourie, "Modern earth masonry: structural properties and structural design," *Structural Engineering*, vol. 90, no. 4, pp. 38–44, 2012.

[7] E. A. Adam and P. J. Jones, "Thermophysical properties of stabilised soil building blocks," *Building and Environment*, vol. 30, no. 2, pp. 245–253, 1995.

[8] D. A. Adesanya, "Evaluation of blended cement mortar, concrete and stabilized earth made from ordinary Portland cement and corn cob ash," *Construction and Building Materials*, vol. 10, no. 6, pp. 451–456, 1996.

[9] L. K. Aggarwal, "Bagasse-reinforced cement composites," *Cement and Concrete Composites*, vol. 17, no. 2, pp. 107–112, 1995.

[10] V. Agopyan, H. Savastano Jr., V. M. John, and M. A. Cincotto, "Developments on vegetable fibre-cement based materials in São Paulo, Brazil: an overview," *Cement and Concrete Composites*, vol. 27, no. 5, pp. 527–536, 2005.

[11] E. A. Basha, R. Hashim, H. B. Mahmud, and A. S. Muntohar, "Stabilization of residual soil with rice husk ash and cement," *Construction and Building Materials*, vol. 19, no. 6, pp. 448–453, 2005.

[12] P. R. Blankenhorn, B. D. Blankenhorn, M. R. Silsbee, and M. DiCola, "Effects of fiber surface treatments on mechanical properties of wood fiber-cement composites," *Cement and Concrete Research*, vol. 31, no. 7, pp. 1049–1055, 2001.

[13] R. S. P. Coutts, "Fibre-matrix interface in air-cured wood-pulp fibre-cement composites," *Journal of Materials Science Letters*, vol. 6, no. 2, pp. 140–142, 1987.

[14] R. Bahar, M. Benazzoug, and S. Kenai, "Performance of compacted cement-stabilised soil," *Cement and Concrete Composites*, vol. 26, no. 7, pp. 811–820, 2004.

[15] J.-C. Morel, A. Pkla, and P. Walker, "Compressive strength testing of compressed earth blocks," *Construction and Building Materials*, vol. 21, no. 2, pp. 303–309, 2007.

[16] J. R. Prusinski and S. Bhattacharja, "Effectiveness of portland cement and lime in stabilizing clay soils," *Transportation Research Record*, vol. 1, no. 1652, pp. 215–227, 1999.

[17] M. Villamizar, V. Araque, C. Reyes, and R. Silva, "Effect of the addition of coal-ash and cassave peels on the engineering properties of compressed earth blocks," *Construction and Building Materials*, vol. 36, pp. 276–286, 2012.

[18] C. Galán-Marín, C. Rivera-Gómez, and J. Petric-Gray, "Effect of animal fibres reinforcement on stabilized earth mechanical properties," *Journal of Biobased Materials and Bioenergy*, vol. 4, no. 2, pp. 121–128, 2010.

[19] S. P. Raut, R. V. Ralegaonkar, and S. A. Mandavgane, "Development of sustainable construction material using industrial and agricultural solid waste: a review of waste-create bricks," *Construction and Building Materials*, vol. 25, no. 10, pp. 4037–4042, 2011.

[20] R. F. Craig, *Craig's Soil Mechanics*, Spon Press, New York, NY, USA, 7th edition, 2004.

[21] I. D. Hay, Z. U. Rehman, and B. H. Rehm, "Membrane topology of outer membrane protein AlgE which is required for alginate production in pseudomonas aeruginosa. Identification of two aflatrem biosynthesis gene loci in aspergillus flavus and metabolic engineering of penicillium paxilli to elucidate their function," *The American Society For Microbiology*, vol. 76, no. 6, pp. 1806–1812, 2010.

[22] I. L. Andresen, O. Skipnes, O. Smidsrod, K. Ostgaard, and P. C. Hemmer, "Some biological functions of matrix components in benthic algae in relation to their chemistry and the composition of seawater," *ACS Symposium Series of the American Chemical Society*, vol. 48, pp. 361–381, 1977.

[23] I. Donati and S. Paoletti, "Materials properties of alginates," in *Alginates: Biology and Applications*, B. H. A. Rehm, Ed., pp. 1–54, Springer, Berlin, Germany, 2009.

[24] G. L. Schneider, "Method of stabilizing soils," USA Patent 4, 002, 633, 1977.

[25] K. Friedemann, F. Stallmach, and J. Kärger, "NMR diffusion and relaxation studies during cement hydration-A non-destructive approach for clarification of the mechanism of internal post curing of cementitious materials," *Cement and Concrete Research*, vol. 36, no. 5, pp. 817–826, 2006.

[26] C. Galán-Marín, C. Rivera-Gómez, and J. Petric, "Clay-based composite stabilized with natural polymer and fibre," *Construction and Building Materials*, vol. 24, no. 8, pp. 1462–1468, 2010.

[27] C. Galán-Marín and G. Rivera-Gómez, "Method for stabilising clay soils with natural organic polymers reinforced with animal fibre," Patent WO2012/101299, 2012.

[28] N. Uddin and R. R. Kalyankar, "Manufacturing and structural feasibility of natural fiber reinforced polymeric structural insulated panels for panelized construction," *International Journal of Polymer Science*, vol. 2011, Article ID 963549, 7 pages, 2011.

[29] G. Design and M. F. Ashby, *CES Edupak: Material Selector*, 2012.

[30] S. Lima, H. Varim, A. Sales, and V. Neto, "Analysis of the mechanical properties of compressed earth block masonry utilizing the sugarcane bagash ash," *Construction and Building Materials*, vol. 35, pp. 829–837, 2012.

[31] A. L. Martínez-Hernández, C. Velasco-Santos, M. de-Icaza, and V. M. Castaño, "Dynamical-mechanical and thermal analysis of polymeric composites reinforced with keratin biofibres from chicken feathers," *Composites B*, vol. 38, pp. 405–410, 2007.

[32] "UNE-EN 583-1/A1. Non-destructive testing. Ultrasonic examination. Part 1: general principles," AENOR, European Committee for Standardization, 2004.

[33] "UNE-EN, 1015-11:2000/A1. Methods of test for mortar for masonry. Part 11: determination of flexural and compressive strength of hardened mortar," AENOR, European Committee for Standardization, 2000.

[34] reparation Protocol System PNT 07LRX0044, CITIUS, "Centro de Investigación, Tecnología e Innovación" (Research Center for Technology and Innovation) of the University of Seville, 2011.

[35] C. Galán-Marín, C. Rivera-Gómez, and F. Bradley, "The mechanical properties and molecular bonding characteristics of clay-based natural composites reinforced with animal fibres," *Journal of Biobased Materials and Bioenergy*, vol. 7, pp. 143–151, 2013.

Effect of Gamma-Irradiated Recycled Low-Density Polyethylene on the High- and Low-Temperature Properties of Bitumen

Perviz Ahmedzade,[1] Alexander Fainleib,[2] Taylan Günay,[3] Olga Starostenko,[2] and Tetyana Kovalinska[4]

[1] *Transportation Division, Department of Civil Engineering, Faculty of Engineering, Ege University, Izmir, Turkey*
[2] *Institute of Macromolecular Chemistry of the National Academy of Sciences of Ukraine (NASU), Kyiv, Ukraine*
[3] *Department of Civil Engineering, Ege University, Izmir, Turkey*
[4] *Institute of Nuclear Research of the NASU, Kyiv, Ukraine*

Correspondence should be addressed to Perviz Ahmedzade; perviz.ahmedzade@ege.edu.tr

Academic Editor: Gonzalo Martínez-Barrera

This paper describes polymer modification of bitumen with gamma-irradiated recycled (γ-LDPE$_R$) low-density polyethylene. The recycled low-density polyethylene (LDPE$_R$) was obtained from greenhouse films exposed to sunlight at least one year. The surface of the LDPE$_R$ was treated by gamma beam irradiation that provided formation of free radicals and some functional groups that may contribute to the creation of strong chemical bonds between polymer modifier and bitumen. Five different samples of bitumen/γ-LDPE$_R$ compositions with the modifier content, wt. %: 1, 3, 5, 7 and 9, were prepared. The effects of the γ-LDPE$_R$ on original and aged bitumen were investigated by means of morphological, chemical, and physical testing program, including FTIR spectroscopy, conventional tests, rotational viscosity (RV), dynamic shear rheometer (DSR), and bending beam rheometer (BBR) tests. Superior performing asphalt pavements (Superpave) specifications were used to analyze mechanical test results as well as to determine the performance grades (PG) of the binders. Optimum usage of the γ-LDPE$_R$ as modifier in bitumen were suggested after testing program. The results reveal the stiffening effect of the γ-LDPE$_R$ on bitumen that provide enhanced temperature susceptibility and also promise better performance grades (PG) with γ-LDPE$_R$ polymer modification.

1. Introduction

The modification of bitumen by means of polymers is the most widespread method in flexible pavement applications. The polymers such as Styrene Butadiene Styrene (SBS), Ethylene vinyl acetate (EVA) are satisfactorily used to enhance temperature susceptibility by increasing stiffness at high temperature and decreasing stiffness at low temperature [1–4]. In addition to SBS, various polymers are utilized in bituminous materials. Polyolefins such as low-density polyethylene (LDPE), high density polyethylene (HDPE), and polypropylene (PP) have been used as modifier, for generally to enhance mechanical properties of bitumen [5–8].

On the other hand, using recycled polymers contributes to reducing amount of waste materials; it has many benefits, such as environmental protection, lower energy consumption, and their affordable cost. Therefore, it is fundamental to find an application area for recycled polymers, in order to increase their commercial use [9–11].

The aim of this work is to study the effect of recycled LDPE (LDPE$_R$) as modifier on bitumen by means of morphologic, chemical, and mechanical test methods. However, there is no chemical reaction between LDPE$_R$ and bitumen which results in a two-phase mixture after the preparation of modified bitumen. In order to provide strong polymer/bitumen bonding, surface of the recycled LDPE was activated by using gamma beam irradiation. The surface activation helps to create double bonds (>C=C<) that can interact with bitumen.

The study contains a comprehensive laboratory evaluation of the γ-irradiated recycled LDPE (γ-LDPE$_R$) modified bitumen (γ-LDPE$_R$/B) in terms of morphology and physical-chemical properties. Fluorescence microscopy has been used

TABLE 1: Physical properties of the base bitumen.

Properties	Standard	160/220
Penetration (25°C; 0.1 mm)	ASTM D5	195.5
Softening point (°C)	ASTM D36	38.7
Ductility (25°C; cm)	ASTM D113	103
Penetration index (PI)	—	−0.73
Specific gravity (25°C; gr/cm^3)	ASTM D70	1.03

to observe morphological changes, while rheological testing have been undertaken by means of dynamic shear rheometer (DSR) and bending beam rheometer (BBR). To understand the influence of γ-irradiation on structure and efficiency of the $LDPE_R$ recycled LDPE, Fourier transform infrared (FTIR) spectroscopy has been performed as well.

2. Experimental

2.1. Materials. The binders were prepared with the type of bitumen having 160/220 penetration grade. Physical properties of the bitumen are given in Table 1. The $LDPE_R$ used as modifier in bitumen was supplied from Sicily, Italy. The $LDPE_R$ was obtained from greenhouse films exposed to sunlight, as well as to other external environmental factors for one year. The waste films were washed, dried, cut into the pieces of dimensions 12 × 12 mm, and finally extruded. The chemical composition of the $LDPE_R$ used was LDPE 65–70%, LLDPE 12–17%, EVA copolymer 12–15%, T_m = 109°C, E = 180 MPa, TS = 16 MPa, EB = 500%, MFI190/2.16 = 0,29 g/10 min, and MFI230/2.16 = 0,95 g/10 min.

2.2. Gamma Irradiation Activation Method. In order to provide a chemical bonding between bitumen and polymer modifier, gamma irradiation method was applied to the $LDPE_R$ [12]. Gamma irradiation method is used in several applications such as food processing, cancer treatments, and some sterilization systems. Gamma irradiation is electromagnetic of high frequency and rays are ionizing radiation. It provides a reformed chemical composition of the materials exposed to decay that might contribute to creating strong bonds between polymer modifier and bitumen. The irradiation source was an electron accelerator—complex electric device, where the transition from regimes of treatment with charged particles (electrons) to regimes of treatment with stream of gamma quanta of a wide range of energies is possible. The $LDPE_R$ granules were γ irradiated with a dose of 20 kGy.

2.3. Preparation of Samples. γ-$LDPE_R$ granules were obtained as pellet shaped. In order to work with smaller pieces, they were milled by using grinder and the particles were sieved with No 50. The amount of additives was selected as %: 1, 3, 5, 7, and 9 by total weight of the binder which is enough to examine the effect of the polymer modified bitumen. High shear mixer was employed for preparing sample. 160/220 penetration grade bitumen was heated for 90 minutes at 163°C and then poured into the mixer flask adjusted to 500 rpm. Subsequently, the γ-$LDPE_R$ was added to bitumen by portions in 15 minutes at certain intervals and then the mixing rate was increased to 1300 rpm and mixing was continued for 150 minutes. After the end of the mixing process, the samples were removed from the flask, divided into small containers, covered with aluminum foil, and stored for various testing.

The different binders were coded as follows:

(i) base bitumen–"B;"

(ii) base bitumen + 1% γ-$LDPE_R$–"B-1-γ-$LDPE_R$;"

(iii) base bitumen + 3% γ-$LDPE_R$–"B-3-γ-$LDPE_R$;"

(iv) base bitumen + 5% γ-$LDPE_R$–"B-5-γ-$LDPE_R$;"

(v) base bitumen + 7% γ-$LDPE_R$–"B-7-γ-$LDPE_R$;"

(vi) base bitumen + 9% γ-$LDPE_R$– "B-9-γ-$LDPE_R$;"

2.4. Testing Program

2.4.1. Fourier Transform Infrared Spectroscopy. Fourier transform infrared (FTIR) spectroscopy was used to characterize various functional groups in asphalt and polymer modifier. FTIR spectra were recorded with a Bruker Tensor 27 DTGS spectrometer between 4000 and 450 cm^{-1} using the attenuated total reflection (ATR) mode. For each spectrum, 32 consecutive scans with a resolution of 4 cm^{-1} were averaged.

2.4.2. Morphology. In order to determine the grade of the dispersion of the modifier within the bitumen, fluorescent microscopy was used. Florescent microscopy allows researchers to study the morphology of bitumen by using a principle in which polymers become swollen after absorbing some of the constituents of the original bitumen [13].

The method of sample preparation for fluorescent microscopy followed the regular procedure that is consisted of heating, homogenizing, and cooling process. The samples were examined under a Carl Zeiss Primo Star generated from a 40 W halogen lamp and magnify up to 1000X.

2.4.3. Conventional Bitumen Tests. The samples of the base and modified bitumen were subjected to conventional tests that are, namely, penetration (ASTM D5), softening point (ASTM D36), and ductility (ASTM D 113). By using penetration and softening point values, penetration index was calculated for each sample of bitumen to investigate the temperature susceptibility of bitumen. A classical approach related to PI calculation has been given in the Shell Bitumen Handbook [14] as shown in the following equation:

$$\mathrm{PI} = \frac{1952 - 500 \times \log(\mathrm{Pen}_{25}) - 20 \times \mathrm{SP}}{50 \times \log(\mathrm{Pen}_{25}) - \mathrm{SP} - 120}, \quad (1)$$

where Pen_{25} is the penetration at 25°C and SP is the softening point temperature of the unmodified and modified bitumen.

2.4.4. Rotational Viscosity Test. The rotational viscometer determines the bitumen viscosity by measuring the torque necessary to maintain a constant rotational speed of a cylindrical spindle submerged in a bitumen specimen held at a constant temperature, described in AASHTO TP48. Unlike the capillary viscometers used with the viscosity-graded method, the rotational viscometer can evaluate modified bitumen binders. Viscosity of bitumen binders can be measured within the range from 0.01 Pa·s (0.1 poise) to 200 Pa·s (2000 poise) [15]. Though this test, which is one of the Superpave Performance Grade (PG) asphalt binder specification, can be conducted at various temperatures for different purposes, the test performed for Superpave PG asphalt binder specification is always conducted at 275°F (135°C). In order to observe the changes in viscosity at various temperatures, a second viscosity reading is taken generally at 165°C. A Brookfield viscometer (DVRV-II Pro Extra) was used in this study for the viscosity tests of the base and the modified bitumens. Viscosity values were measured at 135°C and 165°C, respectively.

2.4.5. Dynamic Shear Rheometer Test. The most commonly used methods for the fundamental rheological characterization of bitumen are dynamic mechanical methods using oscillatory type testing, generally conducted within the region of linear viscoelastic response by using a dynamic shear rheometer (DSR). In order to characterize the viscoelastic behavior of bitumen binders at intermediate and high service temperatures, DSR is used in Superpave specification providing an indication of the rutting resistance of bitumen immediately following the construction. It is also possible to evaluate the resistance to rutting at high service temperatures in the early stages of pavement life and fatigue cracking at immediate service temperatures in the later stages of service life by DSR [16].

The principal viscoelastic parameters obtained from DSR are the complex shear modulus (G^*) and the phase angle (δ). G^* is defined as the ratio of maximum (shear) stress to maximum strain, a measure of the total resistance to deformation when the bitumen is subjected to shear loading. The phase difference between stress and strain in an oscillatory test is named as δ, a measure of the viscoelastic balance of the material behavior [17].

In this study, the principal viscoelastic parameters such as G^*, δ, rutting parameter ($G^*/\sin\delta$), and high-temperature performance grades of the base and modified bitumens were determined by DSR test.

2.4.6. Bending Beam Rheometer Test. The bitumen binder plays a critical role in the thermal cracking potential of hot mix asphalt (HMA) pavement. HMA pavements with a high stiffness modulus at low temperatures are susceptible to thermal cracking. Bitumen binders become harder, or stiffer, as the ambient temperature decreases. Given that the stiffness of an HMA pavement is directly proportional to the stiffness of bitumen binder, high stiffness at low temperatures increases the potential of thermal cracking. The bending beam rheometer (BBR) was developed to determine the stiffness of bitumen binders at low service temperatures and to evaluate the potential of the binder for thermal cracking [16].

Creep tests were conducted at −6, −12, and −18°C which are equivalent to −16, −22, and −28°C, respectively, in low temperature performance grade, using a BBR according to ASTM D6648-01. The bitumen beam (125 mm long, 12.5 mm wide, and 6.25 mm thick) was submerged in a constant-temperature bath and kept at the test temperature for 60 min. After preloading procedure, a constant load of 100 g (980 mN) was then applied to the rectangular beam supported at both ends by stainless steel half-rounds (102 mm apart), and the deflection of center point was measured continuously. Creep stiffness (S) and creep rate (m-value) of the binders were determined at several durations of loading ranging from 8 to 240 s [18]. The creep stiffness is an indicator of the specimen's ability to resist the constant creep load and the creep rate is the rate at which the creep stiffness changes with loading time [16]. For an adequate low-temperature cracking resistance, the creep stiffness must be less than 300 MPa and the m-value must be greater than 0.3 to be in compliance with Superpave specification [19].

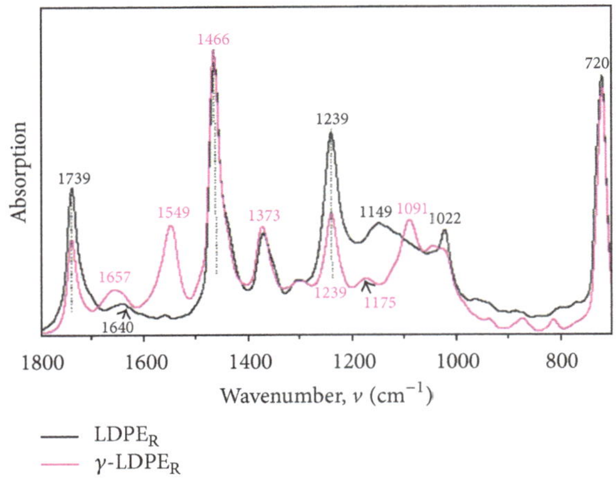

Figure 1: The normalized FTIR spectra of the $LDPE_R$ and γ-$LDPE_R$.

3. Results and Discussions

3.1. FTIR Spectroscopy. In Figure 1, the normalized FTIR spectra (the band at $\nu \approx 720\,cm^{-1}$ has been used as internal standard) of the virgin $LDPE_R$ and γ-$LDPE_R$ are shown. As can be seen from Figure 1, there is no significant difference in position and intensity of the main absorption bands at $\nu \approx$ 720, 1022, 1373, and 1466 cm^{-1}, which are the characteristics for vibrations of C–C and C–H bonds of hydrocarbon chain of virgin LDPE and γ-$LDPE_R$. However, the new peaks are additionally observed at $\nu \approx$ 1091, 1175, 1549, and 1657 cm^{-1} in the spectrum of γ-$LDPE_R$ accompanied by decreasing intensity of the peaks at 1239 cm^{-1} and 1739 cm^{-1}. These new peaks evidence formation of unsaturated bonds in macromolecular chains of gamma-irradiated LDPE (the peak with maximum at $\nu \approx 1657\,cm^{-1}$ corresponds to the non-conjugated C=C bond) that is occurred as a result of high-energy action of the gamma-rays on LDPE. At irradiation treatment free radicals appear at breakage of polymer chains, which can then transform to different groups and bonds including formation of double bonds (>C=C<). Certainly,

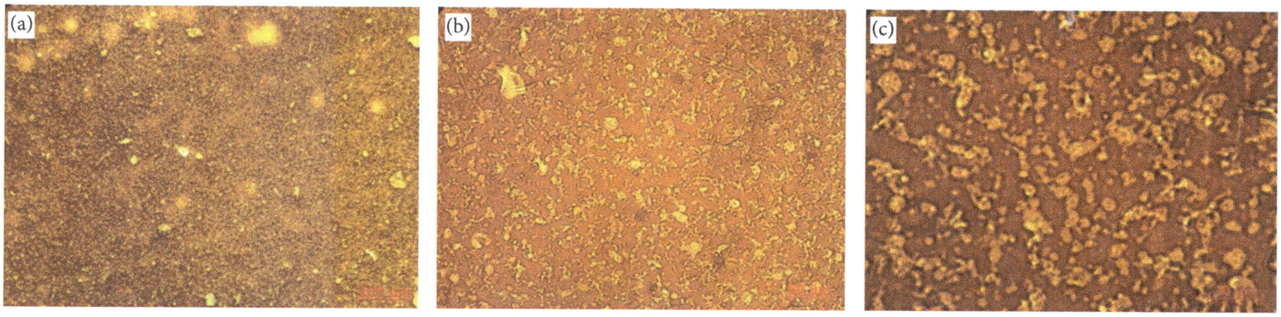

FIGURE 2: (a) 50X magnification, (b) 200X magnification, and (c) 500X magnification.

TABLE 2: Characteristic properties of base and γ-LDPE$_R$ modified bitumens before and after short-term ageing.

Properties	Binder types					
		Modified Bitumens				
	Base	1-γ-LDPE$_R$	3-γ-LDPE$_R$	5-γ-LDPE$_R$	7-γ-LDPE$_R$	9-γ-LDPE$_R$
Penetration (25°C; 0.1 mm)	195.5	168.8	135	102	83.1	57.2
Softening point (°C)	38.7	42.5	48.4	64.2	60.3	70.5
Ductility (cm) at 25°C	103	92	84	77	69	63
Penetration index (PI)	−0.73	0.23	1.31	1.85	2.54	3.28
After ageing						
Change of mass (%)	0.76	0.68	0.65	0.40	0.37	0.35
Penetration (25°C; 0.1 mm)	102	85.6	125	93.8	76.5	52.5
Retained penetration (%)	52	50.7	92.5	91.9	92	92
Softening point (°C)	42.7	45.3	49.5	65.1	61.1	66.5
Change in softening point (°C)	4	2.8	1.1	0.9	0.8	−4

the unsaturated bonds formed in γ-LDPE$_R$ are reactive towards the unsaturated bonds of bitumen components that provides effective reactive compatibilization of the blends of modifier with bitumen through the chemical reactions of the surface functional groups of the activated polyethylene and the double bonds of the bitumen components. In air, the free radicals formed can react with oxygen forming some oxygen-containing functional groups in polymer structure (e.g., C(O)O– group of organic acids or esters with a maximum around $\nu \approx 1549\,\text{cm}^{-1}$ and –C–O–C– bond in ethers with an absorption bands around $\nu \approx 1091, 1175$). These groups can also contribute to physical-chemical compatibilization of γ-LDPE$_R$ with bitumen.

3.2. Morphology. The morphology of γ-LDPE$_R$ modified bitumen was observed with a fluorescent microscope The images given in Figure 2 belong to B-9-γ-LDPE$_R$, the highest polymer content among the all binders. The surface images were taken with using different magnification (i.e. 50X, 200X and 500X). Figure 2(a) shows the texture of the modified bitumen in general, whereas Figure 2(c) shows a detailed image on texture and allows observing the phase structure of the modified bitumen. Two major phases that can be clearly seen (Figures 2(a), 2(b), and 2(c)) are a continuous bitumen-rich phase (brown) and a dispersed polymer phase (yellow). Excluding a few greater particles, the main amount of γ-LDPE$_R$ is mostly ranging from 3 to 10 μm in diameter and is quite well dispersed in bitumen. While modifier particles with a size greater than 30 μm were not seen generally in the modified bitumen samples, it can be said that polymer particles have been swollen by absorbing an amount of the bitumen content. A third phase (mixed phase) can be observed in the image taken at the 500X magnification (Figure 2(c)) formed in interface region obviously due to chemical bonding between bitumen and gamma-irradiated polymer particles.

3.3. Conventional Bitumen Test. In order to understand the effect of γ-LDPE$_R$ additives on physical properties of the bitumen, conventional tests were conducted. The results, including both before and after short-term ageing, are given in Table 2. As can be seen, penetration decreases continuously with increasing amount of γ-LDPE$_R$ polymer content.

Hence, B-9-γ-LDPE$_R$ binder has the lowest penetration value with 57.2. Softening point test results support penetration results, as it increases with increasing γ-LDPE$_R$ polymer content. These physical tests results point out a significant increase in stiffness of the bitumen with increasing amount of the polymer modifier. There is no specification limits for ductility, it is generally expected to be higher than 100 cm. Consequently, decreasing ductility means, binders become more brittle with increasing amount of γ-LDPE$_R$ modifier.

The PI values increase progressively from −0.73 to 3.28, as modifier content increases that signifies γ-LDPE$_R$ polymer

TABLE 3: Viscosity values of the of base and γ-LDPE$_R$ modified bitumens.

Binder types	Rotational viscosity (cP)		$\eta_{modified}/\eta_{pure}$		Temperature range (°C)	
	135°C	165°C	135°C	165°C	Mixing	Compaction
B	202.5	65.5	1	1	135–142	122–128
B-1-γ-LDPE$_R$	250	80	1.23	1.22	142–148	129–135
B-3-γ-LDPE$_R$	380	122.5	1.88	1.87	152–159	139–145
B-5-γ-LDPE$_R$	702.5	208	3.47	3.18	165–171	154–159
B-7-γ-LDPE$_R$	798	330	3.94	5.04	187–195	170–177
B-9-γ-LDPE$_R$	1286	435	6.35	6.64	189–195	175–181

modified bitumen can be used in a wider temperature range than base bitumen.

The tests performed on aged samples reveal that γ-LDPE$_R$ additives have a positive effect on ageing properties of bitumen, as mass loss of the bitumen diminishes. Furthermore, the results of all the samples are within the limitation of change of mass (i.e., 1%). After short-term ageing, the retained penetration tends to increase as γ-LDPE$_R$ content increases which shows that increasing polymer content helps to diminish the effects of ageing on bitumen binders.

3.4. Rotational Viscosity Results. Rotational viscosities performed at 135 and 165°C for base and γ-LDPE$_R$ modified bitumens are shown in Table 3. For all the samples, mixing and compaction temperatures were determined from the viscosity versus temperature graph given in Figure 3. The results reveal a gradual increase in viscosity with modification. As in the penetration and softening point test, this increasing in viscosity gives a clear indication of the stiffening effect of γ-LDPE$_R$ modification. Modification indices, which can be calculated as the viscosity of modified bitumen divided by the viscosity of base bitumen, especially have greater values for B-5-γ-LDPE$_R$, B-7-γ-LDPE$_R$, and B-9-γ-LDPE$_R$ binders. Calculated mixing and compaction temperatures for the base and modified binders tend to increase continuously. Although there are no viscosity values exceeding the Superpave specifications (i.e., 3000 cP) for both base and modified bitumen, this increase in mixing and compaction temperatures means that the workability of bitumen considerably decreases. Besides, mixing temperature of B-7-γ-LDPE$_R$ and B-9-γ-LDPE$_R$ binders reaches 195°C, which can be considered as unfavorable result, because it means that the energy consumption will increase to high levels in applications.

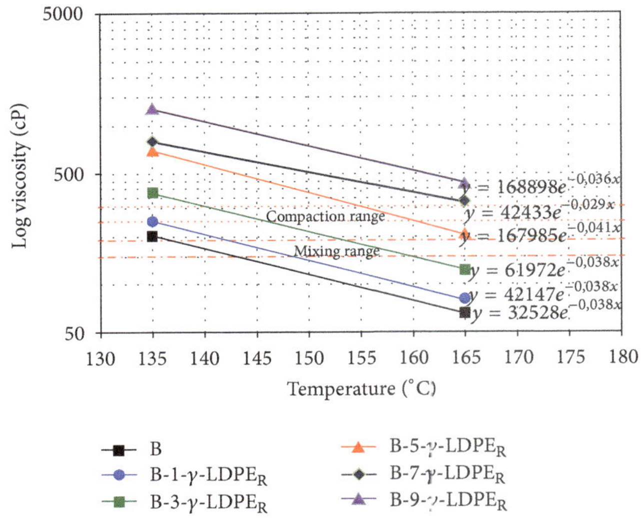

FIGURE 3: The mixing-compaction temperatures for the base and γ-LDPE$_R$ modified bitumens.

3.5. Dynamic Shear Rheometer Test. In this study, high-temperature performances grades of base and modified bitumens were investigated by means of DSR tests. The fundamental rheological parameters for bitumen such as G^*, δ, and $G^*/\sin\delta$ were also determined for original and aged samples in this DSR testing program, and the results are given in Table 4 and Figure 4. Table 4 also contains the high-temperature performance grades (PG) calculated with minimum rutting parameter ($G^*/\sin\delta$) criteria of 1 kPa for original samples and 2.2 kPa for aged samples recommended in Superpave binder specifications. As can be seen, the values of G^*, the measure of the total resistance of bitumen against the deformation subjected to the shear stress, increase significantly for both original and aged samples with increasing γ-LDPE$_R$ modifier content. This result determines that binders become hardened after γ-LDPE$_R$ modification. Therefore, it can be said that the γ-LDPE$_R$ modified binders are more durable than original binders against shear stress deformations.

The phase angle (δ) is generally considered to be more sensitive to the chemical structure than the complex modulus, δ generally tends to decrease as the amount of modifier increases, and, therefore, predominantly viscous behavior with increasing temperatures, γ-LDPE$_R$ modifier improves the elastic response of the modified binders.

The effect of γ-LDPE$_R$ modification on the rheological parameters (G^* and δ) leads to an increased rutting parameter ($G^*/\sin\delta$) of binders that promises an improvement of the rutting deformation. As values of $G^*/\sin\delta$ within the limitations, DSR test is maintained to the next test temperatures.

The increased $G^*/\sin\delta$ parameters also affect the PG of the binders. There is a significant increase in PG with increase of γ-LDPE$_R$ modifier, especially content of 5, 7, and 9%. For instance, high-temperature performance grades step up to

TABLE 4: DSR test results of the original and aged of base and γ-LDPE$_R$ modified bitumens.

Binders		Temperature (°C)	G^* (KPa)	δ (°)	$G^*/\sin\delta$ (KPa)	Specification limits (Pa)	Performance grading
B	Original	52	4.64	86.6	4.65	≥1000	PG 52-Y
		58	1.85	87.4	1.85		
		64	0.766	88.2	0.767		
	Aged	52	3.06	87.1	3.07	≥2200	
		58	1.39	87.2	1.39		
B-1-γ-LDPE$_R$	Original	52	2.75	80.7	2.78	≥1000	PG 58-Y
		58	1.31	77.3	1.35		
		64	0.702	72.7	0.735		
	Aged	52	9.47	82.3	9.56	≥2200	
		58	4.07	84	4.10		
		64	1.88	84.9	1.89		
B-3-γ-LDPE$_R$	Original	52	3.60	85.8	3.61	≥1000	PG 58-Y
		58	1.66	85.5	1.66		
		64	0.843	84.3	0.84		
	Aged	52	10.6	78.7	10.6	≥2200	
		58	4.66	77.9	4.76		
		64	2.42	74.7	2.42		
		70	1.25	71.2	1.32		
B-5-γ-LDPE$_R$	Original	52	5.36	81.6	5.42	≥1000	PG 64-Y
		58	2.47	79.9	2.51		
		64	1.27	77.1	1.30		
		70	0.72	73.4	0.75		
	Aged	52	16.8	80	17.1	≥2200	
		58	7.07	82.4	7.13		
		64	3.14	83.7	3.16		
		70	1.52	84.3	1.53		
B-7-γ-LDPE$_R$	Original	52	7.69	84.5	7.72	≥1000	PG 64-Y
		58	3.34	85.4	3.35		
		64	1.62	85.4	1.63		
		70	0.863	84.1	0.868		
	Aged	52	20.7	77.7	21.2	≥2200	
		58	9.04	80.2	9.18		
		64	4.26	81.7	4.30		
		70	2.13	82.3	2.15		
B-9-γ-LDPE$_R$	Original	52	10.3	71.9	10.8	≥1000	PG 76-Y
		58	5.37	68.4	5.78		
		64	3.13	64.6	3.47		
		70	1.99	60.7	2.28		
		76	1.33	57	1.59		
		82	0.881	53.4	1.10		
	Aged	52	51.4	69.1	55	≥2200	
		58	25	70.9	26.4		
		64	13	71.7	13.7		
		70	5	70.3	5.31		
		76	2.89	68.2	3.11		
		82	1.78	65.2	1.96		

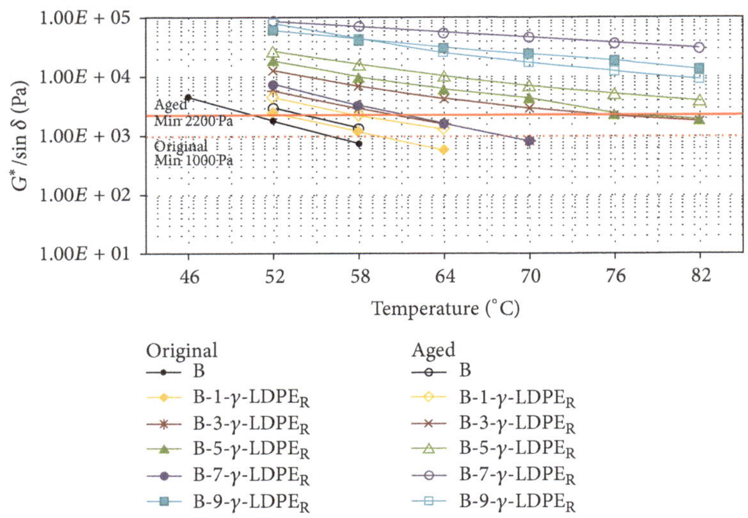

FIGURE 4: Rutting parameter of binders.

TABLE 5: Creep stiffness and m-value of the binders at a loading time of 60 s and different temperatures.

Binder types	Creep stiffness (MPa)			m-value			Performance grading (PG)
	−16°C	−22°C	−28°C	−16°C	−22°C	−28°C	
B	42.13	99.50	199.05	0.479	0.432	0.309	PG Y-28
B-1-γ-LDPE$_R$	40.71	139.69	219.12	0.381	0.343	0.281	PG Y-22
B-3-γ-LDPE$_R$	67.61	146.12	228.33	0.325	0.318	0.274	PG Y-22
B-5-γ-LDPE$_R$	66.93	156.68	248.98	0.324	0.301	0.285	PG Y-22
B-7-γ-LDPE$_R$	78.77	176.58	—	0.311	0.267	—	PG Y-16
B-9-γ-LDPE$_R$	79.39	143.8	—	0.310	0.263	—	PG Y-16

64-Y and 76-Y from 52-Y, when content of the modifier is 5 and 9%, respectively.

3.6. Bending Beam Rheometer Test. In order to study the influence of polymer additives on the low-temperature creep responses of bitumen, the bending beam rheometer test was employed at different loading times of 8, 15, 30, 60, 120, and 240 s and temperatures of −16°C, −22°C, and −28°C. Table 5 contains of compares creep stiffness and m-values obtained at three temperatures and at a loading time of 60 s. The creep stiffness values and m-values of the base and γ-LDPE$_R$ modified bitumens are also graphed in Figures 5 and 6. As can be seen, creep stiffness of binders increases with increasing polymer modifier content. It can be said that this result was expected, as conventional and rotational viscosity test results indicate the effect stiffening of γ-LDPE$_R$ modification. Although it is known that binders with higher creep stiffness are not convenient for cold regions, as we can see from the results, the creep stiffness values of binders obtained do not exceed Superpave specifications (i.e., 300 MPa) at any test temperatures. As to m-value, there is a trend to exceed limitations with increasing polymer modifier content. Consequently, binders of B-7-γ-LDPE$_R$ and B-9-γ-LDPE$_R$ are failed at −22°C to remain within specifications (i.e., 0.3) whereas the rest of the binders step up to next testing temperature. However, any of the binders were not able to provide m-value specification at −28°C excepting base bitumen. Since creep stiffness is within the limitations, m-values determine the low-temperature performance grade of binders. Hence, there is a slightly decreasing trend in low-temperature performance grades as polymer modifier content increases. As can be seen from Table 5, it steps down to −22 and −16 from −28°C, as polymer modified content is increased to 1% and 7%, respectively.

3.7. The Effect of γ-LDPE$_R$ on Performance of Modified Bitumen. As it was discussed above, γ-LDPE$_R$ has positive the effect on high temperature performance grades, while it decreases slightly low temperature performance grades. Thus, in order to clarify the total effects of γ-LDPE$_R$ on the performance and to recommend for optimum usage of γ-LDPE$_R$ in bitumen, Table 6 was prepared to show dependence of the modifier content on failure temperatures of binders for DSR and BBR test. Table 6 also contains the PG of the binders.

The results indicate that, the improvement at high temperatures can be seen clearly at even low polymer modification. For instance, B-1-γ-LDPE$_R$, which is able to be used in regions with a wider temperature range, as improvement of the binder, is 4.53°C. As content of polymer modifier increases, the improvement also continues to increase.

TABLE 6: The effect of γ-$LDPE_R$ on high and low service temperatures and on performance grade (PG).

Binder types	Fails at HT (°C)	Fails at LT (°C)	HT (°C) improvement	LT (°C) decrement	Improvement/decrement	Performance grading (PG)
B	54.3	−28.78	—	—	—	PG 52-28
B-1-γ-$LDPE_R$	61	−26.61	6.7	2.17	4.53	PG 58-22
B-3-γ-$LDPE_R$	63.5	−24.45	9.2	4.33	4.87	PG 58-22
B-5-γ-$LDPE_R$	66.9	−22	12.6	6.78	5.82	PG 64-22
B-7-γ-$LDPE_R$	68.6	−20.5	14.3	8.28	6.02	PG 64-16
B-9-γ-$LDPE_R$	80.5	−17.95	26.2	10.83	15.37	PG 76-16

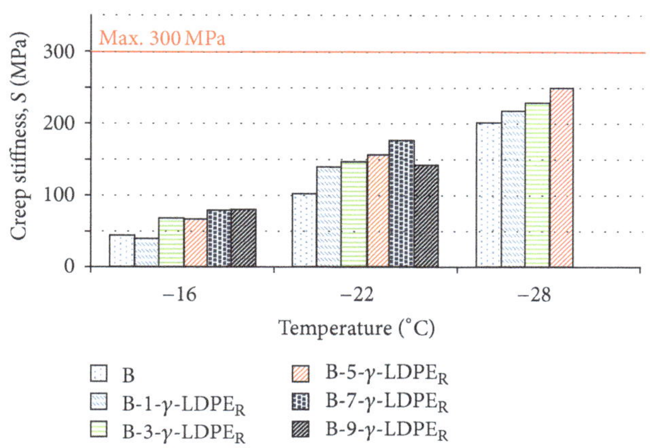

FIGURE 5: Creep stiffness values of the base and γ-$LDPE_R$ modified bitumens.

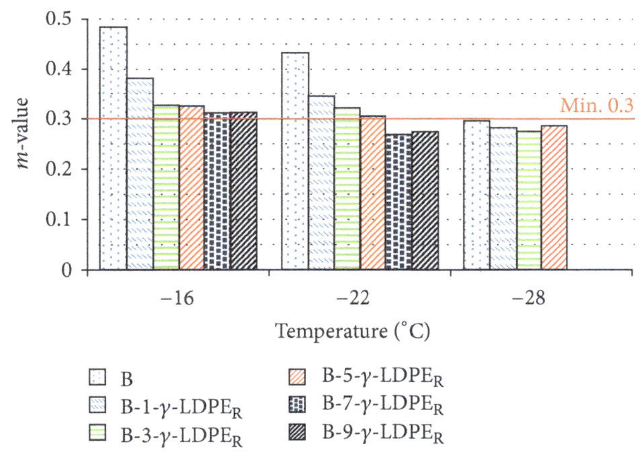

FIGURE 6: m-values of the base and γ-$LDPE_R$ modified bitumens.

Thereby, temperature improvement of B-3-γ-$LDPE_R$ and B-5-γ-$LDPE_R$ binders rises up to 4.87 and 5.82°C, respectively. B-9-γ-$LDPE_R$ has the largest improvement with 15.37°C. However, although failure temperatures are more precise than PG classifications, it is necessary that analyzing the effect of the polymer modifier by using both the failure temperatures and PG.

The PG of the base bitumen was found as PG 52–28, whereas B-1-γ-$LDPE_R$ was a binder of PG 58–22. It can be understood that there is no improvement since high-temperature performance grade steps up to one level (52 to 58) while the decreasing in the low-temperature PG is the same (−28 to −22). However, in fact, as can be seen above there is an improvement in terms of failure temperatures.

B-3-γ-$LDPE_R$ has the same PG (58–22) with B-1-γ-$LDPE_R$, whereas B-5-γ-$LDPE_R$ has PG 64–22, which indicates there is an improvement in terms of both failure temperatures and PG. As for B-7-γ-$LDPE_R$, low-temperature performance grading steps down to two level compared with base bitumen (−16), thereby B-7-γ-$LDPE_R$ has a poorer PG than B-5-γ-$LDPE_R$. The B-9-γ-$LDPE_R$ has the biggest PG (76–16) along all the binders due to vast high-temperature improvement. It might be interpret as a benefit; however, this stiffening effect causes some problems in application while binder is being mixed and compacted. Nevertheless, it is undesirable due to high energy consumptions, as discussed in RV results.

Consequently, in spite of gradual increasing in temperature improvement as polymer content increases, as can be seen B-7-γ-$LDPE_R$ and B-9-γ-$LDPE_R$ have a lower temperature grade of −16 that also indicates that there must be a limitation to the usage of γ-$LDPE_R$. According to the rheological test (DSR and BBR), and considering RV results it can be suggested that the usage of γ-$LDPE_R$ polymers as modifier in bitumen with content of 5% as B-5-γ-$LDPE_R$ has the most stable results due to the previously mentioned reasons.

4. Conclusion

γ-$LDPE_R$ modified bitumen has been investigated by means of conventional, chemical, and mechanical tests. It was found that the rheological properties of bitumen are enhanced with γ-$LDPE_R$ polymer modification as identified by testing program. The surface images of the binders obtained with fluorescence microscopy revealed formation of multiphase system with bitumen continuous phase, γ-$LDPE_R$ dispersed phase, and mixed phase (interfacial layer), which is a sign of the presence of chemical bonding between bitumen and gamma-irradiated polymer particles.

FTIR spectroscopy confirmed the appearance of reactive (>C=C<) bonds in γ-$LDPE_R$ macromolecules after gamma irradiation of the $LDPE_R$. Conventional tests, such as penetration and softening point, indicate a significant increase in stiffness of the bitumen with increasing amount of the γ-$LDPE_R$ modifier.

After short-term ageing by means of RTFOT, all the modified binders are within the Superpave limitations. Furthermore, γ-LDPE$_R$ polymer has a positive effect on bitumen ageing. In addition to the observed stiffening effect of γ-LDPE$_R$ on bitumen, RV test results also suggest that using γ-LDPE$_R$ as modifier more than 5% in bitumen can cause some mixing and compacting problems in applications.

The results obtained from the DSR tests show that γ-LDPE$_R$ modifier provides an increased complex shear modulus (G^*) values and rutting parameters ($G^*/\sin\delta$) and decreased phase angle (δ) values which means the γ-LDPE$_R$ modified binders are more resistant against rutting compared to the base bitumen. This improvement also affects the PG of the binders that increases gradually with increasing modifier content. However, there must be a limitation for the usage of the γ-LDPE$_R$ polymer due to BBR results that indicate an increasing creep stiffness and decreasing m-value as polymer content increases.

Consequently, in this paper, with taking into account the rheological tests and RV results, it was suggested that the optimum usage of γ-LDPE$_R$ is 5% by weight of bitumen.

Acknowledgments

This work was carried out in the framework of the joint international Project no 110M400 of Ege University(Izmir, Turkey) and Institute of Macromolecular Chemistry of the National Academy of Sciences of Ukraine (NASU, Kyiv, Ukraine) under the agreement between TUBITAK (Turkey) and NASU (Ukraine).

References

[1] D. R. Gershkoff, J. Carswell, and J. C. Nicholls, *Rheological Properties of Polymer-Modified Binders for Use in Rolled Asphalt Wearing Course*, Transport Research Laboratory, Crowthorne, UK, 1997.

[2] U. Isacsson, "Laboratory investigation of polymer modified bitumens," *Journal of the Association of Asphalt Paving Technology*, vol. 68, pp. 35–63, 1999.

[3] G. D. Airey, "Rheological evaluation of ethylene vinyl acetate polymer modified bitumens," *Construction and Building Materials*, vol. 16, no. 8, pp. 473–487, 2002.

[4] A. Pérez-Lepe, F. J. Martínez-Boza, C. Gallegos, O. González, M. E. Muñoz, and A. Santamaría, "Influence of the processing conditions on the rheological behaviour of polymer-modified bitumen," *Fuel*, vol. 82, no. 11, pp. 1339–1348, 2003.

[5] M. García-Morales, P. Partal, F. J. Navarro, F. Martínez-Boza, M. R. Mackley, and C. Gallegos, "The rheology of recycled EVA/LDPE modified bitumen," *Rheologica Acta*, vol. 43, no. 5, pp. 482–490, 2004.

[6] G. Polacco, S. Berlincioni, D. Biondi, J. Stastna, and L. Zanzotto, "Asphalt modification with different polyethylene-based polymers," *European Polymer Journal*, vol. 41, no. 12, pp. 2831–2844, 2005.

[7] N. Z. Habib, I. Kamaruddin, M. Napiah, and I. M. Tan, "Rheological properties of polyethylene and polypropylene modified bitumen," *International Journal Civil and Environmental Engineering*, vol. 3, pp. 96–100, 2011.

[8] C. F. Ouyang, Q. Gao, Y. T. Shi, and X. Q. Shan, "Compatibilizer in waste tire powder and low-density polyethylene blends and the blends modified asphalt," *Journal of Applied Polymer Science*, vol. 123, no. 1, pp. 485–492, 2012.

[9] M. García-Morales, P. Partal, F. J. Navarro, and C. Gallegos, "Effect of waste polymer addition on the rheology of modified bitumen," *Fuel*, vol. 85, no. 7-8, pp. 936–943, 2006.

[10] Z. Z. Ismail and E. A. AL-Hashmi, "Use of waste plastic in concrete mixture as aggregate replacement," *Waste Management*, vol. 28, no. 11, pp. 2041–2047, 2008.

[11] F. J. Navarro, P. Partal, M. García-Morales et al., "Bitumen modification with reactive and non-reactive (virgin and recycled) polymers: a comparative analysis," *Journal of Industrial and Engineering Chemistry*, vol. 15, no. 4, pp. 458–464, 2009.

[12] Y. H. Gad, M. M. Magida, and H. H. El-Nahas, "Effect of ionizing irradiation on the thermal blend of waste low density polyethylene/ethylene vinyl acetate/bitumen for some industrial applications," *Journal of Industrial and Engineering Chemistry*, vol. 16, no. 6, pp. 1019–1024, 2010.

[13] V. Wegan, *Sampling for Microscopy-Test Procedure*, Danish Road Institute, 1996.

[14] J. Read and D. Whiteoak, *The Shell Bitumen Handbook*, Shell Bitumen, Thomas Telford, London, UK, 5th edition, 2003.

[15] H. U. Bahia and D. A. Anderson, "Strategic highway research program binder rheological parameters: background and comparison with conventional properties," *Transportation Research Record*, no. 1488, pp. 32–39, 1995.

[16] J. P. Zaniewski and M. E. Pumphrey, *Evaluation of Performance Graded Asphalt Binder Equipment and Testing Protocol*, West Virginia University, 2004.

[17] G. D. Airey, "Rheological properties of styrene butadiene styrene polymer modified road bitumens," *Fuel*, vol. 82, no. 14, pp. 1709–1719, 2003.

[18] S. Liu, W. Cao, J. Fang, and S. Shang, "Variance analysis and performance evaluation of different crumb rubber modified (CRM) asphalt," *Construction and Building Materials*, vol. 23, no. 7, pp. 2701–2708, 2009.

[19] H. I. A. Wahhab, I. A. Al-Dubabe, I. M. Asi, and M. F. Ali, "Performance-based characterization of Arab asphalt," *Building and Environment*, vol. 33, no. 6, pp. 375–383, 1998.

5

Polypropylene Fibers as Reinforcements of Polyester-Based Composites

Gonzalo Martínez-Barrera,[1] Enrique Vigueras Santiago,[1] Susana Hernández López,[1] Osman Gencel,[2] and Fernando Ureña-Nuñez[3]

[1] *Laboratorio de Investigación y Desarrollo de Materiales Avanzados (LIDMA), Facultad de Química, Universidad Autónoma del Estado de México, Km. 12 de la Carretera Toluca-Atlacomulco, 50200 San Cayetano, Mexico*
[2] *Civil Engineering Department, Faculty of Engineering, Bartin University, 74100 Bartin, Turkey*
[3] *Instituto Nacional de Investigaciones Nucleares, Carretera México-Toluca S/N, 52750 La Marquesa Ocoyoacac, Mexico*

Correspondence should be addressed to Gonzalo Martínez-Barrera; gonzomartinez02@yahoo.com.mx

Academic Editor: Carmina Menchaca-Campos

Effects of gamma radiation and the polypropylene fibers on compressive properties of polymer concrete composites (PC) were studied. The PCs had a composition of 30 wt% of unsaturated polyester resin and 70 wt% of marble particles which have three different sizes (small, medium, and large). The PCs were submitted to 200, 250, and 300 kGy of radiation doses. The results show that the compressive properties depend on the combination of the polypropylene fiber concentration and the applied radiation dose. The compressive strength value is highest when using medium particle size, 0.1 vol% of polypropylene fibers and 250 kGy of dose; moreover, the compressive modulus decreases when increasing the particle size.

1. Introduction

Polymer concrete (PC) is a composite material formed by combining mineral aggregates with a thermoset resin. In its elaboration several parameters must be taken into account, such as resin type, initiator, and accelerator concentrations. The unsaturated polyester resin (UPR) is the most widely used due to their balanced mechanical and chemical characteristics, its ease of handling, and low cost; for its polymerization, 2 wt% of methyl-ethyl-ketone peroxide (MEKP), as initiator, and 0.5 wt% of cobalt naphthenate as accelerator are normally used, with at least 40 wt% of styrene.

The composition of polymer concrete is determined by its applications especially loading stress levels and ability to resist corrosive environment. PC is increasingly being used as an alternative to ordinary Portland cement concrete (PCC) in many applications, such as finishing work in cast-in-place applications, precast products, highway pavements, bridge decks, waste water pipes, and even decorative construction panels. In the last 40 years polymer concrete has made tremendous progress and continues to be very promising materials for a wide range of new and innovative applications. Moreover, the use of polymers should be well considered to guarantee better performance and improved sustainability [1–3]. Improvement on mechanical strength and chemical resistance is basic advantages of polymer concrete in comparison to ordinary Portland cement concrete (PCC). Three to five times on the compressive strength, high values for tensile strength (20 MPa), and flexural strength (50 MPa) are still an outstanding advantage of polymer concrete [3, 4].

Mechanical properties of polymer concrete depend on the type of resin and mineral aggregates. In the case of the last, higher specific surface means higher mechanical values, for example, (a) polymer concrete with clean sand have higher values than those with foundry sand [5]; (b) for concrete elaborated with epoxy resin, silica sand, hematite, and colemanite, the mechanical properties depend on the resin and hematite-colemanite concentrations; larger improvement for hematite is done [6]; (c) for concrete with waste marble as aggregate the splitting tensile strength decreases, and improvement on the elasticity modulus is produced [7].

Polymer concrete has relatively low tensile strength compared to its compressive strength. This limits its usefulness for load-bearing applications. Sudden and brittle failures occur in planes where tensile stresses exceed the tensile strength of the polymer concrete. Thus, reinforcement either by synthetic or natural fibers is one option to increasing its strength capacity, ductility, and toughness [5]. For example, glass fibers have been used for reinforcement of polyester polymer concrete, particularly for increasing flexural strength and strain at peak stress. An improvement of 80% in the flexural strength is obtained with respect to unreinforced polymer concrete. The toughness is increased by more than 1440% when 6 wt% of glass fibers is added. The glass fibers were 3 mm long and 0.013 mm of diameter, with a tensile strength of 2.5 GPa and a modulus of 70 GPa [8].

Carbon and glass fibers (1% and 2%, resp.) added to epoxy polymer concrete showed different behaviors. The carbon fibers increase the fracture energy by 340% while glass fibers by 140% when comparing to plain polymer concrete. Nevertheless, the fibers induce failure due to poor adhesion between them and polymer resin including the break of the fibers when compressive test is evaluated [9]. Chopped glass fibers were with random size, while chopped carbon fibers (made from a poly-acrylic-nitrile precursor material) were on average 6 mm.

The fracture behavior of concrete elaborated with polyester resin (18%) and sand and glass fibers (82%) was studied. Stable crack growth prior to peak load was found when adding 4% of glass fibers (13 mm long) [10]. In the case of polymer concrete with polypropylene fibers the compressive strength, splitting tensile strength, and especially flexural strength and elasticity modulus increase while the weight diminishes [11].

Some studies concerned with the use of gamma radiation on polymer/mineral composites have been carried out. The effects of gamma radiation on pure polyester resin and polyester resin/gypsum composites show that the rupture stress is higher for pure resin than that for the composite (6.5 MPa versus 2.0 MPa, at 20 kGy). At highest dose (320 kGy) the values increase up to 9.0 MPa for pure resin and 3.6 MPa for the composite. Such difference is due to the lower tensile strength of gypsum (filler) when comparing to those for resin. Moreover, only the polymer chains (but not the inorganic filler particles) would build cross-links between the chains [12]. In other results, a minimal variation on the hardness is observed at different doses: 91% and 92% for the composite and 89% and 88% for pure resin (at 20 and 320 kGy, resp.) [12].

As it is known, the cross-linking reaction of unsaturated polyester resins (UP) is usually initiated by a thermal or redox initiator, but sometimes the full conversion is not fully achieved, due to the difficulty of mixing with initiators and/or promoters, which reduces the conversion, especially at room temperature [13]. The release of residual styrene of the UP resins creates problems to the environment and is the source of odor in many applications. One alternative route of curing UP resins is by radiation processing [14, 15]. Some differences are found when polyester resin with or without initiators is irradiated. At low dose (11.1 kGy), polyester resin with initiator shows a larger increase in the maximum load (30 N) higher than that for resin without initiator (20 N). Notorious is this value at higher dose (33.3 kGy), 320 N for resin with initiator and 200 N for resin without initiator [16].

2. Materials and Methods

2.1. Specimen Preparation. Before preparing the polymer concrete composites, one set of polypropylene atactic fibers (CONSA, Distrito Federal, Mexico) whose diameters vary from 50 to 60 μm were cut to 10 mm length on the average. For preparing the polymer concrete specimens, marble from a local company (GOSA, Atizapan, Mexico), as the fine aggregate, polypropylene fibers at concentrations of 0.1, 0.2, or 0.3% by volume, and a commercial unsaturated preaccelerated polyester resin (Polylite 32494-00, Reichhold, Atlacomulco, Mexico) were used. The proportion of the polyester resin in the PC was 30% by weight, and the PP fibers were distributed hazardously within the polymer matrix.

Polymer concrete specimens were prepared to be subjected to three different irradiation doses. After mixing, the polymer concrete cubic specimens ($5 \times 5 \times 5$ cm) were placed in a controlled temperature room at $23.0 \pm 3.0^{\circ}$C for 24 hours. Polymer concrete specimens were prepared and subjected to three different irradiation doses (200, 250, and 300 kGy).

2.2. Mechanical Tests. Compressive tests of the polymer concrete specimens were carried out in an Universal Testing Machine model 70-S17C2 (Controls, Cernusco, Italy), located at Laboratory of Research and Development in Advanced Materials (LIDMA) of the Autonomous University of the State of Mexico (UAEM), according to the ASTM C-109M standard.

2.3. Morphological Characterization. First, the fibers were vacuum-coated with carbon (thickness between 3 to 10 nm) with the aid of a Vacuum Evaporator (E. F. Fullam) at 50 mTorr. Then, the fiber surfaces were analyzed by scanning electron microscopy (SEM) in a JEOL model JSM-6510LV microscope in the secondary-electron mode, at 20 keV.

2.4. Irradiation Procedure. Atactic PP fibers and the polymer concrete composites were exposed at various gamma radiation doses using a ^{60}Co source. The fibers were placed in packets of 50 in a capillarity tube. The dosages were 200, 250, and 300 kGy at the dose rate of 3.5 kGy/h; the experiments were performed in air at room temperature. The irradiation was provided by a 651 PT Gammabeam Irradiator manufactured by NORDION (Chalk River, Ontario) and located at the Institute of Nuclear Sciences of the National Autonomous University of Mexico.

3. Results and Discussion

3.1. Compressive Strength. The PCs were fabricated containing different marble-particle sizes and different polypropylene-fiber concentrations. The selection of the particle sizes was in function of the availability of the commercial mesh (sieve), in our case 25, 14, and 8, which correspond to

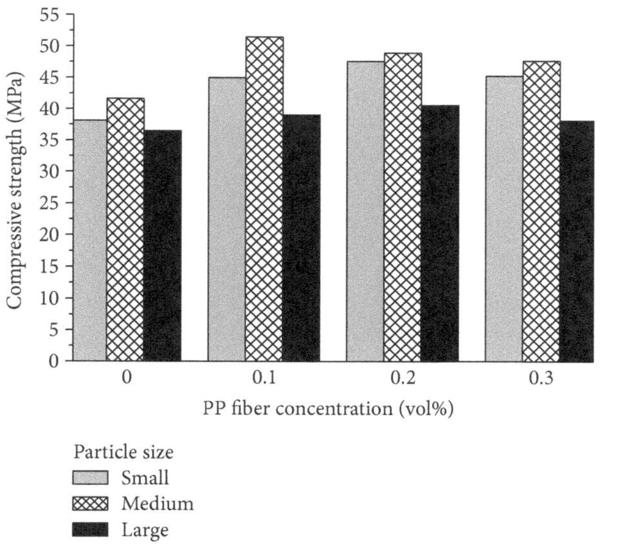

FIGURE 1: Compressive strength of polymer concrete with different marble particle sizes and fiber concentrations.

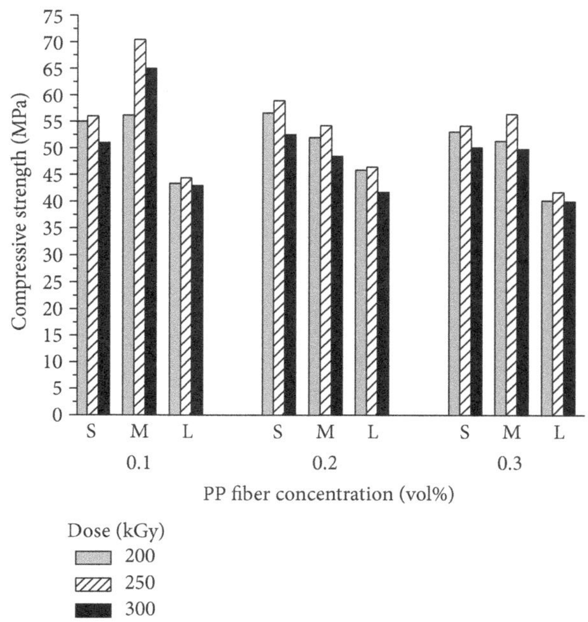

FIGURE 2: Compressive strength of irradiated-polymer concrete with different marble particle sizes and fiber concentrations.

an average particle size of 0.71, 1.4, and 2.36 mm, respectively. Thus, we have approximately 1 : 2 : 3 as the ratio. Through the article we will use the terms small (S), medium (M) or large (L) size, for 0.71, 1.4 and 2.36 mm, respectively.

It can be seen in Figure 1 the compressive strength (σ_c) values for PC elaborated with different particle sizes and polypropylene-fiber concentrations. The σ_c value is highest when using medium particle size and 0.1 vol% of polypropylene fibers, namely, 51 MPa. Conversely, the lowest value (36 MPa) is for those with large particle size and nonpolypropylene fibers.

For plain polymer concrete (without fibers), the σ_c values are in the range from 36 to 41 MPa, while for fiber-reinforced PCs from 38 to 51 MPa; thus, an improvement of 24% is obtained when using PP fibers. Moreover, a high-to-low sequence in the σ_c values for PC with respect to the particle size is as follows: (medium) > (small) > (large), independent of the fiber concentrations.

When comparing the present σ_c values with other PCs, we can see similar results. The σ_c values of the present communication are from 36 to 51 MPa; similar results were obtained as PC with silica sand (49 MPa) [17], but lower than for PC with $CaCO_3$ (86 MPa) [18], or with silica sand + $CaCO_3$ (106 MPa) [19].

The effect of the gamma irradiation is observed in Figure 2. The highest σ_c value is 70 MPa, that is, 58% of improvement in comparison with plain concrete. The present σ_c values are from 46 to 59 MPa.

According to the irradiation dose, different behaviors are well defined for the σ_c values: (a) the highest σ_c values are always obtained at 250 kGy, for all specimens, independent of the fiber concentration and particle size; (b) two well-defined stages are seen for each particle size; the first one consists of an increase of σ_c according to gamma radiation increase up to 250 kGy; then for higher dose the σ_c values decrease; an explanation for such behavior seems related to the radiation effects on the polyester resin; as already noted, the irradiation causes chain scission, but it also produces some crosslinking, chain relaxation, and cage breaking [20, 21].

(c) The maximum values are achieved when using medium particle sizes, and a high-to-low sequence in the σ_c values with respect to the particle size is as follows: (medium) > (small) > (large), independent of the fiber concentrations except for PC with 0.2% of fibers, where (small) > (medium) > (large) sequence is established.

Small size marble particles provide more obstacles to crack propagation in a given amount of concrete. Apparently the use of large particle sizes causes detrimental results due to poor adhesion between the polyester resin and the marble particles. Our PC had 70 wt% of marble particles.

In the case of irradiated PCs, there are significant differences. The present σ_c values were from 40 to 70 MPa and lower than for PC with silica sand (62–86 MPa) [17], with silica sand + $CaCO_3$ (104–112 MPa), [19] or with $CaCO_3$ (126–135 MPa) [18].

Along these lines, the ionizing energy generates more contact points and in consequence larger contact areas between the components: fibers, polyester resin, and marble particles [19, 22]. In turn, an increased number of contact points in the concrete will resist larger loads oriented at various angles relative to the longitudinal axes of the fibers. Eventually, the concrete will split approximately parallel to the dominant axis of the fibers, and the resulting crack will propagate out to the surface. In other words, the energy transfer drops rapidly unless reinforcement is provided to restrain the opening of the splitting crack.

3.2. Compressive Strain at Yield Point. In Figure 3 compressive strain values at yield point are shown. For plain polymer

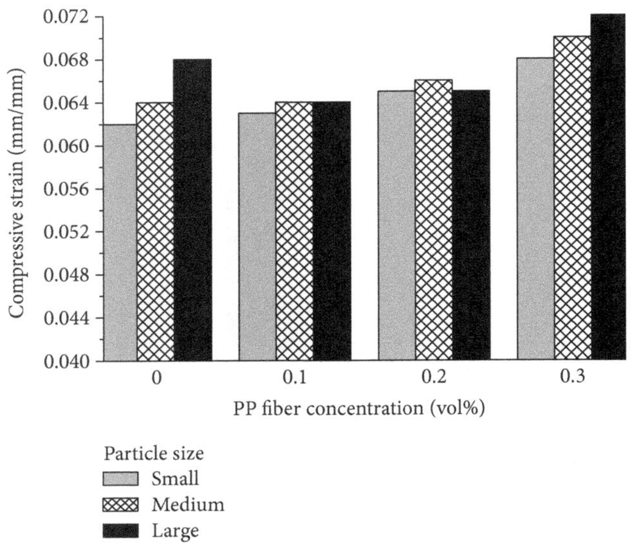

Figure 3: Compressive strain of polymer concrete with different marble particle sizes and fiber concentrations.

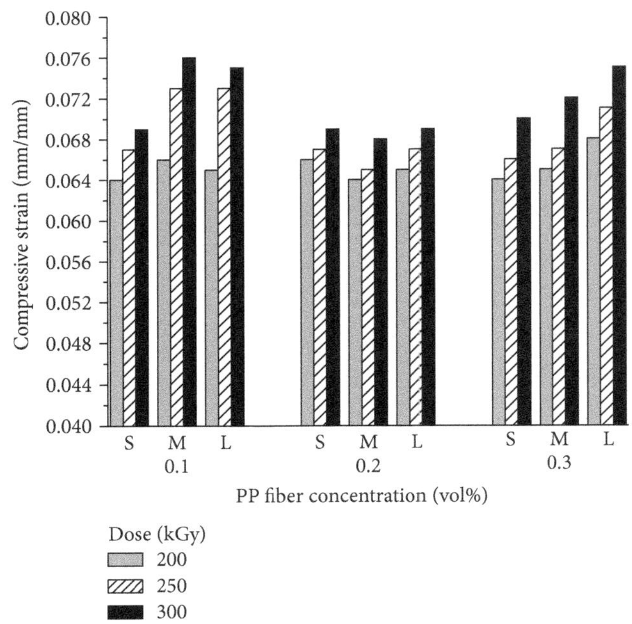

Figure 4: Compressive strain of irradiated-polymer concrete with different marble particle sizes and fiber concentrations.

concrete the values vary from 0.062 to 0.068 mm/mm. The highest values are for PC with 0.3 vol% of fibers (0.068 to 0.073 mm/mm); this means a minimal 7% of difference with respect to polymer concrete without fiber.

The compressive strain increases as the fiber concentration increases. Moreover, the sequence of high-to-low values in terms of the particle sizes is: (large) > (medium) > (small), except for polymer concrete with 0.2 vol% of fibers.

Different behaviors of the compressive strain for irradiated-polymer concrete were observed in Figure 4. (a) the high-to-low sequence on the values in terms of the fiber concentration is (0.1% by volume) > (0.3% by volume) > (0.2% by volume). Thus, an optimal concentration is 0.1% by volume for getting a more ductile concrete and 0.2% by volume, for getting a hard concrete. (b) For all specimens the compressive strain increases when gamma radiation dose increases.

The highest value (0.076 mm/mm) was obtained for polymer concrete with medium particle size and irradiated at 300 kGy; this means 11% of improvement with respect to nonirradiated polymer concrete.

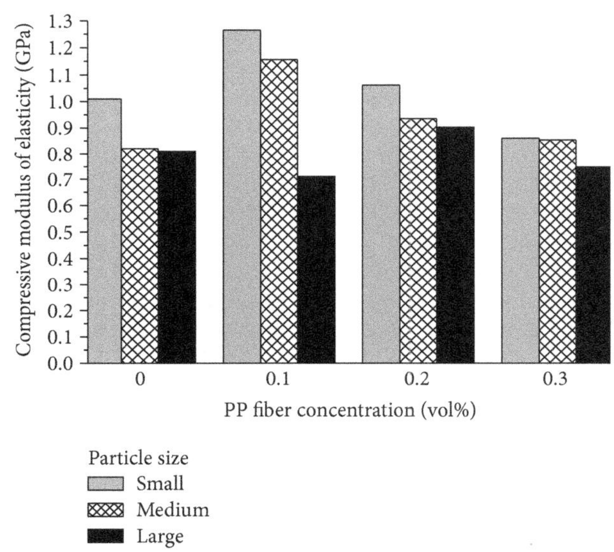

Figure 5: Compressive modulus of elasticity of polymer concrete with different marble particle sizes and fiber concentrations.

3.3. Compressive Modulus of Elasticity. The compressive modulus of elasticity Ec for nonirradiated PCs showed values ranging from 0.71 to 1.26 GPa (Figure 5). Such values exhibit the same behaviour for all specimens: the modulus diminishes as the particle sizes increase, following the sequence: (small) > (medium) > (large); thus a statement can be formulated: the compressive modulus decreases when increasing the particle size. Apparently small particles—while providing good protection against crack propagation—provide more reinforcement than the medium or large ones.

A maximum for PC with small particles size and 0.1 vol% of PP fibers was identified, namely, 1.26 GPa; conversely a minimal value was obtained for PC with large particle size and 0.3 vol% of fibers (0.71 GPa).

When comparing the Ec values for nonirradiated PCs with other PCs elaborated with different mineral aggregates, it was observed that the present value 1.88 GPa is lower than for PC with silica sand + $CaCO_3$ (5.2 GPa) [19], or silica sand + polypropylene fibers (5.7 GPa) [23], or PC with marble + calcium bentonite (6.8 GPa) [22], or for PC with silica sand (7.3 GPa) [17], or PC with $CaCO_3$ (7.6 GPa) [18].

In the case of Ec values for irradiated PCs, several behaviors were observed. The present values are in the range from 0.99 to 1.88 GPa (Figure 6); it means 49% of improvement

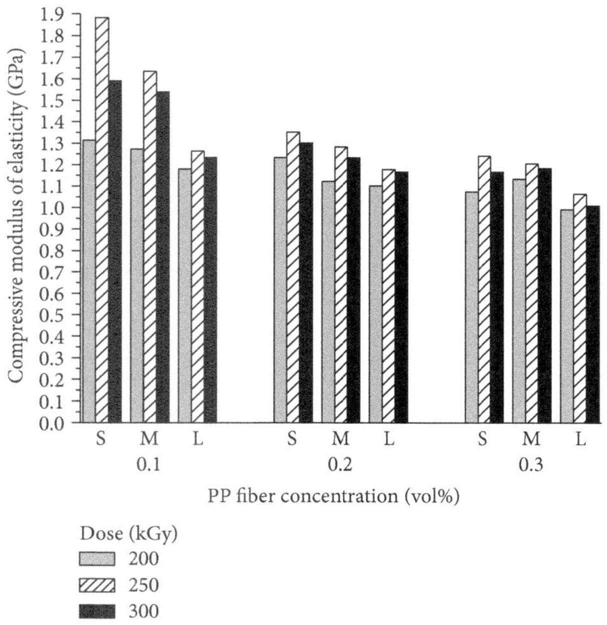

FIGURE 6: Compressive modulus of elasticity of irradiated-polymer concrete with different marble particle sizes and fiber concentrations.

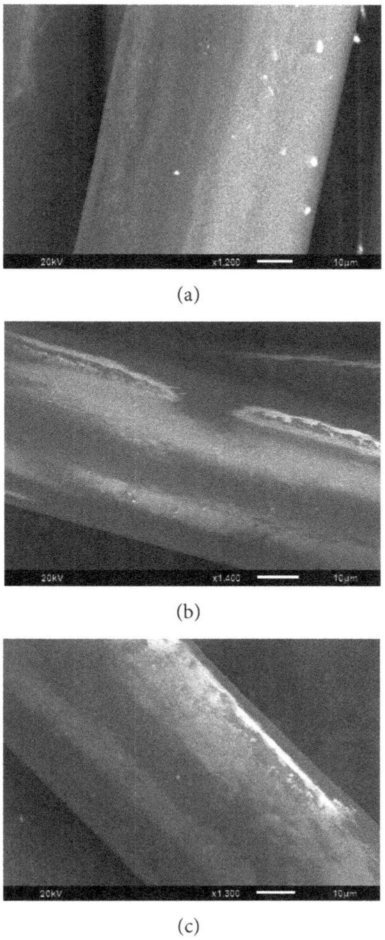

FIGURE 7: SEM images of irradiated polypropylene fibers at (a) 200 kGy, (b) 250 kGy, and (c) 300 kGy.

with respect to Ec values of nonirradiated PCs. In terms of PP fiber concentration, higher values were observed when adding 0.1 vol% of fibers and lower values when adding 0.3 vol% of fibers. Thus, the values follow the sequence: (0.1 vol%) > (0.2 vol%) > (0.3 vol%), and it is established that the compressive modulus decreases when increasing the PP fiber concentration.

With respect to particles size, the Ec values follow the rule: when increasing the particle size, the compressive modulus diminishes; same behavior was observed for nonirradiated PCs. Finally, following the radiation dose, for each particle size a premise is observed: the Ec values show two well-defined stages; the Young modulus increases when the radiation dose increases up to 250 kGy, and after this dose the Young modulus decreases. At 250 kGy the highest value is obtained.

When comparing with another Ec reported in the literature, the highest Ec value (1.88 GPa) is lower than for PC with two mineral aggregates of marble + calcium bentonite (6.3 GPa) [22], or for PC with silica sand + $CaCO_3$ (8.0 GPa) [19], or for PC with silica sand + polypropylene fibers (9.6 GPa) [23], including PCs with one aggregate: silica sand (16.3 GPa) [17] or with $CaCO_3$ (16.1 GPa) [18].

As it was studied the compressive strength and strain as well as compressive modulus depend on: particle size, PP fiber concentration, and gamma dose. The last one involves the modifications of the surface morphology caused by radiation on the polypropylene fibers. SEM images for irradiated fibers at 200, 250, and 300 kGy are shown in Figure 7. Small pieces of scrap of material and well-defined lines at irradiation dose of 200 kGy (Figure 7(a)) can be seen. When increasing the radiation dose at 250 kGy, roughness and deteriorated regions are observed (Figure 7(b)). Such deteriorated regions increase at 300 kGy dose, and more small particles are formed on the surface.

4. Conclusions

As expected, mechanical behavior depends on the combination of polypropylene fiber concentration and the applied radiation dose. Compressive strength values are higher for polymer concrete with medium particle sizes and irradiated at 250 kGy and the lowest elasticity modulus with large particle sizes. Such last behavior suggests generation of a ductile material. Moreover, depending on the latter combination of particle sizes, the polymer concrete needs low or high radiation doses to obtain high deformability with moderate compressive strength.

Conflict of Interests

The authors declare that none of them have a direct financial relationship with the commercial trademarks mentioned in this paper (CONSA, Controls, and GOSA) that might lead to a conflict of interests for any of the authors.

Acknowledgments

Financial support of the Autonomous University of the State of Mexico (UAEM) by Grant UAEM 9017/2013 CAFF and help of Mr. Francisco García Flores for the sample irradiation performed at the Institute of Nuclear Sciences of the National Autonomous University of Mexico (UNAM) are acknowledged.

References

[1] J. M. Laredo Dos Reis, "Mechanical characterization of fiber reinforced Polymer Concrete," *Materials Research*, vol. 8, no. 3, pp. 357–360, 2005.

[2] D. Van Gemert, L. Czarnecki, M. Maultzsch et al., "Cement concrete and concrete-polymer composites: two merging worlds: a report from 11th ICPIC Congress in Berlin, 2004," *Cement & Concrete Composites*, vol. 27, no. 9-10, pp. 926–933, 2005.

[3] L. Czarnecki, "Concrete-polymer composites: trends shaping and future," *International Journal of the Society of Materials Engineering for Resources*, vol. 15, no. 1, pp. 1–5, 2007.

[4] J. M. L. dos Reis, "Effect of textile waste on the mechanical properties of polymer concrete," *Materials Research*, vol. 12, no. 1, pp. 63–67, 2009.

[5] A. J. M. Ferreira, C. Tavares, and C. Ribeiro, "Flexural properties of polyester resin concretes," *Journal of Polymer Engineering*, vol. 20, no. 6, pp. 459–468, 2000.

[6] O. Gencel, W. Brostow, G. Martinez-Barrera, and M. S. Gok, "Mechanical properties of polymer concretes containing different amount of hematite or colemanite," *Polimery*, vol. 57, no. 4, pp. 276–283, 2012.

[7] O. Gencel, C. Ozel, F. Koksal, E. Erdogmus, G. Martínez-Barrera, and W. Brostow, "Properties of concrete paving blocks made with waste marble," *Journal of Cleaner Production*, vol. 21, no. 1, pp. 62–70, 2012.

[8] C. Vipulanandan and S. Mebarkia, "Flexural strength, toughness, and fracture properties of polyester composites," *Journal of Applied Polymer Science*, vol. 50, no. 7, pp. 1159–1168, 1993.

[9] J. M. L. Reis and A. J. M. Ferreira, "A contribution to the study of the fracture energy of polymer concrete and fibre reinforced polymer concrete," *Polymer Testing*, vol. 23, no. 4, pp. 437–440, 2004.

[10] N. Dharmarajan and C. Vipulanandan, "Fracture toughness of particle filled fiber reinforced polyester composites," *Journal of Applied Polymer Science*, vol. 42, no. 3, pp. 601–607, 1991.

[11] O. Gencel, C. Ozel, W. Brostow, and G. Martínez-Barrera, "Mechanical properties of self-compacting concrete reinforced with polypropylene fibres," *Materials Research Innovations*, vol. 15, no. 3, pp. 216–225, 2011.

[12] Z. Ajji, "Preparation of polyester/gypsum/composite using gamma radiation, and its radiation stability," *Radiation Physics and Chemistry*, vol. 73, no. 3, pp. 183–187, 2005.

[13] T. Jurkin and I. Pucić, "Post-irradiation crosslinking of partially cured unsaturated polyester resin," *Radiation Physics and Chemistry*, vol. 75, no. 9, pp. 1060–1068, 2006.

[14] G. Martínez-Barrera and W. Brostow, "Fiber-reinforced polymer concrete: property improvement by gamma irradiation," in *Gamma Radiation Effects on Polymeric Materials and Its Applications*, C. Barrera-Díaz and G. Martínez-Barrera, Eds., pp. 27–44, Research Signpost, Kerala, India, 2009.

[15] G. Martínez-Barrera, C. Menchaca Campos, and F. Ureña-Nuñez, "Gamma radiation as a novel technology for development of new generation concrete," in *Gamma Radiation*, F. Adrovic, Ed., pp. 91–114, InTech, Rijeka, Croatia, 2012.

[16] M. Levitt, D. J. McGahan, and P. R. Hills, "Comparision of concrete polymer composites produced by high energy radiation," *Journal of the Precast/Prestressed Concrete Institute*, vol. 18, no. 3, pp. 35–41, 1973.

[17] G. Martínez-Barrera, U. Texcalpa-Villarruel, E. Vigueras-Santiago, S. Hernández-López, and W. Brostow, "Compressive strength of Gamma-irradiated polymer concrete," *Polymer Composites*, vol. 29, pp. 1210–1217, 2008.

[18] G. Martínez-Barrera, M. E. Espinosa-Pesqueira, and W. Brostow, "Concrete + polyester + $CaCO_3$: mechanics and morphology after gamma irradiation," *e-Polymers*, Article ID 083, 2007.

[19] E. A. Bobadilla-Sánchez, G. Martínez-Barrera, W. Brostow, and T. Datashvili, "Effects of polyester fibers and gamma irradiation on mechanical properties of polymer concrete containing $CaCO_3$ and silica sand," *Express Polymer Letters*, vol. 3, no. 10, pp. 615–620, 2009.

[20] T. Nishiura, S. Nishijima, and T. Okada, "Creep behavior of epoxy resin during irradiation at cryogenic temperature," *Radiation Physics and Chemistry*, vol. 56, no. 5-6, pp. 605–609, 1999.

[21] I. Pucić and F. Ranogajec, "Phase separation during radiation crosslinking of unsaturated polyester resin," *Radiation Physics and Chemistry*, vol. 67, no. 3-4, pp. 415–419, 2003.

[22] G. Martínez-Barrera, L. F. Giraldo, B. L. López, and W. Brostow, "Effects of γ radiation on fiber-reinforced polymer concrete," *Polymer Composites*, vol. 29, no. 11, pp. 1244–1251, 2008.

[23] G. Martínez-Barrera, A. L. Martínez-Hernández, C. Velasco-Santos, and W. Brostow, "Polymer concretes improved by fiber reinforcement and gamma irradiation," *e-Polymers*, vol. 103, pp. 1–14, 2009.

6

Novel Complex Polymers with Carbazole Functionality by Controlled Radical Polymerization

Kazuhiro Nakabayashi and Hideharu Mori

Department of Polymer Science and Engineering, Department of Organic Device Engineering, Graduate School of Science and Engineering, Yamagata University, 4-3-16, Jonan, Yonezawa 992-8510, Japan

Correspondence should be addressed to Hideharu Mori, h.mori@yz.yamagata-u.ac.jp

Academic Editor: Toshifumi Satoh

This review summarizes recent advances in the design and synthesis of novel complex polymers with carbazole moieties using controlled radical polymerization techniques. We focus on the polymeric architectures of block copolymers, star polymers, including star block copolymers and miktoarm star copolymers, comb-shaped copolymers, and hybrids. Controlled radical polymerization of *N*-vinylcarbazole (NVC) and styrene and (meth)acrylate derivatives having carbazole moieties is well advanced, leading to the well-controlled synthesis of complex macromolecules. Characteristic optoelectronic properties, assembled structures, and three-dimensional architectures are briefly introduced.

1. Introduction

Polymers with carbazole groups are of considerable scientific and industrial interest because of their attractive features, such as their hole-transporting, high charge-carrier, and electroluminescent properties [1]. The hole-transporting ability of carbazole-containing polymers makes them especially useful for applications in organic electronics. Numerous studies have been devoted to carbazole-containing polymers as a result of the success of poly(*N*-vinylcarbazole), poly(NVC), in electrophotographic applications [1]. Recent developments in this field are mostly connected to applications in polymeric light-emitting diodes, organic photorefractive materials, and photovoltaic devices. For example, conjugated poly(3,6-carbazole) and poly(2,7-carbazole) derivatives having carbazole moieties in the main chain have been employed for applications in solar cells and white-light-emitting diodes [2–4]. Carbazole-containing polymers can be divided into two general groups: polymers containing carbazole moieties in the main chain, or those containing carbazole moieties in the side chains. Poly(NVC) belongs to the group of polymers having carbazolyl groups in the side chains. A variety of polymers with pendant carbazolyl groups has been reported, including polyacrylate [5, 6], polymethacrylate [7–10], and polystyrene [11, 12] derivatives. Many factors play crucial roles in the manipulation of their properties and practical applications, such as their chemical structures, polymer architectures, conformation, location, and stacking of carbazole units. However, it was difficult to control their molecular weights and architectures, because most of these polymers with pendant carbazolyl groups have been traditionally synthesized by conventional radical polymerization. To manipulate their unique electronic and photonic functions, it is desirable to establish precise synthetic methods of controlling molecular weight, polydispersity, topology, composition, and functions. This can be achieved by using controlled polymerization systems.

In the past few decades, considerable attention has been given to the self-assembly of block copolymers, because of the feasibility of using them to generate nanostructured materials and their numerous potential applications in separation technology, controlled drug delivery and release, and smart catalyst separation technology [13–17]. Traditional amphiphilic block copolymers containing chemically connected hydrophilic and hydrophobic segments provide a great variety of morphologies in selective solvents, for example, water, in which micellar aggregates occur as a result of the association of the insoluble blocks. Block copolymers

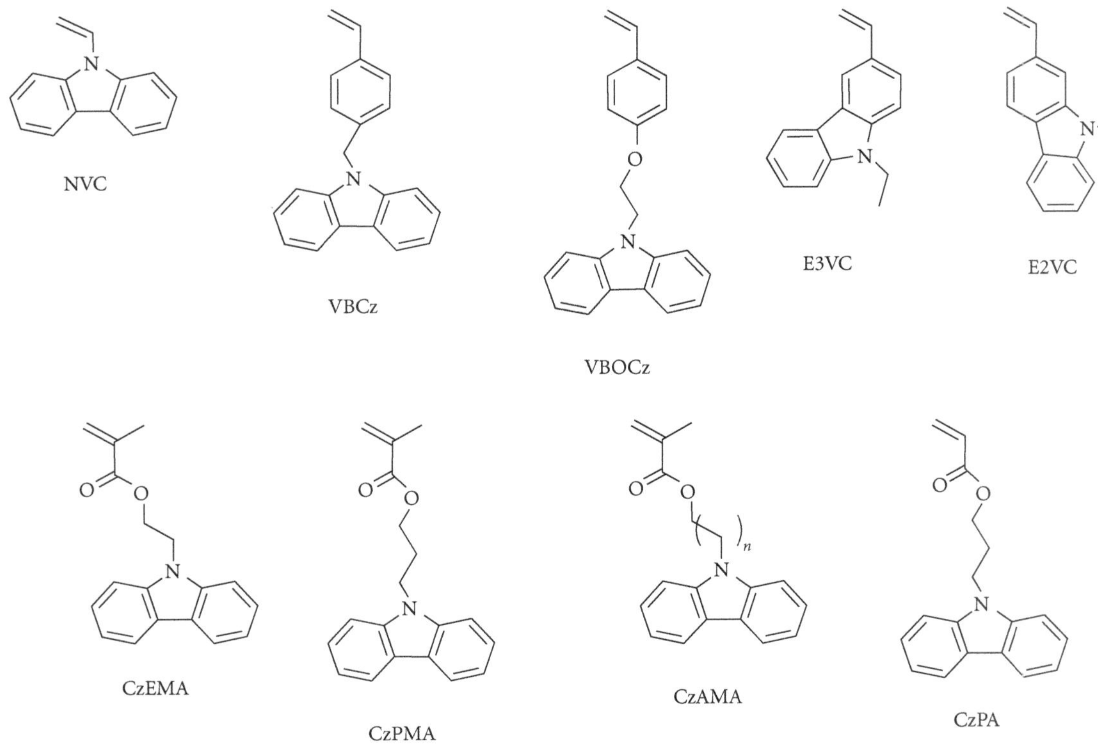

Scheme 1: Representative examples of carbazole-containing vinyl monomers used for controlled radical polymerizations.

incorporating semiconducting segments and characteristic optoelectronic functions have attracted significant research interest [18–23]. Block copolymers are emerging as a promising class of materials for optoelectronic applications, such as organic light-emitting devices, photovoltaics, and organic field-effect transistors, because of their ability to form a variety of ordered structures via self-assembly processes. Depending on the chemical nature of the functional segments and their composition, block copolymers provide a great opportunity to tune their chemical, physical, and optoelectronic properties and assembled structures.

Controlled radical polymerization (aka controlled/"living" radical polymerization or reversible deactivation radical polymerization) combines the benefits of the robust nature of conventional radical polymerization with the capability to prepare well-defined macromolecular architectures common to living polymerization techniques. This method has facilitated the synthesis of various functional polymers with predetermined molecular weights, narrow molecular weight distribution, and controlled architectures, such as star and branched polymers, block and graft copolymers, using a facile approach. The synthesis of functional polymers with controlled molecular weights, low polydispersity, and complex architectures has been possible because of recent significant progress in controlled radical polymerization techniques, such as nitroxide-mediated polymerization (NMP) [24, 25], atom transfer radical polymerization (ATRP) [26, 27], reversible addition-fragmentation chain transfer (RAFT) polymerization [28–41], single electron transfer-living radical polymerization (SET-LRP) [42], organoheteroatom-mediated living radical polymerization [43], and organometallic-mediated radical polymerization [44]. All these systems are based on establishing a rapid dynamic equilibration between a minute amount of growing free radicals and a large majority of dormant species and are more tolerant of functional groups and impurities. Such controlled radical polymerization methods have become key tools for polymer synthesis, especially for synthesizing complex polymers with well-defined structures.

This review highlights recent developments in the design and synthesis of novel complex polymers with carbazole moieties using controlled radical polymerization techniques of carbazole-containing monomers, as shown in Scheme 1. We mainly focus on three classes of carbazole-containing polymers, as follows:

(i) self-assembled block copolymers including amphiphilic block copolymers, block copolymers composed of "more activated" and "less activated" monomers, rod-coil-type block copolymers, and block copolymers possessing a metal ligating functionality

(ii) branched polymers, such as star block copolymers, miktoarm star copolymers, star polymers having a metallocore, and comb-shaped copolymers;

(iii) hybrids obtained by controlled radical polymerization of carbazole-containing monomers.

2. Carbazole-Containing Block Copolymers

Block copolymers incorporating carbazole functionality have attracted significant research interest, mainly as a result of the feasibility of using them to create highly ordered structures and their potential applications in the optoelectronic fields. Depending on the ordered structures and stacking of the carbazole moiety, the three-dimensional hierarchical structures formed by self-organization of the carbazole-containing block copolymers may lead to unique electronic and photonic functions. There has been much research for the synthesis of carbazole-containing block copolymers using controlled radical polymerization of NVC. Styrene and (meth)acrylate derivatives having carbazole moieties have also been used for the controlled synthesis of carbazole-containing block copolymers.

2.1. Controlled Radical Polymerization of N-Vinylcarbazole. Poly(NVC) was the first and most widely studied polymeric photoconductor. NVC, a strong basic monomer, undergoes facile cationic polymerization with almost all cationic initiators to give poly(NVC). The living cationic polymerization of NVC with hydrogen iodine has also been reported [45]. Free radical initiators, such as azo compounds and peroxides, readily polymerize NVC, and radical polymerization was used in manufacturing the polymer [1]. Several attempts to synthesize poly(NVC) using controlled radical polymerization systems have been reported in the literature. For example, Fukuda et al. demonstrated that the control of NVC homopolymerization could not be achieved by a typical NMP, whereas the copolymerization with styrene under the same conditions proceeded in a living fashion [46]. Baethge et al. also reported the controlled NMP of styrene and NVC, while they demonstrated that the synthesis of the homopolymer, poly(NVC), with controlled molecular weights and a narrow polydispersity, was still problematic [47, 48]. Nowakowska et al. reported that the synthesis of poly(sodium styrenesulfonate-*b*-NVC) by NMP was possible only when the polymerization of NVC was conducted in the presence of acetic anhydride as an accelerator [49]. ATRP ($C_{60}Cl_n$/CuCl/2,2′-bipyridine) was also used for the synthesis of poly(NVC) having a narrow polydispersity (M_w/M_n = 1.33). However, the actual structure of the product was a star-like architecture with a C_{60} core because of multiple chlorine atoms in one initiator molecule, $C_{60}Cl_n$, and there was no information on each poly(NVC) and its detachments from the surface of C_{60} [50]. They also demonstrated that the attempt to synthesize the block copolymer, poly(NVC-*b*-styrene), using the ATRP system provided products with bimodal molecular weight distributions [51]. These previous studies failed to provide convincing proof of the controlled character of the homopolymerization of NVC, such as a kinetic investigation, evaluation of the end groups, and a chain extension experiment. The difficulties in adapting controlled radical polymerization to NVC may also be based on the fact that the NVC propagating radical is relatively unstable and thus highly reactive, mainly because of the electron-donating carbazolyl pendant ($Q = 0.26$, $e = -1.29$) [52], leading to a tendency to undergo chain transfer and chain termination reactions. In other words, systems suitable for inducing a fast interconversion between the dormant and the reactive radical species having a high electron density at the radical center derived from NVC were difficult to find.

However, recent advances in the field of controlled radical polymerization have resulted in successful controlled polymerization of NVC. Mori et al. recently synthesized poly(NVC) with predetermined molecular weights (M_n in the range of 3000–48000) and low polydispersities (M_w/M_n in the range of 1.15–1.20) by xanthate-mediated controlled radical polymerization [53]. The RAFT process is generally accomplished by performing a radical polymerization in the presence of a thiocarbonylthio compound, such as a dithioester, dithiocarbamate, trithiocarbonate, or xanthate, all of which act as reversible chain transfer agents (CTAs). When xanthates are employed, the terminology MADIX (macromolecular design via the interchange of xanthates) is frequently used to describe the process [54–56]. In general, the controlled radical polymerization of the *N*-vinyl and *O*-vinyl monomers was considered difficult, since the generated radical species are highly reactive due to their nonconjugated nature and strong electron-donating pendant groups. However, dithiocarbonates (xanthates) were recently reported to be useful for controlling the radical polymerization of highly reactive *O*-vinyl and *N*-vinyl monomers, such as vinyl acetate [57–60], *N*-vinylpyrrolidone [61–63], *N*-vinylcarbazole (NVC) [53], *N*-vinylindole derivatives [64], *N*-vinylphthalimide [65–67], *N*-vinylnaphthalimide [66], and *N*-vinylimidazolium salts [68].

Both the MADIX and RAFT processes are based on the generally accepted reversible addition-fragmentation chain transfer mechanism between an active and a dormant species. To achieve control of the radical polymerization via the RAFT/MADIX process, a delicate balance of the forward and reverse rates of addition (k_{add} and k_{-add}) and fragmentation (k_β and $k_{-\beta}$), together with the rates of reinitiation (k_i) and propagation (k_p), is required. Because the NVC propagating radical is a poor homolytic leaving group, the fragmentation of the RAFT-adduct radical (intermediate radical) is thought to be very slow when dithioesters are used, resulting in insufficient control of the polymerization. In contrast, the xanthate-type CTA is useful for achieving controlled radical polymerization of NVC, because it increases electron density at the radical center, which may lead to the destabilization of the RAFT-adduct radicals relative to the normal dithioester-type CTA and an increase in the fragmentation rate. The electron-donating *O*-alkyl substituents may lead to stabilization of the thiocarbonyl product of fragmentation through their conjugation with the C=S double bond, as shown in Scheme 2. This lowers the rate of addition of the propagating radicals to the sulfur atom, and consequently the overall rate of chain transfer [30, 57, 69–71]; whereas the introduction of electron-withdrawing groups in the Z moiety leads to an increase in the rate of addition [70].

The blocking order is crucial for the synthesis of well-defined block copolymers by the RAFT process. The first dithioester-terminated polymer, S=C(Z)S-A, in which the A block corresponds to the first polymer and Z is the stabilizing

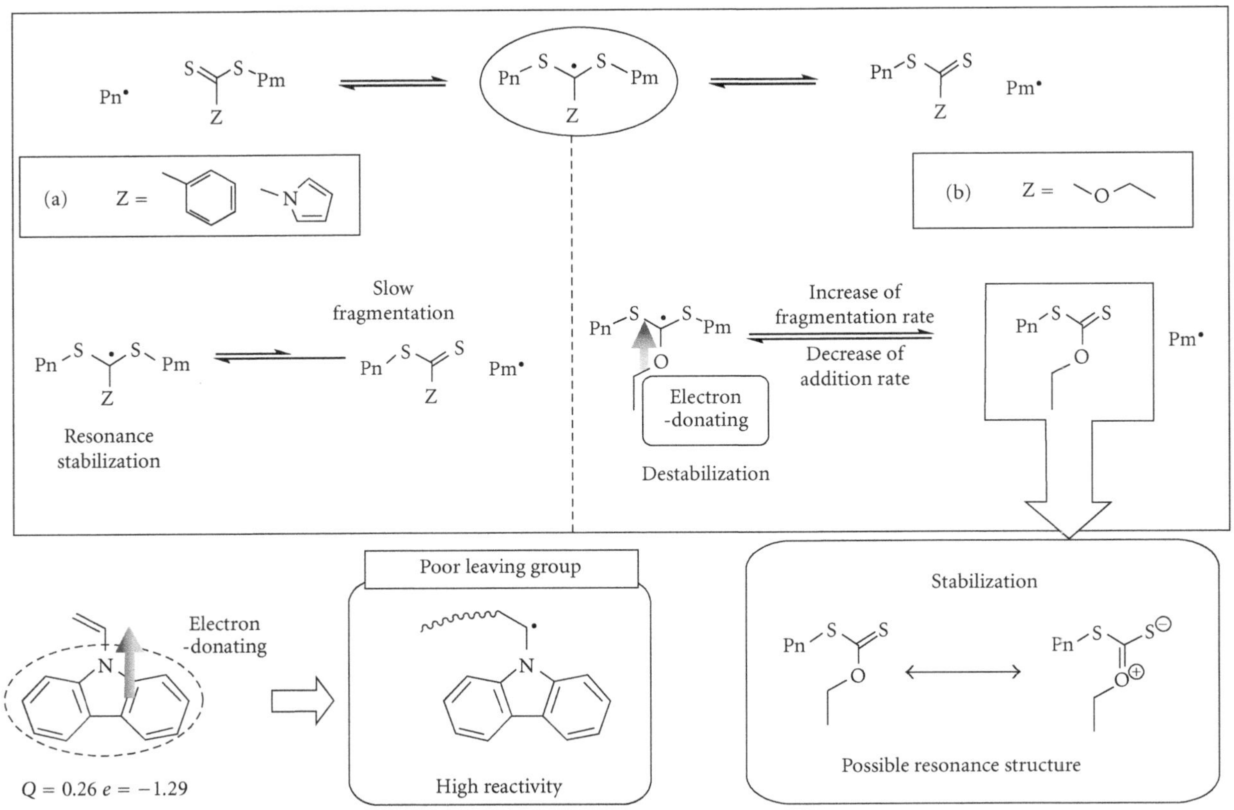

Scheme 2: Proposed mechanism of xanthate-mediated controlled radical polymerization of NVC.

group, should have a high transfer constant in the subsequent polymerization of the second monomers to give the B block [72, 73]. The rapid conversion of macro-CTA to a block copolymer is also required to achieve a block copolymer with low polydispersity, which allows all the second blocks to be initiated at approximately the same time [74, 75].

The xanthate-mediated controlled radical polymerization of NVC was also applied for the synthesis of well-defined block copolymers involving a poly(NVC) segment. For example, a well-defined block copolymer involving a poly(lithium vinyl sulfonate) segment was synthesized by RAFT polymerization of a vinyl sulfonate ester, neopentyl ethenesulfonate, using xanthate-terminated poly(NVC) macro-CTA, followed by deprotection (Scheme 3) [76]. The resulting product can be regarded as an amphiphilic block copolymer, comprising poly(lithium vinyl sulfonate) as a strong anionic polyelectrolyte and poly(NVC) as a hydrophobic segment. Amphiphilic block copolymers, poly(NVC)-*b*-poly(*N*-vinylpyrrolidone), were prepared by xanthate-mediated RAFT polymerization (Scheme 4) [77]. The amphiphilic block copolymers dissolved in several organic solvents; however, depending on their composition, formed either micelles or large aggregates in methanol. The presence of globular aggregates was confirmed by tapping mode atomic force microscopy. The xanthate-mediated RAFT polymerization of NVC was applied for the synthesis of three types of block copolymers, poly(ethylene glycol)-*b*-poly(NVC), poly(NVC)-*b*-poly(vinyl acetate), and poly(NVC)-*b*-poly(vinyl alcohol), as shown in Scheme 5 [78].

Well-defined pH- and thermomultiresponsive fluorescent micelles based on the self-assembly of diblock copolymers, poly(*N*-isopropylacrylamide-*co*-NVC)-*b*-poly[2-(dimethylamino)ethyl acrylate], were obtained by RAFT copolymerization of *N*-isopropylacrylamide and NVC followed by chain extension in the presence of 2-(dimethylamino)ethyl acrylate [79]. The micelles were formed in aqueous solutions in a wide range of temperatures, and their sizes increased from 40 to 65 nm when the pH was varied from basic to acidic. Cross-linking of the poly[2-(dimethylamino)ethyl acrylate)]-containing shell with 1,2-bis(2-iodoethoxy)ethane results in spherical soft nanoparticles. The presence of NVC in concentrations as low as 4% in the core of the micelles was reported to allow the nanoparticles to be tagged by fluorescence, making them well suited to therapeutic applications.

The synthesis of poly(NVC)-based block copolymers functionalized with rhenium diimine complexes or pendant terpyridine ligands was reported [80]. The copolymers were synthesized by RAFT polymerization, and they exhibited interesting morphological properties as a result of the phase separation between different blocks. It was demonstrated that the rhenium complex polymer block could function as a photosensitizer, while the terpyridine-containing polymer

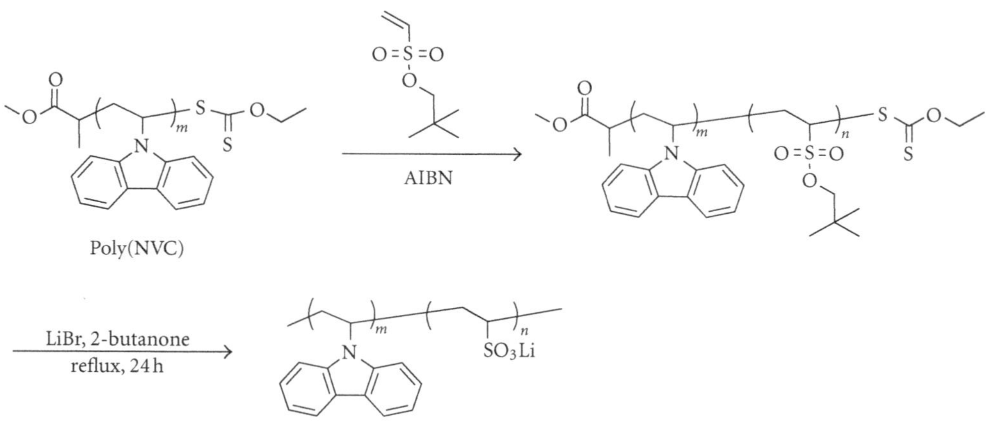

Scheme 3: Synthesis of poly(NVC)-*b*-poly(lithium vinyl sulfonate) by RAFT polymerization of neopentyl ethenesulfonate using poly(NVC) as macro-CTA and subsequent deprotection.

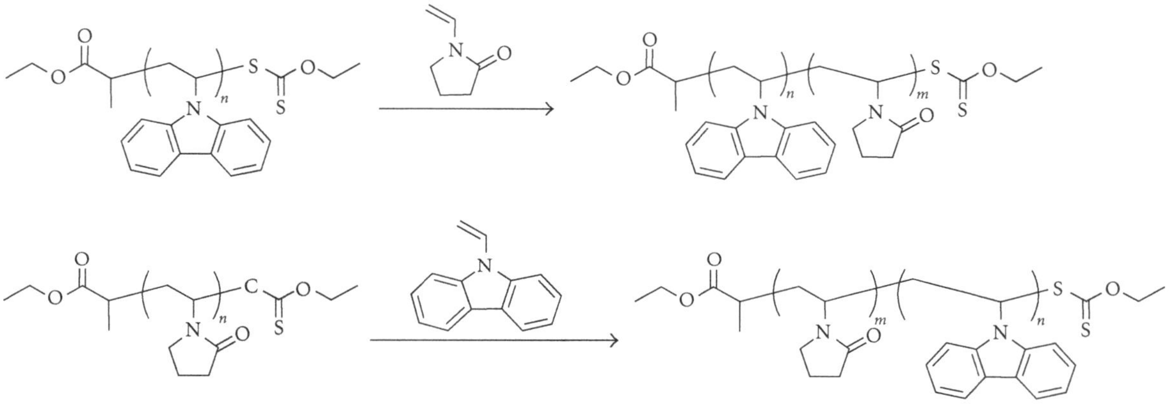

Scheme 4: Synthesis of poly(NVC)-*b*-poly(*N*-vinylpyrrolidone) by xanthate-mediated RAFT polymerization.

block can be used as a template for nanofabrication by selective deposition of zinc complexes.

Generally, the selection of the RAFT agent is dictated by the types of monomer being polymerized. The RAFT agents (dithioesters, trithiocarbonates) suitable for controlling polymerization of "more activated" monomers (e.g., styrene, acrylates, and methacrylates) tend to inhibit polymerization of "less activated" monomers (e.g., NVC, vinyl acetate, and *N*-vinylpyrrolidone). Similarly, RAFT agents suitable for polymerizations of "less activated" monomers tend to give little or poor control over polymerizations of "more activated" monomers. Hence, the synthesis of block copolymers composed of "less activated" and "more activated" monomers was difficult. Recently, switchable RAFT polymerization using *N*-(4-pyridinyl)-*N*-methyldithiocarbamates was found to provide excellent control over polymerization of "less activated" monomers and, after addition of 1 equivelant of a protic or Lewis acid, become effective in controlling polymerization of "more activated" monomers [81]. Switchable RAFT polymerization allowed the synthesis of poly(methyl acrylate)-*b*-poly(NVC) with narrow molecular weight distributions. The preparation of unimodal low dispersity block copolymers of poly(*N*,*N*-dimethylacrylamide) with NVC was also conducted by switchable RAFT polymerization [82]. The controlled radical polymerizations of "less activated" monomers, NVC, vinyl acetate, and *N*-vinylpyrrolidone, were successfully achieved in the presence of a disulfide, isopropylxanthic disulfide, using AIBN as the initiator [83]. The use of RAFT polymerization for the synthesis of optoelectronic polymers was recently reviewed by Moad et al [22]. The polymerization proceeded via MADIX process, where xanthate was formed *in situ* from the reaction of AIBN and isopropylxanthic disulfide. Organoheteroatom-mediated living radical polymerization using organotellurium, organostibine, and organobismuthine chain transfer agents were also useful methods for achieving the controlled character of polymerization of conjugated and nonconjugated monomers [43].

A rod-coil block copolymer consisting of poly(3-hexylthiophene) and poly(NVC) was synthesized by RAFT polymerization of NVC using trichiocarbonate-teminated poly(3-hexylthiophene) as a macro-CTA (Scheme 6) [84]. The rod-coil block copolymer was employed for a white organic electroluminescence device. It is possible to suppress energy transfer from poly(NVC) as wide bandgap units to poly(3-hexylthiophene) as low bandgap blocks by obtaining

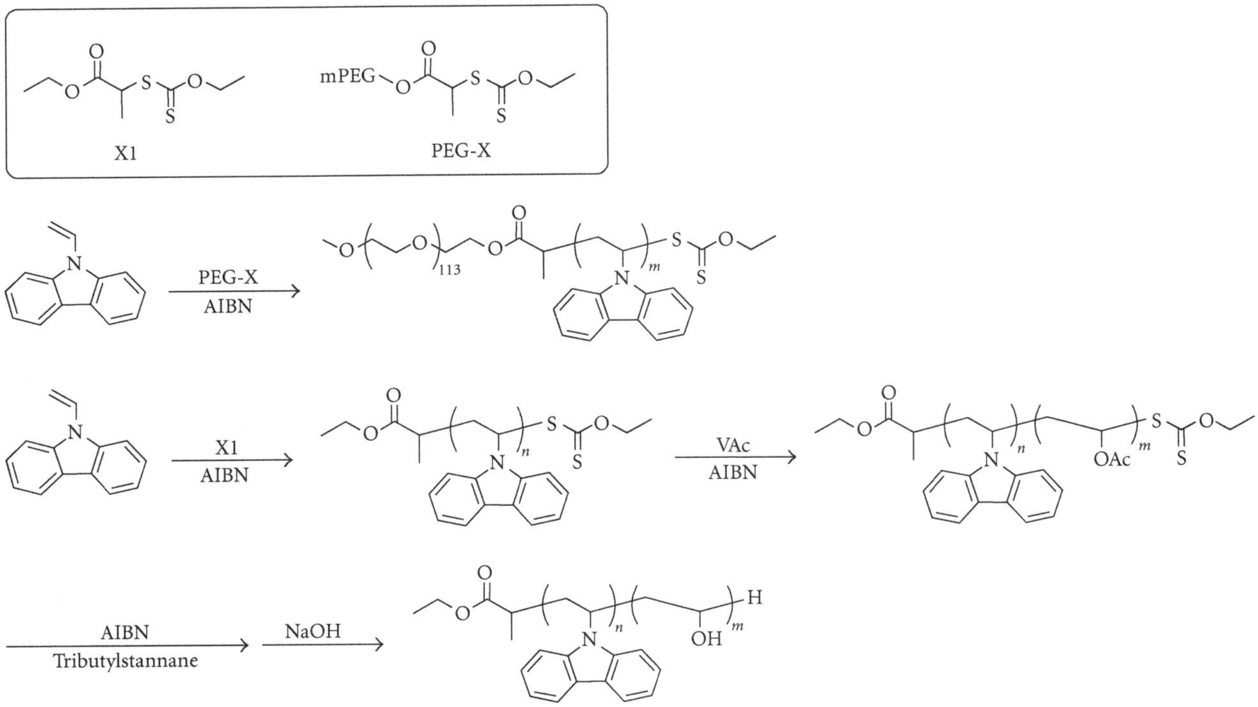

SCHEME 5: Synthesis of poly(ethylene glycol)-*b*-poly(NVC), poly(NVC)-*b*-poly(vinyl acetate), and poly(NVC)-*b*-poly(vinyl alcohol) by xanthate-mediated RAFT polymerization.

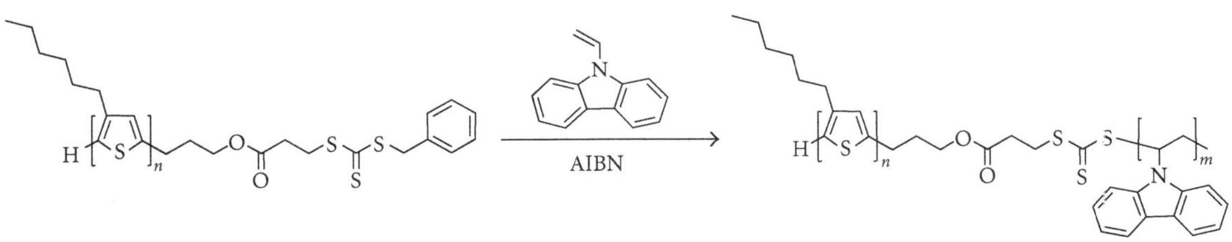

SCHEME 6: Synthesis of poly(3-hexylthiophene)-*b*-poly(NVC) by RAFT polymerization.

the phase-separated domains from thin films of the resulting block copolymer, yielding dual emissions for white electroluminescence with a CIE coordination of (0.34, 0.33).

The homopolymerization of NVC was performed with ATRP with Cu(I)/Cu(II)/2,2′-bipyridine as the catalyst system at 90°C in toluene [85]. *N*-2-Bromoethyl carbazole was used as the initiator, and the optimized ratio of Cu(I) to Cu(II) was found to be 1/0.3. The resulting poly(NVC) was used as a macroinitiator for the ATRP of methyl methacrylate, and this resulted in the formation of a block copolymer. The order of blocking is considered to be also crucial for ATRP or NMP as RAFT process. The polymerization of NVC and 3-(9H-carbazole-9-yl)propyl methacrylate (CzPMA) was carried out using various controlled radical polymerization methods, such as ATRP, single electron transfer-living radical polymerization (SET-LRP), and single electron transfer initiation followed by RAFT (SET-RAFT) [86]. In the case of NVC, well-controlled polymerization with narrow molecular weight distribution was achieved by high-temperature ATRP, while SET-RAFT polymerization was relatively slow and controlled at ambient temperature. In the case of CzPMA, SET-RAFT was more suitable for ambient temperature polymerization. The synthesis of diblock copolymers was achieved from poly(NVC) and poly(CzPMA) macroinitiators using a flavanone-based methacrylate as the second monomer. The emission spectra of these block copolymers showed higher intensities compared with homopolymers.

2.2. Controlled Radical Polymerization of Styrene Derivatives Having Carbazole Moieties. The functional polymer containing carbazole units was successfully prepared via nitroxide-mediated radical polymerization of 9-(4-vinylbenzyl)-9H-carbazole (VBCz) [87]. New nonconjugated random copolymers containing pendant electron-donating poly(VBCz) and electron-accepting 2-phenyl-5-(4-vinylphenyl)-1,3,4-oxadiazole or 2-(4-vinylbiphenyl)-5-(4-phenyl)-1,3,4-oxadiazole were successfully synthesized by

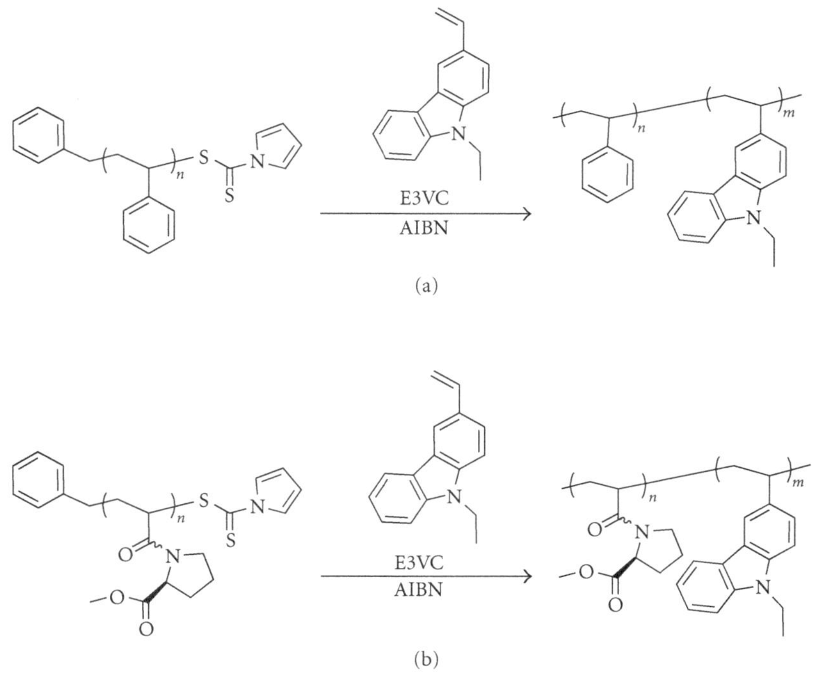

SCHEME 7: Synthesis of block copolymers by RAFT polymerization of E3VC.

the NMP method [88]. The electrical switching behavior, based on the ITO/ polymer/Al device configuration, could be tuned through the donor/acceptor ratio or acceptor trapping ability. The controlled nitroxide-mediated homopolymerization of VBCz and the copolymerization of methyl methacrylate with varying amounts of VBCz were accomplished by using 10 mol % nitroxide relative to 2-({tert-butyl[1-(diethoxyphosphoryl)-2,2-dimethylpropyl]amino}oxy)-2-methylpropionic acid [89]. VBCz was found to be an effective controlling comonomer for NMP of methyl methacrylate, and such low levels of VBCz comonomer ensured transparency in the final copolymer. Well-defined cyclic polymers, cyclic poly(VBCz), with differing molecular weights were efficiently prepared by successive ATRP and a click reaction [90].

N-ethyl-3-vinylcarbazole (E3VC) was also employed as a carbazole-containing monomer, which can be regarded as a styrene derivative. The controlled RAFT polymerization of E3VC was performed using benzyl 1-pyrrolecarbodithioate as a suitable CTA to afford well-defined poly(E3VC) in which the carbazole unit is directly bound to the polymer main chain [91]. Well-defined block copolymers with poly(E3VC) segments have been synthesized using dithiocarbamate-terminated polystyrene as a macro-CTA (Scheme 7(a)). Amino-acid-containing amphiphilic block copolymers composed of poly(E3VC) as a hydrophobic segment and poly(*N*-acryloyl-L-proline methyl ester) [92, 93] as a hydrophilic segment were prepared by RAFT polymerization (Scheme 7(b)) [94]. Formation of micelles consisting of a hydrophobic core of poly(E3VC) and a hydrophilic shell of poly(*N*-acryloyl-L-proline methyl ester) was observed in methanol, which is a selective solvent for the poly(*N*-acryloyl-L-proline methyl ester) segment. The chiroptical properties were affected by the assembled structure induced by the selective solvent and were evaluated by fluorescence and circular dichroism spectra.

N-Ethyl-2-vinylcarbazole (E2VC) is a monomer structurally similar to the styrenes, which are known to polymerize well by NMP [95]. Polymerization of E2VC from a unimolecular alkoxyamine was reported to show a molecular weight that increases steadily with time, whereas termination or decomposition of the "living" radical ends was clearly evident. Nevertheless, a sufficient fraction of the chains remained living that poly(E2VC)-*b*-polystyrene could be formed by chain extension of the poly(E2VC) macroinitiator with styrene.

2.3. Controlled Radical Polymerization of (Meth)Acrylate Derivatives Having Carbazole Moieties. Photorefractive active block copolymers were successfully synthesized via ATRP of acrylate containing liquid crystalline moieties and hole transporting carbazole moieties [96]. The photorefractive characteristics of the copolymers indicated a higher diffraction efficiency and larger coupling gain for the block copolymer, compared to the corresponding statistical random copolymer. The controlled radical polymerization of 2-(*N*-carbazolyl)ethyl methacrylate (CzEMA) and 4-(5-(4-*tert*-butylphenyl-1,3,4-oxadiazol-2-yl)phenyl) methacrylate via the RAFT process has been studied for the synthesis of functional polymers with hole- or electron-transfer ability [97]. RAFT polymerization of these monomers using cumyl dithiobenzoate as a suitable CTA and AIBN as an initiator afforded well-defined polymethacrylates having hole- and electron-transporting groups. Monomers containing a

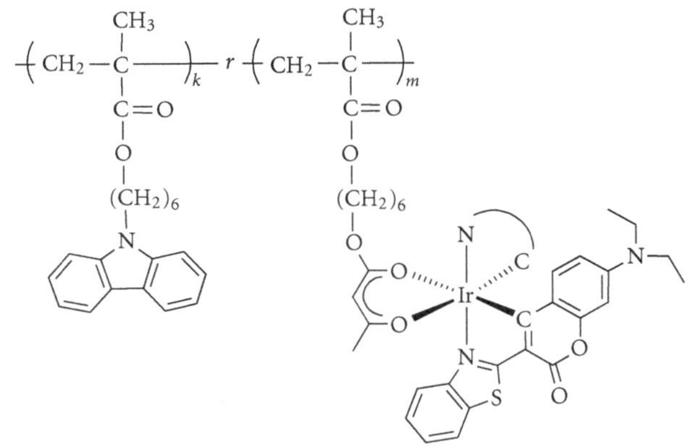

SCHEME 8: Methacrylate copolymers containing phosphorescent Ir(III)-complex and carbazole moiety.

chromophore group, that is, 9-anthrylmethyl methacrylate and CzEMA, were copolymerized with methyl methacrylate using ATRP, resulting in two series of copolymers with various amounts of included chromophore units [98]. Luminescence comparison demonstrated more intense light emission by polymethacrylates with carbazolyl groups than by copolymers with anthryl groups. The wavelength of emitted light also differed in the range of blue–violet and blue–green fluorescence for carbazolyl and anthryl copolymers, respectively.

A methacrylate-functionalized phosphorescent Ir(III)-complex was copolymerized with a methacrylate-functionalized carbazole-derivative using ATRP (Scheme 8), with the aim of applying this type of host-guest-system to organic light-emitting devices [99]. In the system, the emission originated almost exclusively from the complex, suggesting efficient intrachain energy transfer.

3. Carbazole-Containing Branched Polymers

3.1. Star Polymer/Star Block Copolymer/Miktoarm Star Copolymer. There are two basic routes for synthesizing star polymers [100]: the "core first" method (polymerization from multifunctional initiators or microgels), and the "arm first" method, in which growing polymer chain ends are reacted with a multifunctional terminating agent or a divinyl compound. The use of multifunctional initiators or terminators leads to stars with a well-known (though often, low) number of arms; while the use of microgels or divinyl compounds leads to a rather broad arm number distribution, with a high average arm number. The design and synthesis of novel star copolymers having characteristic architectures, such as star block copolymers and heteroarm or miktoarm (mixed) star polymers, are attracting significant attention because of their interesting structures and properties [17, 101–106]. Amphiphilic star block copolymers and miktoarm star polymers can form a variety of superstructures as a result of self-organization, and the resulting assembled structures should be governed by the branched architecture, as well as the chemical nature of the components, their composition, and molecular weight.

The xanthate-mediated RAFT polymerization of NVC was also applied for the synthesis of star polymers [107]. RAFT polymerization was employed for the purpose, using three different xanthate-type tetrafunctional CTAs: two *Z*-designed CTAs having different *R* (leaving) groups in which the *Z* (stabilizing) groups are linked to the core, and one *R*-designed CTA. The *R*-group approach was found to be the most efficient for the controlled synthesis of four-arm poly(NVC) stars having low polydispersities and controlled molecular weights. Amphiphilic star block copolymers were synthesized by the polymerization of NVC using the poly(acrylic acid) star, which was prepared from the R-designed tetrafunctional CTA (Scheme 9) [108]. DLS measurements of the amphiphilic stars in THF solution, which is a good solvent for both components, suggested the existence of unimolecular star micelles. In contrast, the formations of the micelles and inverse micelles were observed in water, which is a good solvent only for the poly(acrylic acid) segment, and in $CHCl_3$, which is a good solvent only for poly(NVC) segment. The absorbance and fluorescence spectra indicated that specific conformations of amphiphilic stars, like micelles and inverse micelles, affected the characteristic optoelectronic properties.

Star polymers containing chemically different arms are designated heteroarm or miktoarm (mixed arm) star polymers when they comprise two or more different kinds of arms, respectively. Their unique architectures cause them to reveal interesting properties in the solid state and in solution [109]. A novel amphiphilic A_3B miktoarm star copolymer, poly(*N*-isopropylacrylamide)$_3$-poly(NVC), was successfully synthesized by a combination of single-electron transfer living radical polymerization and RAFT polymerization (Scheme 10) [110]. First, the well-defined three-armed poly(*N*-isopropylacrylamide) was prepared via SET-LRP of *N*-isopropylacrylamide using a tetrafunctional bromoxanthate iniferter (Xanthate-Br_3) as the initiator and Cu(0)/PMDETA as a catalyst system. Secondly, the target amphiphilic A_3B miktoarm star copolymer was prepared via RAFT polymerization of NVC employing three-armed poly(*N*-isopropylacrylamide) having a xanthate moiety as the macro-CTA. The amphiphilic A_3B miktoarm star

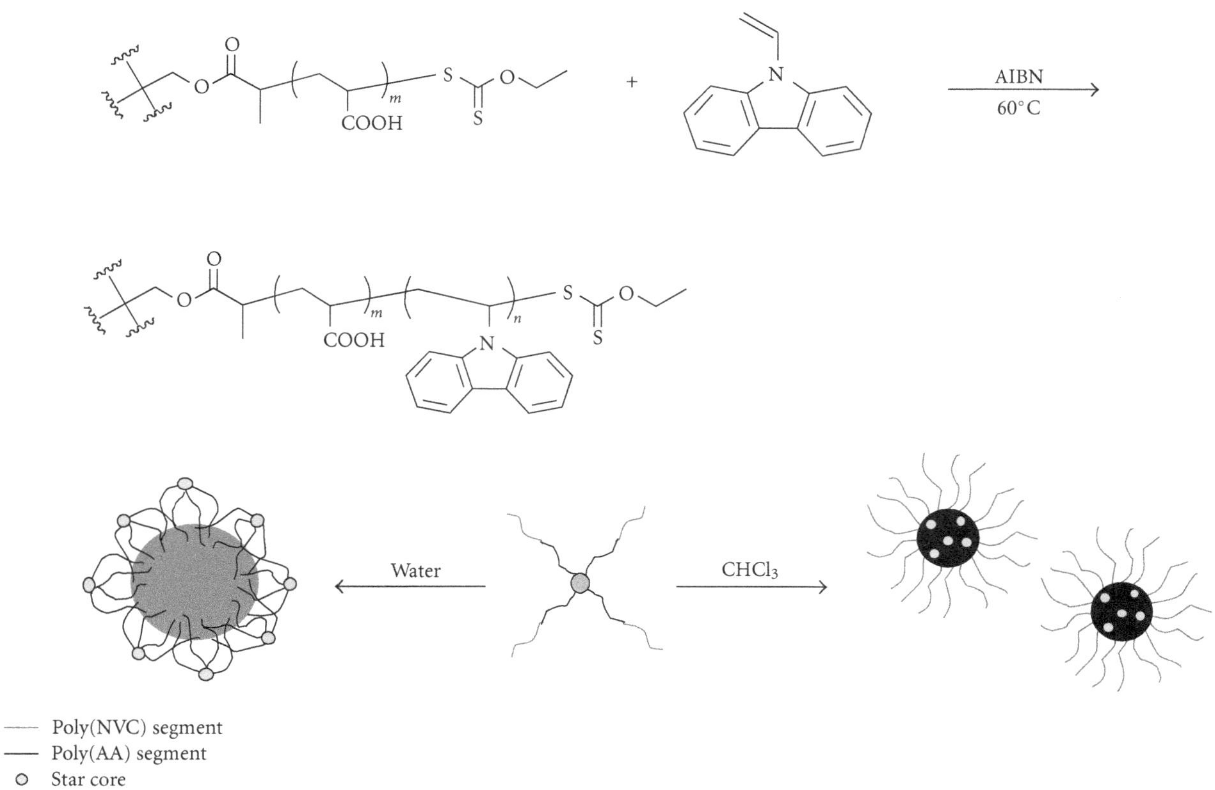

SCHEME 9: Synthesis and assembled structures of amphiphilic star block copolymers.

copolymer showed an increase in the fluorescence intensity of micelle with temperature and had good temperature reversibility.

Styrene-type monomer, VBCz, and methacrylate-type monomer, 2-(9H-carbazole-9-yl)ethyl methacrylate (CzEMA), were polymerized to star polymers via ATRP using zinc 5,10,15,20-tetrakis(4-(2-methyl-2-bromopropoxy) phenyl) porphyrin as an initiator (Scheme 11) [111]. The emission spectra of the two star polymers, poly(VBCz) and poly (CzEMA) stars, displayed red light emission in the solid state, while those of the two monomers showed blue light emission, suggesting the effective energy transfer from the carbazole to the Zn porphyrin core. It was demonstrated that the flexible arms of star polymers can effectively hinder π-π interaction of the porphyrin cores and present aggregation, which would lead to fluorescent self-quenching in the solid state.

A derivative Ru(II) complex with radical initiating sites was employed in the ATRP of functional N-(ω'-alkylcarbazoly) methacrylates to provide linear metallopolymers with the metal chromophores at one terminus of the polymer chain [112]. Similar Ru(II) complexes with one and three initiating sites were also employed as metalloinitiators for ATRP of N-(ω'-alkylcarbazoly) methacrylates with $NiBr_2(PPh_3)_2$ as a catalyst (Scheme 12) [113]. Star-like three-armed polymers could be obtained from the metalloinitiators with three initiating sites.

3.2. Comb-Shaped Copolymers. There have been very few reports of the synthesis of comb-shaped copolymers with carbazole-containing segments. These comb-shaped copolymers can be regarded as cylindrical polymer brushes having the same number of side chains as the degree of polymerization of the main chain. These cylindrical polymer brushes with carbazole-containing segments are architecturally interesting for both experimental and theoretical chemists because of the possibility of forming extended chain conformations, based on the intramolecular excluded-volume interactions between side chains densely grafted to the backbone. The homopolymerization of macromonomers, "grafting onto" and "grafting from" can be used in much the same way as conventional graft copolymers and comb-shaped copolymers. The "grafting from" method was mainly employed for the synthesis of comb-shaped copolymers with carbazole-containing segments.

The comb-shaped copolymer having poly(VBCz) side chains was prepared via NMP of VBCz using a macro-TEMPO agent, which was synthesized by anion ring-opening polymerization of 4-glycidyloxy-2,2,6,6-tetramethylpiperidine1oxyl (Scheme 13) [114]. The fluorescence, ultraviolet intensities, and cyclic voltammetries of the comb-shaped copolymers with different molecular weights showed a regular order. It was reported that the highest occupied molecular orbital (HOMO) and lowest unoccupied molecular orbital (LUMO) energy levels accompanied by bandgaps

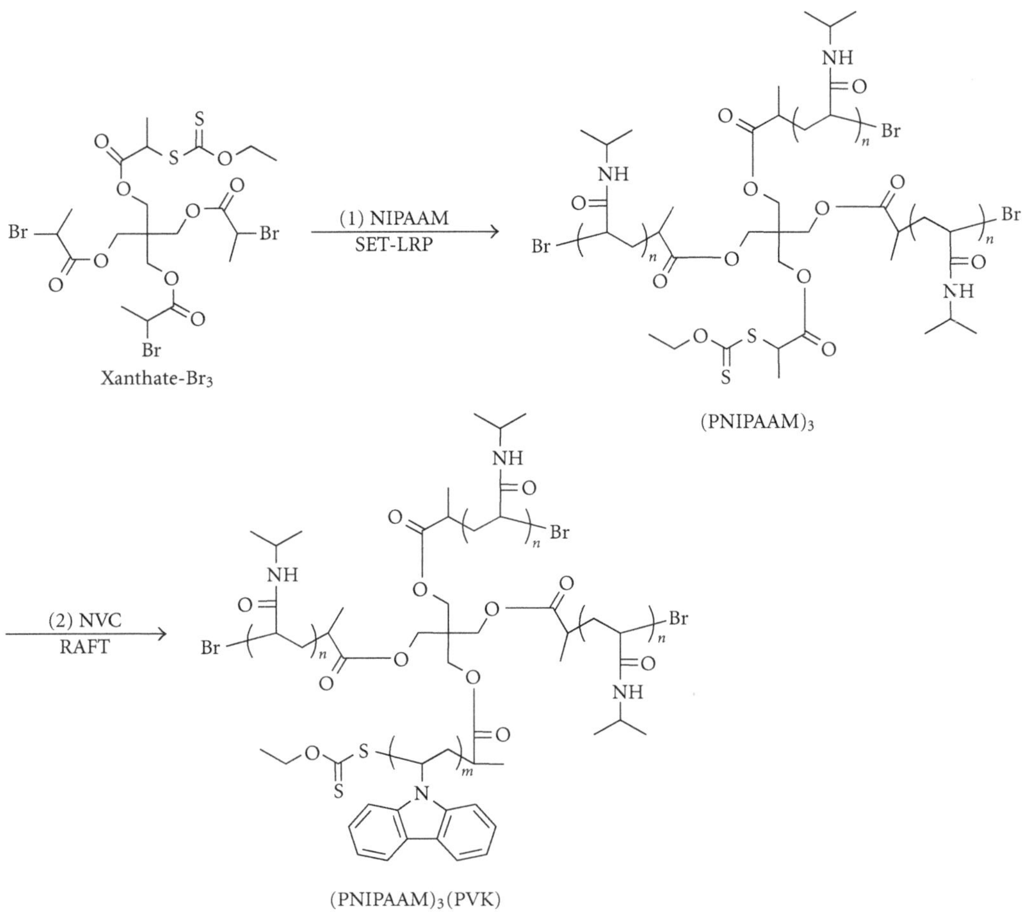

Scheme 10: Synthesis of A_3B miktoarm star copolymer, poly(*N*-isopropylacrylamide)$_3$-poly(NVC).

were adjustable via molecular weights of the polymers, which should be favorable to the application of these polymers in electronic devices.

A new light-emitting and hole transporting comb-shaped copolymer was prepared via ATRP of an acrylate monomer, 3-(*N*-carbazolyl)propyl acrylate (CzPA), initiated by a macroinitiator with ATRP initiating sites distributed along a polythiophene backbone (Scheme 14). This polymer can be regarded as a rod-coil-type comb-shaped copolymer, in which the polythiophene rod is located in the backbone and the poly(CzPA) coil is located in the side chain. The energy transfer process from the poly(CzPA) side chains to the polythiophene backbone was evidenced in both the solution and film states. Electroluminescence devices fabricated from the copolymer demonstrated that the copolymer could act as both a light emissive and hole-transporting material [115].

4. Carbazole-Containing Hybrids

Carbazole-containing polymer chains attached to planar and spherical surfaces have recently attracted much interest as candidates for various optoelectronic industrial applications. Electroactive substrates, such as carbon nanotubes, fullerene, graphene, and quantum dots, were mainly employed. The so-called "grafting onto," and "grafting from" methods can be used for the synthesis of carbazole-containing hybrids. In the "grafting from" process, the side chains of the brush are formed via controlled radical polymerizations involving ATRP and NMP initiated by the pendant initiating groups on the surface. When RAFT polymerization was employed, the CTA moiety was attached to the surface. Well-defined polymer brushes with high grafting density and rather narrow distributions can be obtained using this method. The "grafting onto" technique was frequently used for the synthesis of the hybrids, in which the carbazole-containing polymers were attached to the substrates by adsorption or chemical reaction of end-functionalized polymers or block copolymers onto the surfaces.

The poly(NVC)-grafted multiwalled carbon nanotubes hybrid materials were synthesized by RAFT polymerization of NVC in the presence of a trithiocarbonate-type CTA functionalized covalently onto multiwalled carbon nanotubes (Scheme 15) [116]. Incorporation of poly(NVC) onto the surfaces of the multiwalled carbon nanotubes can considerably improve their solubility and processability. Hybrid materials that are suitable candidates for viable optical limiting devices exhibit significant nonlinear optical

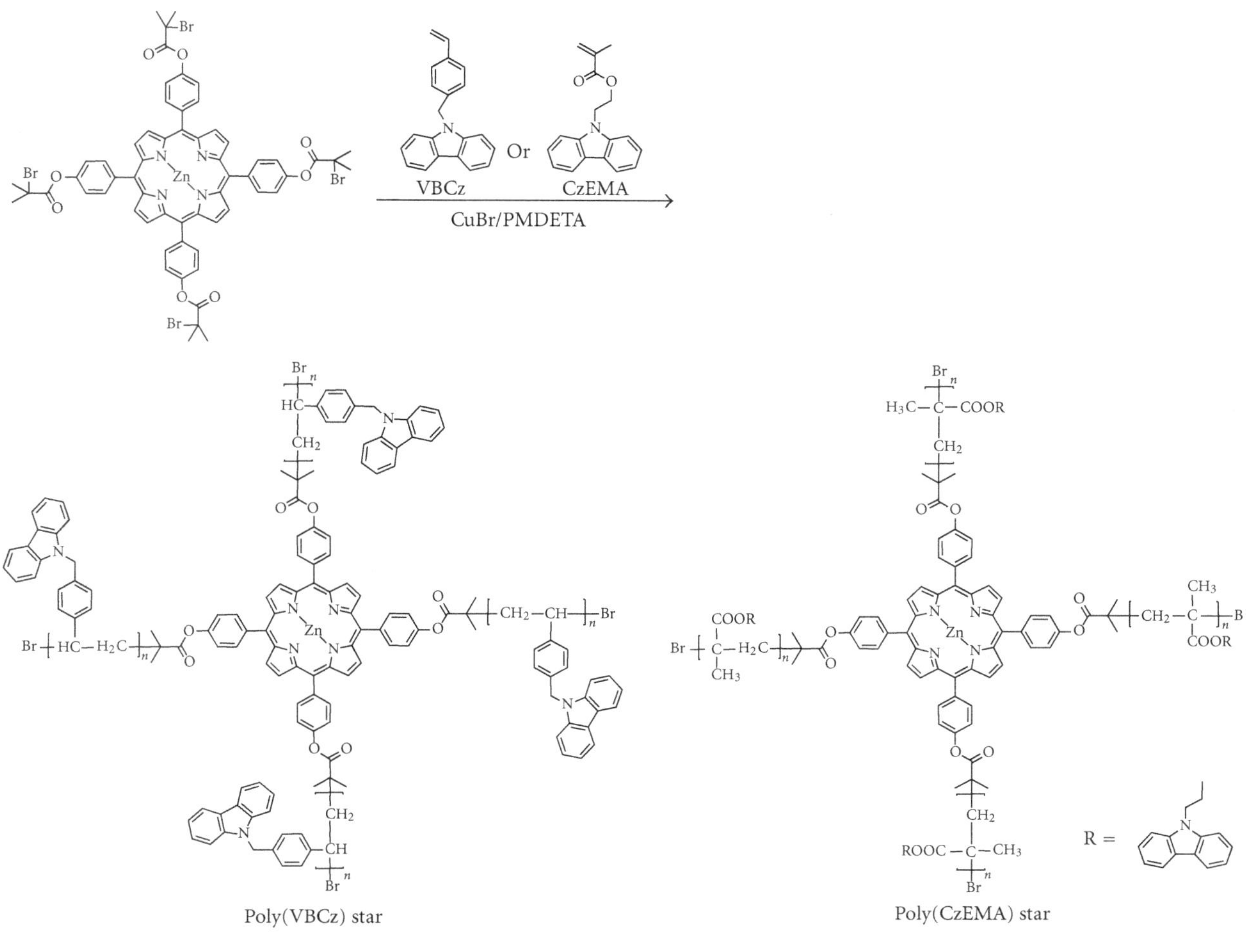

SCHEME 11: Synthesis of star polymers via ATRP using zinc porphyrin as a tetrafunctional initiator.

responses. Poly(NVC) chemically modified graphene oxide was obtained by RAFT polymerization of a trithiocarbonate type-CTA functionalized on graphene oxide [117, 118]. The resulting hybrid material showed good solubility in organic solvents and a significant energy bandgap in the vicinity of 2.49 eV [118]. The poly(NVC)-C_{60} composites were synthesized by ATRP ($C_{60}Cl_n$/CuCl/2,2′-bipyridine) of NVC [50, 119, 120]. The photoconducting properties of poly(NVC)-C_{60} composites were also investigated.

The CdSe-polymer composite was prepared via ATRP of NVC on functionalized CdSe quantum dots (Scheme 16) [121, 122]. It was revealed that grafting poly(NVC) onto the surface of CdSe nanocrystals would reduce the bandgap of poly(NVC) and cause a red shift in the emission peak. The nanocrystal-polymer nanocomposite was employed as the electron acceptor in polymer bulk heterojunction solar cells. Ultrasound-assisted bulk synthesis of CdS-poly(NVC) nanocomposites was reported via RAFT polymerization [123]. Poly(methyl methacrylate)/zinc oxide (ZnO) or carbazole-containing polymers, poly(CzEMA)/ZnO nanocomposites, which are composed of high molecular weight polymers with narrow molecular weight distributions and ZnO quantum dots, were prepared by ATRP [124]. 2-Bromo-2-methylpropionyl group was introduced onto the ZnO nanoparticle surfaces, which was employed for the surface-initiated ATRP.

The styrene derivative having a carbazole moiety (VBCz) was employed for the synthesis of bifunctional nanoparticles with fluorescence and magnetism by surface-initiated iron(III)-mediated ATRP with activators generated by electron transfer (AGET ATRP) [125]. VBCz was grafted from magnetic nanoparticles (ferroferric oxide) via AGET ATRP, using $FeCl_3*6H_2O$ as the catalyst, tris(3,6-dioxaheptyl)amine as the ligand, and ascorbic acid as the reducing agent. Another interesting system involves the synthesis of multistimuli-response hybrid nanoparticles with magnetic cores and thermoresponsive fluorescence-labeled shells by surface-initiated RAFT polymerization using carbazole-containing CTA-functionalized magnetic silica nanoparticles [126].

Polymer brushes refer to an assembly of polymer chains, which are tethered by one end to a surface or an interface. Surface-initiated polymerization has the advantage of allowing easy modification of surface properties by varying the composition of the polymer brush, grafting density, and the degree of polymerization. Electrochemically

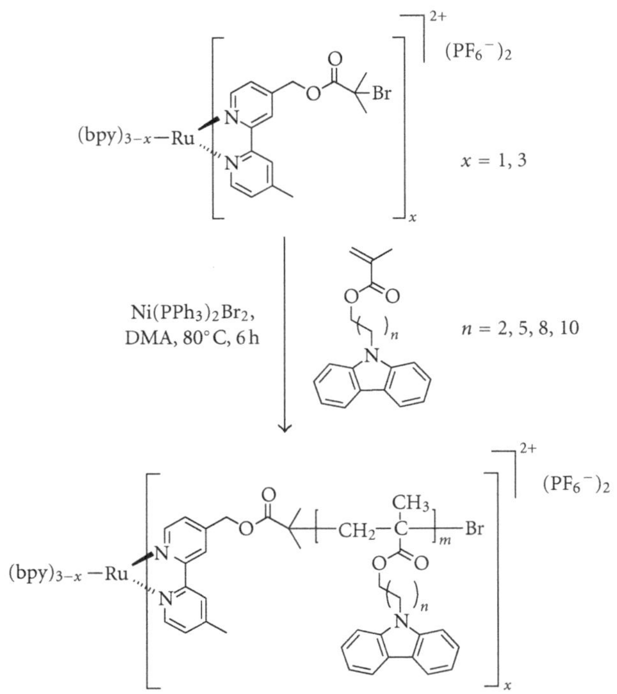

Scheme 12: Synthesis of three-armed polymer by ATRP of the carbazole-containing methacrylates with a metalloinitiator.

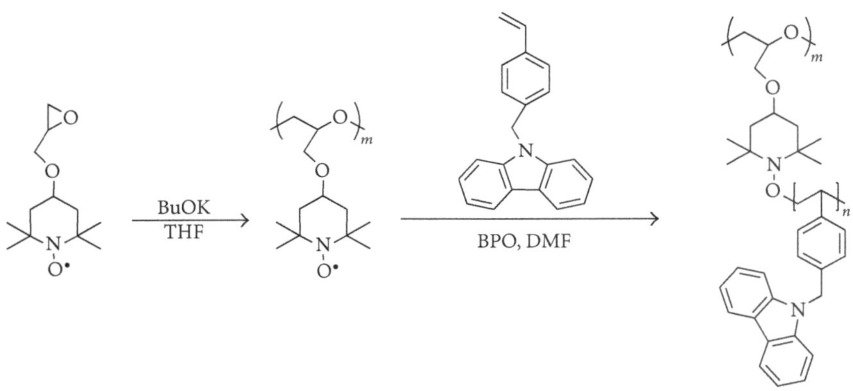

Scheme 13: Synthesis of a comb-shaped copolymer having poly(VBCz) side chains.

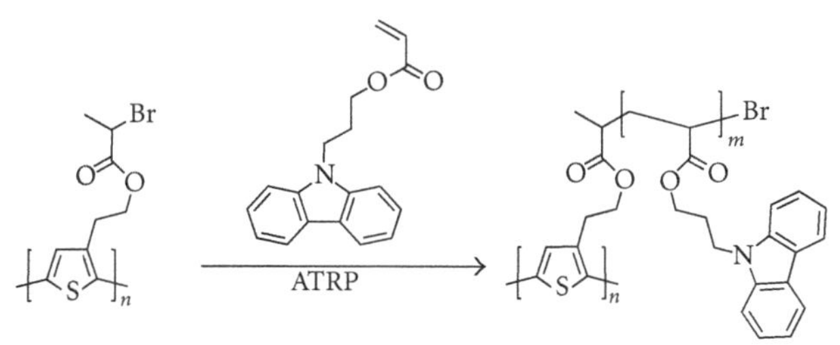

Scheme 14: Synthesis of a comb-shaped copolymer by ATRP of CzPA from a polythiophene backbone.

crosslinked surface-grafted poly(NVC) brushes have been demonstrated as hole transport layers on a photovoltaic device using surface-initiated polymerization [127, 128]. The pendant carbazole group made the poly(NVC) an electrochemically cross-linkable precursor, capable of forming conjugated polymer network films. The covalent linkage of the poly(NVC) brush allowed for direct electroluminescent device preparation on modified ITO, giving the advantage of strong adhesion to ITO with possible long-term stability against acid dopants and oxygen. Surface-initiated ATRP of

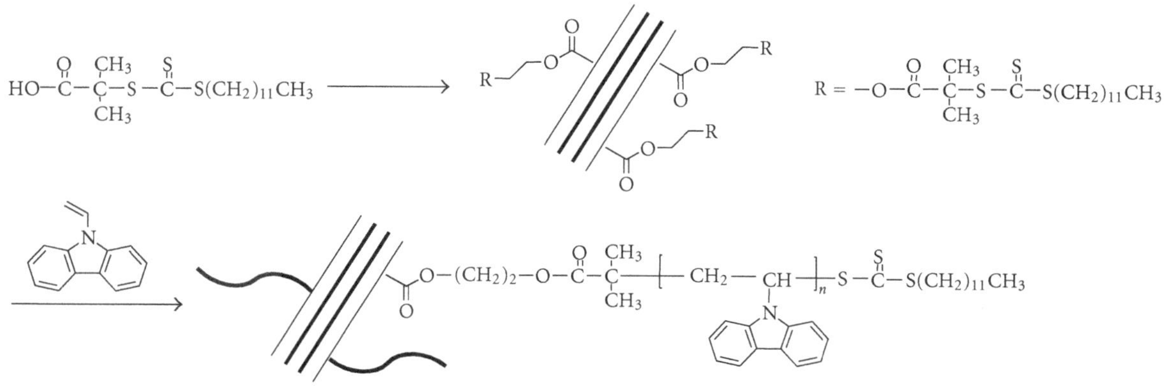

SCHEME 15: Synthesis of poly(NVC)-grafted multiwalled carbon nanotubes by RAFT polymerization of NVC.

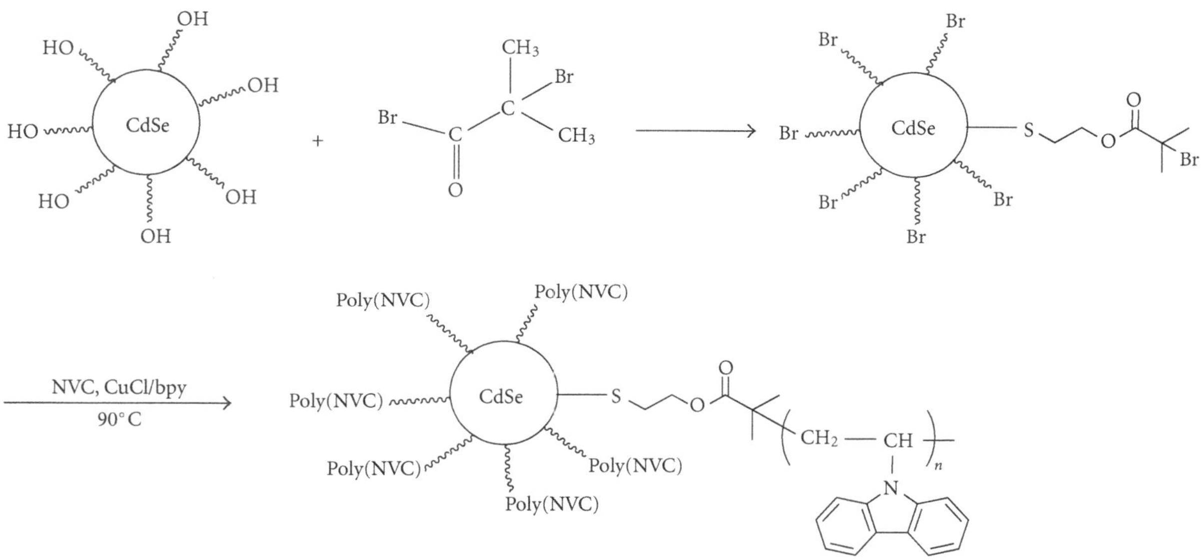

SCHEME 16: Synthesis of CdSe-polymer composite by ATRP of NVC.

9-(2-(4-vinyl(benzyloxy)ethyl)-9H-carbazole (VBOCz) was also employed for the synthesis of carbazole-based polymer brushes on silicon surfaces [129]. Conductance switching at a voltage of about 2.1 V was observed in the memory device based on the brushes.

5. Conclusion

This review has summarized the approaches to synthesis of a variety of tailor-made polymers containing carbazole functionality. The combination of different polymerization mechanisms, self-organization of block copolymers, and surface-initiated polymerization have allowed great advances in the design and synthesis of various nano- and mesoscale polymeric materials with sophisticated structures, in addition to recent progress in controlled radical polymerization techniques. The synthetic methodologies for novel architectures, characteristic bulk, and solution properties related to the self-organization process, and a wide range of applications of block copolymers containing poly(NVC) segments have been extensively investigated in the last decade. Controlled radical polymerization of NVC, styrene, and (meth)acrylate derivatives having the carbazole moiety were well advanced, leading to the well-controlled synthesis of complex macromolecules, such as star polymers, star block copolymers, and polymer brushes with the carbazole functionality. The carbazole-containing complex macromolecules can be combined with a broad range of functional groups, such as fluorophores, electroactive groups, dyes, and other biorelated materials. These carbazole-containing polymers with well-defined architectures can provide viable tailored materials with unique electronic and photonic properties for a wide range of applications, including polymeric light-emitting materials, and organic photorefractive materials.

References

[1] J. V. Grazulevicius, P. Strohriegl, J. Pielichowski, and K. Pielichowski, "Carbazole-containing polymers: synthesis,

properties and applications," *Progress in Polymer Science*, vol. 28, no. 9, pp. 1297–1353, 2003.

[2] P. L. T. Boudreault, S. Beaupré, and M. Leclerc, "Polycarbazoles for plastic electronics," *Polymer Chemistry*, vol. 1, no. 2, pp. 127–136, 2010.

[3] S. Beaupré, P. L. T. Boudreault, and M. Leclrec, "Solar-energy production and energy-efficient lighting: photovoltaic devices and white-light-emitting diodes using poly(2,7-fluorene), poly(2,7-carbazole), and poly(2,7-dibenzosilole) derivatives," *Advanced Materials*, vol. 22, no. 8, pp. E6–E27, 2010.

[4] J. Li and A. C. Grimsdale, "Carbazole-based polymers for organic photovoltaic devices," *Chemical Society Reviews*, vol. 39, no. 7, pp. 2399–2410, 2010.

[5] T. Uryu, H. Ohkawa, and R. Oshima, "Synthesis and High Hole Mobility of Isotactic Poly(2-*N*-carbazolylethyl acrylate)," *Macromolecules*, vol. 20, p. 705, 1987.

[6] C. J. Hu, R. Oshima, S. Sato, and M. Seno, "Synthesis and photoinduced discharge characteristics of polyacrylates with pendant carbazole group," *Journal of Polymer Science Part C*, vol. 26, no. 10, pp. 441–446, 1988.

[7] M. Keyanpour-Rad, A. Ledwith, A. Hallam et al., "Some photophysical properties of five new carbazole-containing methacrylate polymers," *Macromolecules*, vol. 11, no. 6, pp. 1114–1118, 1978.

[8] F. S. Du, Z. C. Li, W. Hong, Q. Y. Cao, and F. M. Li, "Vinyl monomers bearing chromophore moieties and their polymers. XI. Synthesis and photochemical behavior of carbazole-containing methacrylic monomers and their polymers," *Journal of Polymer Science Part A*, vol. 38, no. 4, pp. 679–688, 2000.

[9] A. Ledwith, N. J. Rowley, and S. M. Walker, "Fluorescence emission from poly[2-(9-ethyl)carbazolyl-methylmethacrylate]," *Polymer*, vol. 22, no. 4, pp. 435–436, 1981.

[10] S. Barik and S. Valiyaveettil, "Synthesis and self-assembly of copolymers with pendant electroactive units," *Macromolecules*, vol. 41, no. 17, pp. 6376–6386, 2008.

[11] T. Kanbara, Y. Yokokawa, and K. Hasegawa, "Palladium-catalyzed modification of poly(p-bromostyrene) with carbazole and related heteroarenes containing an N-H bond and their properties," *Journal of Polymer Science Part A*, vol. 38, no. 1, pp. 28–34, 2000.

[12] Y. S. Cho, S. W. Kim, C. S. Ihn, and J. S. Lee, "Anionic polymerization of 4-(9-carbazolyl)methylstyrene," *Polymer*, vol. 42, no. 18, pp. 7611–7616, 2001.

[13] S. Förster and T. Plantenberg, "From self-organizing polymers to nanohybrid and biomaterials," *Angewandte Chemie International Edition*, vol. 41, no. 5, pp. 688–714, 2002.

[14] T. Liu, C. Burger, and B. Chu, "Nanofabrication in polymer matrices," *Progress in Polymer Science*, vol. 28, no. 1, pp. 5–26, 2003.

[15] D. E. Discher and A. Eisenberg, "Polymer vesicles," *Science*, vol. 297, no. 5583, pp. 967–973, 2002.

[16] M. Sauer and W. Meier, "Polymer nanocontainers with controlled permeability," *Australian Journal of Chemistry*, vol. 54, no. 3, pp. 149–151, 2001.

[17] G. Riess, "Micellization of block copolymers," *Progress in Polymer Science*, vol. 28, no. 7, pp. 1107–1170, 2003.

[18] F. J. M. Hoeben, P. Jonkheijm, E. W. Meijer, and A. P. H. J. Schenning, "About supramolecular assemblies of π-conjugated systems," *Chemical Reviews*, vol. 105, no. 4, pp. 1491–1546, 2005.

[19] J. Liu, E. Sheina, T. Kowalewski, and R. D. McCullough, "Tuning the electrical conductivity and self-assembly of regioregular polythiophene by block copolymerization: nanowire morphologies in new di-and triblock copolymers," *Angewandte Chemie International Edition*, vol. 41, no. 2, pp. 329–332, 2002.

[20] R. A. Segalman, B. McCulloch, S. Kirmayer, and J. J. Urban, "Block copolymers for organic optoelectronics," *Macromolecules*, vol. 42, no. 23, pp. 9205–9216, 2009.

[21] I. Botiz and S. B. Darling, "Optoelectronics using block copolymers," *Materials Today*, vol. 13, no. 5, pp. 42–51, 2010.

[22] G. Moad, M. Chen, M. Häussler, A. Postma, E. Rizzardo, and S. H. Thang, "Functional polymers for optoelectronic applications by RAFT polymerization," *Polymer Chemistry*, vol. 2, no. 3, pp. 492–519, 2011.

[23] A. de Cuendias, R. C. Hiorns, E. Cloutet, L. Vignau, and H. Cramail, "Conjugated rod-coil block copolymers and optoelectronic applications," *Polymer International*, vol. 59, no. 11, pp. 1452–1476, 2010.

[24] C. J. Hawker, A. W. Bosman, and E. Harth, "New polymer synthesis by nitroxide mediated living radical polymerizations," *Chemical Reviews*, vol. 101, no. 12, pp. 3661–3688, 2001.

[25] R. B. Grubbs, "Nitroxide-mediated radical polymerization: limitations and versatility," *Polymer Reviews*, vol. 51, no. 2, pp. 104–137, 2011.

[26] K. Matyjaszewski and J. Xia, "Atom transfer radical polymerization," *Chemical Reviews*, vol. 101, no. 9, pp. 2921–2990, 2001.

[27] M. Kamigaito, T. Ando, and M. Sawamoto, "Metal-catalyzed living radical polymerization," *Chemical Reviews*, vol. 101, no. 12, pp. 3689–3745, 2001.

[28] J. Chiefari, Y. K. Chong, F. Ercole et al., "Living free-radical polymerization by reversible addition—fragmentation chain transfer: the RAFT process," *Macromolecules*, vol. 31, no. 16, pp. 5559–5562, 1998.

[29] C. Barner-Kowollik, T. P. Davis, J. P. A. Heuts, M. H. Stenzel, P. Vana, and M. Whittaker, "RAFTing down under: tales of missing radicals, fancy architectures, and mysterious holes," *Journal of Polymer Science Part A*, vol. 41, no. 3, pp. 365–375, 2003.

[30] G. Moad, E. Rizzardo, and S. H. Thang, "Living radical polymerization by the RAFT process," *Australian Journal of Chemistry*, vol. 58, no. 6, pp. 379–410, 2005.

[31] C. L. McCormick and A. B. Lowe, "Aqueous RAFT polymerization: recent developments in synthesis of functional water-soluble (Co)polymers with controlled structures," *Accounts of Chemical Research*, vol. 37, no. 5, pp. 312–325, 2004.

[32] S. Perrier and P. Takolpuckdee, "Macromolecular design via reversible addition-fragmentation chain transfer (RAFT)/ xanthates (MADIX) polymerization," *Journal of Polymer Science Part A*, vol. 43, no. 22, pp. 5347–5393, 2005.

[33] A. Favier and M. T. Charreyre, "Experimental requirements for an efficient control of free-radical polymerizations via the reversible addition-fragmentation chain transfer (RAFT) process," *Macromolecular Rapid Communications*, vol. 27, no. 9, pp. 653–692, 2006.

[34] G. Moad, E. Rizzardo, and S. H. Thang, "Living radical polymerization by the RAFT process—a first update," *Australian Journal of Chemistry*, vol. 59, no. 10, pp. 669–692, 2006.

[35] A. B. Lowe and C. L. McCormick, "Reversible addition-fragmentation chain transfer (RAFT) radical polymerization and the synthesis of water-soluble (co)polymers under homogeneous conditions in organic and aqueous media," *Progress in Polymer Science*, vol. 32, no. 3, pp. 283–351, 2007.

[36] C. Barner-Kowollik, M. Buback, B. Charleux et al., "Mechanism and kinetics of dithiobenzoate-mediated RAFT polymerization. I. The current situation," *Journal of Polymer Science Part A*, vol. 44, no. 20, pp. 5809–5831, 2006.

[37] M. L. Coote, E. H. Krenske, and E. I. Izgorodina, "Computational studies of RAFT polymerization-mechanistic insights and practical applications," *Macromolecular Rapid Communications*, vol. 27, no. 7, pp. 473–497, 2006.

[38] L. Barner, T. P. Davis, M. H. Stenzel, and C. Barner-Kowollik, "Complex macromolecular architectures by reversible addition fragmentation chain transfer chemistry: theory and practice," *Macromolecular Rapid Communications*, vol. 28, no. 5, pp. 539–559, 2007.

[39] A. Goto and T. Fukuda, "Kinetics of living radical polymerization," *Progress in Polymer Science*, vol. 29, no. 4, pp. 329–385, 2004.

[40] G. Moad, E. Rizzardo, and S. H. Thang, "Radical addition-fragmentation chemistry in polymer synthesis," *Polymer*, vol. 49, no. 5, pp. 1079–1131, 2008.

[41] C. Barner-Kowollik, *Handbook of RAFT Polymerization*, Wiley-VCH, Weinheim, Germany, 2007.

[42] B. M. Rosen and V. Percec, "Single-electron transfer and single-electron transfer degenerative chain transfer living radical polymerization," *Chemical Reviews*, vol. 109, no. 11, pp. 5069–5119, 2009.

[43] S. Yamago, "Precision polymer synthesis by degenerative transfer controlled/living radical polymerization using organotellurium, organostibine, and organobismuthine chain-transfer agents," *Chemical Reviews*, vol. 109, no. 11, pp. 5051–5068, 2009.

[44] M. Hurtgen, C. Detrembleur, C. Jerome, and A. Debuigne, "Insight into organometallic-mediated radical polymerization," *Polymer Reviews*, vol. 51, no. 2, pp. 188–213, 2011.

[45] M. Sawamoto, J. Fujimori, and T. Higashimura, "Living cationic polymerization of N-vinylcarbazole initiated by hydrogen iodide," *Macromolecules*, vol. 20, no. 5, pp. 916–920, 1987.

[46] T. Fukuda, T. Terauchi, A. Goto, Y. Tsujii, T. Miyamoto, and Y. Shimizu, "Well-defined block copolymers comprising styrene-acrylonitrile random copolymer sequences synthesized by "living" radical polymerization," *Macromolecules*, vol. 29, no. 8, pp. 3050–3052, 1996.

[47] H. Baethge, S. Butz, C. H. Han, and G. Schmidt-Naake, "Rate enhancement of the N-oxyl-controlled free radical copolymerization of styrene with N-vinylcarbazole," *Die Angewandte Makromolekulare Chemie*, vol. 267, pp. 52–56, 1999.

[48] H. Baethge, S. Butz, and G. Schmidt-Naake, "'Living' free radical copolymerization of styrene and N-vinylcarbazole," *Macromolecular Rapid Communications*, vol. 18, no. 10, pp. 911–916, 1997.

[49] M. Nowakowska, S. Zapotoczny, and A. Karewicz, "Polymeric photosensitizers. Part 4. Synthesis of poly(sodium styrenesulfonate-block-N-vinylcarbazole) by nitroxide-mediated free radical polymerization," *Polymer*, vol. 42, no. 5, pp. 1817–1823, 2001.

[50] J. Hua, D. B. Chen, Y. L. Yu et al., "Preparation of C60 bonded poly(N-vinylcarbazole) with C60Cln/CuCl/Bpy catalyst system," *Polymer Bulletin*, vol. 48, no. 2, pp. 135–141, 2002.

[51] J. Hua, D. Chen, X. Jing, L. Xu, Y. Yu, and Y. Zhang, "Preparation and photoconducting property of C60Cln-m-bonded poly(N-vinylcarbazole) with C60Cln/CuCl/Bpy catalyst system," *Journal of Applied Polymer Science*, vol. 87, no. 4, pp. 606–609, 2003.

[52] J. Brandrup and E. H. Immergut, *Polymer Handbook*, John Wiley & Sons, New York, NY, USA, 3rd edition, 1991.

[53] H. Mori, H. Ookuma, S. Nakano, and T. Endo, "Xanthate-mediated controlled radical polymerization of N-vinylcarbazole," *Macromolecular Chemistry and Physics*, vol. 207, no. 12, pp. 1005–1017, 2006.

[54] D. Charmot, P. Corpart, H. Adam, S. Z. Zard, T. Biadatti, and G. Bouhadir, "Controlled radical polymerization in dispersed media," *Macromolecular Symposia*, vol. 150, pp. 23–32, 2000.

[55] D. Taton, A. Z. Wilczewska, and M. Destarac, "Direct synthesis of double hydrophilic statistical di-and triblock copolymers comprised of acrylamide and acrylic acid units via the MADIX process," *Macromolecular Rapid Communications*, vol. 22, no. 18, pp. 1497–1503, 2001.

[56] M. Destarac, D. Taton, S. Z. Zard, T. Saleh, and Y. Six, "On the importance of xanthate substituents in the MADIX process," in *Advances in Controlled/Living Radical Polymerization*, K. Matyjaszewski, Ed., vol. 854 of *ACS Symposinum Series*, chapter 37, pp. 536–550, American Chemical Society, Washington, DC, USA, 2003.

[57] M. H. Stenzel, L. Cummins, G. E. Roberts, T. P. Davis, P. Vana, and C. Barner-Kowollik, "Xanthate mediated living polymerization of vinyl acetate: a systematic variation in MADIX/RAFT agent structure," *Macromolecular Chemistry and Physics*, vol. 204, no. 9, pp. 1160–1168, 2003.

[58] M. L. Coote and L. Radom, "Substituent effects in xanthate-mediated polymerization of vinyl acetate: Ab initio evidence for an alternative fragmentation pathway," *Macromolecules*, vol. 37, no. 2, pp. 590–596, 2004.

[59] M. H. Stenzel, T. P. Davis, and C. Barner-Kowollik, "Poly (vinyl alcohol) star polymers prepared via MADIX/RAFT polymerisation," *Chemical Communications*, vol. 10, no. 13, pp. 1546–1547, 2004.

[60] M. Destarac, D. Charmot, X. Franck, and S. Z. Zard, "Dithiocarbamates as universal reversible addition-fragmentation chain transfer agents," *Macromolecular Rapid Communications*, vol. 21, no. 15, pp. 1035–1039, 2000.

[61] T. L. U. Nguyen, K. Eagles, T. P. Davis, C. Barner-Kowollik, and M. H. Stenzel, "Investigation of the influence of the architectures of poly(vinyl pyrrolidone) polymers made via the reversible addition-fragmentation chain transfer/macromolecular design via the interchange of xanthates mechanism on the stabilization of suspension polymerizations," *Journal of Polymer Science Part A*, vol. 44, no. 15, pp. 4372–4383, 2006.

[62] D. Wan, K. Satoh, M. Kamigaito, and Y. Okamoto, "Xanthate-mediated radical polymerization of N-vinylpyrrolidone in fluoroalcohols for simultaneous control of molecular weight and tacticity," *Macromolecules*, vol. 38, no. 25, pp. 10397–10405, 2005.

[63] R. Devasia, R. L. Bindu, R. Borsali, N. Mougin, and Y. Gnanou, "Controlled radical polymerization of N-vinylpyrrolidone by reversible addition-fragmentation chain transfer process," *Macromolecular Symposia*, vol. 229, pp. 8–17, 2005.

[64] Y. Maki, H. Mori, and T. Endo, "Xanthate-mediated controlled radical polymerization of N-vinylindole derivatives," *Macromolecules*, vol. 40, no. 17, pp. 6119–6130, 2007.

[65] Y. Maki, H. Mori, and T. Endo, "Controlled RAFT polymerization of N-vinylphthalimide and its hydrazinolysis to poly(vinyl amine)," *Macromolecular Chemistry and Physics*, vol. 208, no. 24, pp. 2589–2599, 2007.

[66] Y. Maki, H. Mori, and T. Endo, "Synthesis of well-defined alternating copolymers by RAFT copolymerization of

N-vinylnaphthalimide," *Macromolecules*, vol. 41, no. 22, pp. 8397–8404, 2008.

[67] Y. Maki, H. Mori, and T. Endo, "Synthesis of amphiphilic and double-hydrophilic block copolymers containing polyvinyl amine segments by RAFT polymerization of N-vinylphthalimide," *Macromolecular Chemistry and Physics*, vol. 211, no. 1, pp. 45–56, 2010.

[68] H. Mori, M. Yanagi, and T. Endo, "RAFT polymerization of N-vinylimidazolium salts and synthesis of thermoresponsive ionic liquid block copolymers," *Macromolecules*, vol. 42, no. 21, pp. 8082–8092, 2009.

[69] G. Moad, J. Chiefari, Y. K. Chong et al., "Living free radical polymerization with reversible addition—fragmentation chain transfer (the life of RAFT)," *Polymer International*, vol. 49, no. 9, pp. 993–1001, 2000.

[70] M. Destarac, W. Bzducha, D. Taton, I. Gauthier-Gillaizeau, and S. Z. Zard, "Xanthates as chain-transfer agents in controlled radical polymerization (MADIX): structural effect of the O-alkyl group," *Macromolecular Rapid Communications*, vol. 23, no. 17, pp. 1049–1054, 2002.

[71] A. Favier, C. Barner-Kowollik, T. P. Davis, and M. H. Stenzel, "A detailed on-line FT/NIR and 1H NMR spectroscopic investigation into factors causing inhibition in xanthate-mediated vinyl acetate polymerization," *Macromolecular Chemistry and Physics*, vol. 205, no. 7, pp. 925–936, 2004.

[72] B. Y. K. Chong, T. P. T. Le, G. Moad, E. Rizzardo, and S. H. Thang, "More versatile route to block copolymers and other polymers of complex architecture by living radical polymerization: the RAFT process," *Macromolecules*, vol. 32, no. 6, pp. 2071–2074, 1999.

[73] R. T. A. Mayadunne, E. Rizzardo, J. Chiefari et al., "Living polymers by the use of trithiocarbonates as reversible addition-fragmentation chain transfer (RAFT) agents: ABA triblock copolymers by radical polymerization in two steps," *Macromolecules*, vol. 33, no. 2, pp. 243–245, 2000.

[74] Y. A. Vasilieva, D. B. Thomas, C. W. Scales, and C. L. McCormick, "Direct controlled polymerization of a cationic methacrylamido monomer in aqueous media via the RAFT process," *Macromolecules*, vol. 37, no. 8, pp. 2728–2737, 2004.

[75] P. Vana, T. P. Davis, and C. Barner-Kowollik, "Kinetic analysis of reversible addition fragmentation chain transfer (RAFT) polymerizations: conditions for inhibition, retardation, and optimum living polymerization," *Macromolecular Theory and Simulations*, vol. 11, no. 8, pp. 823–835, 2002.

[76] H. Mori, E. Kudo, Y. Saito, A. Onuma, and M. Morishima, "RAFT polymerization of vinyl sulfonate esters for the controlled synthesis of poly(lithium vinyl sulfonate) and sulfonated block copolymers," *Macromolecules*, vol. 43, no. 17, pp. 7021–7032, 2010.

[77] C. F. Huang, J. A. Yoon, and K. Matyjaszewski, "Synthesis of N-vinylcarbazole-N-vinylpyrrolidone amphiphilic block copolymers by xanthate-mediated controlled radical polymerization," *Canadian Journal of Chemistry*, vol. 88, no. 3, pp. 228–235, 2010.

[78] N. Hu, W. X. Ji, Y. Y. Tong, Z. C. Li, and E. Q. Chen, "Synthesis of diblock copolymers containing poly(N-vinylcarbazole) by reversible addition-fragmentation chain transfer polymerization," *Journal of Polymer Science Part A*, vol. 48, no. 20, pp. 4621–4626, 2010.

[79] N. Suchao-in, S. Chirachanchai, and S. Perrier, "pH-and thermo-multi-responsive fluorescent micelles from block copolymers via reversible addition fragmentation chain transfer (RAFT) polymerization," *Polymer*, vol. 50, no. 17, pp. 4151–4158, 2009.

[80] W. Y. Tam, C. S. K. Mak, A. M. C. Ng, A. B. Djurišič, and W. K. Chan, "Multifunctional poly(N-vinylcarbazole)-based block copolymers and their nanofabrication and photosensitizing properties," *Macromolecular Rapid Communications*, vol. 30, no. 8, pp. 622–626, 2009.

[81] M. Benaglia, J. Chiefari, Y. K. Chong, G. Moad, E. Rizzardo, and S. H. Thang, "Universal (Switchable) RAFT agents," *Journal of the American Chemical Society*, vol. 131, no. 20, pp. 6914–6915, 2009.

[82] D. J. Keddie, C. Guerrero-Sanchez, G. Moad, E. Rizzardo, and S. H. Thang, "Switchable reversible addition-fragmentation chain transfer (raft) polymerization in aqueous solution, n, n -dimethylacrylamide," *Macromolecules*, vol. 44, no. 17, pp. 6738–6745, 2011.

[83] Y. Yan, W. Zhang, Y. Qiu et al., "Universal xanthate-mediated controlled free radical polymerizations of the "less activated" vinyl monomers," *Journal of Polymer Science Part A*, vol. 48, no. 22, pp. 5206–5214, 2010.

[84] M. Heo, J. Kim, J. Y. Kim, and C. Yang, "A first approach to white organic electroluminescence device from a single rod-coil poly[thiophene-block-(N-vinylcarbazole)] diblock copolymer," *Macromolecular Rapid Communications*, vol. 31, no. 23, pp. 2047–2052, 2010.

[85] A. S. Brar and S. Kaur, "Atom transfer radical polymerization of N-vinyl carbazole: optimization to characterization," *Journal of Polymer Science Part A*, vol. 44, no. 5, pp. 1745–1757, 2006.

[86] N. Haridharan and R. Dhamodharan, "Controlled polymerization of carbazole-based vinyl and methacrylate monomers at ambient temperature: a comparative study through ATRP, SET, and SET-RAFT polymerizations," *Journal of Polymer Science Part A*, vol. 49, no. 4, pp. 1021–1032, 2011.

[87] W. Zhang, Y. Yan, N. Zhou et al., "Controlled synthesis and fluorescent properties of poly(9-(4-vinylbenzyl)-9H-carbazole) via nitroxide-mediated living free-radical polymerization," *European Polymer Journal*, vol. 44, no. 10, pp. 3300–3305, 2008.

[88] Y. K. Fang, C. L. Liu, and W. C. Chen, "New random copolymers with pendant carbazole donor and 1,3,4-oxadiazole acceptor for high performance memory device applications," *Journal of Materials Chemistry*, vol. 21, no. 13, pp. 4778–4786, 2011.

[89] B. Lessard, E. J. Y. Ling, M. S. T. Morin, and M. Marić, "Nitroxide-mediated radical copolymerization of methyl methacrylate controlled with a minimal amount of 9-(4-vinylbenzyl)-9H-carbazole," *Journal of Polymer Science Part A*, vol. 49, no. 4, pp. 1033–1045, 2011.

[90] X. Zhu, N. Zhou, Z. Zhang et al., "Cyclic polymers with pendent carbazole units: enhanced fluorescence and redox behavior," *Angewandte Chemie International Edition*, vol. 50, no. 29, pp. 6615–6618, 2011.

[91] H. Mori, S. Nakano, and T. Endo, "Controlled synthesis of poly(N-ethyl-3-vinylcarbazole) and block copolymers via RAFT polymerization," *Macromolecules*, vol. 38, no. 20, pp. 8192–8201, 2005.

[92] H. Mori, H. Iwaya, and T. Endo, "Structures and chiroptical properties of thermoresponsive block copolymers containing L-proline moietiesa," *Macromolecular Chemistry and Physics*, vol. 208, no. 17, pp. 1908–1918, 2007.

[93] H. Mori, H. Iwaya, and T. Endo, "Controlled synthesis of thermoresponsive polymer via RAFT polymerization of an acrylamide containing l-proline moiety," *Reactive and Functional Polymers*, vol. 67, no. 10, pp. 916–927, 2007.

[94] H. Mori and S. Okabayashi, "Synthesis, assembled structure, and chiroptical properties of amino acid-based amphiphilic block copolymers containing carbazole moiety," *Reactive and Functional Polymers*, vol. 69, no. 7, pp. 441–449, 2009.
[95] J. D. Quinn and R. A. Register, "Nitroxide-mediated radical polymerization of N-ethyl-2-vinylcarbazole," *Polymers for Advanced Technologies*, vol. 19, no. 6, pp. 556–559, 2008.
[96] K. Ogino, T. Goma, D. Kageyama, H. Sato, and N. Yonezawa, "Multifunctional block copolymers for organic photorefractive materials," *Journal of Photopolymer Science and Technology*, vol. 19, no. 3, pp. 419–424, 2006.
[97] P. Zhao, Q. D. Ling, W. Z. Wang, J. Ru, S. B. Li, and W. Huang, "Reversible addition-fragmentation chain transfer polymerization of methacrylates containing hole-or electron-transporting groups," *Journal of Polymer Science Part A*, vol. 45, no. 2, pp. 242–252, 2007.
[98] D. Neugebauer, D. Charasim, A. Swinarew et al., "Polymethacrylates with anthryl and carbazolyl groups prepared by atom transfer radical polymerization," *Polymer Journal*, vol. 43, no. 5, pp. 448–454, 2011.
[99] C. Ulbricht, C. R. Becer, A. Winter, D. Veldman, and U. S. Schubert, "Copolymers containing phosphorescent iridium(III) complexes obtained by free and controlled radical polymerization techniques," *Macromolecular Rapid Communications*, vol. 29, no. 24, pp. 1919–1925, 2008.
[100] N. Hadjichristidis, M. Pitsikalis, S. Pispas, and H. Iatrou, "Polymers with complex architecture by living anionic polymerization," *Chemical Reviews*, vol. 101, no. 12, pp. 3747–3792, 2001.
[101] A. Hirao, M. Hayashi, S. Loykulnant et al., "Precise syntheses of chain-multi-functionalized polymers, star-branched polymers, star-linear block polymers, densely branched polymers, and dendritic branched polymers based on iterative approach using functionalized 1,1-diphenylethylene derivatives," *Progress in Polymer Science*, vol. 30, no. 2, pp. 111–182, 2005.
[102] H. Mori and A. H. E. Müller, "New polymeric architectures with (meth)acrylic acid segments," *Progress in Polymer Science*, vol. 28, p. 1403, 2003.
[103] N. Hadjichristidis, H. Iatrou, M. Pitsikalis, S. Pispas, and A. Avgeropoulos, "Linear and non-linear triblock terpolymers. Synthesis, self-assembly in selective solvents and in bulk," *Progress in Polymer Science*, vol. 30, no. 7, pp. 725–782, 2005.
[104] K. V. Bernaerts and F. E. du Prez, "Dual/heterofunctional initiators for the combination of mechanistically distinct polymerization techniques," *Progress in Polymer Science*, vol. 31, no. 8, pp. 671–722, 2006.
[105] Y. Yagci and M. A. Tasdelen, "Mechanistic transformations involving living and controlled/living polymerization methods," *Progress in Polymer Science*, vol. 31, no. 12, pp. 1133–1170, 2006.
[106] D. Taton, Y. Gnanou, R. Matmour et al., "Controlled polymerizations as tools for the design of star-like and dendrimer-like polymers," *Polymer International*, vol. 55, no. 10, pp. 1138–1145, 2006.
[107] H. Mori, H. Ookuma, and T. Endo, "Synthesis of star polymers based on xanthate-mediated controlled radical polymerization of N-vinylcarbazole," *Macromolecular Symposia*, vol. 249-250, pp. 406–411, 2007.
[108] H. Mori, H. Ookuma, and T. Endo, "Poly(N-vinylcarbazole) star polymers and amphiphilic star block copolymers by xanthate-mediated controlled radical polymerization," *Macromolecules*, vol. 41, no. 19, pp. 6925–6934, 2008.
[109] N. Hadjichristidis, S. Pispas, M. Pitsikalis, H. Iatrou, and C. Vlahos, "Asymmetric star polymers: synthesis and properties," *Advances in Polymer Science*, vol. 142, pp. 71–127, 1999.
[110] W. Zhang, W. Zhaing, Z. Zhang et al., "Thermo-responsive fluorescent micelles from amphiphilic A3B miktoarm star copolymers prepared via a combination of SET-LRP and RAFT polymerization," *Journal of Polymer Science Part A*, vol. 48, no. 19, pp. 4268–4278, 2010.
[111] Y. Tao, Q. Xu, N. Li, J. Lu, L. Wang, and X. Xia, "Synthesis and photoluminescent property of star polymers with carbzole pendent and a zinc porphyrin core by ATRP," *Polymer*, vol. 52, no. 19, pp. 4261–4267, 2011.
[112] A. A. Farah and W. J. Pietro, "Atom transfer radical polymerization of N-(ω'-alkylcarbazolyl) methacrylates via the use of novel heteroleptic Ru(II) polypyridyl initiator," *Inorganica Chimica Acta*, vol. 357, no. 13, pp. 3813–3824, 2004.
[113] A. A. Farah and W. J. Pietro, "Synthesis and characterization of multifunctional polymers via atom transfer radical polymerization of N-(ω'-alkylcarbazolyl) methacrylates initiated by Ru(II) polypyridyl chromophores," *Journal of Polymer Science Part A*, vol. 43, no. 23, pp. 6057–6072, 2005.
[114] C. Chang, J. Zhu, Z. Zhang, N. Zhou, Z. Cheng, and X. Zhu, "Synthesizing and characterization of comb-shaped carbazole containing copolymer via combination of ring opening polymerization and nitroxide-mediated polymerization," *Polymer*, vol. 51, no. 9, pp. 1947–1953, 2010.
[115] J. Shen, H. Masaoka, K. Tsuchiya, and K. Ogino, "Synthesis and properties of a novel brush-type copolymers bearing thiophene backbone and 3-(N-carbazolyl)propyl acrylate side chains for light-emitting applications," *Polymer Journal*, vol. 40, no. 5, pp. 421–427, 2008.
[116] B. Zhang, J. Wang, Y. U. Chen et al., "Multiwalled carbon nanotubes covalently functionalized with poly(N-vinylcarbazole) via RAFT polymerization: synthesis and imonliner optical properties," *Journal of Polymer Science Part A*, vol. 48, no. 14, pp. 3161–3168, 2010.
[117] B. Zhang, Y. U. Chen, X. Zhuang et al., "Poly(N-vinylcarbazole) chemically modified graphene oxide," *Journal of Polymer Science Part A*, vol. 48, no. 12, pp. 2642–2649, 2010.
[118] B. Zhang, Y. Chen, L. Xu et al., "Growing poly(N-vinylcarbazole) from the surface of graphene oxide via RAFT polymerization," *Journal of Polymer Science Part A*, vol. 49, no. 9, pp. 2043–2050, 2011.
[119] J. Hua, D. Chen, X. Jing, L. Xu, Y. Yu, and Y. Zhang, "Preparation and Photoconducting Property of $C_{60}Cl_{n-m}$-Bonded Poly(*N*-vinylcarbazole) with $C_{60}Cl_n$/CuCl/Bpy Catalyst System," *Journal of Applied Polymer Science*, vol. 84, 2003.
[120] J. Hua, W. Yang, and Y. Zhu, "Optical-limiting effect of C_{60} bonded poly(*N*-vinylcarbazole) initiated with $C_{60}Cl_n$/CuCl/Bpy catalyst system," *Materials Letters*, vol. 59, p. 644, 2005.
[121] T.-L. Wang, C.-H. Yang, Y.-T. Shieh, A.-C. Yeh, L.-W. Juan, and H. C. Zeng, "Synthesis of new nanocrystal-polymer nanocomposite as the electron acceptor in polymer bulk heterojunction solar cells," *European Polymer Journal*, vol. 46, p. 634, 2009.
[122] T. L. Wang, C. H. Yang, Y. T. Shieh, and A. C. Yeh, "Synthesis of CdSe-Poly(N-vinylcarbazole) nanocomposite by atom transfer radical polymerization for potential optoelectronic applications," *Macromolecular Rapid Communications*, vol. 30, no. 19, pp. 1679–1683, 2009.
[123] M. Feng, Y. Chen, N. He et al., "Ultrasound-assisted bulk synthesis of CdS-PVK nanocomposites via RAFT

polymerization," *Journal of Polymer Science Part A*, vol. 46, no. 16, pp. 5702–5707, 2008.

[124] M. Sato, A. Kawata, S. Morito, Y. Sato, and I. Yamaguchi, "Preparation and properties of polymer/zinc oxide nanocomposites using functionalized zinc oxide quantum dots," *European Polymer Journal*, vol. 44, no. 11, pp. 3430–3438, 2008.

[125] J. Liu, W. He, L. Zhang et al., "Bifunctional Nanoparticles with Fluorescence and Magnetism via Surface-Initiated AGET ATRP Mediated by an Iron Catalyst," *Langmuir*, vol. 27, pp. 12684–12692, 2011.

[126] Q. Li, L. Zhang, L. Bai et al., "Multistimuli-responsive hybrid nanoparticles with magnetic core and thermoresponsive fluorescence-labeled shell via surface-initiated RAFT polymerization," *Soft Matter*, vol. 7, no. 15, pp. 6958–6966, 2011.

[127] T. M. Fulghum, P. Taranekar, and R. C. Advincula, "Grafting hole-transport precursor polymer brushes on ITO electrodes: surface-initiated polymerization and conjugated polymer network formation of PVK," *Macromolecules*, vol. 41, no. 15, pp. 5681–5687, 2008.

[128] M. C. Tria, K. S. Liao, N. Alley, S. Curran, and R. Advincula, "Electrochemically crosslinked surface-grafted PVK polymer brushes as a hole transport layer for organic photovoltaics," *Journal of Materials Chemistry*, vol. 21, no. 28, pp. 10261–10264, 2011.

[129] Y. Wei, D. Gao, L. Li, and S. Shang, "Memory effect in polymer brushes containing pendant carbazole groups," *Polymer*, vol. 52, no. 6, pp. 1385–1390, 2011.

7

Design and Synthesis of Functional Silsesquioxane-Based Hybrids by Hydrolytic Condensation of Bulky Triethoxysilanes

Hideharu Mori

Department of Polymer Science and Engineering, Graduate School of Science and Engineering, Yamagata University, 4-3-16, Jonan, Yonezawa 992-8510, Japan

Correspondence should be addressed to Hideharu Mori, h.mori@yz.yamagata-u.ac.jp

Academic Editor: Kimihiro Matsukawa

This paper presents a short overview of recent advances in the design and synthesis of organic-inorganic hybrids using silsesquioxane-based nanoparticles having nanometer size, relatively narrow size distribution, high functionalities, and various characteristic features, mainly focusing on our recent researches related to the subject. A highlight of this paper is the water-soluble silsesquioxane-based nanoparticles, including hydroxyl-functionalized and cationic silsesquioxanes, which were synthesized via the one-step condensation of the bulky triethoxysilane precursors. The design and synthesis of R-$SiO_{1.5}$/SiO_2 and R-$SiO_{1.5}$/TiO_2 hybrids by hydrolytic cocondensation of a triethoxysilane precursor and metal alkoxides are briefly introduced. This paper also deals with recent results in stimuli-responsive hybrids based on the water-soluble silsesquioxane nanoparticles and fluorinated and amphiphilic silsesquioxane hybrids.

1. Introduction

The silsesquioxane family is now recognized to have an enormous potential as a building block for various advanced materials, and their applications can be found in the areas of catalysis, coordination chemistry, and material science, such as organic-inorganic nanocomposites [1–9]. Silsesquioxane is a family of compounds characterized by a ratio of 1.5 between the silicon and oxygen atoms [4], and the structures can be expressed in the general formula: $(\text{R–SiO}_{1.5})_n$ (n = even number) [1]. In particular, much interest has been paid to cubic T_8 silsesquioxane $(\text{R–SiO}_{1.5})_8$, consisting of a rigid, crystalline silica-like core that is perfectly defined spatially (0.5–0.7 nm) and that can be linked covalently to eight R groups [10, 11], because of their possible applications in optics, electronics, engineering, and biosciences.

Although some T_8 compounds are commercially available, the synthesis of polyhedral oligomeric silsesquioxane (also known as T_8 silsesquioxane or POSS) is plagued by relatively low yields of multistage syntheses and time-consuming procedures. One alternative to exploit them for practical applications is to use a more easily accessible mixture of silsesquioxanes. The hydrolysis and polycondensation of substituted alkoxysilanes, $\text{R–Si(OR}')_3$, containing a nonhydrolyzable Si–C bond or chlorosilanes are known to be a representative method to afford a variety of silsesquioxanes with various substituent groups and cage structures [2–4]. In the synthesis of silsesquioxanes via hydrolytic condensation of $\text{R–Si(OR}')_3$, it is known that the reactions and the resulting structures are influenced by many factors, such as the nature of the R′ and R groups, solvent, concentration, addition rate, quality of water, temperature, pH, and reaction time [3, 4]. Among them, water related with the hydrolysis reaction to give trisilanol and the solvent related with the solubilities of a precursor and resulting hybrid may play a relevant role in influencing the synthesis.

A variety of silsesquioxanes having various functional groups, different numbers of substituent groups, and cage structures has been developed. Completely condensed silsesquioxanes, $(\text{R–SiO}_{1.5})_n$ with $n = 4, 6, 8, 10$, and 12, have been synthesized and characterized [4]. Completely condensed structures, $(\text{R–SiO}_{1.5})_n$ with $n > 12$, are not so common, but some examples have been reported [3, 12, 13]. In addition to the fully condensed structures, $(\text{R–SiO}_{1.5})_n$,

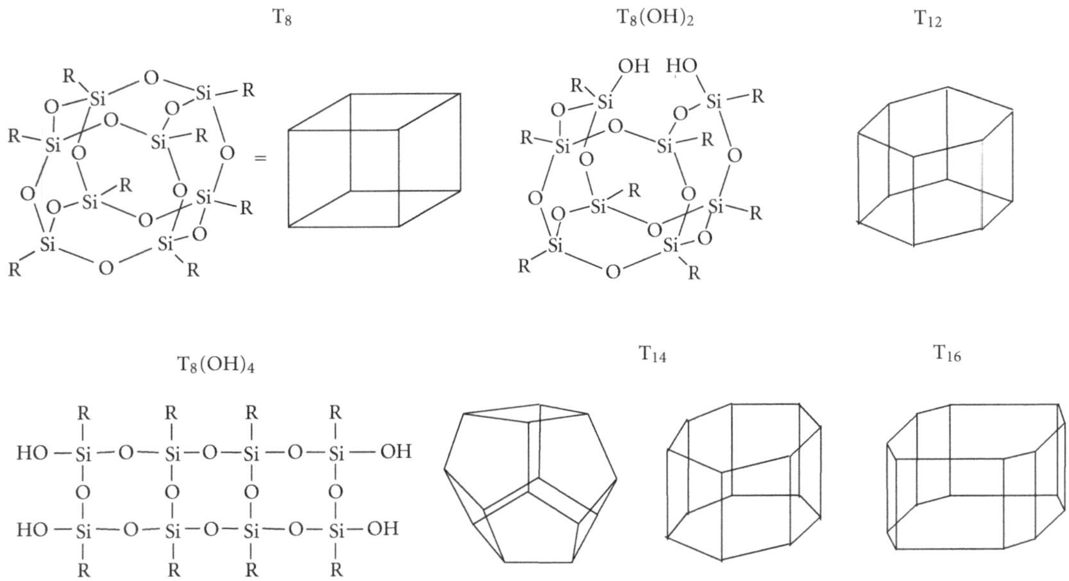

FIGURE 1: Schematic representation of various structures, ladder ($T_8(OH)_4$), partially cage ($T_8(OH)_2$), and cage (T_8–T_{16}) structures.

which is denoted as T_n (n = even number), incompletely condensed structures containing Si–OH groups are known, which have generic formula; $[R\text{–}SiO_{1.5-x}(OH)_{2x}]_n$ or $T_n(OH)_m$ [1, 4, 14]. The structures include perfect polyhedra, incompletely condensed polyhedra (species with 1–3 OH per molecule), ladder-type structure (species with 4 OH per molecule), open structure (species with more than 5 OH per molecule), linear structure, and all their possible combinations. Examples of possible structures are shown in Figure 1.

Preferable formation of cubic T_8 species over the T_{10}, T_{12}, and other fully condensed species is known to be due to the stability of the Si_4O_4 ring structures. Even if a variety of cubic silsesquioxanes having various functional groups has been developed by the hydrolysis/condensation process, it does have certain inherent disadvantages, such as long reaction times, up to three months in some cases, and low yields less than 50% [5]. The low yield is one of the most important problems to afford cubic T_8 species by the hydrolysis/condensation. This is mainly attributed to the formation of the mixtures of the fully condensed products with byproducts of ladder and other nonpolyhedral silsesquioxanes and formation of oligomeric products having higher contents of silicon atoms, such as T_{10}, T_{12}. The difficulty in the separation of the desired cubic T_8 species from the byproducts is another reason for low yields.

Organic-inorganic hybrid nanoparticles have attracted a great deal of attention due to their potential applications in a wide range of fields. Further advances of such organic-inorganic nanocomposite materials require fine-tuning of the sizes, structures, compositions, topologies, and spatial assembly of individual constitutes and their interfaces. Several efforts have been directed at the preparation of novel colloids and nanoparticles based on the hydrolytic condensation of monosilanes, and the resulting products are often called polysilsesquioxanes colloids or polyorganosiloxane nanoparticles. For example, Bronstein et al. reported the synthesis of a new family of polysilsesquioxanes colloids based by hydrolytic condensation of N-(6-aminohexyl)aminopropyltrimethoxysilane [15]. They claimed that the well-defined colloids (30–50 nm) were obtained at neutral initial pH. The synthesis of nanoparticles of poly(phenyl/methylsilsesquioxane) was conducted by polymerization of the cohydrolyzate from a mixture of trichlorophenylsilane and trichloromethylsilane in aqueous solution [16]. By changing the initial conditions, the average size of the resultant particles could be controlled from 30 to 250 nm in diameter. Other examples involve the syntheses of poly(phenylsilsesquioxane)s with particle sizes from 30 to 110 nm by polycondensation of phenylsilanetriol formed in aqueous solution [17] and polyorganosiloxane nanoparticles by using alkylalkoxysilanes, in which methyltrimethoxysilane was used as a network-forming monomer and diethoxydimethylsilane was employed as a chain-forming monomer in the presence of the surfactant [18].

This paper highlights recent developments in the design and synthesis of organic-inorganic hybrids using silsesquioxane-based nanoparticles having nanometer size, relatively narrow size distribution, high functionalities, and various characteristic features (Figure 2). Because the synthesis of T_8 silsesquioxane requires complicated and time-consuming procedures, much attention has been paid in developing facile synthetic methods for silsesquioxane-based nano-objects having uniform size and characteristic properties. This paper initially deals with the water-soluble silsesquioxane-based nanoparticles, including hydroxyl-functionalized silsesquioxanes (Figure 2(a)) and cationic silsesquioxanes (Figure 2(c)), which were synthesized via the one-step condensation of the bulky triethoxysilane precursors. An attractive feature of this system is that it makes it feasible to create a variety of hybrid materials

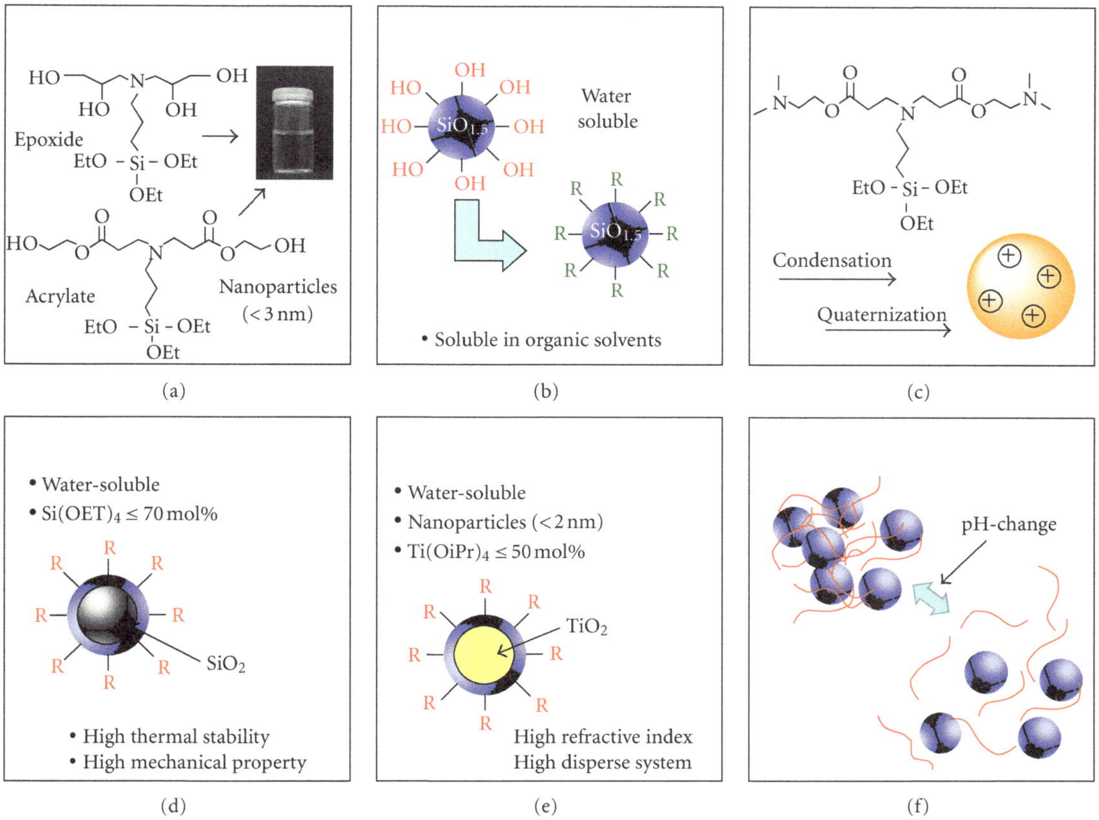

FIGURE 2: Summary of water-soluble silsesquioxane-based nanoparticles. (a) Water-soluble hybrids synthesized by hydrolytic condensation of hydroxyl-functionalized triethoxysilanes; (b) end-group modifications of the hydroxyl groups of the water-soluble hybrids; (c) cationic hybrids obtained by hydrolytic condensation, followed by quaternization; (d) R–$SiO_{1.5}$/SiO_2 hybrids; (e) R–$SiO_{1.5}$/TiO_2 hybrids prepared by hydrolytic cocondensation of the hydroxyl-functionalized triethoxysilane and metal alkoxides; (f) smart hybrids based on the complexation of the silsesquioxane nanoparticles and weak polyelectrolyte.

having different functional organic components through the design of appropriate triethoxysilane precursors, because various epoxy compounds and vinyl monomers can be employed as starting materials. A facile synthetic method with short synthetic steps of high yield is another advantage of this approach, an advantage which is a crucial in practical applications. Other examples involve the synthesis of R–$SiO_{1.5}$/SiO_2 hybrids (Figure 2(d)) and R–$SiO_{1.5}$/TiO_2 hybrids (Figure 2(e)) by hydrolytic cocondensation of the hydroxyl-functionalized triethoxysilane precursor and metal alkoxides. The design and synthesis of stimuli-responsive hybrids based on the silsesquioxane nanoparticles (Figure 2(f)) are briefly introduced. Examples of fluorinated and amphiphilic silsesquioxane hybrids are also given (Figure 3).

2. Hydroxyl-Functionalized Silsesquioxane-Based Hybrids

Research on water-soluble metal/metal oxide nanoparticles with narrow size distributions and characteristic properties has been extensively conducted, due to their potential applications, particularly in biorelated fields. Effective use of water-soluble nanoparticles in a given application requires fine-tuning of two factors; the physical properties of the particles itself (size, topology, composition, and nature of the particle itself) and the chemical properties of their interfaces to avoid unfavorable aggregation of individual nanoparticle and to promote specific interactions with target molecules. During recent years, there has been increasing attention paid to water-soluble silsesquioxane-based materials with tunable properties and well-defined multidimensional architectures [19–23].

Water-soluble silsesquioxane-based nanoparticles (diameter < 3.0 nm) were synthesized by hydrolytic condensation of hydroxyl-functionalized triethoxysilanes under mild conditions (Figure 2(a)). The first example is the synthesis of water-soluble nanoparticles by hydrolytic condensation of a functionalized precursor, *N*,*N*-di(2,3-dihydroxypropyl)aminopropyltriethoxysilane, which was prepared by addition reaction of (3-aminopropyl)triethoxysilane and glycidol (Figure 4(a)) [24, 25]. After the solvents (methanol and ethanol) were evaporated in vacuum, the product was obtained as a glassy solid at room temperature. In addition to tertiary amino groups, the resulting nanoparticle should have hydroxyl groups on the outermost surface, which lead to water-soluble property. The resulting silsesquioxane-based nanoparticle is soluble directly in water, methanol,

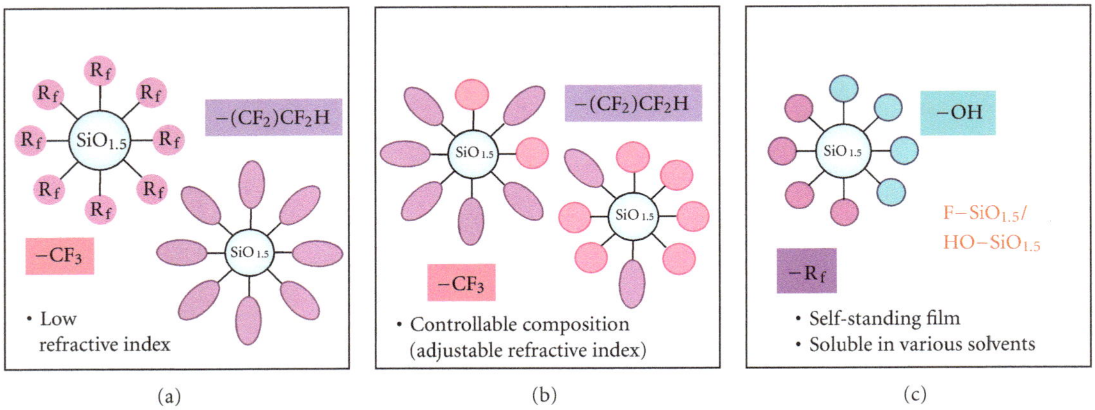

Figure 3: Summary of fluorinated silsesquioxane-based nanoparticles. (a) Fluorinated hybrids prepared by hydrolytic condensation of triethoxysilane precursors derived from fluoroalkyl acrylates; (b) cocondensation products obtained by hydrolytic cocondensation of TFEA-based and OFPA-based triethoxysilane precursors; (c) amphiphilic hybrids based on hydrolytic cocondensation of the hydroxyl-functionalized triethoxysilane precursor and fluorinated triethoxysilane precursors.

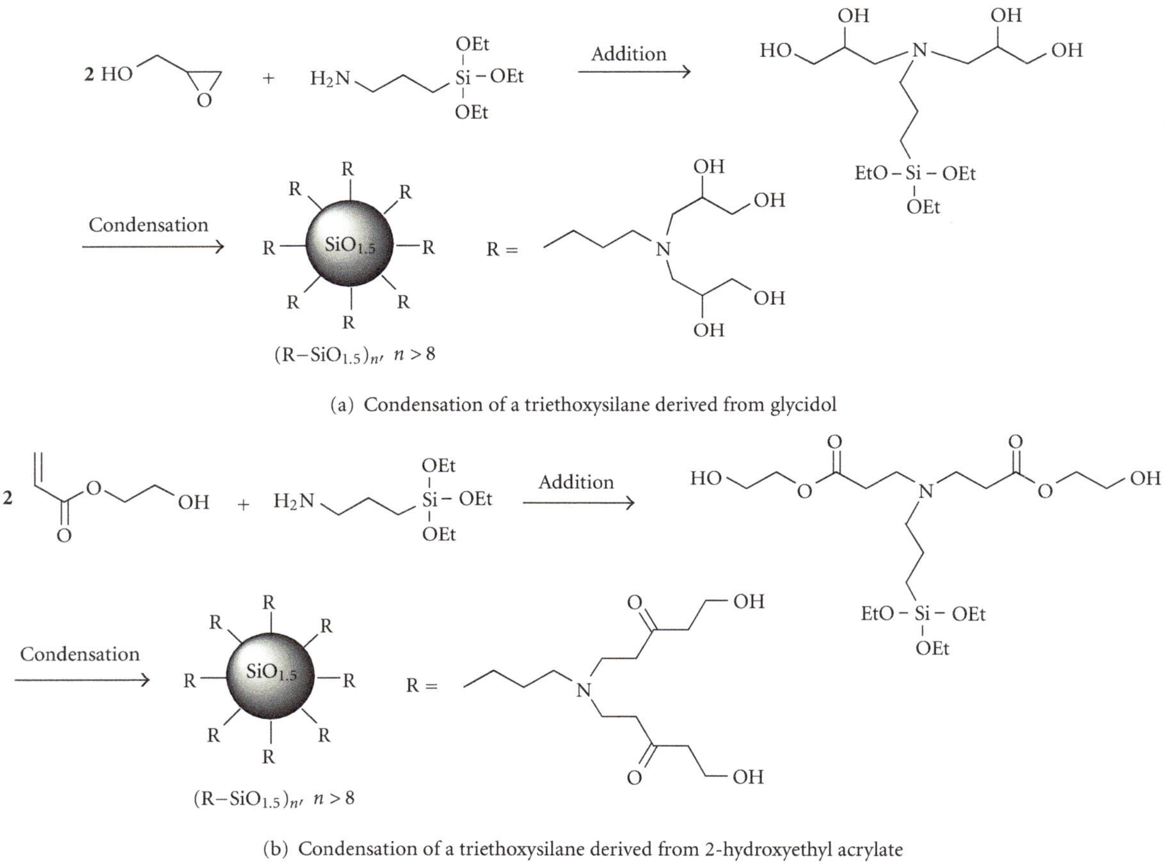

Figure 4: Synthesis of water-soluble silsesquioxane-based nanoparticles by hydrolytic condensation of hydroxyl-functionalized triethoxysilanes.

DMF, and DMSO, while insoluble in most organic solvents, such as dichloromethane, acetone, and dioxane, and so forth. The resulting particles have relatively narrow size distribution with average particle diameter less than 3.0 nm, as confirmed by transmission electron microscopy (TEM) and scanning force microscopy (SFM). Due to the tiny size and high functionality, the nanoparticles can be uniformly dispersed in water and behave as single dissolved molecules to form a transparent colloidal solution. MALDI-TOF MS analysis indicated that the product consists of many species having 12–18 Si atoms with different numbers of intramolecular cyclizations, and the Si–O–C bonds are formed through the reaction of SiOH (or SiOEt) groups with the hydroxyl functionalities of an organic moiety bonded to a Si atom. The species having high number of intermolecular cyclization were predominantly detected, suggesting that product mainly consists of complete and incomplete cage-like structures. Note that most of functional groups are considered to exist on the surface of the silsesquioxane-based materials, there is a possibility for some functional groups to exist within the materials. Nevertheless, this facile synthetic method with short synthetic steps with high yield endowed the nanoparticles with characteristic properties such as high functionalities, good solubility in aqueous media, nanometer size, and narrow size distribution. The silsesquioxane-based nanoparticles with a high density of hydroxyl groups on the surface were used as functional cores for the syntheses of organic/inorganic hybrid stars via "grafting from" polymerization methods [26, 27], amphiphilic silsesquioxanes having various molar ratios of hydrophilic and hydrophobic terminal groups [28], smart nanohybrid based on the complexation with amphiphilic block copolymer micelles [29], and polyurethane-silsesquioxane hybrids [30].

Although the mechanism of the formation is not yet fully understood, the feasibility to create the nanoparticles having narrow size distribution by one step condensation of the organotriethoxysilane is of interest, as it offers the possibility of large-scale production without tedious and time-consuming process. Similar behavior was also reported by another group, in which hydrolytic condensation of triethoxysilanes bearing bulky organic substituents with hydroxyl groups produced cage-type silsesquioxane having a sharp distribution with 8, 9, and 10 Si atoms [31, 32]. They demonstrated that the formation of the closed structures seems to be originated by the presence of (β-hydroxyl) tertiary amine groups in the starting organotrialkoxysilanes, in addition to the bulkiness of the organic group attached to silicon atoms [33, 34].

A bulky functionalized triethoxysilane precursor prepared by addition reaction of (3-aminopropyl)triethoxysilane and 2-hydroxyethyl acrylate (HEA) could be also employed for the preparation of water-soluble nanoparticles (Figure 4(b)) [35]. This can be regarded as the synthesis of a new family of the silsesquioxane-based nanoparticles having uniform size distribution and characteristic water-soluble property by hydrolytic condensation of the hydroxyl-functionalized triethoxysilane precursor. As shown in Figure 4(b), the first step is the addition reaction of aminopropyltriethoxysilane and HEA, followed by acidic condensation of the addition product to afford silsesquioxane-based nanoparticles. The condensation of the triethoxysilane precursor proceeded as a homogeneous system in methanol in the presence of HF (3.2%) at ambient temperature for 2 h to afford the water-soluble hybrids quantitatively. In addition to tertiary amino groups, the resulting silsesquioxane-based nanoparticles should have hydroxyl groups on the outermost surface, which leads to water soluble property. The size distribution was relatively small, and the average particle size was less than 2 nm, as confirmed by X-ray diffraction (XRD) and SFM measurements (Figure 5). The narrow polydispersity ($M_w/M_n = 1.08$) and a reasonable molecular weight ($M_n = 3300$), corresponding to species having 6–12 silicon atoms, were also confirmed by size-exclusion chromatography (SEC) measurements. For SEC measurement, the hydroxyl groups of the water-soluble silsesquioxane-based nanoparticles were converted into isobutyl ester form by esterification reaction (Figure 2(b)). The investigation of the hydrolytic condensation under various conditions suggested that the presence of the hydroxyl groups in the alkyl chain facilitates the internal cyclization, and the homogeneous reaction systems in alcohol may be prerequisite for the nanoparticle formation. In this system, the triethoxysilane precursor, $R–Si(OCH_2CH_3)_3$, $R = –CH_2CH_2CH_2N(CH_2CH_2COOCH_2CH_2OH)_2$, has a bulky organic group attached to a silicon atom, but the position of the hydroxyl group is far from the tertiary amino group. The bulkiness of the organic group and the distance between the hydroxyl group and the tertiary amino group may have some influence on the internal cyclization, resulting in the formation of closed structures and avoidance of gelation.

In this synthetic procedure, the reaction system is homogeneous during the hydrolysis/condensation reaction, since both the precursor and the resulting silsesquioxane hybrid are soluble in methanol in the presence of a small amount of HF aqueous solution used as acidic catalyst. This is a quite different feature from conventional systems, in which the reaction is heterogeneous, because precursors containing a H or an alkyl chain bonded to a Si atom are only soluble in organic solvents. In this system, the usage of the functional triethoxysilane precursor having two hydroxyl groups and ester groups may support the formation of the functionalized silsesquioxane hybrid that has nanometer size, uniform size distribution, and good solubility in many solvents. These results suggest the feasibility of creating the silsesquioxane hybrid having a high density of chemically bonded peripheral hydroxyl groups on the outermost surface, via the one-step condensation of the bulky triethoxysilane precursor, which can be achieved through the careful choice of the organic structure and condensation conditions.

3. Water-Soluble $R–SiO_{1.5}/SiO_2$ Hybrid Nanoparticles by Cocondensation

Cocondensation of a trialkoxysilane and a tetraalkoxysilane compound provides organic-inorganic hybrid materials, in which thermal and mechanical properties can be manipulated by the tetraalkoxysilane content. The increase in a tetraalkoxysilane compound in the feed may lead to the formation of glassy hybrids with higher inorganic component. Further, the cocondensation can be regarded as a convenient and promising process to produce novel functional hybrids, which have inorganic silica-like properties and exhibit characteristic solubility in appropriate solvents, if the presence of a particular trialkoxysilane prevents the gelation effectively during the hydrolytic cocondensation of the mixtures.

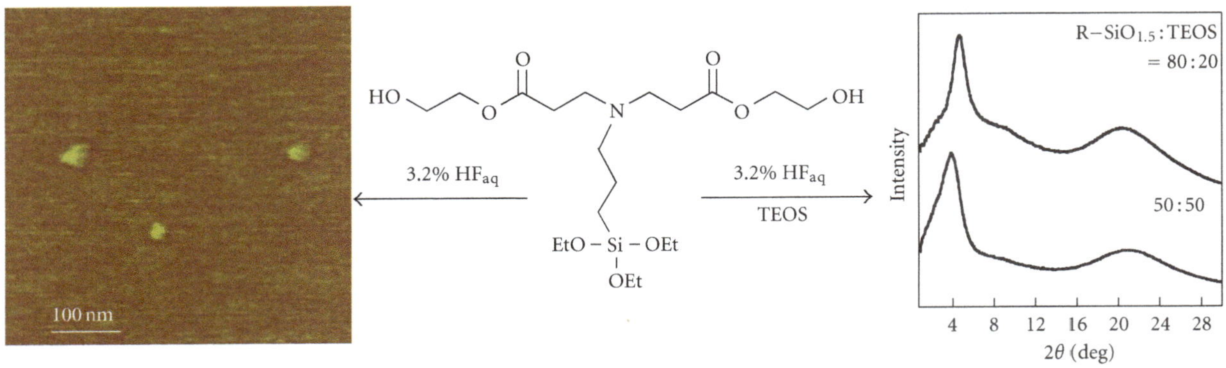

Figure 5: SFM height image and XRD spectra of water-soluble silsesquioxane-based nanoparticles synthesized by hydrolytic (co-)condensation of the hydroxyl-functionalized triethoxysilane precursor derived from 2-hydroxyethyl acrylate (HEA). Reprinted with permission from [35]. Copyright 2007 American Chemical Society.

The cocondensation of tetraethoxysilane (TEOS) with the hydroxyl-functionalized triethoxysilane precursor provided water-soluble nanoparticles having similar sizes (ca, 2 nm) with higher thermal property, depending on the composition in the feed (Figure 6(a)) [35]. Cocondensation was carried out under different feed ratios in methanol at room temperature, and the water-soluble products were obtained in the cases of TEOS molar ratio up to 70% (Figure 2(d)). Thermal stability and the char yield were found to increase with increasing the TEOS content in the feed, as determined by thermogravimetric analysis. The isolated nanoparticles distributed homogeneously without any aggregation were visualized by SFM, when the product was prepared at TEOS/the triethoxysilane = 50/50 mol%. The formation of the nanoparticles by hydrolytic cocondensation was also confirmed by the XRD studies (Figure 5). In the cocondensation process, the content of inorganic component, solubility, thermal stability, and size of the hybrid materials could be manipulated, depending on the triethoxysilane precursor/tetraethoxysilane ratio in the feed. Another attractive feature of this system is the feasibility to create a variety of hybrid materials, because various metal alkoxides can be employed as starting materials.

4. Water-Soluble R–$SiO_{1.5}$/TiO_2 Hybrid Nanoparticles by Cocondensation

Titania-silica mixed oxides have attracted significant research interest, because of their wide range of applications, such as glasses with a low thermal coefficient [36] or high strength [37], super-hydrophilic surfaces [38, 39], implant coatings for direct tissue attachment [40, 41], high-k dielectric materials [42], optical sensors [43–45], antireflection coating for solar cells [46], optical planar waveguide [47–49], and heterogeneous catalysts [50–54]. Various kinds of TiO_2–SiO_2 composites have been developed, which involve spherical particles [55, 56], thin films [39, 44, 46], fibers [57, 58], and porous materials [50, 54, 59]. Manipulation of the size, shape, microstructures, and surface area is crucial to improve characteristic properties, such as catalytic activity, photoactivity, chemical durability, and optical and thermal properties.

Novel water-soluble R–$SiO_{1.5}$/TiO_2 hybrid nanoparticles were synthesized by hydrolytic cocondensation of titanium alkoxides ($Ti(OR')_4$ R′ = ethyl, isopropyl, and butyl) with the hydroxyl-functionalized triethoxysilane precursor (Figure 2(e)) [60]. Cocondensation of a titanium alkoxide with the triethoxysilane precursor was investigated at different feed ratios, suggesting that water-soluble nanoparticles were obtained only at less than 30% of $Ti(OEt)_4$ molar ratio in the feed. In contrast, the cocondensation of titanium tetraisopropoxide, $Ti(O^iPr)_4$, with the triethoxysilane precursor in the presence of acetylacetone proceeded as a homogeneous system until 70% of $Ti(O^iPr)_4$ molar ratio to afford water-soluble organic-inorganic hybrid nanoparticles containing titania-silica mixed oxides (Figure 6(b)), as confirmed by NMR, FT-IR, elemental, and ICP analyses. SFM measurements of the product prepared at $Ti(O^iPr)_4$/the triethoxysilane = 50/50 mol% with acetylacetone indicated the formation of the nanoparticles having relatively narrow size distribution with average particle diameter less than 2.0 nm without aggregation. The refractive index of the hybrid nanoparticle was 1.571. The isolated nanoparticles distributed homogeneously were visualized by TEM, and the size of the hybrid nanoparticle (1.9 nm) was determined by XRD.

5. Cationic Hybrid Nanoparticles

Cationic silsesquioxane-based materials with tunable properties and well-defined multidimensional architectures have become of special interest, because of their industrial importance and scientifically interesting properties. The potential applications involve biocompatible drug carrier [61], detection of conformation transformation of double-stranded DNA [62], probe for the detection of DNA [63], light-harvesting unimolecular nanoparticle for fluorescence amplification in cellular imaging [64], modification of montmorillonite [65], and preparation of nanocomposite thin

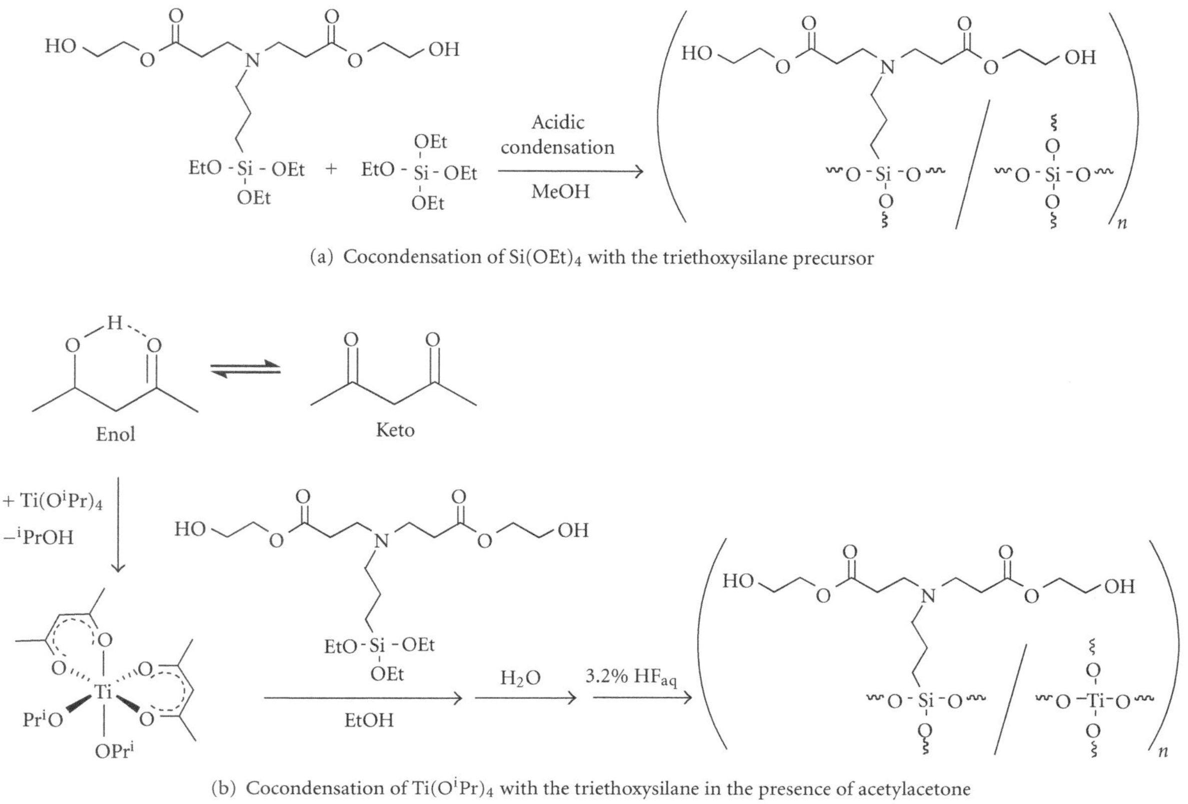

Figure 6: Synthesis of hybrid nanoparticles by cocondensation of the hydroxyl-functionalized triethoxysilane precursor derived from 2-hydroxyethyl acrylate (HEA).

films via layer-by-layer electrostatic self-assembly [22, 66]. Most of these cationic silsesquioxanes have been prepared from octafunctional polyhedral oligomeric silsesquioxanes, consisting of a rigid, crystalline silica-like core that is perfectly defined spatially (0.5–0.7 nm) and that can be linked covalently to eight R groups, such as octavinyl- and octaamino-substituted ones.

Recently, the synthesis of novel cationic silsesquioxane hybrids having uniform size distribution and characteristic water-soluble property by hydrolytic condensation of a triethoxysilane precursor derived from 2-(dimethylamino)ethyl acrylate (DMAEA) was reported, as shown in Figure 2(c) [67]. The development of an easily accessible mixture of silsesquioxanes with ionic or ionizable groups is one option for their exploitation in practical applications. As shown in Figure 7(a), the first step is the addition reaction of aminopropyltriethoxysilane and DMAEA, followed by hydrolytic condensation of the addition product to afford silsesquioxane hybrid. Acidic condensation of the DMAEA-based triethoxysilane precursor proceeded as a homogeneous system in methanol at room temperature to afford the water-soluble hybrid almost quantitatively. In addition to the ester groups, the resulting silsesquioxane hybrid exhibits a high density of chemically bonded peripheral tertiary amino groups on the outermost surface, which leads to water-solubility and various characteristic properties. The relatively low polydispersity ($M_w/M_n = 1.33$) and a reasonable molecular weight ($M_n = 2700$) were confirmed by SEC measurement. The size of the silsesquioxane hybrid (1.7 nm) was also determined by XRD. Quaternization reaction of the tertiary amine-containing hybrids with methyl iodide led to cationic silsesquioxane hybrids containing quaternized amine functionalities, which showed good solubility in polar solvents. SFM measurements indicated the formation of the cationic silsesquioxane hybrids having relatively narrow size distribution with average particle diameter (about 2.0 nm) without aggregation. Hence, it is entirely fair to say that the product obtained by the hydrolytic condensation of the functional precursor derived from DMAEA is a mixture of silsesquioxanes having different silicon atoms, which can be called silsesquioxane-based nanoparticles, because their structures are topologically equivalent to a sphere. Note that polyhedral silsesquioxanes are often referred to as spherosiloxanes, since polyhedral structures are topologically equivalent to a sphere. Cocondensation of TEOS with the triethoxysilane precursor was carried out under different feed ratios, and water-soluble products were obtained in the cases of TEOS molar ratio up to 40% (Figure 7(b)). This convenient synthetic approach may lead to further development of novel organic-inorganic hybrids, because of the characteristic properties of the cationic silsesquioxane hybrid, such as high functionalities, solubility in aqueous medium, nanometer size, and narrow size distribution.

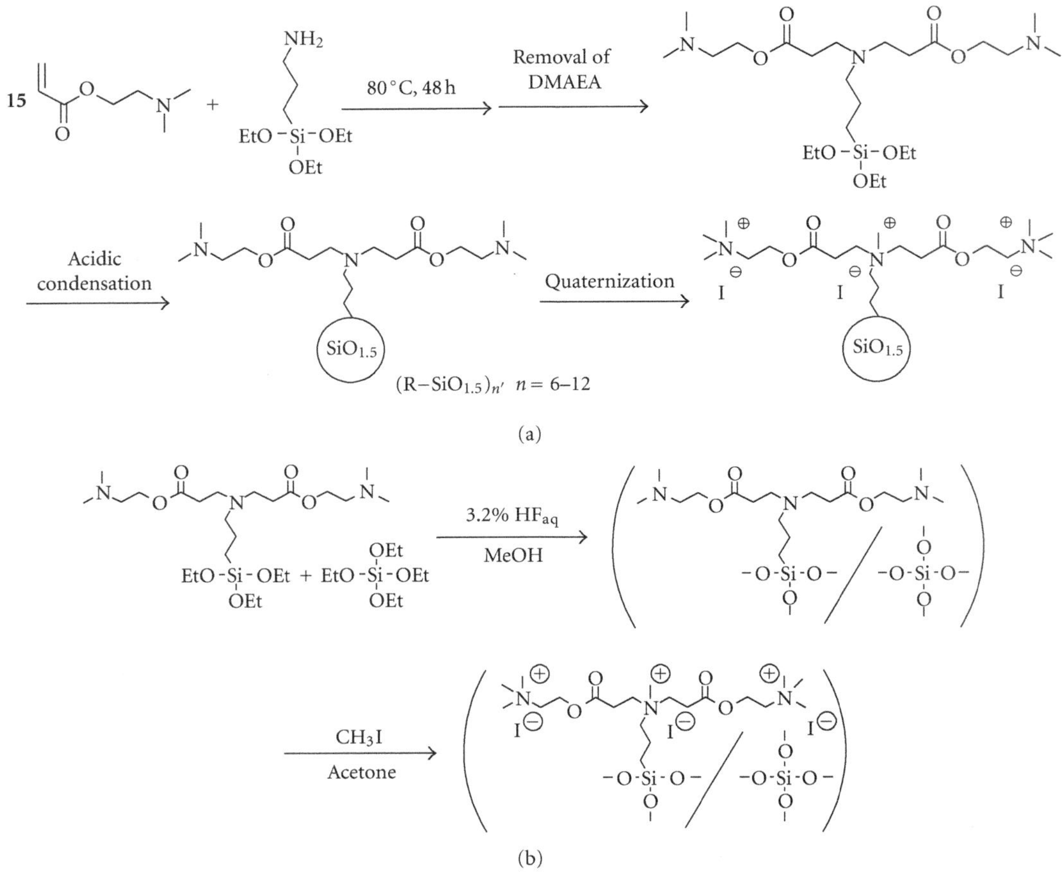

FIGURE 7: Synthetic routes for cationic silsesquioxane hybrids by (a) hydrolytic condensation of a functionalized precursor derived from 2-(dimethylamino)ethyl acrylate (DMAEA), followed by quaternization, and (b) hydrolytic cocondensation of tetraethoxysilane (TEOS) with the DMAEA-based triethoxysilane precursor.

6. Fluorinated Silsesquioxane-Based Hybrids

In recent years, increasing attention has been paid to fluorinated silsesquioxanes, which consist of a silicon-oxygen core framework and a fluoroalkyl shell. For example, fluorinated polyhedral oligomeric silsesquioxane molecules, in which the rigid cage is surrounded by perfluoroalkyl groups, were employed in the design of superoleophobic surfaces [68]. In this system, the fluorinated silsesquioxanes exhibited limited solubility because they were synthesized by basic condensation of triethoxysilane derivatives having hydrophobic and oleophobic fluoroalkyl groups such as 1*H*,1*H*,2*H*,2*H*-heptadecafluorodecyl and 1*H*,1*H*,2*H*,2*H*-tridecafluorooctyl groups [68, 69]. A number of fluorinated polyhedral oligomeric silsesquioxane structures possessing a high degree of hydrophobicity has been prepared via a facile corner-capping methodology using various fluoroalkyl trichlorosilanes [70, 71]. A fluorinated polyhedral oligomeric silsesquioxane was also prepared by hydrosilylation of octakis(dimethylsiloxy)silsesquioxane with a fluorinated allyl ether derivative [72]. Another example is the synthesis of a heteroleptic silsesquioxane consisting of perfluoro, isooctyl, and amino (or alkoxy) groups by basic hydrolysis of the corresponding trialkoxysilane precursors [73, 74]. Fluorinated silsesquioxane/polymer hybrids have recently inspired additional research efforts in which the silsesquioxane component has been either blended or covalently linked with a polymer [75–79]. Various fluorinated silsesquioxane/polymer hybrids have been also employed for modification of surface dewettability [80–82] and in lithography [83, 84].

Recently, the synthesis of novel low-refractive-index fluorinated silsesquioxane-based hybrids that have uniform size distribution, good solubility, and a tunable refractive index by hydrolytic condensation of triethoxysilane precursors derived from fluoroalkyl acrylates was demonstrated (Figure 3(a)) [85]. As shown in Figure 8(a), the first step is the addition reaction of aminopropyltriethoxysilane and fluoroalkyl acrylates, followed by acidic condensation of the addition products. Two acrylates having different fluoroalkyl chains—1*H*,1*H*,5*H*-octafluoropentyl acrylate (OFPA) and 2,2,2-trifluoroethyl acrylate (TFEA)—were employed for the preparation of the fluorinated triethoxysilane precursors ($R\text{–}Si(OCH_2CH_3)_3$). The condensation of the fluorinated triethoxysilane precursors in the presence of a small amount of HF aqueous solution proceeded as a homogeneous system

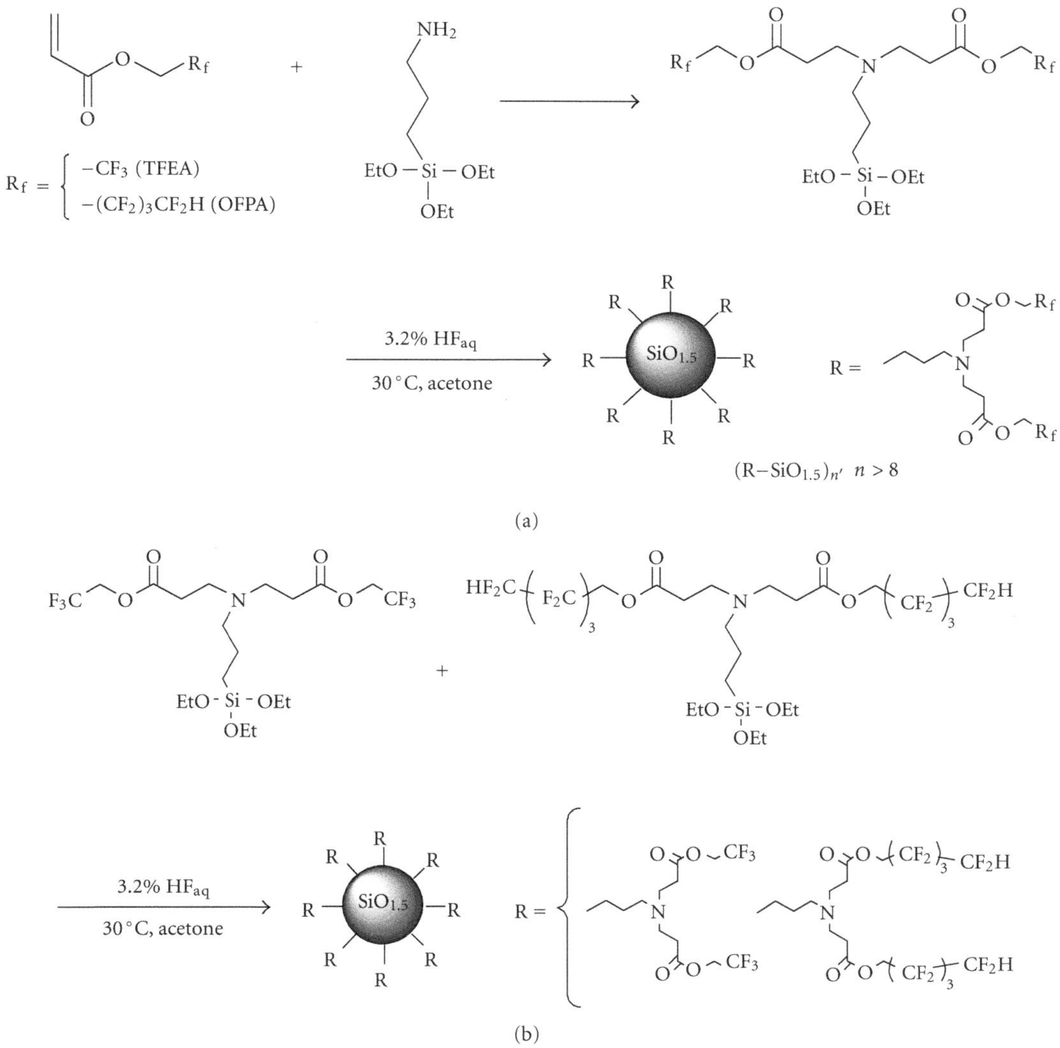

FIGURE 8: Synthetic routes for fluorinated silsesquioxane-based hybrids obtained via (a) hydrolytic condensation of triethoxysilane precursors derived from fluoroalkyl acrylates, and (b) hydrolytic cocondensation of TFEA-based and OFPA-based triethoxysilane precursors.

in acetone at 30°C until the end of the reaction. After the solvents (acetone and ethanol) were evaporated in a vacuum, the products were obtained quantitatively. The products were soluble in a variety of organic solvents, including $CHCl_3$, THF, and acetone, but were insoluble in hexane and water. The resulting silsesquioxane-based hybrids exhibited a high density of chemically bonded peripheral fluoroalkyl groups, which led to various characteristic properties, including a low refractive index. The low polydispersities and reasonable molecular weights of the resulting fluorinated silsesquioxanes ($M_n = 4800$, $M_w/M_n = 1.01$; and $M_n = 4300$, $M_w/M_n = 1.07$ for the OFPA- and TFEA-based products, resp.) were confirmed by SEC. The formation of spherical hybrids having relatively narrow size distributions (average particle diameter < 3.0 nm) without aggregation was confirmed by SFM measurements. The XRD and DSC measurements indicated that the fluorinated silsesquioxane hybrids can be regarded as amorphous glass having low glass transition temperatures. Different from the cubic silsesquioxane crystals, some silsesquioxanes having long and/or specific substituent groups were amorphous, which showed broad XRD peaks [86–88]. Similar to such amorphous silsesquioxanes, the fluorinated silsesquioxane hybrids [85], hydroxyl-functionalized silsesquioxane hybrids [35], and cationic silsesquioxane hybrids [67] synthesized via the one-step condensation of the bulky triethoxysilane precursors can be regarded as amorphous glasses, which are mainly due to the long alkyl groups. The refractive indexes of the TFEA- and OFPA-based silsesquioxane hybrids were 1.43 and 1.40, respectively.

Cocondensation of the TFEA- and OFPA-based triethoxysilane precursors afforded a series of fluorinated hybrids whose refractive index and various other properties can be manipulated by varying the composition of the feed (Figure 8(b)). The formation of hybrids having spherical structures via hydrolytic cocondensation was also confirmed

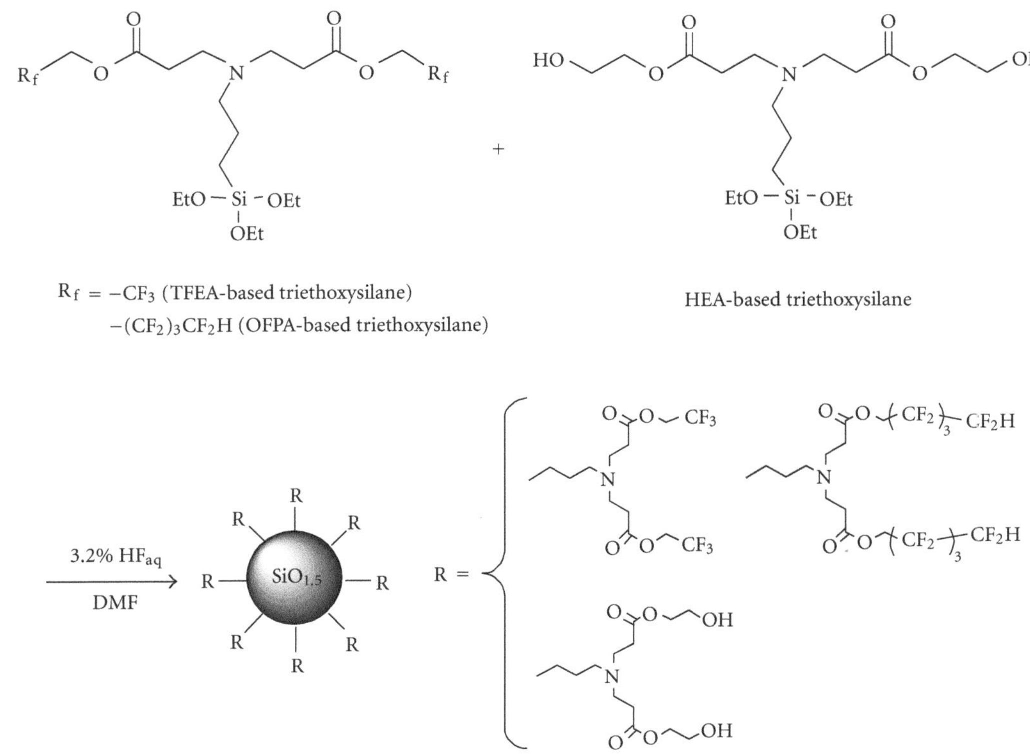

Figure 9: Synthetic route for the amphiphilic silsesquioxane-based hybrids based on hydrolytic cocondensation of the hydroxyl-functionalized triethoxysilane precursor and fluorinated triethoxysilane precursors.

by XRD and SFM measurements. Low refractive indexes of the fluorinated silsesquioxane-based hybrids could be tuned by adjusting the feed ratio of the TFEA-based and OFPA-based triethoxysilane precursors in the cocondensation (Figure 3(b)).

7. Amphiphilic Silsesquioxane-Based Hybrids

Silsesquioxanes having two or more different organic groups immobilized to an inorganic core have recently attracted considerable attention, because of their intriguing phase behavior and enormous potential as a building block for various advanced materials [5]. For example, Gunawidjaja et al. reported bulk and surface assembly of amphiphilic silsesquioxane compounds with various hydrophilic and hydrophobic terminal group compositions [28]. For the synthesis of silsesquioxanes having different substituent groups, several methods have been employed, which involve cohydrolysis/cocondensation of chlorosilanes and alkoxysilanes, corner-capping method, and synthetic modification of a preexisting silsesquioxane compound [5]. During recent years, there has been increasing attention paid to silsesquioxanes having fluoroalkyl chain as a hydrophobic component and another organic group. The synthesis of heteroleptic silsesquioxane consisting of perfluoro, isooctyl, and amino (or alkoxy) groups was conducted by basic hydrolysis of corresponding trialkoxysilane precursors [73, 74]. A number of fluorinated polyhedral oligomeric silsesquioxanes structures possessing another organic group has been prepared via a facile corner-capping methodology [70, 71]. Additionally, fluorinated polyhedral oligomeric silsesquioxanes possessing one reactive functional group by the corner-capping method were employed for preparation of various fluorinated silsesquioxanes/polymer hybrids [89–91].

Amphiphilic silsesquioxanes-based hybrids were synthesized by hydrolytic cocondensation of a hydroxyl-functionalized triethoxysilane precursor derived from 2-hydroxyethyl acrylate (HEA) and fluorinated triethoxysilane precursors derived from 1*H*,1*H*,5*H*-octafluoropentyl acrylate (OFPA) and 2,2,2-trifluoroethyl acrylate (TFEA), as shown in Figure 9 [92]. The OFPA-based triethoxysilane precursor has a bulky fluoroalkyl group attached to a silicon atom, whereas the TFEA-based precursor has a shorter fluoroalkyl group. The bulkiness of the fluoroalkyl group may have some influence on the internal cyclization, resulting in the formation of closed structures and avoidance of gelation. On the basis of the preparation method (hydrolytic cocondensation of two different triethoxysilanes, R_f–Si(OEt)$_3$ and R_{OH}–Si(OEt)$_3$), the general structure of the hybrids is expected to be $([R_f\text{-}co\text{-}R_{OH}]-SiO_{1.5})_n$, where each silicon atom is bound to an average of one-and-a-half oxygens and to one alkyl chain involving the fluoroalkyl groups or hydroxyalkyl groups. The resulting amphiphilic silsesquioxane hybrids should have fluoroalkyl groups and hydroxyalkyl groups on the outermost surface, in addition to the tertiary amino groups and ester groups, which lead to various characteristic properties.

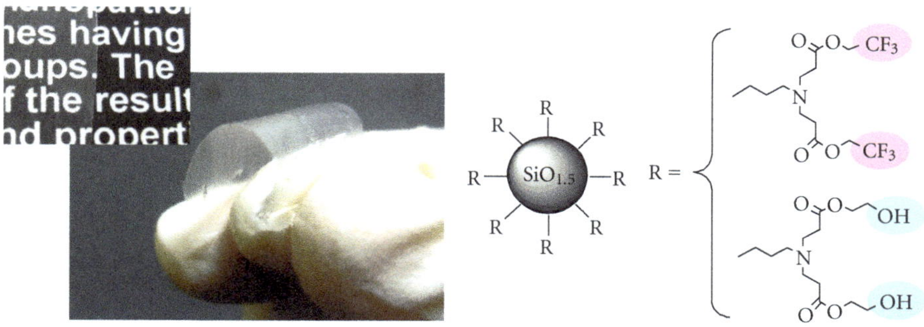

FIGURE 10: Photograph of flexible film of the TFEA-*co*-HEA silsesquioxane hybrid prepared by hydrolytic cocondensations. Reprinted with permission from [92]. Copyright 2012 Elsevier.

Hydrolytic cocondensations of two functionalized triethoxysilane precursors proceeded as homogeneous systems in *N,N*-dimethylformamide (DMF) to afford amphiphilic silsesquioxanes hybrids, which were soluble in a variety of solvents, depending on the composition. The structure of the constitutional unit of the hybrids was confirmed by the results of NMR and FT-IR measurements. SFM and XRD measurements indicated the formation of spherical hybrids having relatively narrow size distributions (average particle diameter < 3.0 nm) without aggregation. Cocondensation of the hydroxyl-functionalized precursor (HEA-based triethoxysilane) and fluorinated precursors (OFPA-based and TFEA-based triethoxysilane) provided amphiphilic hybrid materials, in which solubility, amphiphilicity, refractive index, film forming, and various properties could be manipulated by the composition in the feed (Figure 3(c)). As can be seen in Figure 10, flexible semitransparent films were obtained from the amphiphilic silsesquioxane hybrids having hydrophilic and hydrophobic chains connected chemically to an inorganic core, which were prepared by hydrolytic cocondensations at suitable feed ratios [92]. Note that self-standing hybrid films were obtained from the amphiphilic silsesquioxane hybrids in this system without any addition of cross-linker and polymer, although most of film-forming nanocomposites were prepared from organic-inorganic hybrid systems comprising of polymer chains and inorganic particles. In the systems, the formation of the flexible semitransparent films would be the result of specific interactions between the hydroxyl groups, tertiary amino groups, fluoroalkyl groups, and ester groups in the amphiphilic silsesquioxane hybrids. The presence of intermolecular interactions between these functional groups of the hybrids may contribute to achieving good film-forming property, in which physically cross-linked polymer-like materials can be easily fabricated into the self-standing hybrid films.

8. Stimuli-Responsive Organic-Inorganic Hybrids

Stimuli-responsive organic-inorganic hybrids have recently attracted considerable interest, because combination with inorganic materials offers the opportunity to develop new nanosized "intelligent" or "smart" hybrids [25, 93–97]. Manipulation of specific intermolecular interactions, such as hydrogen-bonding, acid-base interactions, and oppositely charged ionic interactions, is crucial for the design and development of stimuli-responsive organic-inorganic hybrid structures with nanometer dimensions [98–101]. The water soluble silsesquioxane-based nanoparticles obtained from glycidol (Figure 4(a)) could be used as a component for characteristic intelligent colloidal hybrids, in which the complexation of tertiary amine-containing nanoparticles and a weak anionic polyelectrolyte can be manipulated simply by pH change in aqueous solution (Figure 2(f)) [25, 97]. Poly(acrylic acid), PAA, was selected as a weak polyelectrolyte, because the degree of ionization of carboxylic acids can be easily controlled by the pH value. In this system, both PAA and the silsesquioxane nanoparticles formed visually transparent solutions in water, while a white turbid dispersion was obtained just after mixing the two solutions at room temperature. The complex formation in water was strongly affected by the pH value, and the pH-induced association-dissociation behavior was a reversible and rapid process. The reversible pH-induced colloid formation due to the complexation of the inorganic-organic nanomaterials can provide a viable route to the production of tailored materials with unique properties for various applications.

The nature of the interaction of biomolecules, such as proteins, peptides, and amino acids, with inorganic materials is a subject of extraordinary relevance due to increasing interest in biointerfaces for medical, diagnostic, and biotechnology applications. In recent years, increasing attention has also been paid to developing silsesquioxane/biomolecule hybrids [102–109]. A series of researches in the design and synthesis of amino acid-based polymers including stimuli-responsive polymers, such as pH-responsive, thermoresponsive, and dual-stimuli-responsive block copolymers, and self-assembled block copolymers having tunable chiroptical properties, have been reported, which were obtained by reversible addition-fragmentation chain transfer (RAFT) polymerization [110].

Recently, smart amino acid-based polymer/silsesquioxane hybrids was developed, in which the complexation of water-soluble silsesquioxane nanoparticles [35], derived from HEA-based triethoxysilane (Figure 11(a)), and amino

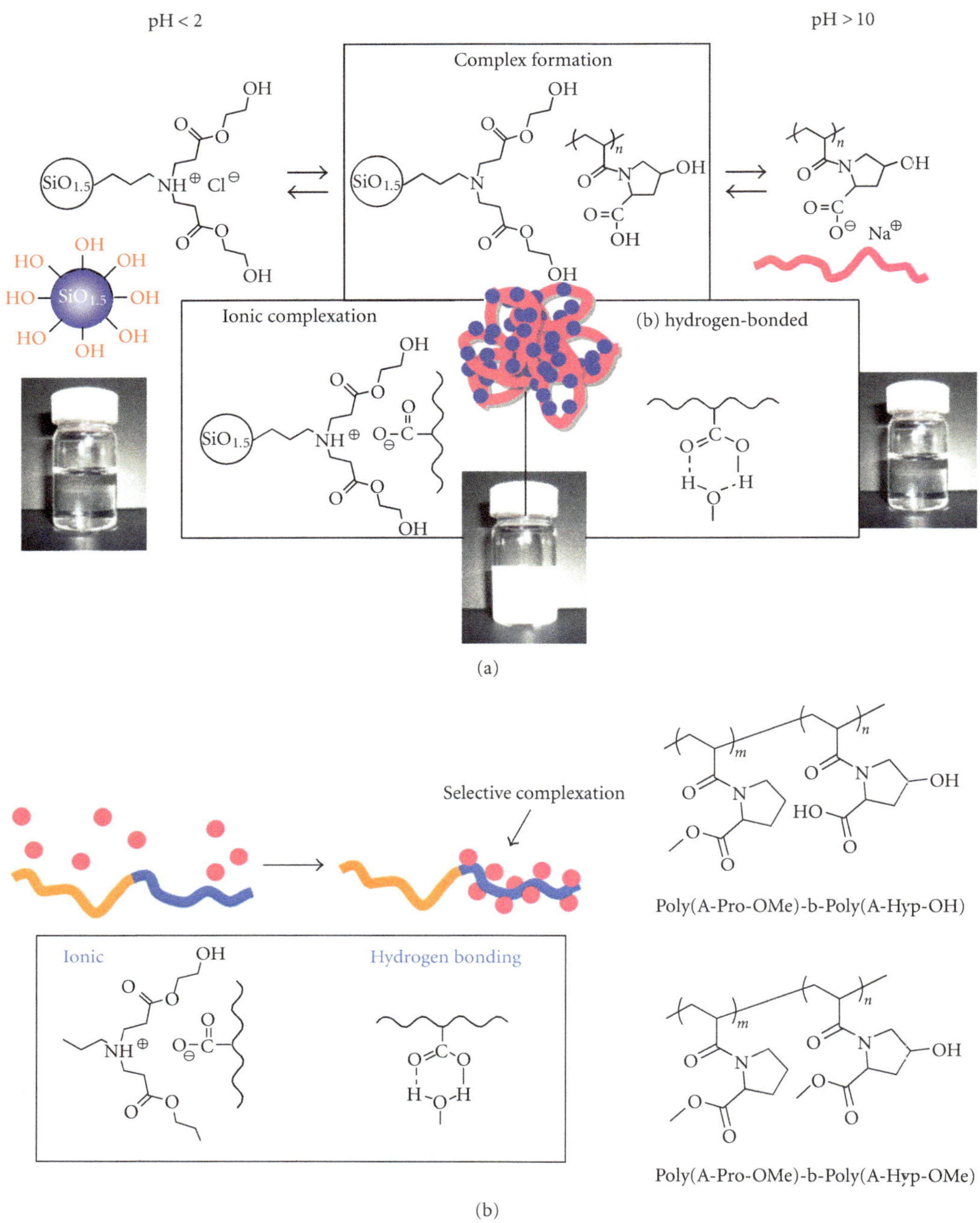

FIGURE 11: Smart organic-inorganic hybrids based on (a) complexation of amino acid-based polymer and water-soluble silsesquioxane nanoparticles, and (b) selective complexation of the silsesquioxane nanoparticles and block copolymer.

acid-based polymer, can be manipulated by pH changes in aqueous media [111]. The nanoparticles obtained from HEA-based triethoxysilane should have many ester groups on the surface, which may contribute hydrogen bond and therefore affect stimuli-responsive complexation of tertiary amine-containing nanoparticles and a weak anionic polyelectrolyte. Simple mixing of aqueous solutions of the amino acid-based polymer obtained from *N*-acryloyl-4-*trans*-hydroxy-l-proline (A-Hyp-OH) [112] and the silsesquioxane nanoparticles led to the straightforward formation of the pH-responsive hybrids, in which a white turbid dispersion was observed at pH = 5–8, whereas transparent solutions were obtained at pH = 2 and 10 (Figure 11(a)). The methylated sample, poly(A-Hyp-OMe), exhibited a characteristic soluble-insoluble transition at around 49.5°C [112], and thermoresponsive hybrids were obtained by complexation of the silsesquioxane nanoparticles with poly(A-Hyp-OMe) [111]. In contrast, poly(*N*-acryloyl-l-proline methyl ester) and poly(A-Pro-OMe), exhibited a relatively lower phase-separation temperature (around 18°C) in neutral water (pH = 7) [113–115], and the thermoresponsive polymer showed no specific interaction with the silsesquioxane nanoparticles [111].

Self-assembly of block copolymer/nanoparticle hybrids has also generated significant research interest, because the nanoparticles can be spatially organized in the formed aggregates [29, 116–120]. Depending on the chemical nature of the functional segments and their composition, the block copolymers afford a great opportunity for tuning chemical and physical properties as well as assembled structures. Additionally, the size, shape, and nature of the inorganic nanoparticles and specific interactions between the organic and inorganic components act as crucial elements to provide an effective route for the controlled self-ordering of nanoparticles with polymers and for the endowment of characteristic properties. Smart organic-inorganic hybrids were prepared using noncovalent interactions between water-soluble silsesquioxane nanoparticles containing tertiary amine moieties and two amino acid-based block copolymers prepared by RAFT polymerization (Figure 11(b)) [121]. A dual thermoresponsive block copolymer displaying LCST and UCST was employed, in which only the poly(A-Hyp-OH) segment could interact with the silsesquioxane nanoparticles, whereas another poly(A-Pro-OMe) segment showed a characteristic thermoresponsive property without any interaction with the nanoparticles. The simple mixing procedure of two transparent aqueous solutions led to the formation of smart organic-inorganic hybrids through the selective complexation of the silsesquioxane nanoparticles and the poly(A-Hyp-OH) segment in the block copolymer.

9. Conclusion

This paper has summarized recent development of the silsesquioxanes-based nanoparticles prepared by hydrolytic condensation of bulky triethoxysilane precursors derived from functional acrylate derivatives. The addition reaction of aminopropyltriethoxysilane with 2-hydroxyethyl acrylate (HEA), 2-(dimethylamino)ethyl acrylate (DMAEA), 1*H*,1*H*,5*H*-octafluoropentyl acrylate (OFPA), and 2,2,2-trifluoroethyl acrylate (TFEA) afforded functional triethoxysilane precursors. The hydrolytic condensation of the addition products proceeded as homogeneous systems under suitable conditions to afford the functional silsesquioxane hybrids almost quantitatively. The development of an easily accessible mixture of silsesquioxanes with a variety of functional groups is one option for their exploitation in practical applications. This convenient synthetic approach have allowed great advances in the further development of novel organic-inorganic hybrids, because of the characteristic properties of the functional silsesquioxane hybrid, such as high functionalities, solubility in aqueous medium, nanometer size, and narrow size distribution. Depending on the chemical nature of the functional trietoxysilane precursors, additional component (trialkoxysilanes and metal alkoxides used for the cocondensation), and their composition, the silsesquioxane hybrids afforded great flexibility for tuning functional groups in the organic part, components of the inorganic parts, and various properties as well as size and shape of the hybrids. The manipulation of noncovalent interactions such as hydrogen bonding and interelectrolyte interaction was also crucial for providing organic-inorganic smart hybrids having characteristic stimuli-responsive properties. The multiscale ordering of such functional nanomaterials is a powerful technique for the creation of tailored hybrid materials with unique properties for various applications.

Acknowledgments

The author acknowledges Professor Takeshi Endo (Director of Molecular Engineering Institute, Kinki University and Vice President of Kinki University) and Professor Koichiro Yonetake (Yamagata University) for helpful suggestions, discussions, and collaborations.

References

[1] P. Eisenberg, R. Erra-Balsells, Y. Ishikawa et al., "Cagelike precursors of high-molar-mass silsesquioxanes formed by the hydrolytic condensation of trialkoxysilanes," *Macromolecules*, vol. 33, no. 6, pp. 1940–1947, 2000.

[2] F. J. Feher and T. A. Budzichowski, "Silasesquioxanes as ligands in inorganic and organometallic chemistry," *Polyhedron*, vol. 14, no. 22, pp. 3239–3253, 1995.

[3] M. G. Voronkov and V. I. Lavrent'yev, "Polyhedral oligosilsesquioxanes and their homo derivatives," *Topics in Current Chemistry*, vol. 102, pp. 199–236, 1982.

[4] P. P. Pescarmona and T. Maschmeyer, "Oligomeric silsesquioxanes: synthesis, characterization and selected applications," *Australian Journal of Chemistry*, vol. 54, no. 9-10, pp. 583–596, 2001.

[5] D. B. Cordes, P. D. Lickiss, and F. Rataboul, "Recent developments in the chemistry of cubic polyhedral oligosilsesquioxanes," *Chemical Reviews*, vol. 110, no. 4, pp. 2081–2173, 2010.

[6] S. Sulaiman, A. Bhaskar, J. Zhang, R. Guda, T. Goodson III, and R. M. Laine, "Molecules with perfect cubic symmetry as nanobuilding blocks for 3-D assemblies. Elaboration of octavinylsilsesquioxane. Unusual luminescence shifts may indicate extended conjugation involving the silsesquioxane core," *Chemistry of Materials*, vol. 20, no. 17, pp. 5563–5573, 2008.

[7] R. Y. Kannan, H. J. Salacinski, P. E. Butler, and A. M. Seifalian, "Polyhedral oligomeric silsesquioxane nanocomposites: the next generation material for biomedical applications," *Accounts of Chemical Research*, vol. 38, no. 11, pp. 879–884, 2005.

[8] R. M. Laine, "Nanobuilding blocks based on the $[OSiO_{1.5}]_x$ (x = 6, 8, 10) octasilsesquioxanes," *Journal of Materials Chemistry*, vol. 15, no. 35-36, pp. 3725–3744, 2005.

[9] P. D. Lickiss and F. Rataboul, "Chapter 1 fully condensed polyhedral oligosilsesquioxanes (POSS): from synthesis to application," *Advances in Organometallic Chemistry*, vol. 57, pp. 1–116, 2008.

[10] A. Provatas and J. G. Matisons, "Silsesquioxanes: synthesis and applications," *Trends in Polymer Science*, vol. 5, pp. 327–332, 1997.

[11] C. Zhang and R. M. Laine, "Hydrosilylation of allyl alcohol with $[HSiMe_2OSiO_{1.5}]_8$: octa(3-hydroxypropyldimethylsiloxy)octasilsesquioxane and its octamethacrylate derivative as potential precursors to hybrid nanocomposites,"

Journal of the American Chemical Society, vol. 122, no. 29, pp. 6979–6988, 2000.

[12] P. A. Agaskar, V. W. Day, and W. G. Klemperer, "A new route to trimethylsilylated spherosilicates: synthesis and structure of $[Si_{12}O_{18}](OSiMe_3)_{12}$, D_{3h}-$[Si_{14}O_{21}](OSiMe_3)_{14}$, and C_{2v}-$[Si_{14}O_{21}](OSiMe_3)_{14}$," *Journal of the American Chemical Society*, vol. 109, no. 18, pp. 5554–5556, 1987.

[13] C. L. Frye and W. T. Collins, "The oligomeric silsesquioxanes, $(HSiO_{3/2})_n$," *Journal of the American Chemical Society*, vol. 92, no. 19, pp. 5586–5588, 1970.

[14] R. J. J. Williams, R. Erra-Balsells, Y. Ishikawa, H. Nonami, A. N. Mauri, and C. C. Riccardi, "UV-MALDI-TOF and ESI-TOF mass spectrometry characterization of silsesquioxanes obtained by the hydrolytic condensation of (3-glycidoxypropyl)-trimethoxysilane in an epoxidized solvent," *Macromolecular Chemistry and Physics*, vol. 202, no. 11, pp. 2425–2433, 2001.

[15] L. M. Bronstein, C. N. Linton, R. Karlinsey et al., "Controlled synthesis of novel metalated poly(aminohexyl)-(aminopropyl)silsesquioxane colloids," *Langmuir*, vol. 19, no. 17, pp. 7071–7083, 2003.

[16] C. Ma, I. Taniguchi, M. Miyamoto, and Y. Kimura, "Formation of stable nanoparticles of poly(phenyl/methylsilsesquioxane) in aqueous solution," *Polymer Journal*, vol. 35, no. 3, pp. 270–275, 2003.

[17] C. Ma and Y. Kimura, "Preparation of nano-particles of poly(phenylsilsesquioxane)s by emulsion polycondensation of phenylsilanetriol formed in aqueous solution," *Polymer Journal*, vol. 34, no. 9, pp. 709–713, 2002.

[18] N. Jungmann, M. Schmidt, and M. Maskos, "Characterization of polyorganosiloxane nanoparticles in aqueous dispersion by asymmetrical flow field-flow fractionation," *Macromolecules*, vol. 34, no. 23, pp. 8347–8353, 2001.

[19] P. Maitra and S. L. Wunder, "Oligomeric poly(ethylene oxide)-functionalized silsesquioxanes: interfacial effects on Tg, Tm, and δHm," *Chemistry of Materials*, vol. 14, no. 11, pp. 4494–4497, 2002.

[20] R. Knischka, F. Dietsche, R. Hanselmann, H. Frey, R. Mülhaupt, and P. J. Lutz, "Silsesquioxane-based amphiphiles," *Langmuir*, vol. 15, no. 14, pp. 4752–4756, 1999.

[21] K. Y. Mya, X. Li, L. Chen, X. Ni, J. Li, and C. He, "Core-corona structure of cubic silsesquioxane-poly(ethylene oxide) in aqueous solution: fluorescence, light scattering, and TEM studies," *Journal of Physical Chemistry B*, vol. 109, no. 19, pp. 9455–9462, 2005.

[22] G. Wu and Z. Su, "Polyhedral oligomeric silsesquioxane nanocomposite thin films via layer-by-layer electrostatic self-assembly," *Chemistry of Materials*, vol. 18, no. 16, pp. 3726–3732, 2006.

[23] B. L. Frankamp, N. O. Fischer, R. Hong, S. Srivastava, and V. M. Rotello, "Surface modification using cubic silsesquioxane ligands. facile synthesis of water-soluble metal oxide nanoparticles," *Chemistry of Materials*, vol. 18, no. 4, pp. 956–959, 2006.

[24] H. Mori, M. G. Lanzendörfer, A. H. E. Müller, and J. E. Klee, "Silsesquioxane-based nanoparticles formed via hydrolytic condensation of organotriethoxysilane containing hydroxy groups," *Macromolecules*, vol. 37, no. 14, pp. 5228–5238, 2004.

[25] H. Mori, A. H. E. Müller, and J. E. Klee, "Intelligent colloidal hybrids via reversible pH-induced complexation of polyelectrolyte and silica nanoparticles," *Journal of the American Chemical Society*, vol. 125, no. 13, pp. 3712–3713, 2003.

[26] S. Muthukrishnan, F. Plamper, H. Mori, and A. H. E. Müller, "Synthesis and characterization of glycomethacrylate hybrid stars from silsesquioxane nanoparticles," *Macromolecules*, vol. 38, no. 26, pp. 10631–10642, 2005.

[27] J. Xu and W. Shi, "Synthesis and crystallization kinetics of silsesquioxane-based hybrid star poly(εlunate-caprolactone)," *Polymer*, vol. 47, no. 14, pp. 5161–5173, 2006.

[28] R. Gunawidjaja, F. Huang, M. Gumenna et al., "Bulk and surface assembly of branched amphiphilic polyhedral oligomer silsesquioxane compounds," *Langmuir*, vol. 25, no. 2, pp. 1196–1209, 2009.

[29] M. Schumacher, M. Ruppel, J. Yuan et al., "Smart organic-inorganic nanohybrids based on amphiphilic block copolymer micelles and functional silsesquioxane nanoparticles," *Langmuir*, vol. 25, no. 6, pp. 3407–3417, 2009.

[30] V. N. Bliznyuk, T. A. Tereshchenko, M. A. Gumenna et al., "Structure of segmented poly(ether urethane)s containing amino and hydroxyl functionalized polyhedral oligomeric silsesquioxanes (POSS)," *Polymer*, vol. 49, no. 9, pp. 2298–2305, 2008.

[31] D. P. Fasce, R. J. J. Williams, F. Méchin, J. P. Pascault, M. F. Llauro, and R. Pétiaud, "Synthesis and characterization of polyhedral silsesquioxanes bearing bulky functionalized substituents," *Macromolecules*, vol. 32, no. 15, pp. 4757–4763, 1999.

[32] D. P. Fasce, R. J. J. Williams, R. Erra-Balsells, Y. Ishikawa, and H. Nonami, "One-step synthesis of polyhedral silsesquioxanes bearing bulky substituents: UV-MALDI-TOF and ESI-TOF mass spectrometry characterization of reaction products," *Macromolecules*, vol. 34, no. 11, pp. 3534–3539, 2001.

[33] I. E. dell'Erba, D. P. Fasce, R. J. J. Williams, R. Erra-Balsells, Y. Fukuyama, and H. Nonami, "Poly(silsesquioxanes) derived from the hydrolytic condensation of organotrialkoxysilanes containing hydroxyl groups," *Journal of Organometallic Chemistry*, vol. 686, no. 1-2, pp. 42–51, 2003.

[34] D. P. Fasce, I. E. Dell'Erba, and R. J. J. Williams, "Synthesis of a soluble functionalized-silica by the hydrolysis and condensation of organotrialkoxysilanes bearing (β-hydroxy) tertiary amine groups with tetraethoxysilane," *Polymer*, vol. 46, no. 17, pp. 6649–6656, 2005.

[35] H. Mori, Y. Miyamura, and T. Endo, "Synthesis and characterization of water-soluble silsesquioxane-based nanoparticles by hydrolytic condensation of triethoxysilane derived from 2-hydroxyethyl acrylate," *Langmuir*, vol. 23, no. 17, pp. 9014–9023, 2007.

[36] Z. Deng, E. Breval, and C. G. Pantano, "Colloidal sol/gel processing of ultra-low expansion TiO_2/SiO_2 glasses," *Journal of Non-Crystalline Solids*, vol. 100, no. 1–3, pp. 364–370, 1988.

[37] S. Kirtay, E. Oktay, and V. Gunay, "Glass strengthening by SiO_2-TiO_2 organically modified silica coating," *Thin Solid Films*, vol. 515, no. 4, pp. 2145–2152, 2006.

[38] M. Houmard, D. Riassetto, F. Roussel et al., "Morphology and natural wettability properties of sol-gel derived TiO_2-SiO_2 composite thin films," *Applied Surface Science*, vol. 254, no. 5, pp. 1405–1414, 2007.

[39] S. Permpoon, M. Houmard, D. Riassetto et al., "Natural and persistent superhydrophilicity of SiO_2/TiO_2 and TiO_2/SiO_2 bi-layer films," *Thin Solid Films*, vol. 516, no. 6, pp. 957–966, 2008.

[40] V. Ääritalo, S. Areva, M. Jokinen, M. Lindén, and T. Peltola, "Sol-gel-derived TiO_2-SiO_2 implant coatings for direct tissue attachment. Part I: Design, preparation and

characterization," *Journal of Materials Science*, vol. 18, no. 9, pp. 1863–1873, 2007.

[41] S. Areva, V. Ääritalo, S. Tuusa, M. Jokinen, M. Lindén, and T. Peltola, "Sol-Gel-derived TiO_2-SiO_2 implant coatings for direct tissue attachment. Part II: Evaluation of cell response," *Journal of Materials Science*, vol. 18, no. 8, pp. 1633–1642, 2007.

[42] D. K. Sarkar, D. Brassard, M. A. E. Khakani, and L. Ouellet, "Dielectric properties of sol-gel derived high-k titanium silicate thin films," *Thin Solid Films*, vol. 515, no. 11, pp. 4788–4793, 2007.

[43] M. Zaharescu, A. Barau, L. Predoana et al., "TiO_2-SiO_2 sol-gel hybrid films and their sensitivity to gaseous toluene," *Journal of Non-Crystalline Solids*, vol. 354, no. 2–9, pp. 693–699, 2008.

[44] A. Łukowiak, R. Dylewicz, S. Patela, W. Stręk, and K. Maruszewski, "Optical properties of SiO_2-TiO_2 thin film waveguides obtained by the sol-gel method and their applications for sensing purposes," *Optical Materials*, vol. 27, no. 9, pp. 1501–1505, 2005.

[45] X. Shu, Y. Chen, H. Yuan, S. Gao, and D. Xiao, "H_2O_2 sensor based on the room-temperature phosphorescence of nano TiO_2/SiO_2 composite," *Analytical Chemistry*, vol. 79, no. 10, pp. 3695–3702, 2007.

[46] S. Y. Lien, D. S. Wuu, W. C. Yeh, and J. C. Liu, "Tri-layer antireflection coatings (SiO_2/SiO_2-TiO_2/TiO_2) for silicon solar cells using a sol-gel technique," *Solar Energy Materials and Solar Cells*, vol. 90, no. 16, pp. 2710–2719, 2006.

[47] W. Chen, X. Tao, J. Zhang, Q. Fang, and J. Yang, "Sol-gel-processed SiO_2/TiO_2/methylcellulose composite materials for optical waveguides," *Journal of the American Ceramic Society*, vol. 88, no. 11, pp. 2998–3002, 2005.

[48] N. D. Afify, R. Grisenti, G. Dalba et al., "Short-range order around Er^{3+} in silica waveguides containing aluminium, titanium and hafnium," *Optical Materials*, vol. 28, no. 6-7, pp. 864–867, 2006.

[49] A. C. Marques and R. M. Almeida, "Raman spectra and structure of multicomponent oxide planar waveguides prepared by sol-gel," *Journal of Sol-Gel Science and Technology*, vol. 40, no. 2-3, pp. 371–378, 2006.

[50] W. Dong, Y. Sun, C. W. Lee et al., "Controllable and repeatable synthesis of thermally stable anatase nanocrystal-silica composites with highly ordered hexagonal mesostructures," *Journal of the American Chemical Society*, vol. 129, no. 45, pp. 13894–13904, 2007.

[51] A. O. Bouh, G. L. Rice, and S. L. Scott, "Mono- and dinuclear silica-supported titanium(IV) complexes and the effect of TiOTi connectivity on reactivity," *Journal of the American Chemical Society*, vol. 121, no. 31, pp. 7201–7210, 1999.

[52] J. Jarupatrakorn and T. D. Tilley, "Silica-supported, single-site titanium catalysts for olefin epoxidation. A molecular precursor strategy for control of catalyst structure," *Journal of the American Chemical Society*, vol. 124, no. 28, pp. 8380–8388, 2002.

[53] M. Fujiwara, H. Wessel, H. S. Park, and H. W. Roesky, "A sol-gel method using tetraethoxysilane and acetic anhydride: immobilization of cubic μ-oxo Si-Ti complex in a silica matrix," *Chemistry of Materials*, vol. 14, no. 12, pp. 4975–4981, 2002.

[54] A. R. Oki, Q. Xu, B. Shpeizer et al., "Synthesis, characterization and activity in cyclohexene epoxidation of mesoporous TiO_2-SiO_2 mixed oxides," *Catalysis Communications*, vol. 8, no. 6, pp. 950–956, 2007.

[55] A. Matsuda, Y. Higashi, K. Tadanaga, and M. Tatsumisago, "Hot-water treatment of sol-gel derived SiO_2-TiO_2 microparticles and application to electrophoretic deposition for thick films," *Journal of Materials Science*, vol. 41, no. 24, pp. 8101–8108, 2006.

[56] J. W. Lee, S. Kong, W. S. Kim, and J. Kim, "Preparation and characterization of SiO_2/TiO_2 core-shell particles with controlled shell thickness," *Materials Chemistry and Physics*, vol. 106, no. 1, pp. 39–44, 2007.

[57] S. W. Lee, Y. U. Kim, S. S. Choi, T. Y. Park, Y. L. Joo, and S. G. Lee, "Preparation of SiO_2/TiO_2 composite fibers by sol-gel reaction and electrospinning," *Materials Letters*, vol. 61, no. 3, pp. 889–893, 2007.

[58] S. Zhan, D. Chen, X. Jiao, and Y. Song, "Mesoporous TiO_2/SiO_2 composite nanofibers with selective photocatalytic properties," *Chemical Communications*, no. 20, pp. 2043–2045, 2007.

[59] J. Jiao, Q. Xu, and L. Li, "Porous TiO_2/SiO_2 composite prepared using PEG as template direction reagent with assistance of supercritical CO_2," *Journal of Colloid and Interface Science*, vol. 316, no. 2, pp. 596–603, 2007.

[60] H. Mori, Y. Miyamura, and T. Endo, "Synthesis and characterization of water-soluble $SiO_{1.5}$/TiO_2 hybrid nanoparticles by hydrolytic co-condensation of triethoxysilane containing hydroxyl groups," *Materials Chemistry and Physics*, vol. 115, no. 1, pp. 287–295, 2009.

[61] C. McCusker, J. B. Carroll, and V. M. Rotelo, "Cationic polyhedral oligomeric silsesquioxane (POSS) units as carriers for drug delivery processes," *Chemical Communications*, no. 8, pp. 996–998, 2005.

[62] L. Cui, D. Chen, and L. Zhu, "Conformation transformation determined by different self-assembled phases in a DNA complex with cationic polyhedral oligomeric silsesquioxane lipid," *ACS Nano*, vol. 2, no. 5, pp. 921–927, 2008.

[63] Q. C. Zou, Q. J. Yan, G. W. Song, S. L. Zhang, and L. M. Wu, "Detection of DNA using cationic polyhedral oligomeric silsesquioxane nanoparticles as the probe by resonance light scattering technique," *Biosensors and Bioelectronics*, vol. 22, no. 7, pp. 1461–1465, 2007.

[64] K. Y. Pu, K. Li, and B. Liu, "Cationic oligofluorene-substituted polyhedral oligomeric silsesquioxane as light-harvesting unimolecular nanoparticle for fluorescence amplification in cellular imaging," *Advanced Materials*, vol. 22, no. 5, pp. 643–646, 2010.

[65] F. Zhao, C. Wan, X. Bao, and B. Kandasubramanian, "Modification of montmorillonite with aminopropylisooctyl polyhedral oligomeric silsequioxane," *Journal of Colloid and Interface Science*, vol. 333, no. 1, pp. 164–170, 2009.

[66] Y. C. Li, S. Mannen, J. Schulz, and J. C. Grunlan, "Growth and fire protection behavior of POSS-based multilayer thin films," *Journal of Materials Chemistry*, vol. 21, no. 9, pp. 3060–3069, 2011.

[67] H. Mori and M. Yamada, "Synthesis and characterization of cationic silsesquioxane hybrids by hydrolytic condensation of triethoxysilane derived from 2-(dimethylamino)ethylacrylate," *Colloid and Polymer Science*. In press.

[68] A. Tuteja, W. Choi, M. Ma et al., "Designing superoleophobic surfaces," *Science*, vol. 318, no. 5856, pp. 1618–1622, 2007.

[69] J. M. Mabry, A. Vij, S. T. Iacono, and B. D. Viers, "Fluorinated polyhedral oligomeric silsesquioxanes (F-POSS)," *Angewandte Chemie*, vol. 47, no. 22, pp. 4137–4140, 2008.

[70] S. T. Iacono, A. Vij, W. Grabow, D. W. Smith Jr., and J. M. Mabry, "Facile synthesis of hydrophobic fluoroalkyl

functionalized silsesquioxane nanostructures," *Chemical Communications*, no. 47, pp. 4992–4994, 2007.

[71] K. Koh, S. Sugiyama, T. Morinaga et al., "Precision synthesis of a fluorinated polyhedral oligomeric silsesquioxane-terminated polymer and surface characterization of its blend film with poly(methyl methacrylate)," *Macromolecules*, vol. 38, no. 4, pp. 1264–1270, 2005.

[72] Y. S. Ye, Y. C. Yen, W. Y. Chen, C. C. Cheng, and F. C. Chang, "A simple approach toward low-dielectric polyimide nanocomposites: blending the polyimide precursor with a fluorinated polyhedral oligomeric silsesquioxane," *Journal of Polymer Science A*, vol. 46, no. 18, pp. 6296–6304, 2008.

[73] I. Jerman, M. Koželj, and B. Orel, "The effect of polyhedral oligomeric silsesquioxane dispersant and low surface energy additives on spectrally selective paint coatings with self-cleaning properties," *Solar Energy Materials and Solar Cells*, vol. 94, no. 2, pp. 232–245, 2010.

[74] I. Jerman, B. Orel, A. S. Vuk, M. Koželj, and J. Kovač, "A structural and corrosion study of triethoxysilyl and perfluorooctyl functionalized polyhedral silsesquioxane nanocomposite films on AA 2024 alloy," *Thin Solid Films*, vol. 518, no. 10, pp. 2710–2721, 2010.

[75] S. T. Iacono, S. M. Budy, J. M. Mabry, and D. W. Smith Jr., "Synthesis, characterization, and surface morphology of pendant polyhedral oligomeric silsesquioxane perfluorocyclobutyl aryl ether copolymers," *Macromolecules*, vol. 40, no. 26, pp. 9517–9522, 2007.

[76] H. Hussain, B. H. Tan, C. S. Gudipati et al., "Synthesis and characterization of organic/inorganic hybrid star polymers of 2,2,3,4,4,4-hexafluorobutyl methacrylate and octa(aminophenyl)silsesquioxane nano-cage made via atom transfer radical polymerization," *Journal of Polymer Science A*, vol. 46, no. 22, pp. 7287–7298, 2008.

[77] H. Hussain, B. H. Tan, K. Y. Mya, Y. Liu, C. B. He, and T. P. Davis, "Synthesis, micelle formation and bulk properties of poly(ethylene glycol)b-poly(pentafluorostyrene)-g-polyhedral oligomeric silsesquioxane amphiphilic hybrid copolymers," *Journal of Polymer Science A*, vol. 48, no. 1, pp. 152–163, 2010.

[78] A. G. Kannan, N. R. Choudhury, and N. Dutta, "Fluoro-silsesquioxane-urethane hybrid for thin film applications," *ACS Applied Materials & Interfaces*, vol. 1, no. 2, pp. 336–347, 2009.

[79] H. Fujiwara, T. Narita, and H. Hamana, "Novel fluorinated hybrid polymer preparation from silsesquioxanes by radical polyaddition," *Journal of Fluorine Chemistry*, vol. 125, no. 9, pp. 1279–1285, 2004.

[80] S. T. Iacono, S. M. Budy, D. W. Smith, and J. M. Mabry, "Preparation of composite fluoropolymers with enhanced dewetting using fluorinated silsesquioxanes as drop-in modifiers," *Journal of Materials Chemistry*, vol. 20, no. 15, pp. 2979–2984, 2010.

[81] R. Misra, R. D. Cook, and S. E. Morgan, "Nonwetting, nonrolling, stain resistant polyhedral oligomeric silsesquioxane coated textiles," *Journal of Applied Polymer Science*, vol. 115, no. 4, pp. 2322–2331, 2010.

[82] K. Zeng and S. Zheng, "Nanostructures and surface dewettability of epoxy thermosets containing hepta(3,3,3-trifluoropropyl) polyhedral oligomeric silsesquioxane-capped polyethylene oxide," *Journal of Physical Chemistry B*, vol. 111, no. 50, pp. 13919–13928, 2007.

[83] A. M. Douvas, F. Van Roey, M. Goethals et al., "Partially fluorinated, polyhedral oligomeric silsesquioxane-functionalized (meth)acrylate resists for 193 nm bilayer lithography," *Chemistry of Materials*, vol. 18, no. 17, pp. 4040–4048, 2006.

[84] C. Pina-Hernandez, P. F. Fu, and L. J. Guo, "Easy duplication of stamps using UV-cured fluoro-silsesquioxane for nanoimprint lithography," *Journal of Vacuum Science and Technology B*, vol. 26, no. 6, pp. 2426–2429, 2008.

[85] H. Mori, C. Sada, T. Konno, and K. Yonetake, "Synthesis and characterization of low-refractive-index fluorinated silsesquioxane-based hybrids," *Polymer*, vol. 52, pp. 5452–5463, 2011.

[86] Y. C. Sheen, C. H. Lu, C. F. Huang, S. W. Kuo, and F. C. Chang, "Synthesis and characterization of amorphous octakis-functionalized polyhedral oligomeric silsesquioxanes for polymer nanocomposites," *Polymer*, vol. 49, no. 18, pp. 4017–4024, 2008.

[87] R. M. Laine and M. F. Roll, "Polyhedral phenylsilsesquioxanes," *Macromolecules*, vol. 44, no. 5, pp. 1073–1109, 2011.

[88] C. Liu, Y. Liu, Z. Shen, P. Xie, D. Dai, R. Zhang et al., "Synthesis and characterization of novel alcohol-soluble ladderlike poly(silsesquioxane)s containing side-chain hydroxy groups," *Macromolecular Chemistry and Physics*, vol. 202, pp. 1576–1580, 2001.

[89] K. Zeng, L. Wang, S. Zheng, and X. Qian, "Self-assembly behavior of hepta(3,3,3-trifluoropropyl) polyhedral oligomeric silsesquioxane-capped poly(ε-caprolactone) in epoxy resin: nanostructures and surface properties," *Polymer*, vol. 50, no. 2, pp. 685–695, 2009.

[90] K. Zeng, Y. Liu, and S. Zheng, "Poly(ethylene imine) hybrids containing polyhedral oligomeric silsesquioxanes: preparation, structure and properties," *European Polymer Journal*, vol. 44, no. 12, pp. 3946–3956, 2008.

[91] K. Zeng and S. Zheng, "Synthesis and characterization of organic/inorganic Polyrotaxanes from Polyhedral oligomeric silsesquioxane and poly(ethylene oxide)/α- cyclodextrin polypseudorotaxanes via click chemistry," *Macromolecular Chemistry and Physics*, vol. 210, no. 9, pp. 783–791, 2009.

[92] H. Mori, C. Sada, T. Konno, R. Koizumi, and K. Yonetake, "Film-forming amphiphilic silsesquioxane hybrids prepared by hydrolytic co-condensation of hydroxyl-functionalized and fluorinated triethoxysilanes," *Polymer*, vol. 53, pp. 3849–3860, 2012.

[93] J. Rodríguez-Hernández, F. Chécot, Y. Gnanou, and S. Lecommandoux, "Toward 'smart' nano-objects by self-assembly of block copolymers in solution," *Progress in Polymer Science*, vol. 30, no. 7, pp. 691–724, 2005.

[94] H. Mori and A. H. E. Müller, "New polymeric architectures with (meth)acrylic acid segments," *Progress in Polymer Science*, vol. 28, no. 10, pp. 1403–1439, 2003.

[95] S. Förster, V. Abetz, and A. H. E. Müller, "Polyelectrolyte block copolymer micelles," *Advances in Polymer Science*, vol. 166, pp. 173–210, 2004.

[96] A. O. Moughton and R. K. O'Reilly, "Using metallo-supramolecular block copolymers for the synthesis of higher order nanostructured assemblies," *Macromolecular Rapid Communications*, vol. 31, no. 1, pp. 37–52, 2010.

[97] H. Mori, M. G. Lanzendörfer, A. H. E. Müller, and J. E. Klee, "Organic-inorganic nanoassembly based on complexation of cationic silica nanoparticles and weak anionic polyelectrolytes in aqueous and alcohol media," *Langmuir*, vol. 20, no. 5, pp. 1934–1944, 2004.

[98] A. K. Boal, F. Ilhan, J. E. Derouchey, T. Thurn-Albrecht, T. P. Russell, and V. M. Rotello, "Self-assembly of nanoparticles into structured spherical and network aggregates," *Nature*, vol. 404, no. 6779, pp. 746–748, 2000.

[99] B. L. Frankamp, O. Uzun, F. Ilhan, A. K. Boal, and V. M. Rotello, "Recognition-mediated assembly of nanoparticles into micellar structures with diblock copolymers," *Journal of the American Chemical Society*, vol. 124, no. 6, pp. 892–893, 2002.

[100] M. Hara, J. T. Lean, and T. E. Mallouk, "Photocatalytic oxidation of water by silica-supported tris(4,4′-dialkyl-2,2′-bipyridyl)ruthenium polymeric sensitizers and colloidal iridium oxide," *Chemistry of Materials*, vol. 13, no. 12, pp. 4668–4675, 2001.

[101] R. Tamaki and Y. Chujo, "Synthesis of polystyrene and silica gel polymer hybrids utilizing ionic interactions," *Chemistry of Materials*, vol. 11, no. 7, pp. 1719–1726, 1999.

[102] A. R. Bassindale, A. Codina-Barrios, N. Frascione, and P. G. Taylor, "The use of silsesquioxane cages and phage display technology to probe silicone-protein interactions," *New Journal of Chemistry*, vol. 32, no. 2, pp. 240–246, 2008.

[103] N. Alobaid, H. J. Salacinski, K. M. Sales et al., "Nanocomposite containing bioactive peptides promote endothelialisation by circulating progenitor cells: an in vitro evaluation," *European Journal of Vascular and Endovascular Surgery*, vol. 32, no. 1, pp. 76–83, 2006.

[104] A. De Mel, G. Punshon, B. Ramesh et al., "In situ endothelialisation potential of a biofunctionalised nanocomposite biomaterial-based small diameter bypass graft," *Bio-Medical Materials and Engineering*, vol. 19, no. 4-5, pp. 317–331, 2009.

[105] T. L. Kaneshiro, X. Wang, and Z. R. Lu, "Synthesis, characterization, and gene delivery of poly-L-iysine octa(3-aminopropyl)silsesquioxane dendrimers: nanoglobular drug carriers with precisely defined molecular architectures," *Molecular Pharmaceutics*, vol. 4, no. 5, pp. 759–768, 2007.

[106] T. L. Kaneshiro and Z. R. Lu, "Targeted intracellular codelivery of chemotherapeutics and nucleic acid with a well-defined dendrimer-based nanoglobular carrier," *Biomaterials*, vol. 30, no. 29, pp. 5660–5666, 2009.

[107] S. W. Kuo, H. F. Lee, W. J. Huang, K. U. Jeong, and F. C. Chang, "Solid state and solution self-assembly of helical polypeptides tethered to polyhedral oligomeric silsesquioxanes," *Macromolecules*, vol. 42, no. 5, pp. 1619–1626, 2009.

[108] S. W. Kuo and H. T. Tsai, "Control of peptide secondary structure on star shape polypeptides tethered to polyhedral oligomeric silsesquioxane nanoparticle through click chemistry," *Polymer*, vol. 51, no. 24, pp. 5695–5704, 2010.

[109] S. Fabritz, D. Heyl, V. Bagutski et al., "Towards click bioconjugations on cube-octameric silsesquioxane scaffolds," *Organic & Biomolecular Chemistry*, vol. 8, no. 9, pp. 2212–2218, 2010.

[110] H. Mori and T. Endo, "Amino acid-based block copolymers by RAFT polymerization," *Macromolecular Rapid Communications*, vol. 33, pp. 1090–1107, 2012.

[111] H. Mori and S. Saito, "Smart organic-inorganic hybrids based on the complexation of amino acid-based polymers and water-soluble silsesquioxane nanoparticles," *Reactive and Functional Polymers*, vol. 71, no. 10, pp. 1023–1032, 2011.

[112] H. Mori, I. Kato, M. Matsuyama, and T. Endo, "RAFT polymerization of acrylamides containing proline and hydroxyproline moiety: controlled synthesis of water-soluble and thermoresponsive polymers," *Macromolecules*, vol. 41, no. 15, pp. 5604–5615, 2008.

[113] H. Mori, H. Iwaya, A. Nagai, and T. Endo, "Controlled synthesis of thermoresponsive polymers derived from L-proline via RAFT polymerization," *Chemical Communications*, no. 38, pp. 4872–4874, 2005.

[114] H. Mori, H. Iwaya, and T. Endo, "Controlled synthesis of thermoresponsive polymer via RAFT polymerization of an acrylamide containing l-proline moiety," *Reactive and Functional Polymers*, vol. 67, no. 10, pp. 916–927, 2007.

[115] H. Mori, H. Iwaya, and T. Endo, "Structures and chiroptical properties of thermoresponsive block copolymers containing L-proline moieties," *Macromolecular Chemistry and Physics*, vol. 208, no. 17, pp. 1908–1918, 2007.

[116] S. Luo, J. Xu, Y. Zhang, S. Liu, and C. Wu, "Double hydrophilic block copolymer monolayer protected hybrid gold nanoparticles and their shell cross-linking," *Journal of Physical Chemistry B*, vol. 109, no. 47, pp. 22159–22166, 2005.

[117] B. J. Kim, J. J. Chiu, G. R. Yi, D. J. Pine, and E. J. Kramer, "Nanoparticle-induced phase transitions in diblock-copolymer films," *Advanced Materials*, vol. 17, no. 21, pp. 2618–2622, 2005.

[118] Y. Kang and T. A. Taton, "Controlling shell thickness in core-shell gold nanoparticles via surface-templated adsorption of block copolymer surfactants," *Macromolecules*, vol. 38, no. 14, pp. 6115–6121, 2005.

[119] B. S. Kim and T. A. Taton, "Multicomponent nanoparticles via self-assembly with cross-linked block copolymer surfactants," *Langmuir*, vol. 23, no. 4, pp. 2198–2202, 2007.

[120] Y. Kang and T. A. Taton, "Core/shell gold nanoparticles by self-assembly and crosslinking of micellar, block-copolymer shells," *Angewandte Chemie*, vol. 44, no. 3, pp. 409–412, 2005.

[121] H. Mori, S. Saito, and K. Shoji, "Complexation of amino-acid-based block copolymers with dual-thermoresponsive properties and water-soluble silsesquioxane nanoparticles," *Macromolecular Chemistry and Physics*, vol. 212, pp. 2558–2572, 2011.

8

Scaffolds for Growth Factor Delivery as Applied to Bone Tissue Engineering

Keith A. Blackwood,[1] Nathalie Bock,[1,2,3] Tim R. Dargaville,[2] and Maria Ann Woodruff[1]

[1] *Biomaterials and Tissue Morphology Group, Institute of Health and Biomedical Innovation, Queensland University of Technology, 60 Musk Avenue, Kelvin Grove, QLD 4059, Australia*
[2] *Tissue Repair and Regeneration Program, Institute of Health and Biomedical Innovation, Queensland University of Technology, 60 Musk Avenue, Kelvin Grove, QLD 4059, Australia*
[3] *Regenerative Medicine Group, Institute of Health and Biomedical Innovation, Queensland University of Technology, 60 Musk Avenue, Kelvin Grove, QLD 4059, Australia*

Correspondence should be addressed to Maria Ann Woodruff, mia.woodruff@qut.edu.au

Academic Editor: Christopher Batich

There remains a substantial shortfall in the treatment of severe skeletal injuries. The current gold standard of autologous bone grafting from the same patient has many undesirable side effects associated such as donor site morbidity. Tissue engineering seeks to offer a solution to this problem. The primary requirements for tissue-engineered scaffolds have already been well established, and many materials, such as polyesters, present themselves as potential candidates for bone defects; they have comparable structural features, but they often lack the required osteoconductivity to promote adequate bone regeneration. By combining these materials with biological growth factors, which promote the infiltration of cells into the scaffold as well as the differentiation into the specific cell and tissue type, it is possible to increase the formation of new bone. However due to the cost and potential complications associated with growth factors, controlling the rate of release is an important design consideration when developing new bone tissue engineering strategies. This paper will cover recent research in the area of encapsulation and release of growth factors within a variety of different polymeric scaffolds.

1. Introduction

Skeletal injuries are some of the most debilitating types of injury known and can lead to severe long-term effects without surgical intervention. These injuries may occur via trauma, disease (degenerative diseases and tumour excisions), or growth defects (such as cleft palette). Cost of treatment in the US exceeded $17 billion in 2005 and is predicted to rise to $25 billion per year by 2025 [1]. Importantly, demographic data reveals that, due to the ageing population, complications associated with the musculoskeletal system will increase over the coming years [2]. It is therefore vital to have the best technologies available to orthopaedic surgeons to assist bone repair.

Over the last 30 years extensive work has been carried out on the development of biomaterial scaffolds for the promotion of bone growth and to assist in regenerating wounds that would otherwise not heal. Advancements have led to the development of composite scaffolds, consisting of materials that have good structural characteristics combined with materials which promote bone growth, and one group of molecules for this enhancement is growth factors (GFs).

"Growth factor" is a term for several families of intracellular signalling proteins which have a number of different defined roles, such as recruitment of cells, as well as the promotion of migration [3] and differentiation [4, 5] of these cells. Specific GFs are responsible for the formation, maturation, and maintenance of tissues throughout the body.

Several GF families play a role in the generation and repair of bone: some stimulate pathways specific for the formation of bone, such as promoting osteoinduction [6], and others such as vascular endothelial growth factor (VEGF) [3]

and platelet-derived growth factor (PDGF) promote angiogenesis, the sprouting of new blood vessels from existing ones.

While many GFs play an extensive role in the *de novo* formation of bone, in the field of tissue engineering there has been particular interest in using bone morphogenetic proteins (BMPs) for the treatment of bone defects [4, 7, 8]. BMPs are a family of GFs closely associated with the growth, maturation, and regulation of bone, and they are considered the most potent stimulator of new bone formation.

Currently, the most commonly used clinical technique using BMPs is the relatively straightforward method of combining reconstituted BMP with collagen scaffolds just prior to implantation [9, 10]. The scaffold and BMP are supplied separately, and the scaffold is soaked in the GF solution before implantation. This technique, while clinically accepted, exhibits poor GF release characteristics, eluting BMP rapidly and presenting the wound site with a BMP concentration far above those found in normal physiological conditions. Due to the short half-life of GFs *in situ*, the high concentrations are needed to maintain a therapeutical release profile. This technical shortfall not only increases the cost of the therapy, but also generates serious concerns over long-term complications associated with the high dosage of GFs [11]. It would be more favourable to have a smaller therapeutic dose delivered, at appropriate physiological levels over the time period of the wound-healing process.

Several techniques have been employed to control the release profiles of GFs from scaffolds. Many of these techniques involve encapsulating the GFs within degradable polymer networks, such that they are slowly released from the degrading scaffold into the wound site. By understanding the release kinetics, this process has the potential to maintain a therapeutic dose for longer than the currently available rapidly releasing scaffolds. Microspheres and hydrogels are the most commonly studied matrices for this type of encapsulation.

Delivery systems for GFs [12, 13] as well as other therapeutics [14] have already been reviewed several times over the last decade. Thus this paper will focus on providing an in-depth discussion on the most recent strategies and techniques being utilised for the production of GF-eluting scaffolds providing *sustained* release of GFs, thus moving towards providing a more physiologically relevant environment for the promotion of bone regeneration.

2. Basic Structure of Bone

The skeletal structure of mammals plays many important roles throughout the body, ranging from structural support, protection of organs, and the maintenance of homeostasis via the release and storage of calcium [15]. Human bone can be classified as two forms: cancellous and cortical bone. Cancellous bone, also known as spongy bone, is formed from a highly porous (50%–90%) 3D lattice and compromises approximately 20% of all bone found in the body. Cancellous bone is predominantly found in the ribs and the epiphysis of long bones. Cortical bone, also known as compact bone, is a much more dense (10% porosity) making up the remaining 80% of bone found in the human body. Cortical bone itself is surrounded by the periosteum, a thin fibrous membrane.

Bone has a remarkable ability to regenerate damaged regions back to full functionality with no scarring. However if the trauma is too large, defect healing cannot occur without surgical intervention, with the sheer size of the defect exceeding the bones' natural healing ability. This is often the case during complex fractures and large tumour excision [16]. Defects occurring in different regions of the body can lead to a number of different complications. For example, cranial defects do not naturally heal well and often require surgical intervention for satisfactory healing [17]. When dealing with a cranial defect, issues arise due to contour irregularities, but when dealing with spinal fractures another set of fundamental problems arise revolving around the need to provide more robust mechanical support, whilst maintaining movement [18].

Autologous bone grafting (transplant from the same patient) is the "gold standard" treatment for bone defects. Cancellous bone sourced from the iliac crest is usually used and has the potential for facilitating *de novo* bone formation due to the natural presence of GFs which promote angiogenesis and osteogenesis [19]. While effective, the use of autologous bone has severe limitations, mainly revolving around the need for a donor site from the same patient which causes a secondary wound site, in addition to limited available supply, and risk of infection [20]. In a recent review, Dimitriou et al. revealed that 19.4% of iliac crest donor sites have complications ranging from infection, fracture, nerve injuries, chronic donor site pain, hernias, and haematoma [21]. It should be noted that techniques are being developed to reduce the rate of complications with a reamer/irrigator/aspirator system developed by Synthes Inc. (West Chester, PA, USA) which has shown reduction in complications of the donor site down to >6%. However the total number of completed studies with this system is still small [21].

Allografting (transplant from a different patient) has been successfully used, predominantly in hip surgery where major femoral bone loss has occurred [22], but this technique has similar limitations to autografting, with difficulty in sourcing the grafting materials, as well as the potential for infection from donor to host.

Due to the shortcomings of the current clinical approaches, there remains an important niche to fill, such as, the field of tissue engineering has emerged to develop methods to bridge the existing gap in clinical treatments for severe bone loss.

3. Tissue Engineering of Bone

Tissue engineering is a multidisciplinary research field which combines aspects of conventional biology, engineering, medicine, and chemistry to create functional solutions for the treatment of clinical problems. Due to the aforementioned problems associated with allografting and autografting, there remains a need for an improved clinical solution for the treatment of bone trauma to improve on the current gold standards. Tissue-engineering strategies have thus emerged as alternative treatments, and most approaches involve the combination of a scaffold/matrix with living cells and/or biologically active molecules (biomolecules).

Tissue-engineered scaffolds often comprise polymers to provide structural support to the defect/wound site during repair. Polymers are often used due to ease of synthesis, adaptability to a wide number of structural conformations, biodegradability, and low associated immune response. Although both synthetic [23] and natural polymers [24–26] have been used in the development of tissue-engineered solutions for bone healing, using synthetic matrices is advantageous; it avoids the problems associated with grafts and allows for tailoring of the physicochemical properties required for a particular application. The main problem associated with these synthetic matrices, however, is their lack of biological recognition, since the matrix is often "tissue-conductive" and not "tissue-inductive," consequently impairing complete morphogenesis inside the matrix [27], which is why the incorporation of growth factors in matrices has been proposed.

Despite extensive research into the field of bone regeneration over the last 30 years, there still remains a requirement for an efficient tissue-engineered substitute for autologous bone grafting, one which provides both an adequate environment for influx of tissue growth and maturation of bone formation, whilst concomitantly providing structural support as these processes take place.

Polyesters are commonly used in tissue engineering. These plastics provide structural support in a similar manner to commonly used metallic implants. Unlike steel or titanium, these polymers are bioresorbable [28, 29], breaking down into soluble oligomeric chains, and eventually principle monomers, which are biologically metabolised via the citric acid cycle. This gives the potential to produce a scaffold which will be mechanically robust for a certain time period, gradually breaking down at a rate whereby it can be replaced with newly formed bone. The most commonly used polyesters are polylactide derived. Polylactide (PLA) is a polymer of the chimerical lactide monomer which has a well-established history in the field of tissue engineering spanning back to the 1960s [30]. Polylactide glycolide (PLGA) is an FDA-approved polymer and has a long history of use as a biomaterial owing to its biodegradability profile on a timescale commensurate with tissue remodelling and nontoxic degradation products. It is often used in preference to pure PLLA, P(D,L)LA, and PGA because there is a greater degree of control of the degradation rate by altering the ratio of glycolide to lactide, with increased ratio of glycolide increasing the speed of breakdown. Several studies have investigated the biocompatibility and degradation rates of PLGAs both *in vitro* and *in vivo* with good results [29].

Many types of polyesters are FDA approved and have seen clinical use for treatment of bone defects/wounds although they were initially introduced to the market in the form of resorbable sutures. For example Lactosorb, a lactide-rich PLGA cranial plate substitute to titanium implants [31] has had over 50,000 surgical procedures to date (source Lactosorb website). Lactosorb is just one of a number of biodegradable plate/screw systems in current clinical use for maxillofacial applications, other examples include Inion [31], MacroPore [32], Biofix [33], Resorb X [34], and Rapisorb [31].

These polymers offer potentially superior clinical outcomes to conventional materials, as they degrade within a time frame of 12 to 24 months, a rate dependent upon the specific composition of the polyester. However while these polymers are generally good as plates and rods, they lack any osteoinductive capacity to make them suitable for bone replacement. Current common materials for promoting osteoinduction are based on hydroxyapatite (HA) and β-tri-calcium-phosphate (TCP)—see Kamitakahara et al. for a review on the area [35]. These products, ceramic in nature, are quite brittle when used alone. Compared to polymeric materials such as polyesters ceramics lack structural robustness but promote bone formation. As such combining osteoinductive materials with structurally supportive polymers is a popular avenue of research.

Composite scaffolds, consisting of a structural support phase and a second phase for promoting cellular infiltration and differentiation, offer the potential to facilitate more rapid wound healing. Several different composite scaffolds are currently being investigated. Examples include the previously mentioned combination of TCP and HA to promote mineralisation with a structural support scaffold [36]. Another common composite scaffold approach is the addition of GFs, signalling molecules found in natural bone and wound sites. This is the focus of this paper.

In the field of tissue-engineering bone substitutes, the most commonly used growth factors are BMPs, specifically BMP-2 and BMP-7. It should, however, be noted that BMPs are not bony tissue specific [37]—all mammalian cells have receptors for BMPs, and the effects of BMP stimulation on a large number of cells types are unknown. Therefore there is need for caution when delivering BMPs to defect/wound sites to ensure that the dosage is not adversely affecting areas away from the trauma site, as BMP is a potent inducer of new bone and has been shown to stimulate ectopic bone formation (outside the expected area of bone) [11].

BMPs are a dimeric subset of cytokines from the TGF-β superfamily [39]. It has been established that BMPs have a wide variety of different signalling roles, most prominently in *de novo* bone formation [40] but are also involved in cartilage [41] and neural crest formation [42]. More than 20 different BMPs have been identified, but, in the field of tissue engineering, BMP-2, usually made by recombinant human technology of mammalian cell lines, most frequently derived from Chinese hamster ovaries, is by far the most commonly studied and is commercially available for surgical use on a collagen sponge carrier [43] for a variety of different bone-related applications. BMP-2 is known to promote the differentiation of cells into the osteoblast lineage and upregulate the calcification of bone defects thereby increasing bone formation [4, 39].

Current scaffolds for BMP delivery are either based upon loaded collagen sponges or direct injection into wound sites, a technique which has been reviewed by Geiger and coworkers in 2003 [24]. The two most common commercial products have been available for about a decade now for lumbar repair are Medtronic's Infuse (a BMP-2) and Olympus Biosciences OP1 (BMP-7), both are strikingly similar products which use a collagen sponge into which a BMP solution is soaked into prior to implantation. BMP-7 (OP-1) was developed as a commercial product by Stryker but is now licensed by Olympus Biosciences, and it comprises a mix of OP-1 in a purified bovine collagen type I forming a putty which can be used for nonunion bone injuries and spinal fusion. The kit comprises BMP-2, a collagen sponge and a surrounding support structure. Recently both companies have released products for the treatment of critical-sized long-bone defects.

The fundamental principles for the use of BMPs in bone regeneration and repair have recently been outlined in a review in 2009 by Haidar et al. in terms of the current challenges faced [44]. BMPs play a central role in most bone regeneration strategies [45] and are involved in the entire complex cascade of bone formation, from migration of mesenchymal stem cells to differentiation into osteoblasts. Tissue-engineering strategies for supplying BMPs to wound sites revolve around delivering the GF in gels or solid scaffolds in a ceramic, polymeric (natural or synthetic) or composite carrier. Excess quantities of BMP can cause a phenomena whereby bone grows beyond its conventional boundaries (ectopic bone formation) and can cause serious long-term complications, especially in the treatment of the spinal cord [11]. Due to this danger, controlling the levels of BMP in an implant and the rate of release of BMP from the implant is a critical step.

Collagen sponges loaded with GFs such as BMP-2 and BMP-7 exhibit poor release kinetics of adsorbed molecules making them inefficient [24, 46]. These scaffolds have a high initial burst release of GF, leading to a poor long-term sustained release, and use a very simple method of incorporating BMPs into the scaffold, namely, pH-induced anionic binding. This is a very inefficient delivery system as only a small percentage of the BMP which is released interacts with the appropriate cells. The remainder is simply washed away from the defect site via the circulation or becomes inactive after a certain period of time in contact with physiological fluid.

To summarise, the current FDA-approved systems only involve the delivery of recombinant BMPs (rhBMPs) from a simple reconstituted collagen matrix (OP-1 and INFUSE) and raise safety and cost concerns owing to supraphysiological doses present in these products. By using a degradable polymeric system to encapsulate BMPs, we can potentially overcome this issue by delivering overall smaller, more effective doses over longer time periods, preventing high doses leaking from the wound site and therefore targeting only the tissue of interest.

4. Desirable Characteristics of Growth Factor-Eluting Scaffolds

Because natural bone has a number of different roles within the body, scaffolds for bone tissue engineering require a number of different characteristics in order to effectively function as a scaffold in a bone environment. For several years the basic desirable characteristics for biomaterial scaffolds have been well defined [24] and are summarised in Table 1. All tissue-engineering scaffolds should exhibit a low immunogenic and antigenic response, while still facilitating the rapid infiltration of the host cells throughout the margins of the scaffold to the centre. When loading GFs into a scaffold, both high encapsulation efficiency as well as a release rate allowing a sustained therapeutic dose is also required. Additionally these scaffolds should break down into nonharmful products at a rate where host tissue can successfully take over the mechanical features.

5. Techniques for Assessing the Potential of Growth Factor-Eluting Scaffolds

It is important to develop a fundamental understanding of how a GF may be loaded and released from a scaffold prior to further in detail studies. Encapsulation efficiency (the actual amount of GF incorporated into the scaffold compared to the amount of GF used) plays a key role in GF delivery strategies, and this is driven by their high cost.

The controlled release of GFs from implanted scaffolds is one of the most important fields of study within the context of GF therapies. Release kinetics can be studied in a variety of different ways. The release of the GF is frequently measured by enzyme-linked immunosorbent assay (ELISA). This method has the advantage of being simple to setup, provided that the level of released GF is readily detectable. Other techniques include measuring autofluorescence of the GF [47] or measuring fluorescently-tagged GFs [48, 49]. Another advantage of release kinetic experiments is that cheaper analogues of growth factors can be used as a model system. For example bovine serum albumin has often been used as a substitute to model release of proteins from scaffolds [49–51], as has soya bean trypsin inhibitor [52].

Beyond basic release kinetics, *in vitro* cellular models allow for more detailed study of GF release profiles. They enable us to determine bioactivity of the released GF and to gain insight into appropriate therapeutic dosage of the GF by the inclusion and monitoring of relevant cells into the system in response to released GF. This gives the ability to evaluate any potential damage to the GF which may have occurred during the formation of the scaffold as we can assume that the GF is damaged if there is no measurable cell response [53].

Conventional bioactivity assays such as the (3-(4,5-dimethylthiazol-2-yl)-2,5-diphenyltetrazolium bromide, a yellow tetrazole), (MTT) [50, 54, 55], (3-(4,5-dimethylthiazol-2-yl)-5-(3-carboxymethoxyphenyl)-2-(4-sulfophenyl)-2H-tetrazolium), (MTS) [56], and Alamar blue [57], assays which all measure the metabolic activity of a cell population, have been commonly used to assess the ability for cells to be cultured in the presence of GF-loaded scaffolds, thereby assessing any cytotoxicity effects of either the scaffold or the released GFs. More specific assays test for the effects GFs have upon differentiation of certain cell types.

TABLE 1: Desirable characteristics of drug-eluting scaffolds adapted and updated from [24].

Desirable characteristics of growth factor-eluting scaffolds (adapted from [24])
Low immunogenic and antigenic response.
Controllable breakdown into benign biodegradable products occurring contemporaneously with new bone formation.
Good mechanical properties during breakdown, while retaining a porous structure capable of cellular infiltration
Good "biocompatibility"*.
Ability to load a growth factor with high efficiency and elute a therapeutical dosage over a sustained period of time relevant to the application.
Ability to be readily sterilised without either loss of mechanical function or denaturation of delivered drug.
Adaptability to wound site, surgical malleability.
Relatively long shelf-life, allowing for ease of access to a surgeon.

*We use the Williams definition of "biocompatibility" found in [38].

For example, when assessing BMP-2 activity, monitoring the upregulation of alkaline phosphatise (ALP), a cell surface enzyme associated with the maintenance of bone turnover, is a common tool [48, 55, 57–60]. C2C12 myofibroblast cells have routinely been used as they are prone to BMP-2 mediated osteoinduction [4, 48, 49, 54, 60] due to their low baseline ALP activity. Stimulation via BMP-2 will cause the upregulation of ALP thus showing bioactivity of the GF. Other osteoinductable cells such as MC3T3-E1 [56, 61] and bone marrow stromal W20-17 [55] have also been used. ALP activity is often used in conjunction with Alizarin Red assays [57, 60, 61] as a collective measurement of matrix mineralization, a key component in the maturation of new bone. Another technique used to measure osteoinduction is quantitative real-time PCR [52].

While *in vitro* studies allow for the demonstration of release kinetics and enable basic biocompatibility testing of the scaffolds with different cell populations, they cannot hope to model the extreme complexity of a wound site in sufficient depth to provide a complete picture of the scaffold's bone regenerative potential. As such animal models are still very much an important part of preclinical research and are essential prerequisite for the translation of any scaffold technology into the clinical.

A critical-sized bone defect is defined as "the smallest size intraosseous wound that will not spontaneously heal completely with bone tissue, or the defect will heal by connective tissue during the lifetime of the animal" [62]. Typically, models for bone regeneration will utilise bone defects, but occasionally people have worked on simpler *in vivo* models such as subcutaneous regions in the flanks of rats [63] or muscles [64]. The ectopic model can provide insight on the effect GFs can have on the differentiation of tissue, as well as immune responses, away from the ultimate intended site of implantation. For example Hulsart-Billström et al. used a rat thigh muscle model to determine the ectopic bone formation of a number of rhBMP-2-loaded hydrogels [64]. Briefly, an aldehyde-modified hyaluronic acid and hydrazide-modified polyvinyl alcohol were mixed to form a stable injectable hydrogel. The hydrogel was first loaded with rhBMP-2 and supplemented with a number of variations of HA prior to injection in a rat thigh. Ectopic bone formation was then assessed using radiographs and histology.

A good blood supply is a critical step for all wound healing; this is one of the principle reasons for undertaking ectopic bone formation studies in muscle as opposed to a subcutaneous pocket. However another model involving the arteriovenous loop has been used in bone tissue engineering [65]. This technique involves looping an artery around the implant and connecting it to a vein, allowing for the infiltration of blood vessels via angiogenesis [65–67].

Several animal models have been used to study *in situ* bone regeneration, and these revolve around different models to test different types of injuries, such as maxillofacial and cranial defects where rats are frequently used [49, 55, 68–70]. Lumbar spine, where sheep are commonly used [26, 71], along with goats [72] and long bone, where rats are the most commonly used model [25, 46, 73–79]. Smaller animals such as mice and rats have been commonly used as models for tissue-engineering applications due to their size making them more economical than the larger animals while still providing valuable information which cannot be gained via *in vitro* models. However there is a push towards using more relevant models in bone regeneration, such as the sheep model, which have closer characteristics to the human body in terms of physical size and structure than smaller animals. An effective 3 cm tibial defect was, for instance, established, and now being, extensively, reproducibly used, by Reichert et al. to study scaffolds for large bone defect repair [16, 80].

6. Current Strategies for Delivering Bone Growth Factors

The current methods for the fabrication of GF-loaded scaffolds can be divided into two main categories: attachment of the GF (physical and nonphysical) to the scaffold or entrapment of the GF within the scaffold, see Figure 1 for a summary. The attachment may be achieved by the adsorption of the GF onto the scaffold or through chemical cross-linking of the GF to the polymer scaffold. Fibrin-based matrices are popular biomaterials for this purpose. For instance, GFs with

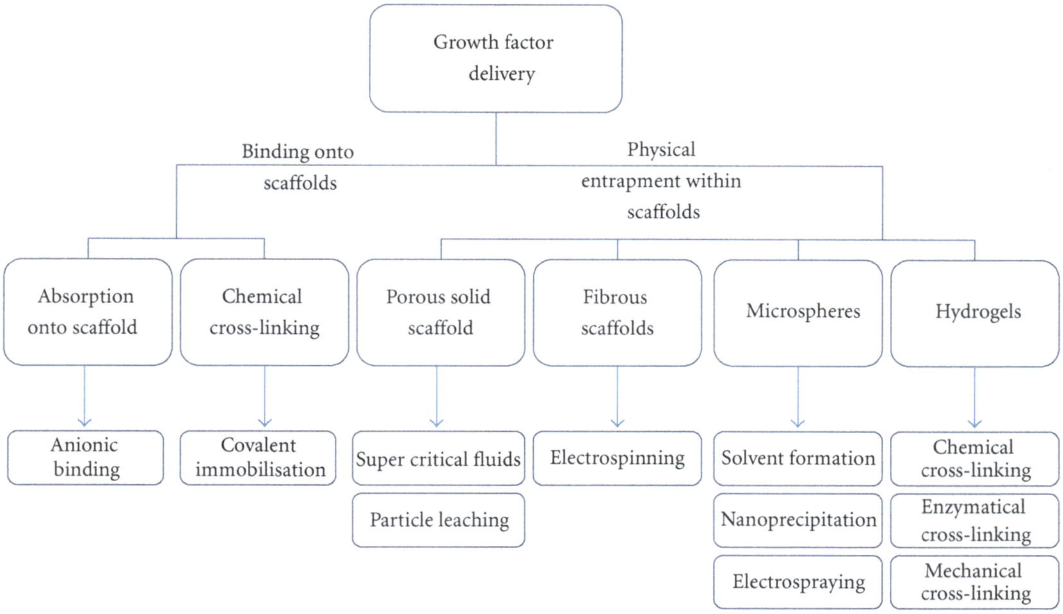

Figure 1: Current methods for fabrication of GF-loaded scaffolds (adapted from [84]).

heparin binding sequences like basic FGF (bFGF) [81, 82] and BMP-2 [83] have been covalently cross-linked to fibrin during the enzymatical coagulation of fibrin. This process has been further improved by the addition of heparin or hyaluronic acid to the fibrinogen mix prior to enzymatic coagulation. The release then occurs passively and through simultaneous fibrin/heparin degradation.

It is, however, often desirable to release GFs in a slower manner; as such, many methods of encapsulation have been investigated over the past decade. Encapsulation is a simple method of slowing the release, by creating a physical barrier, blocking the ability of a GF to diffuse away until the capsule in which the GF is contained has been sufficiently degraded; see Figure 2. By altering the composition of the capsule, the rate of degradation and the subsequent release of the entrapped GF can be somewhat controlled [85]. As previously mentioned bioresorbable polymers, most commonly aliphatic polyesters, are used for this purpose [28]. These polymers first undergo random hydrolytic breakdown of their more amorphous regions. As these polymers continue to breakdown, they become soluble oligomers and undergo surface degradation. During this phase the surface of the polymer begins to erode away, and GFs close to the surface begin to be exposed and release into the environment; ergo the rate of breakdown of the polymer is directly related to the release rate of the entrapped GF.

Other approaches to supply GFs to a wound site include seeding a population of cells which have been genetically modified to elute specific GFs into their local environment. As an example, a recent work by Saito et al. investigated the effect of adding BMP-7-transduced human gingival fibroblasts to phase deposition modelled PLLA and PLGA 50 : 50 scaffolds followed by subcutaneous implantation to assess bone formation [86]. The focus of the study was to investigate the effect pore size and polymer degradation rate had upon bone volume, and, as such, a rat subcutaneous pocket model was used to investigate this. Briefly two million transduced fibroblasts were seeded onto scaffolds using a fibrin gel; three different scaffold configurations were used per polymer (small pore, medium pores, and large pores), prior to implantation then assessed for *de novo* bone formation via μCT. All samples were shown to produce new bone, suggesting that the fibroblasts had successfully survived the transfer procedure and were able to express BMP-7 which led to the differentiation of local progenitor cells into bone-forming cells. As this method of cell transplantation digresses from the topic of this paper, which aims to review scaffolds with entrapped GFs rather than scaffolds with entrapped cells which express GFs, we will focus the remainder of the paper to this area.

6.1. Porous Solid Scaffolds. Entrapment of GFs has been achieved in porous solid scaffolds, through solvent casting/particulate leaching and supercritical fluid (SCF) processing [87–89]. SCF processing is attractive for GF delivery, since it does not require the use of harmful solvents which could adversely affect the GFs. In fact, protein denaturation during scaffold fabrication is the main problem associated with the incorporation of proteins in various systems and leads to a considerable loss of their activity. As a solution, SCF processing allows the preservation of biomolecules, and supercritical CO_2 is typically used as a solvent. Howdle et al. were able to successfully retain close to 100% enzymatic activity of ribonuclease processed via SCF within a PLLA scaffold [90]. Briefly, SCFs like CO_2 are used above their critical point and enable the glass transition of the exposed polymer to be depressed. GFs are initially mixed within the polymer powder within a pressure chamber which can

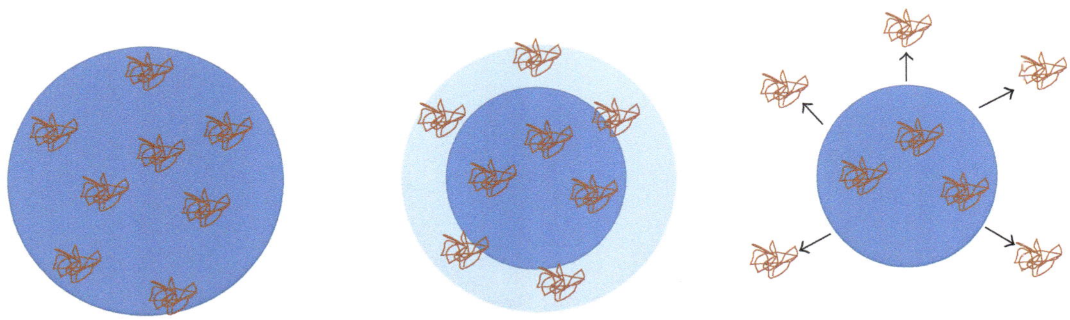

Figure 2: Schematic illustrating the breakdown and release of GFs from polymeric encapsulation.

maintain the temperature and pressure conditions required for supercritical fluid formation; for CO_2 this is 31°C and 73.8 bar, respectively [90]. Entrapment of GF then occurs during depressurization which under the reducing pressure also nucleates gas bubbles as the gas solvent attempts to escape from the polymer, leading to the formation of a foamed structure. Several studies show good GF entrapment using this method, incorporating VEGF [87, 88] or BMP-2 [89] into PLGA and its derivative polymers.

In a preliminary study by Kanczler et al. entrapment of VEGF within a P(D,L)LA scaffold via supercritical CO_2 demonstrated almost 100% entrapment rate, as assessed by ELISA, with a release rate of approximately 0.8% of the encapsulated VEGF after 21 days. The viability of the released VEGF was assessed by successful stimulation of human umbilical vascular endothelial cells (HUVECs) to form enclosed endothelial cell networks. The group followed this work by combining supercritical CO_2 scaffolds with an alginate hydrogel to allow for two independent GF release curves in a single implant [91]; this work is discussed in detail later in the review in Section 6.6 "Combining different loading techniques for delivering multiple GFs."

Another technique for binding GFs onto a porous scaffold is via layer-by-layer (LbL) polyelectrolyte multilayer films [92]. These films are created by dip-coating scaffolds with a system of positively and negatively charged molecules, and GFs can be introduced into a layer depending on its charge. For example MacDonald et al. used a LbL system comprised of a negatively charged Poly (β-aminoester) 2 layer, then a positively charged chondroitin sulphate layer, followed by a negatively charged BMP-2 and finally another chondroitin sulphate. This tetra-layer structure was repeated 100 times onto polycaprolactone/β-tricalcium phosphate printed scaffolds. Release profiles showed that 80% of the BMP-2 attached eluted within 2 days, with the remaining 20% being released over 2 weeks.

6.2. Fibrous Scaffolds. Similar to the SCF technique, biomolecules can also be directly incorporated into electrospun nanofibres. Electrospinning is a popular technique for the rapid fabrication of either microfibres or nanofibres. Briefly, the electrospinning concept lies with the interaction of an applied electric field and the surface charge of a polymer solution in a capillary. Mutual charge repulsion in the polymer solution generates a force directly opposite to the surface tension of the polymer solution. Once the electric charge is sufficient to overcome the surface tension of the polymer solution, a polymer jet is ejected and undergoes instabilities due to elongations and solvent evaporation before it hits the collector, leading to the formation of nanofibre electrospun scaffolds. GF may be incorporated into the produced fibers during the electrospinning process by mixing the biomolecules into a polymer solution prior to electrospinning the mixture [93], or by using coaxial electrospinning wherein a secondary polymer solution containing the biomolecules is electrospun within the core of the forming nanofibre surrounded by a shell polymer [94, 95].

Nanofibre scaffolds applied to drug delivery have predominantly been focused upon the loading of antibiotics and anticancer agents [96], and only relative few studies report the incorporation of GFs into these types of scaffolds [60, 68, 95, 97–103].

For bone regeneration purposes, BMP-2 has been incorporated into a number of different polymers and electrospun, including PLLA [68, 102, 103], PLGA [97, 99] PCL [46], cellulose [60], and silk [98]. PLLA nanofibers loaded with BMP-2 have been achieved by dissolving BMP-2 in solvents such as acetic acid and then dispersing this mixture into a PLLA solution dissolved in dichloromethane (DCM) [68, 102, 103] before the mixture was loaded into a syringe and electrospun as loaded fibres.

The type of solvent used in the electrospinning process is important for the structural properties of the resultant scaffold, as shown in a study comparing the strength of PLLA when spun in DCM compared with hexafluoroisopropanol (HFIP). The results indicated that the use of HFIP solvent produced physically stronger scaffolds compared to those produced by DCM, with an increase in maximum load (~ 30%), tensile strain (~900%), and Young's modulus (~25%). However, the addition of BMP-2 reduced the strength of fibres so, significantly, they were unable to be tested mechanically [102]. For this particular experiment BMP-2 loading was achieved by creating an emulsion of BMP-2 in acetic acid with the PLLA in DCM. A similar effect was observed with the addition of collagen to the PLLA fibres, suggesting that the tertiary structure of the polymer was significantly affected by the addition of "impurities." Culturing MG36 cells on the PLLA fibrous meshes for 22 days maintained cell viability, but once again demonstrated a loss of mechanical strength in scaffolds prepared with BMP-2 or collagen versus pure PLLA fibres [102].

Electrospun PLLA scaffolds loaded with rhBMP-2 have been implanted for 12 weeks into critical-sized cranial defects in a mouse model. Results showed a significant increase in the overall *de novo* bone formation with 45% of the defect filled with new bone within the BMP-2-loaded scaffold versus 10% with the unloaded PLLA scaffold. The upregulation of osteogenic markers, Smad-5 and osteocalcin was also observed at the 12-week time point in the BMP-2-loaded implant compared to both the unloaded BMP-2 and the collagen control [68]. μCT scans showed more mineralisation of the PLLA/BMP-2 scaffold than compared to the blank control, collagen, and unloaded PLLA scaffolds (Figure 3).

rhBMP-2 has been incorporated into electrospun scaffolds in association with hydroxyapatite (HA) in order to mimic bone extracellular matrix [97–99]. The studies involved silk [98] or PLGA [97, 99], and, in the silk study by Li et al., poly(ethylene) oxide (silk : PEO 80% to 20% wt) was used to generate a stable and continuous spinning jet, which was not possible using silk alone. Five different groups of scaffolds with similar morphology were produced. A control group of silk/PEO and a second control group of silk with the PEO extracted from the fibre prior to seeding were used along with scaffolds containing either rhBMP-2, HA nanoparticles, or both. After 31 days of *in vitro* culture with human bone marrow-derived mesenchymal stem cells, all scaffolds supported proliferation and differentiation into bone-forming cells, as shown by a calcium extraction assay (trichloroacetic acid), and total DNA content assay (Pico green). The calcium deposition was significantly higher in the rhBMP-2+HA group than either rhBMP-2 or HA alone, but these groups were both higher than either control. However, no information on the loading and release profiles were mentioned for this system, which would be valuable additions to the study and would add value to the sustainability of the release system [98].

In more detailed studies, Nie et al. and later Fu et al. examined the release profiles of BMP-2 from PLGA/HA scaffolds using both *in vitro* and *in vivo* assessment [97, 99]. Encapsulation efficiencies of scaffolds ranged from 49 to 66%. It was observed that increasing the contents of HA nanoparticles accelerated the release rate of rhBMP-2 from the scaffold. This might be explained by the hydrophilic nature of HA which served to increase the overall hydrophilicity of the scaffold, rendering it more easily degradable. However, the surface of scaffolds seemed weaker, with cracks and holes occurring when HA was incorporated [97]. Interestingly, *in vivo* studies (tibial bone defect in nude mice) demonstrated that PLGA scaffolds loaded only with BMP-2 had no significant effect on bone healing. However, when HA nanoparticles were incorporated into the PLGA during electrospinning, the BMP-2, released from either the scaffolds where BMP-2 was encapsulated inside the fibres or just coated on the surface of fibres, resulted in a greater bone healing [99]. These results suggested that the presence of HA was more determinant on sustained release of BMP-2 than the initial loading location of the GF.

6.3. Nano-/Microspheres. The term microspheres or nanospheres is a term used to describe small particles which are in the micro- or nanoscale. The use of microspheres loaded with GFs in bone TE has been investigated in various ways: microspheres can be incorporated into different types of preformed matrices (solid, soft, polymeric, and ceramic) or formed during the matrix fabrication, to generate composite scaffolds containing microspheres. They can also be fused together to form a microporous scaffold in itself. In all these techniques, the loading generally happens during the formation of microspheres. It is important to note that independent incorporation of microspheres in a preformed matrix holds great potential, since the particles and matrix can be made from different materials. This option enables the use of scaffold materials which may be manufactured to closely fit their intended anatomical site and affords the inclusion of separate microsphere systems which may contain different and site-specific GFs for the specific application. While nanospheres can be used in a similar manner, research into GF loading for bone TE has been extremely limited, compared to microspheres, with few groups looking at nanospheres [104].

In bone tissue engineering, the scaffold needs to provide a structural support which often requires mechanical integrity, meaning that slow degradation is essential so as to maintain mechanical integrity whilst the tissue regenerates [14]. Bone scaffold materials are thus not optimal for direct loading of growth factors since their release will only be afforded through the degradation of the scaffold matrix, which will be very slow. This supposes that growth factors are released when the scaffold starts to degrade, that is, also losing its mechanical integrity. However, in order to maximise efficiency of bone tissue regeneration using a scaffold, GFs should preferentially be readily and constantly released whilst the scaffold can still provide sufficient mechanical support. This provides an interesting composite option: a material providing faster degradation for the nano-/microspheres with a material presenting lower degradation profiles for the scaffold. Furthermore, encapsulating GFs in microspheres rather than directly into the scaffolds leads to a better control of the GFs release, since microspheres can be optimized and tailored independently. Their release profiles can thus be studied before incorporation to the scaffolds and further optimized for incorporation [85, 105–107]. Additionally it is possible to incorporate various GFs in differentially degrading polymer microspheres and give a "zonal" release of GF over time by including slow and faster degrading nano-/microspheres, a condition which is relevant in tissue regeneration where there is usually a temporal expression of GFs during healing [108].

Many different polymers, both natural and synthetic, have been used in the production of GF-loaded microspheres. The most commonly used natural material for microspheres is gelatine, often cross-linked with genipin [52, 72], while the most universally used synthetic polymers

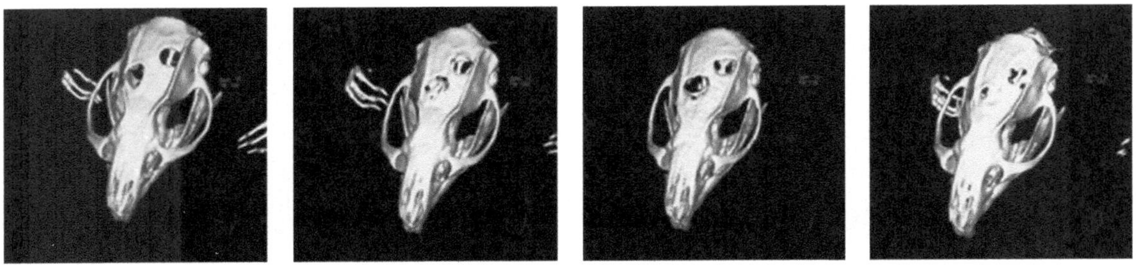

(a) (b) (c) (d)

FIGURE 3: 3D reconstructions of cranial CT scans of defects used for quantification of radiological bone density in critical sized defects after 12 weeks either; left blank (a), collagen control filler (b), electrospun PLLA (c), or electrospun PLLA/BMP-2 (d) [68].

for the production of microspheres are PLGAs, commonly a 50 : 50 ratio of comonomers [109–111]. PLGA has widely been used for a number of reasons including FDA approval, ease of the formation of microspheres, reasonable encapsulation efficiencies, and desirable release characteristics [112].

In recent years several GFs have been incorporated into microspheres for bone TE applications, the most studied GF is recombinant human BMP-2. See Table 2 for a summary of recent publications utilising microspheres to deliver GFs which have a role in bone TE.

6.3.1. Loading of Growth Factors into Microspheres. Incorporating GFs into microspheres is usually performed during the formation of the microsphere. Table 2 summarises current research in microspheres, the techniques used to produce the microspheres, and the size of microspheres beings used. As shown by Table 2, the w/o/w continues to be the most popular technique for GF delivery.

The classical ways of forming microspheres are based on single or double emulsions: oil-in water (o/w), solid-in-oil-in-water (s/o/w), oil-oil (o/o), and water-in-oil-in-water (w/o/w) approaches [112, 123, 124]. Of these techniques, the w/o/w is the most commonly used method. As shown in Figure 4, the first step involves mixing an aqueous solution containing the GFs with an organic solution (generally ethyl acetate or dichloromethane [81]) containing the dissolved polymer intended for the matrix material. This leads to the formation of the primary water-in-oil (w/o) emulsion. This emulsion is then dispersed in an aqueous medium under continuous mechanical agitation to form the secondary oil-in-water emulsion (o/w). The aqueous medium generally contains stabilizers such as PVA or PEG to promote the formation of microdroplets. The subsequent solvent evaporation allows the microdroplets to turn into microspheres. Finally, microspheres are harvested by the centrifugation or filtration of the solution followed by being washed and freeze dried [123, 125–127]. This technique is commonly used for the incorporation of GFs mainly within PLGA copolymer matrices [127].

A common problem with these methods of producing microspheres with encapsulated GFs is that they require exposure of the GF to organic solvents, which may cause denaturation. The addition of protective agents, such as bovine serum albumin (BSA) (acting as a carrier protein) [117, 128], and polymers such as polyethylene glycol (PEG) [129] are often incorporated to reduce denaturation of growth factors via organic solvents. Acidic degradation is another issue to be considered. When polyesters like PLGA degrade, they do so into monomers which can potentially lower the local environmental pH if not properly buffered, Crotts and Park determined that the localised environment in the centre of PLGA microspheres could be as low as pH 4.0 [130].

A variant of the w/o/w double emulsion is the solid-in-oil-in-water (s/o/w) emulsion, also called the single or spontaneous emulsion. This variant skips the first w/o emulsion involving the dissolution of GFs directly in the organic solution containing the dissolved polymer. Microdroplets are then formed by adding the organic solution to the aqueous medium under mechanical agitation to form the o/w emulsion directly, and then the same steps as for the w/o/w are followed [123, 125, 126].

In terms of techniques, the s/o/w technique—although quite similar to the w/o/w approach—is generally used less frequently for GF encapsulation. However, there is an upward trend in literature showing that s/o/w might be a better option than w/o/w. For example, when encapsulating brain-derived neurotrophic factor (BDNF) in PLGA microspheres (50 : 50, 10 kDa), it was observed that 2.4 times more GFs were released from microspheres prepared by s/o/w than prepared by w/o/w with a more sustained delivery over the period of the study (60 days) [131]. A variant of the s/o/w method has also been shown to enhance encapsulation efficiency by micronization of a model protein with PEG before emulsion [132]. This same approach was also used for the encapsulation of trypsin, horseradish peroxidise (HRP), and rhBMP-2. In this study, the results for the "model proteins" demonstrated encapsulation efficiencies and retention activities of more than 75% with s/o/w compared to 25% and 13% for w/o/w although it should be mentioned that encapsulation efficiency for rhBMP-2 was not discussed [133].

Another method of producing microspheres which is available but not commonly used is microprecipitation. An example is the production of silk fibroin microspheres (Cao et al.) [134]. Briefly regenerated silk fibroin solution was prepared by first washing the silk in Na_2CO_3 solution at 95°C, followed by dissolving in LiBr solution. This solution was then dialysed against pure water to remove the salt. Finally

Table 2: Summary of the growth factor, polymer used, and size of microspheres studies recently released.

Polymer	Growth factor	Formation technique	Microsphere size	Reference
PEG-PLGA diblock	BMP-2	w/o/w	37–67 μm	[48]
PLGA/PLGA-PEG-PLGA	BMP-2	w/o/w	67–127 μm	[61]
Gelatine cross-linked with genipin	BMP-2	w/o	undisclosed	[72]
Gelatine cross-linked with genipin	BMP-2	w/o	2–6 μm	[52]
PLGA	BMP-2	w/o/w	100–150 μm	[63]
PLGA	BMP-2	w/o/w	Average of 50 μm	[113]
PLGA	BMP-2	w/o/w	Undisclosed	[114]
PLGA	VEGF	w/o/w	Undisclosed	[115]
PLGA	VEGF	w/o/w	20–53 μm	[116]
PLGA	VEGF + FGF-2	w/o/w	undisclosed (fused microsphere bridge)	[108]
PLGA	VEGF + FGF-2	s/o/w	Undisclosed	[67]
Alginate	VEGF	w/o/w	300–400 μm	[47]
PEG-CAM-PEG	VEGF + BMP-2	s/o/w	5–30 μm	[117]
PLGA	VEGF, FGF-2, VEGF + FGF-2	w/o/w	~20–100 μm	[65]
PLGA	PDGF	w/o/w	7–31 μm	[118] microsphere formation in [119]
PLGA	BMP-2 + TGFβ	Ultrasound o/w	Undisclosed (fused microsphere bridge)	[120]
PLGA	LTB_4	w/o/w	5–10 μm	[107]
Hyaluronic acid	BMP-2	w/o/w	710 nm–1.7 μm	[121]
PLGA	BMP-2	w/o/w	120–470 μm	[122]
$P(TMC\text{-}CL)_2$-PEG	BMP-6 + TGFβ3	Electrospraying o/w	~65-80 μm	[51]
Gelatine cross-linked with glutaraldehyde	VEGF + BMP-2	w/o	50–100 μm	[70]
Silk fibroin	BMP-2 BMP-9 or BMP-14	Nanoprecipitation	2.4–3 μm	[58]
PLLA	BMP-2	w/o	35–55 nm	[104]

the solution was centrifuged, and the supernatant was then collected and diluted as required. The GF was then added to this solution at room temperature under agitation. At this point a nonmiscible solution (in Cao's example absolute ethanol, which causes precipitation of regenerated silk fibroin), was added dropwise under gentle agitation to form microspheres. Following this step the samples were first frozen and then lyophilised to produce dry microspheres. By using different ratios of the two solutions, particle size may be altered; by increasing the ratio of ethanol to fibroin solution from 1/10 to 1/4 the average particle size was decreased from ~1100 to 400 nm in diameter.

In addition to the classical techniques for the formation of microspheres, electrospraying has emerged as an effective technique for the formation of microparticles [50]. Electrospraying is a similar technique to electrospinning, using identical apparatus. When compared to electrospinning, electrospraying requires polymer solutions with lower viscosity; these can be obtained by using lower molecular weight polymers, or more often using lower polymer to solvent concentrations. GFs can be incorporated into the polymer solution prior to electrospraying resulting in particles with GFs trapped throughout. There are several advantages to electrospraying compared to the classical techniques including that the use of an emulsion is optional but not required. This is very advantageous since the interface between the organic and aqueous phases may result in GF denaturation and aggregation, which is established as the main drawback of all emulsion-based methods [135]. In addition, in electrospraying there is no use of high temperature such as in spray drying, there is no further drying step required since particles are dried during the spray process, and there is an enhanced control over the size distribution of particles produced in classical ways, with the possibility of producing near monodispersion of particles by choosing optimal processing parameters. For instance relative standard deviations as narrow as 2% of the average size have been obtained when encapsulating inhalation drugs into electrosprayed PLLA submicron particles [136]. This is advantageous in drug delivery since degradation rates and diffusion of drugs from the polymeric matrix can be controlled to a better precision for sizes of particles within a narrow distribution, allowing a desired application to be obtained. However, this requirement is rarely achieved for microparticles made from emulsion fabrication methods, known to provide very broad distributions with relative standard deviation of up to 110% [137].

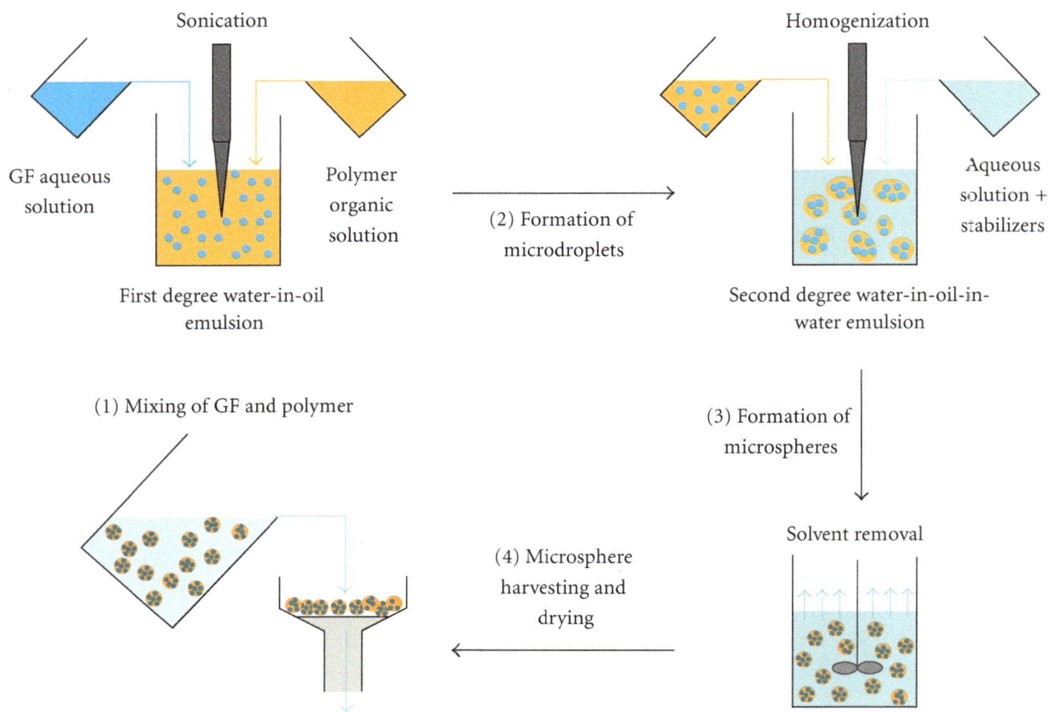

Figure 4: Schematic showing the four principal steps in GF-loaded microsphere preparation by the water-in-oil-in-water emulsion (followed by solvent extraction) technique.

To date, the encapsulation of GFs in electrosprayed particles has been successfully achieved in the production of PLGA microcapsules loaded with IGF-1 via coaxial electrospraying [138], and in electrosprayed droplets comprising a hyaluronic hydrogel containing VEGF and PDGF-BB [139]. Both studies showed that the released growth factors were bioactive and that they enhanced angiogenesis *in vitro*. Recently Sukarto and Amsden [51] trialled an exotic electrosprayed poly(1,3-trimethylene carbonate-*co*-ε-caprolactone)-*b*-poly(ethylene glycol)-*b*-poly(1,3-trimethylenecarbonate-*co*-ε-caprolactone), $P(TMC\text{-}CL)_2$-PEG polymer which was tested for dual release of 2 GFs associated with bone formation; BMP-6 and TGF-*β*. This polymer was chosen due to biodegradability, slow breakdown, and limited production of acidic byproducts, and due to the low melting point of the polymer (in the region of 40°C). Briefly the 2 GFs were mixed with BSA in a sucrose solution, which was mixed with the $P(TMC\text{-}CL)_2$-PEG polymer prior to electrospraying. The solution was then electrosprayed into an ethanol collection bath. They report two release profiles, with 80% of the TGF-*β* being released within 20 days, while only 40% of the BMP-6 was released over the same time, taking over 80 days for 80% to be released.

6.3.2. Growth Factor Delivery from Microsphere-Based Scaffolds. Nano- and microspheres containing BMPs are rarely used in isolation for bone regeneration purposes. Usually they are immobilised within/on a matrix to form a composite scaffold containing microspheres, or they are fused together to form a microsphere scaffold. Most scaffolds incorporating microspheres showed significant differences in bone formation compared to having the BMPs adsorbed on the scaffold surface. For instance, one study showed that rhBMP-7 delivery from PLGA nanospheres incorporated into a PLLA scaffold induced significant ectopic bone formation while failure of bone formation was observed with passive adsorption of rhBMP-7 onto the scaffold, likely due to significant loss of biological activity of the GF and diffusion away from the implantation site [140].

The benefit of incorporating nano-/microspheres into scaffolds is less evident when compared to direct entrapment of the GF in the bulk scaffold matrix as discussed in [141]. In an interesting study, PLGA microspheres containing rhBMP-2 were incorporated into a polyurethane (PU) scaffold, and the release kinetics and bone formation in a rat femoral defect were compared with a PU scaffold containing free rhBMP-2. It was shown that rhBMP-2 delivery from the microsphere-containing scaffold presented a reduced burst release compared to free rhBMP-2 in PU, but showed less new bone formation at the early time points, in particular for smaller microspheres (1.3 μm compared to 114 μm). This phenomenon was compensated later between weeks 2 and 4 after implantation, with the PLLA rhBMP-2 microspheres provoking similar bone formation when compared to the rhBMP-2 freely loaded into PU [141]. This study linked the burst release to new bone formation, suggesting that an initial burst release followed by sustained release is actually preferential for promoting new bone formation. This can be explained by the fact that BMP-2 plays an important role during the healing process to promote cell differentiation but is equally important at the early stages in the healing processes for the initial migration of progenitor cells and triggering the fracture healing cascade [8, 142]. As a consequence, burst release is implied to have an initial beneficial effect on early bone formation. This burst release was controlled by the size of microspheres as shown by different sizes of microspheres enabling tailoring of the GF release for enhanced bone formation [141].

It is important to note that care should be taken when translating *in vitro* results to *in vivo* studies. For instance, BMP-2 has been incorporated into PLGA microspheres through w/o/w and further cross-linked to various scaffolds: a gelatin hydrogel, a poly(propylene fumarate) PPF scaffold, and a PPF scaffold surrounded by a gelatin hydrogel [53]. All the delivery systems used microspheres to deliver BMP-2 and were compared to a hydrogel matrix simply cross-linked with BMP-2. In terms of release, the latter case showed an *in vitro* burst release of 33% compared to less than 10% with all microsphere systems (Figure 5(a)). However, the burst release of the free BMP-2 system was increased to 92% *in vivo* while the microspheres systems showed no burst release but exhibited a much faster BMP-2 release upon implantation (Figure 5(b)). The bone in growth, assessed by μCT, is presented in Figure 5(c). Microspheres/PPF scaffolds gave significantly better results with more bone formation observed via μCT, when compared to other gelatin scaffolds at 6 weeks. But this result was statistically equivalent to the microsphere/PPF surrounded by gelatin at 12 weeks as shown in Figures 5(c) and 5(d). Similarly to other studies [141], the use of microspheres with gelatin did not provide significantly better results than the BMPs directly loaded in the gelatine hydrogel. Only the use of another material (PPF) alone (no gelatine hydrogel) was able to provide significantly more bone formation after 6 weeks of implantation [53] when compared to PFF with gelatine-loaded implant.

Recent work has included microspheres dispersed within a hydrogel or in bone cements [26, 72] or combined PLGA/PLGA-PEG-PLGA microspheres with a fused deposition PCL and 20% w/w β-tricalcium phosphate scaffold [61]. Other materials used as the mechanical support matrix include PFF [78, 143], brushite [122], and coralline hydroxyapatite [113].

Combining microspheres with other scaffolds is a common approach, but it is also possible to fuse microspheres into solid scaffolds themselves with functional mechanical properties via sintering effects [56, 108, 120]. The general technique of creating a sintered scaffold is to compress microspheres in a mold and chemically sinter.

One interesting study by Dormer et al. established a gradient of GF loading (TGF-β) and BMP-2 into PLGA 50 : 50 microspheres where a high concentration of TGF-β was present at the top of the implant to promote chondrogenesis; the formation of cartilage, and a high concentration of BMP-2 at the bottom to promote osteogenesis; the formation of bone [120]. Microspheres suspended in water were introduced into a mold via syringe pumps, creating a BMP-2-rich region, a transition gradient region leading from a high concentration of BMP-2 and a low concentration of TGF-β to a low concentration of BMP-2 and a high concentration of TGF-β. Ethanol was used as the sintering agent to fuse the microspheres together, and the scaffold was tested with and without cells *in vivo* in a rabbit osteochondral knee defect versus a sham (empty) control, and an unloaded scaffold. The infiltration of cellular tissue was assessed using histology, mineralisation was assessed using alizarin red and von kossa staining. Although in this study the sham control arguably had the most superior tissue infiltration and superior bone mineralisation, it was theorised that this result was due to the rate of degradation of the microspheres. They concluded that a more rapidly degrading scaffold would have a superior cell infiltration and tissue maturation. However it should be noted that they were able to create zonal differentiation of cells into cartilage and bone within their loaded scaffold.

Loading of BMPs onto microspheres may be approached after the production of the microspheres by dipping into a BMP solution. PLGA/HA microspheres for potential use as an injectable system for bone defects have been prepared in this manner, by treating the surface of the microspheres with an alkali treatment followed by dipping into rhBMP-2 to allow better GF adsorption. The system released 80% of rhBMP-2 after 3 weeks *in vitro* but did not show a significant effect on cell attachment and proliferation although it was shown to affect osteoblast differentiation [109]. Another study by Woo et al. involved the preparation of PLGA microspheres through a w/o/w process and further treatment in an rhBMP-2 solution [144]. The GFs attached through adsorption and the loaded microspheres were lyophilised before incorporation into a carboxymethyl cellulose for implantation into rat calvarial defects. Several immediate and sustained release systems were compared, and the best bone regeneration was observed with a sustained-release implant able to heal 75–79% of the defect in 6 weeks as compared to the 45% recovery from the immediate-release implant. *In vitro*, the microspheres used in the sustained-release implant first showed a 45 to 55% burst release over 24 hours, before steadily reaching 70% release after 3 weeks. This supports the theory that both an initial burst and sustained release are required for optimal bone formation. However, it must be kept in mind, in this case, that 63% of rhBMP-2 was not physically bound to the microspheres with this method of loading, that is, that a consequential amount of GF was still free in the matrix [144].

6.4. Hydrogels. Hydrogels are widely used within tissue engineering due to the highly versatile nature of their construc tion and the ability to entrap molecules within their hydrated matrix with relative ease. A hydrogel is a cross-linked, insoluble hydrophilic polymer, and highly swollen in water. Hydrogels swell in contact with water, forming an insoluble 3D network, and as such certain polymers can be surgically injected into a wound site in a partially dehydrated state and undergo swelling *in situ*, and other techniques can use pH and temperature-sensitive polymers which will gel following injection [145]. Hydrogels are commonly used in tissue engineering for drug delivery due to the ease of incorporating soluble and insoluble drugs into the hydrogel, often only requiring simple mixing the drug and hydrogel precursors [145]. The cross-linking of hydrogels can be chemical like PEG cross-linking [146], enzymatically such as hydrazone cross-linking

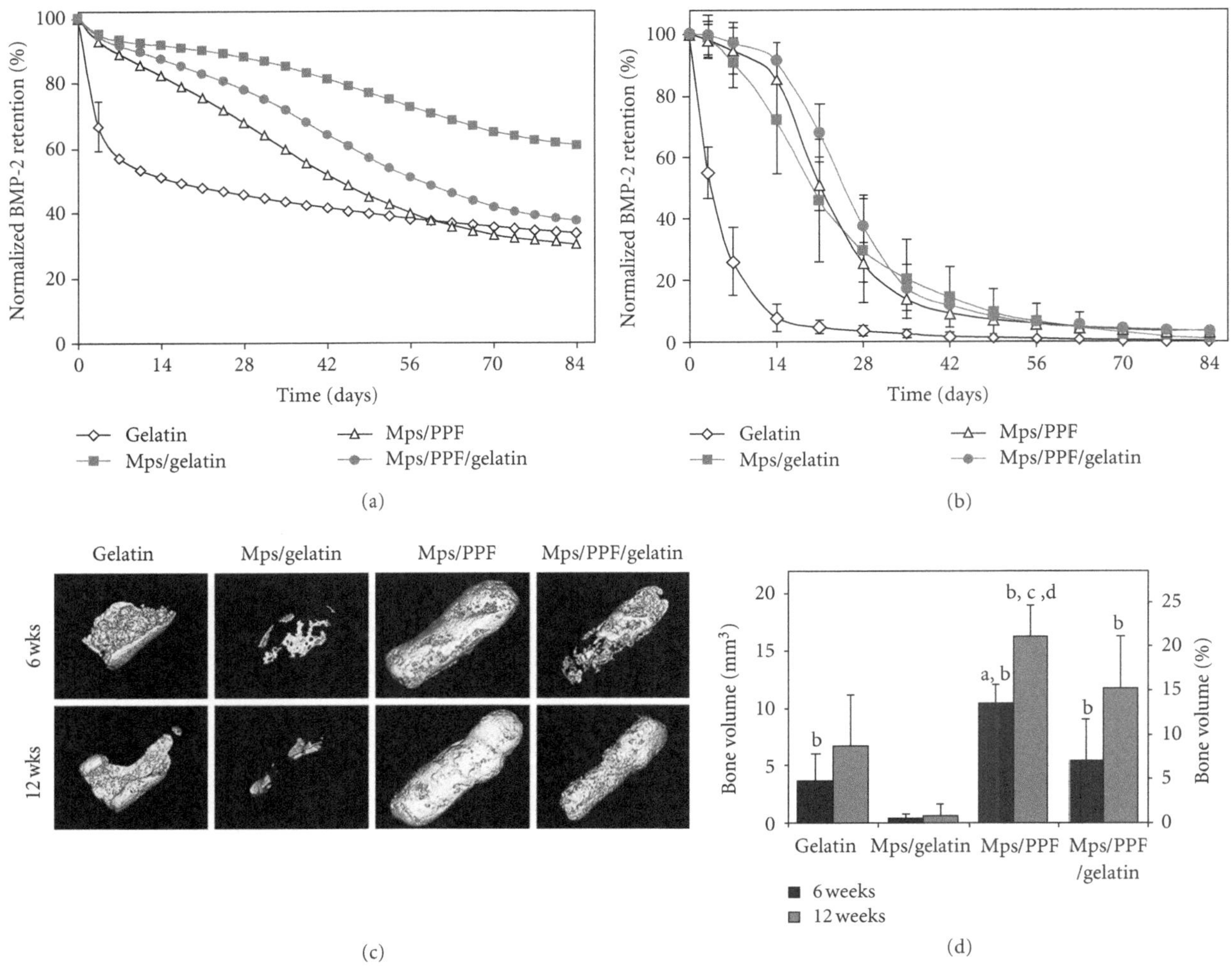

Figure 5: Normalized release profiles of ^{125}I-BMP-2 from the four different implants (% of the initial loading): (a) *in vitro* (b) *in vivo* in a rat subcutaneous implantation model. (c) New bone formation after 6 and 12 weeks of implantation. Image reconstruction was derived from μCT scans using standardized thresholds. (d) Average volumes of newly formed bone in the four different scaffolds after 6 and 12 weeks of implantation (left axis), and normalized to an 8 mm long, 3.5 mm diameter cylindrical scaffold (right axis). Significant difference relative to (a) all other implants, (b) Mps/gelatin implant, and (c) gelatin implant at the same time point. (d) Significant increase in bone formation relative to the previous time point [53].

of hyaluronic acid [55], or physical, such as the sonification of silk fibroin [147] in nature.

Like microspheres, many different polymers have been used in the production of hydrogels including natural polymers like silk [147], chitosan [79, 148], and alginate [148], as well as synthetic ones such as PEG [149]. Diblock and triblock copolymers have also been studied as they can provide useful physical properties such as temperature-dependent physical characteristics [145].

6.4.1. Loading Growth Factors into Hydrogels. The most common method of loading GFs into hydrogels is to simply mix the GF into the hydrogel precursor. As the polymerisation occurs, the GFs become entangled in the polymer chains and are physically trapped within the hydrogel network. As the hydrogel begins to degrade, via hydrolysis or proteolytic degradation, GFs are able to freely move out and elute from the matrix, starting with GFs loaded on the outside of the gel and moving inwards (see Figure 6). Hydrogels pore size is often larger than the size of the GF, and hydrogels often release the majority of the contained GF within a few days. Hydrogels with nanosized pores are known as nanogels and provide increased retention of GFs over conventional hydrogels, with some gels retaining GFs for months, dependent upon the degradation of the polymer, although limited research has been conducted in the field [147, 150].

In addition to pore size affecting the release rate of GFs from the hydrogel, other factors such as the concentration of the polymer in the gel and the degree of crosslinking can have an effect upon the release rate of GFs. The polymers constituting hydrogels can also be modified to include specific cell-binding sites such as covalent attachment of the peptide sequence arginine-glycine-aspartic acid (RGD) [151, 152] to the polymer during synthesis to improve cell attachment.

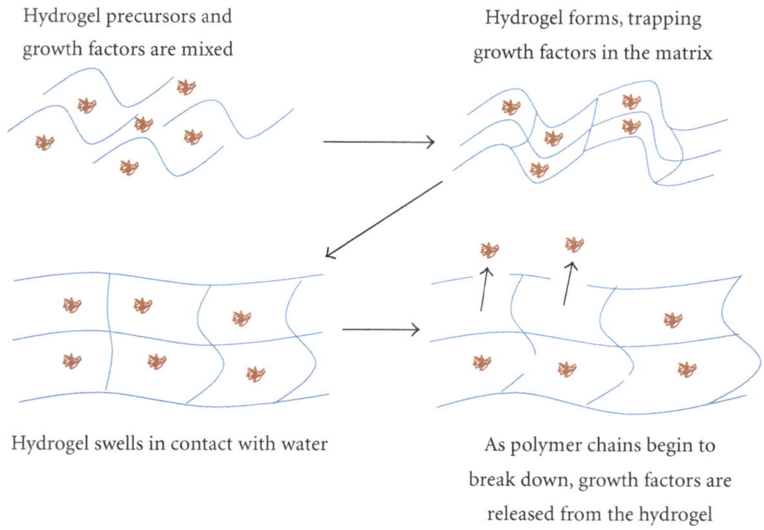

FIGURE 6: Schematic showing the formation of growth factor loaded hydrogel formation, breakdown and release.

Zieris et al. combined amino end-functionalized four-arm star PEG with EDC/sulfo-NHS-activated carboxylic acid groups of heparin in addition to RGD functionalization of the carboxylic acid heads of the PEG chains to improve GF binding [153]. See Table 3 for recent investigations into GF loading hydrogels.

6.5. Growth Factor Delivery from Hydrogel-Based Scaffolds. The release of BMP-2 from hydrogels has been widely studied as shown in Table 3. Other GFs such as VEGF and PDGF have received less attention but have also been investigated to some extent [26, 139, 147].

The material used to compose the hydrogel has a significant effect upon the formation of new bone; a study by Ludmila et al. showed that an rhBMP-2-loaded hyaluronic acid hydrogel produced a more osteoinductive response when compared to a chitosan one of similar design [148]. However, it should be noted that although more bone formation was observed in the hyaluronic acid hydrogel, the bone formed in the chitosan was described as being in a more mature state after three weeks with significantly higher levels of calcification, as observed by Masons trichrome staining and μCT imaging in a rat ectopic muscle model [148]. This increase in maturation within the chitosan hydrogels was attributed to the more rapid breakdown of the hydrogel leading to a more rapid release of the rhBMP-2. A further study conducted by the same group supplemented the rhBMP-2 chitosan hydrogel with β-tricalcium phosphate (β-TCP) which improved the osteoinduction at the cost of slower mineralisation, as shown histologically with more overall bone growth observed, but less mineralisation as assessed by a Masons trichrome staining. They attributed this effect to slower release of rhBMP-2 from the scaffold due to the affinity of rhBMP-2 to bind to β-TCP [79]. Therefore they were able to reduce the release rate of the rhBMP-2 while maintaining the more rapid hydrogel breakdown.

Hänseler et al. compared the loading efficiency of a PEG hydrogel versus deproteinized bovine bone matrix (DBBM) for the release of rhBMP-2, both in a glycosylated and non-glycosylated form [154]. The glycosylation of rhBMP-2 occurs during the production within an animal cell, such as transfected Chinese hamster ovary cells (CHO), while non-glycosylated rhBMP-2 is sourced from a prokaryotic expression system in modified *Escherichia coli*. Eukaryotic cells glycosylate disulphide bonds, while proteins derived from *E. coli* source do not. However studies have shown that nonglycosylated BMP-2 can stimulate bone growth in humans. RhBMP-2 was loaded by mixing thiol PEG in a 0.1 M triethanolamine solution in the presence of 25 μg of rhBMP-2 per 50 μL of unreacted PEG gel. It should be noted that this simplistic method of encapsulating nonglycosylated rhBMP-2 produced significantly lower levels of encapsulated rhBMP-2 than expected, with only ~25% total active rhBMP-2. It was theorised that the thiol groups on the linear PEG could attack the disulfide bonds of the rhBMP-2 dimers, causing them to lose affinity for each other and thus inactivate the rhBMP-2. The use of glycosylated rhBMP-2 improved the encapsulation to ~68% within the PEG hydrogel alone; this value was still below the encapsulation level when DBBM was incorporated into the hydrogel. The release rate showed a burst over 3 days whereby nearly all of the rhBMP-2 both, glycosylated and nonglycosylated, was shown to be released from the gels.

Kolambkar et al. [25] used an alginate hydrogel loaded with 500 ng rhBMP-2 which was injected into an electrospun PCL mesh tubes and the same tubes containing perforations, as a method for long bone regeneration. Day 0 evaluation suggested a ~55% encapsulation efficiency with ~275 ng of active rhBMP-2 detected. However *in vitro* release studies detected a burst release profile of bioactive rhBMP-2 over 7 days whereby ~70 ng of rhBMP-2 was released as detected by ELISA, very little additional rhBMP-2 was detected as releasing after 21 days (~71 ng total), and only 27 ng of the active

TABLE 3: Recent publications of GF-loaded hydrogels.

Growth factor	Polymer	Quantity loaded	Release	Reference
rhBMP-2 and VEGF	Silk	40 μg/mL VEGF 60 μg/mL BMP-2	~15% VEGF after 28 days ~15% BMP-2 after 28 days	[147]
rhBMP-2	PEG	50 mg/mL	~56% after 15 days for glycosylated rhBMP-2, ~35% after 15 days for nonglycosylated	[154]
BMP-2	Alginate	Up to 25 μg/mL	~85% after 14 days	[46]
BMP-2	Alginate	40 μg/mL	~98% after 7 days	[25]
BMP-2	Hyaluronic acid	150 μg/mL	Not disclosed	[55]
BMP-2	Hyaluronic acid	150 μg/mL	Not conducted (animal study)	[64]
BMP-2	Chitosan	0.75 mg/mL	Not conducted (animal study)	[79]
BMP-2	Chitosan or hyaluronic acid	0.75 mg/mL	Not conducted (animal study)	[148]
BMP-2	Collagen-hydroxyapatite composite PEG	200 μg/mL	~72% after 3 days, ~95% after 7 days	[149]
BMP-2 and VEGF	Hyaluronic acid	0.01 mg/mL VEGF 0.01 mg/mL BMP-2	~100% after 4 days for fast degrading hydrogels ~100% after 7 days for slow degrading hydrogels (VEGF releasing slightly slower than BMP-2)	[49]
BMP-2	Hyaluronic acid	200 μg/mL	~10% after 24 hours ~25% after 35 days	[66]
VEGF-A and PDGF-BB	Alginate	30 μg/mL PDGF-BB and/or 3 μg/mL VEGF-A_{165}	~50% VEGF after 24 hours, ~80% after 30 days ~15% PDGF-BB after 24 hours, ~75% after 30 days	[155]
FGF and VEGF	PEG heparin	1 or 5 μg/mL	~10% after 4 days	[153]

rhBMP-2 was still detectable within the gel. The remaining undetected rhBMP-2 was theorised to have degraded during the 21 days of the release study. An *in vivo* model in critical sized rat femoral defects showed superior *de novo* bone formation in rhBMP-2 loaded hydrogels, when compared to unloaded alginate gels as examined by X-ray, μCT, and torsion testing. An interesting result was observed when the electrospun meshes were perforated, an increase in the bone formation occurred versus the unperforated mesh. It was hypothesised that the perforations allowed for superior infiltration of cellular material into the defect and increased vascularisation which in turn increased wound healing.

In a similar study, Boerckel et al. used rhBMP-2 (quantities of 0.1 μg to 5 μg) loaded alginate hydrogels combined with an electrospun PCL scaffold to provide physical support and retain alginate at the wound site and compared the rate of healing of an 8 mm femur defect when compared to a collagen sponge [46]. In order to improve the rate of cellular migration into the alginate hydrogel, RGD-binding sites were added to the alginate. However, μCT imaging showed that RGD binding sites were not sufficient, in isolation, to promote the formation of bone (Figure 7). As shown by the scaffold without rhBMP-2 and the 0.1 μg rhBMP-2-loaded scaffold failing to elicit successful defect closure after 12 weeks. In contrast the 1 μg loading of rhBMP-2 in the alginate/mesh composite scaffold showed good bone formation within the defect, but not the collagen mesh. It was theorised that the addition of the slowly degrading alginate hydrogel acted as a diffusion barrier, allowing slower release of rhBMP-2 than the collagen sponge. It is also worth noting that alginate is negatively charged and in theory should provide superior binding to the positively charged rhBMP-2, thus retaining more rhBMP-2 in the defect site compared to collagen [46].

Another recent study examining the release of rhBMP-2 in a dental application looked at two methods of delivering rhBMP-2, either bound to a sandblasted, acid-etched surface (SLA) titanium implant, which was then surrounded by an unloaded scaffold, or loaded onto the scaffold before being placed around the implant. The scaffold test groups were PEG hydrogels, collagen/HA (Col/HA) sponges and combination of the PEG hydrogel and Col/HA sponge. This study found that the latter group, the PEG hydrogel, and Col/HA exhibited what was deemed to be the best release characteristics reducing the level of burst release and providing a constant sustained release [149]. Using a single loaded PEG hydrogel led to a rapid initial burst of rhBMP-2 from the scaffold, while Col/HA prevented a large initial burst release of rhBMP-2, and the scaffold started to break down rapidly and rapidly released rhBMP-2 after 10 days in an *in vitro* model. The combined Col/HA PEG hydrogel revealed a slow initial release leading to a long-sustained release, which was

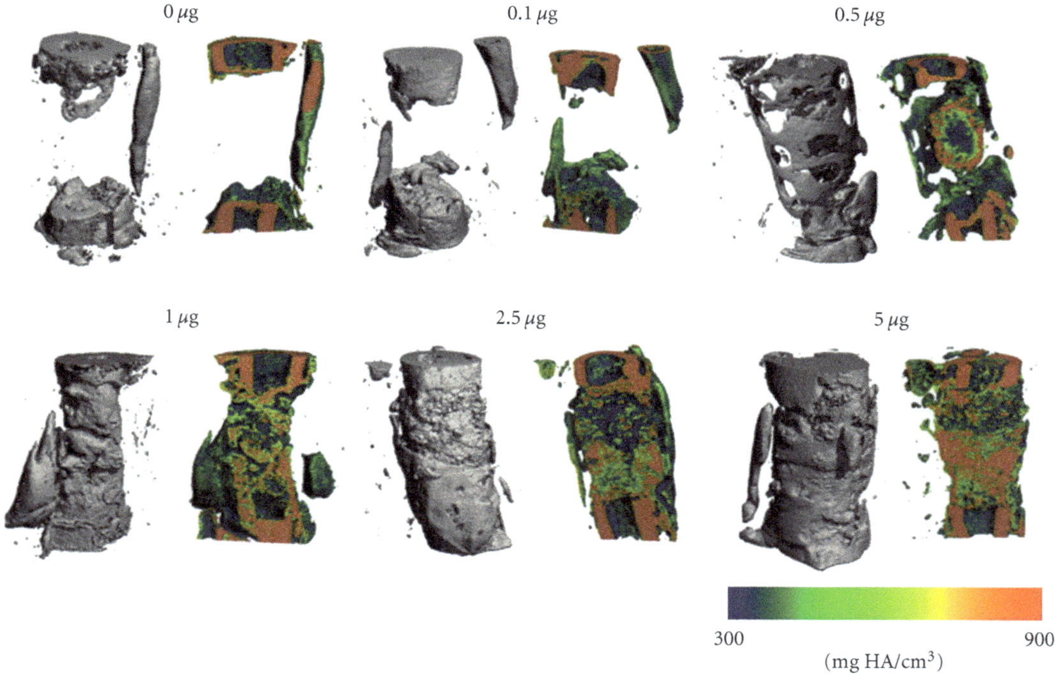

FIGURE 7: μCT reconstructions of 3D structure and sagittal cross-sections illustrating local mineral density mapping of critical sized femur defects of an alginate hydrogel rhBMP-2 delivery system. Dose of BMP-2 delivered per implant as shown [46].

attributed to the rhBMP-2 bound to both the Col/HA and the PEG hydrogel. In the *in vivo* study, a rat mandible model, the Col/HA PEG hydrogel had approximately twice the *de novo* bone formation than that of the Col/HA alone, as measured by μCT (Figure 8). Although the rhBMP-2 loaded on the SLA implant surrounded by the PEG gel itself produced the most *de novo* bone formation in the study, when the rhBMP-2 was loaded into the PEG hydrogel itself it produced the least, suggesting that it was mechanically unsuitable for the purpose.

6.6. Combining Different Loading Techniques for Delivering Multiple Growth Factors. Natural wound healing occurs as a part of a cascade involving multiple different cell types and GFs. While many studies have investigated the potential for individual GFs to facilitate the reconstruction of damaged tissues, it stands to reason that complementary GF loading may create a superior scaffold than loading a single GF. Multiple GF delivery has frequently been considered for bone formation studies, since several GFs are known to be involved in the complex process of bone repair/regeneration, and the strategy to mimic this delivery seems logical, albeit not trivial.

Alginate scaffolds, loaded with TGF-β3 and BMP-2, and transplanted with progenitor cells induced enhanced bone formation in mice compared to scaffolds loaded only with single GFs [156]. Similarly, bone formation was enhanced by incorporating TGF-β1 and IGF-1 in the PDLLA coating of titanium wires which were implanted in rats, showing synergistic effects of GFs [157]. However direct loading into scaffolds does not allow a precise control of the release kinetics and often when GFs are exposed to the medium, this may result in loss of bioactivity. Using micro- or nanoparticles overcomes this problem by protecting the GF, and allowing the use of various populations of particles with different degradation properties and thus different release kinetics of the encapsulated GF [105].

Basmanav et al. prepared polyelectrolyte complexes for encapsulation of BMP-2 and BMP-7 before further introduction into PLGA scaffolds and undertook *in vitro* testing with bone marrow-derived stem cells [158]. A release study was also undertaken using BSA as the BMP-2 substitute during the release assays. The authors selected the populations of microspheres that would rapidly release the BMP-2 and then release the BMP-7 much later. It was shown that the delivery system had no effect on proliferation but enhanced osteogenic differentiation to a higher degree than with single administration of either BMP-2 or BMP-7 alone [158]. In another study of dual delivery applied to cartilage regeneration, different populations of PLGA microspheres containing IGF-I and TGF-β1 were fused together to provide sequential release [159]. It was shown that the materials and additives used during processing had significant influences on burst release. The incorporation of BSA in the IGF-I formulation decreased the initial burst release from 80% to 20% while using uncapped PLGA containing a carboxylic acid terminal group for the TGF-β1 formulation did not produce any burst release compared to a 60% burst release using uncapped PLGA alone. Three types of scaffolds were further fabricated with the different microsphere formulations providing tailored release kinetics (either simultaneous, initial release of TGF-β1 followed by IGF-I, or initial release of IGF-I followed

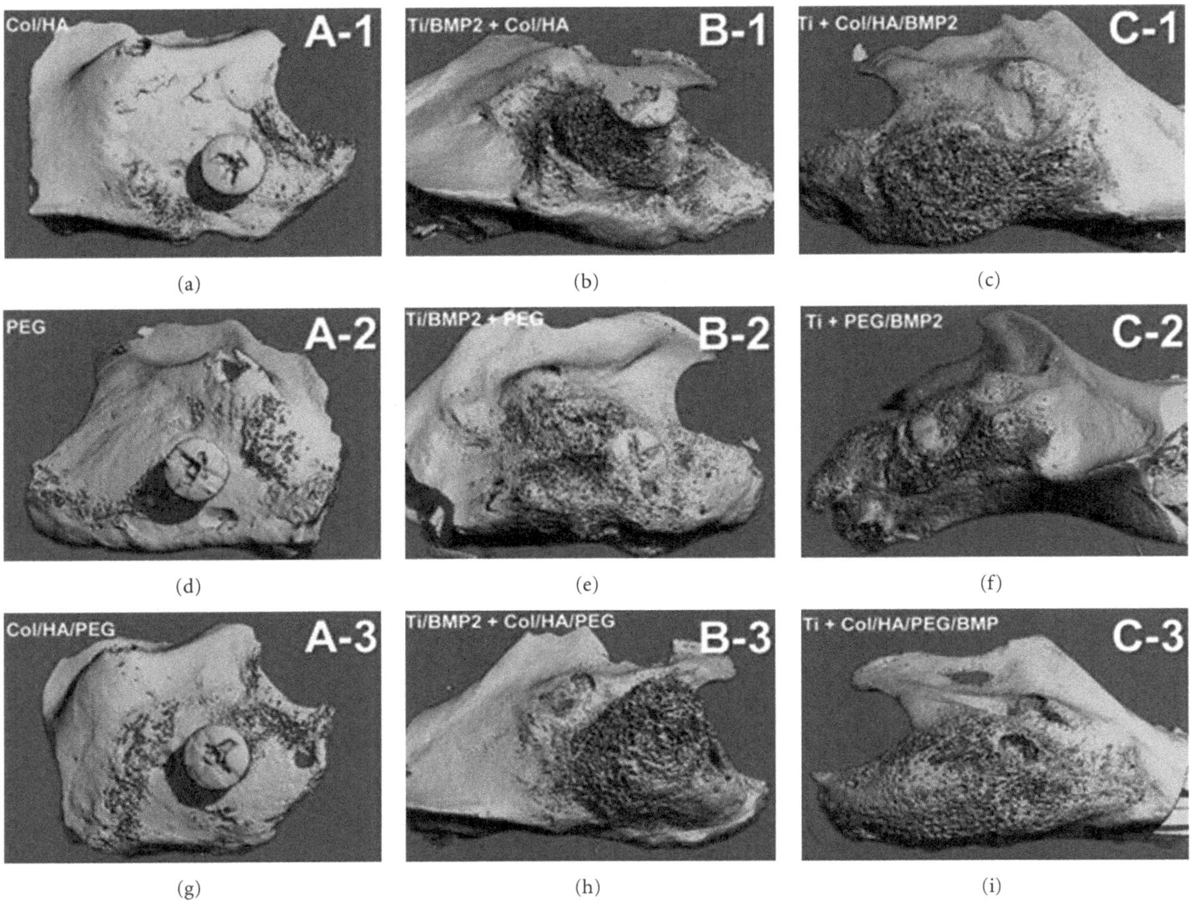

FIGURE 8: μCT images of *de novo* bone growth around a titanium screw implant [149], showing either unloaded scaffolds (A), BMP-2 loading onto the screw (B), or BMP-2 loaded onto the surrounding implant (C).

by TGF-β1). However neither *in vitro* nor *in vivo* studies were performed for assessment of the potential of these scaffolds for tissue regeneration [159].

BMPs are central to bone regeneration due to their osteoinduction potential. However, the formation of blood vessels (angiogenesis) is also essential for successful bone regeneration and in this context, vascular endothelial growth factor (VEGF) is known to be the key GF to regulate angiogenesis. VEGF has been shown to interact with BMPs, and both types of GFs were shown to stimulate the proliferation of either endothelial cells or osteoblasts, respectively, and differentiation of osteoprogenitor cells [7, 160]. VEGF has been successfully encapsulated in several types of devices to obtain a controlled release of the GFs such as PLLA scaffolds using super critical CO_2 ($scCO_2$) processing [87, 161], PLGA scaffolds using gas foaming/particulate leaching process [88], and in microspheres (s/o/w processed PLLA microspheres [162, 163]) suggesting potential as part of a dual delivery system incorporating BMPs. For instance, Kempen et al. recently produced a dual delivery system and tested it ectopically (subcutaneously) and orthotopically (critical sized femoral defect) in rats [45]. BMP-2 was loaded in PLGA microspheres through the w/o/w process (leading to an 85% encapsulation efficiency) and further embedded into a polypropylene fumarate rod by photo-cross-linking. In parallel, a gelatin hydrogel was processed as a tubular implant and impregnated with VEGF before association with the PPF rod previously loaded with microspheres (see Figure 9). This system was aimed at giving a sequential release of the entrapped GF: an initial peak of VEGF, followed by a sustained release of BMP-2, since the same pattern has been observed in normal bone healing [142]. As expected, the *in vivo* profiles were faster than *in vitro*, showing a rapid release of VEGF in the first 2 weeks and a more sustained release of BMP-2 over the full 8-week period. In ectopic defects, the presence of VEGF with scaffolds loaded with BMP-2 was shown to significantly enhance bone formation compared to the BMP-2 only scaffolds. However, there were no significant differences in orthotopic defects. It was hypothesized that the peak release of VEGF occurred too early for orthotopic defects while it was sufficient for ectopic formation, since circulation could have enhanced blood flow and thus accessibility of the implantation site for cells [142].

In a similar study, BMP-2 and VEGF were loaded by diffusion into gelatin microspheres and further incorporated into PPF scaffolds for implantation into a rat cranial critical sized defect [164]. The dual release group showed significantly higher bone formation than other groups at 4 weeks

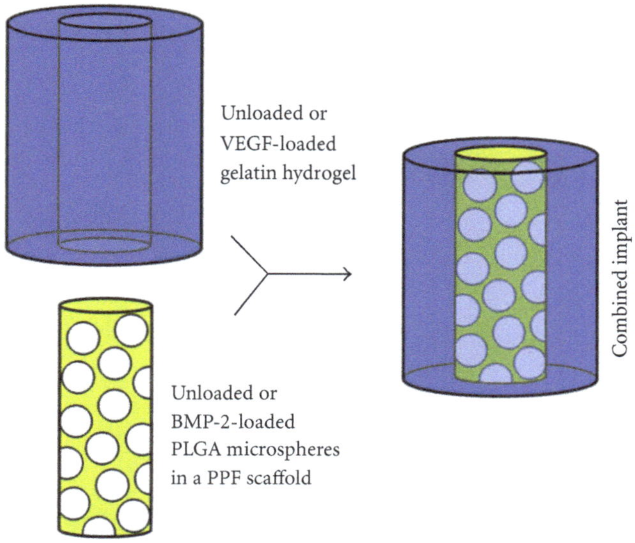

FIGURE 9: Schematic showing the composite scaffolds for the sequential release of VEGF and BMP-2. The PPF rod and the gelatin shell were prepared separately and combined just before implantation [45].

but was equivalent to the BMP-2 group at 12 weeks, suggesting only a synergistic effect of the GFs for early bone regeneration. However, the dual delivery system showed enhanced bone bridging and union of the defect compared to the BMP-2 group at 12 weeks as detected by μCT.

As previously discussed, the ability to inject a hydrogel into a wound site creates a system which is surgically minimally invasive. Zhang et al. have investigated the potential of silk fibroin as dual delivery system for rhBMP-2 and $VEGF_{165}$ [147]. The experimental design compared the effect of loading either rhBMP-2 or $VEGF_{165}$ or both GFs into the hydrogel then injecting them into a maxillary sinus floor model in rats. An *in vitro* release kinetics study on 100 μL of the dual loaded gel showed a steady sustained release over the 28-day study. No noticeable initial burst of both rhBMP-2 and $VEGF_{165}$ was detected (ELISA), with approximately 70% as much $VEGF_{165}$ being released as rhBMP-2. The presence of rhBMP-2 in the hydrogel increased the level of sinus floor by an average of 11.5 mm ($n = 6$) compared to the unloaded silk hydrogel (7.5 mm). While rhBMP-2 elevated the sinus floor as much as the dual loaded scaffold (11.5 mm), μCT suggested that the dual loaded hydrogel had produced a more dense bone mineralisation although this difference was not statistically significant.

Kanczler et al. used $scCO_2$ to produce P(D,L)LA scaffolds to load rhBMP-2 and combined this with an alginate hydrogel loaded with $VEGF_{165}$ [91]. This technique was able to produce two independent release curves, with an initial release of $VEGF_{165}$ from day 0 occurring in a linear fashion of $\sim$100 pg/day. In comparison, there was very limited initial BMP-2 released from the PLLA implant; however, the final release profile was exponential with more BMP-2 released than $VEGF_{165}$ after 7 days. Further *in vitro* testing showed that both GFs maintained bioactivity in culture, and the implant was able to stimulate an osteogenic response from the BMP-2 in C2C12 cells and an angiogenic response from the $VEGF_{165}$ in HUVECs. However, it should be noted that *in vivo* work on mouse critical-sized femoral defects showed no significant effect of the loaded GFs in inducing bone formation without the addition of human bone marrow stromal cells to the implant [91].

Certainly, these studies underline the complexity of dual delivery of VEGF and BMPs. VEGF remains an interesting option for complete bone formation and is certainly a key GF for angiogenesis in general, as shown in a dual delivery of w/o/w-processed PLGA microspheres containing PDGF incorporated in a scaffold containing VEGF, for enhanced generation of mature vascular network [165]. The selection of the correct GF cocktail and the control of the relationship between concentration and timing considerations remain two major challenges in GF delivery [158]. However, the specific BMP-2 GF has multiple roles in bone healing, shown to play a role in both early and late time points; therefore, it may be easier to achieve the optimization of its release through a single GF delivery perspective rather than a dual approach and provide a considerable advance in bone TE with this optimized release of a single BMP. Nevertheless, such a release might not be easily achieved with the traditional encapsulation processes, and new techniques need to be studied and optimized as better alternatives.

7. Conclusion

The currently available clinical treatments for bone-related deficiencies are not ideal for the healing of large bone defects. Tissue engineering has widely been acknowledged as a promising approach to tackle this issue, although with nearly 30 years of research, while much insight has been gained, an optimal solution has yet to be realised. A tissue engineering scaffold must provide a wide range of functions; including physical robustness, as well as promoting *de novo* bone formation. Polymeric scaffolds offer promising physical characteristics, but alone they have proven to lack sufficient osteoinductive capacity. GFs such as rhBMP-2 have shown the ability to promote bone formation in a clinical environment, but, due to short active lifespan in the defect environment, there remains a challenge of delivery. The ability to entrap within polymeric scaffolds offers a potential solution to this problem, producing a scaffold which would have both the structural capacity and osteogenic potential required for successful treatment.

While no particular individual strategy has proven so far to be sufficient; see Table 4 for a summary of advantages and drawbacks of the most popular techniques, entrapping GFs into scaffolds shows much promise.

8. Future Perspectives

Wound healing is a complex process, and many current studies still only focus on the release of a single growth factor.

TABLE 4: Current strategies for growth factor delivery in bone tissue engineering and the associated advantages and drawbacks.

Delivery system	Release characteristics	Advantages	Drawbacks
Supercritical CO_2	Sustained release	Good encapsulation efficiency Very high bioactivity retention Solvent less formation	Limited available scaffold conformations suitable for bone tissue engineering Complex equipment requirement for manufacture
Electrospinning	Varying release	Rapid fabrication of scaffold Large number of polymers available for electrospinning	Limited available scaffold conformations suitable for bone tissue engineering Requires organic solvents which can denature GFs and reduce bioactivity Limited information on encapsulation efficiency Limited information on release characteristics
Nano/microspheres	Initial burst then slow sustain	Several techniques available for fabrication Easy to combine with other scaffold fabrication techniques	Limited available scaffold conformations suitable for bone tissue engineering Requires organic solvents which can denature GFs and reduce bioactivity Classical formation techniques have poor encapsulation efficiency
Hydrogels	Short-term release	Can be injected into wound sites (*in situ* formation) Good encapsulation efficiency Solvent less formation Easy to combine with other techniques	Rapid degradation of most commonly used hydrogels Unsuitable mechanical characteristics for bone tissue engineering
Functionalisation of surfaces, for example, polyelectrolyte systems nanocoating, and so forth.	Short-term release	Can be applied to scaffolds with good mechanical properties High bioactivity retention	Technique is time intensive Alters existing surface chemistry

Advancing knowledge through combining scaffolds with different release profiles gives the potential to load different growth factors into a single constructed scaffold, which, when implanted will lead to a cascade of temporal GF release, for example, providing VEGF early to a wound site, followed by a slower release of rhBMP2 after an established blood supply has formed. Such scaffold design shows promise to be a superior single step solution for the treatment of critical sized bone defects and may ultimately negate the need for autografting and permanent implants.

While the biological activity of a scaffold can be enhanced by GFs, there are other techniques which when employed with GF loading could produce more rapid cellular infiltration and differentiation than either technique alone. The scope for codelivery of GFs with other therapeutic agents such as short-interfering RNA sequences and other osteogenic promoters such as hydroxyl-apatite is another avenue which has great potential.

For a complete clinical solution, a tissue-engineered scaffold will have to be able to fulfil a number of criteria, including mechanical suitability and biological activity. GF loading into polymeric scaffolds is a promising avenue of research. The technique both provides a mechanically robust scaffold, in addition to a system for controlling the release of GF into the defect, but there is still much work to be done.

The rate of entrapment, denaturation rate, and release of the GFs from the scaffold must all be carefully considered when designing such systems; many recent studies are moving towards utilising far less overall GF than those currently used in the clinic and also provide the opportunity to maintain a therapeutic level for longer. This drive to avoid the supraphysiological loadings present in clinical setting will hopefully provide a safer, cheapers and more efficacious treatment.

Conflict of Interests

All authors would like to declare that they do not have a direct financial relationship with any of the commercial products mentioned within the paper.

Authors' Contribution

K. A. Blackwood and N. Bock have contributed equally to this work.

Acknowledgment

The authors would like to thank the Australian Research Council (ARC) for the Linkage Grants LP110200082 and LP100200084.

References

[1] R. Burge, B. Dawson-Hughes, D. H. Solomon, J. B. Wong, A. King, and A. Tosteson, "Incidence and economic burden of osteoporosis-related fractures in the United States, 2005–2025," *Journal of Bone and Mineral Research*, vol. 22, no. 3, pp. 465–475, 2007.

[2] K. M. Sanders, G. C. Nicholson, A. M. Ugoni, J. A. Pasco, E. Seeman, and M. A. Kotowicz, "Health burden of hip and other fractures in Australia beyond 2000. Projections based on the Geelong Osteoporosis Study," *Medical Journal of Australia*, vol. 170, no. 10, pp. 467–470, 1999.

[3] N. Ferrara, H. P. Gerber, and J. LeCouter, "The biology of VEGF and its receptors," *Nature Medicine*, vol. 9, no. 6, pp. 669–676, 2003.

[4] T. Katagiri, A. Yamaguchi, M. Komaki et al., "Bone morphogenetic protein-2 converts the differentiation pathway of C2C12 myoblasts into the osteoblast lineage," *The Journal of Cell Biology*, vol. 127, no. 6, pp. 1755–1766, 1994.

[5] M. Inada, T. Katagiri, S. Akiyama et al., "Bone morphogenetic protein-12 and -13 inhibit terminal differentiation of myoblasts, but do not induce their differentiation into osteoblasts," *Biochemical and Biophysical Research Communications*, vol. 222, no. 2, pp. 317–322, 1996.

[6] C. A. Kirker-Head, "Potential applications and delivery strategies for bone morphogenetic proteins," *Advanced Drug Delivery Reviews*, vol. 43, no. 1, pp. 65–92, 2000.

[7] M. M. L. Deckers, R. L. van Bezooijen, G. van der Horst et al., "Bone morphogenetic proteins stimulate angiogenesis through osteoblast-derived vascular endothelial growth factor A," *Endocrinology*, vol. 143, no. 4, pp. 1545–1553, 2002.

[8] J. Fiedler, G. Röderer, K. P. Günther, and R. E. Brenner, "BMP-2, BMP-4, and PDGF-bb stimulate chemotactic migration of primary human mesenchymal progenitor cells," *Journal of Cellular Biochemistry*, vol. 87, no. 3, pp. 305–312, 2002.

[9] W. Friess, H. Uludag, S. Foskett, R. Biron, and C. Sargeant, "Characterization of absorbable collagen sponges as rhBMP-2 carriers," *International Journal of Pharmaceutics*, vol. 187, no. 1, pp. 91–99, 1999.

[10] J. O. Hollinger, J. M. Schmitt, D. C. Buck et al., "Recombinant human bone morphogenetic protein-2 and collagen for bone regeneration," *Journal of Biomedical Materials Research*, vol. 43, no. 4, pp. 356–364, 1998.

[11] E. J. Carragee, E. L. Hurwitz, and B. K. Weiner, "A critical review of recombinant human bone morphogenetic protein-2 trials in spinal surgery: emerging safety concerns and lessons learned," *Spine Journal*, vol. 11, no. 6, pp. 471–491, 2011.

[12] R. Vasita and D. S. Katti, "Growth factor-delivery systems for tissue engineering: a materials perspective," *Expert Review of Medical Devices*, vol. 3, no. 1, pp. 29–47, 2006.

[13] R. R. Chen and D. J. Mooney, "Polymeric growth factor delivery strategies for tissue engineering," *Pharmaceutical Research*, vol. 20, no. 8, pp. 1103–1112, 2003.

[14] J. R. Porter, T. T. Ruckh, and K. C. Popat, "Bone tissue engineering: a review in bone biomimetics and drug delivery strategies," *Biotechnology Progress*, vol. 25, no. 6, pp. 1539–1560, 2009.

[15] M. Noda, *Cellular and Molecular Biology of Bone*, Academic Press, New York, NY, USA, 1993.

[16] J. C. Reichert, S. Saifzadeh, M. E. Wullschleger et al., "The challenge of establishing preclinical models for segmental bone defect research," *Biomaterials*, vol. 30, no. 12, pp. 2149–2163, 2009.

[17] C. Szpalski, J. Barr, M. Wetterau, P. B. Saadeh, and S. M. Warren, "Cranial bone defects: current and future strategies," *Neurosurgical Focus*, vol. 29, no. 6, p. E8, 2010.

[18] A. R. Vaccaro, K. Singh, R. Haid et al., "The use of bioabsorbable implants in the spine," *Spine Journal*, vol. 3, no. 3, pp. 227–237, 2003.

[19] P. N. Soucacos, E. O. Johnson, and G. Babis, "An update on recent advances in bone regeneration," *Injury*, vol. 39, supplement 2, pp. S1–S4, 2008.

[20] E. M. Younger and M. W. Chapman, "Morbidity at bone graft donor sites," *Journal of orthopaedic trauma*, vol. 3, no. 3, pp. 192–195, 1989.

[21] R. Dimitriou, G. I. Mataliotakis, A. G. Angoules, N. K. Kanakaris, and P. V. Giannoudis, "Complications following autologous bone graft harvesting from the iliac crest and using the RIA: a systematic review," *Injury*, vol. 42, supplement 2, pp. S3–S15, 2011.

[22] B. A. Rogers, A. Sternheim, D. Backstein, O. Safir, and A. E. Gross, "Proximal femoral allograft for major segmental femoral bone loss: a systematic literature review," *Advances in Orthopedics*, vol. 2011, Article ID 257572, 7 pages, 2011.

[23] M. A. Woodruff and D. W. Hutmacher, "The return of a forgotten polymer—polycaprolactone in the 21st century," *Progress in Polymer Science*, vol. 35, no. 10, pp. 1217–1256, 2010.

[24] M. Geiger, R. H. Li, and W. Friess, "Collagen sponges for bone regeneration with rhBMP-2," *Advanced Drug Delivery Reviews*, vol. 55, no. 12, pp. 1613–1629, 2003.

[25] Y. M. Kolambkar, K. M. Dupont, J. D. Boerckel et al., "An alginate-based hybrid system for growth factor delivery in the functional repair of large bone defects," *Biomaterials*, vol. 32, no. 1, pp. 65–74, 2011.

[26] Y. Gu, L. Chen, H. L. Yang, Z. P. Luo, and T. S. Tang, "Evaluation of an injectable silk fibroin enhanced calcium phosphate cement loaded with human recombinant bone morphogenetic protein-2 in ovine lumbar interbody fusion," *Journal of Biomedical Materials Research A*, vol. 97, no. 2, pp. 177–185, 2011.

[27] E. S. Place, N. D. Evans, and M. M. Stevens, "Complexity in biomaterials for tissue engineering," *Nature Materials*, vol. 8, no. 6, pp. 457–470, 2009.

[28] M. Vert, S. M. Li, G. Spenlehauer, and P. Guerin, "Bioresorbability and biocompatibility of aliphatic polyesters," *Journal of Materials Science*, vol. 3, no. 6, pp. 432–446, 1992.

[29] A. U. Daniels, M. K. Chang, and K. P. Andriano, "Mechanical properties of biodegradable polymers and composites proposed for internal fixation of bone," *Journal of Applied Biomaterials*, vol. 1, no. 1, pp. 57–78, 1990.

[30] R. K. Kulkarni, K. C. Pani, C. Neuman, and F. Leonard, "Polylactic acid for surgical implants," *Archives of Surgery*, vol. 93, no. 5, pp. 839–843, 1966.

[31] J. Y. Paeng, J. Hong, C. S. Kim, and M. J. Kim, "Comparative study of skeletal stability between bicortical resorbable and titanium screw fixation after sagittal split ramus osteotomy for mandibular prognathism," *Journal of Cranio-Maxillofacial Surgery*. In press.

[32] R. E. Holmes, S. R. Cohen, G. B. Cornwall, K. A. Thomas, K. K. Kleinhenz, and M. Z. Beckett, "MacroPore resorbable devices in craniofacial surgery," *Clinics in Plastic Surgery*, vol. 31, no. 3, pp. 393–406, 2004.

[33] N. Ashammakhi and P. Rokkanen, "Absorbable polyglycolide devices in trauma and bone surgery," *Biomaterials*, vol. 18, no. 1, pp. 3–9, 1997.

[34] B. C. Kim, B. L. Padwa, H. S. Park, and Y. S. Jung, "Stability of maxillary position after le fort i osteotomy using self-reinforced biodegradable poly-70L/30DL-lactide miniplates and screws," *Journal of Oral and Maxillofacial Surgery*, vol. 69, no. 5, pp. 1442–1446, 2011.

[35] M. Kamitakahara, C. Ohtsuki, and T. Miyazaki, "Behavior of ceramic biomaterials derived from tricalcium phosphate in physiological condition," *Journal of Biomaterials Applications*, vol. 23, no. 3, pp. 197–212, 2008.

[36] E. S. Place, L. Rojo, E. Gentleman, J. P. Sardinha, and M. M. Stevens, "Strontium-and zinc-alginate hydrogels for bone tissue engineering," *Tissue Engineering A*, vol. 17, no. 21-22, pp. 2713–2722, 2011.

[37] Z. S. Haidar, R. C. Hamdy, and M. Tabrizian, "Delivery of recombinant bone morphogenetic proteins for bone regeneration and repair. Part B: delivery systems for BMPs in orthopaedic and craniofacial tissue engineering," *Biotechnology Letters*, vol. 31, no. 12, pp. 1825–1835, 2009.

[38] D. Williams, "Revisiting the definition of biocompatibility," *Medical Device Technology*, vol. 14, no. 8, pp. 10–13, 2003.

[39] D. Chen, M. Zhao, and G. R. Mundy, "Bone morphogenetic proteins," *Growth Factors*, vol. 22, no. 4, pp. 233–241, 2004.

[40] A. H. Reddi, "Bone morphogenetic proteins: from basic science to clinical applications," *Journal of Bone and Joint Surgery A*, vol. 83, supplement 1, part 1, pp. S1–S6, 2001.

[41] M. I. Menendez, D. J. Clark, M. Carlton et al., "Direct delayed human adenoviral BMP-2 or BMP-6 gene therapy for bone and cartilage regeneration in a pony osteochondral model," *Osteoarthritis and Cartilage*, vol. 19, no. 8, pp. 1066–1075, 2011.

[42] J. H. Christiansen, E. G. Coles, and D. G. Wilkinson, "Molecular control of neural crest formation, migration and differentiation," *Current Opinion in Cell Biology*, vol. 12, no. 6, pp. 719–724, 2000.

[43] R. W. Haid Jr., C. L. Branch Jr., J. T. Alexander, and J. K. Burkus, "Posterior lumbar interbody fusion using recombinant human bone morphogenetic protein type 2 with cylindrical interbody cages," *Spine Journal*, vol. 4, no. 5, pp. 527–538, 2004.

[44] Z. S. Haidar, R. C. Hamdy, and M. Tabrizian, "Delivery of recombinant bone morphogenetic proteins for bone regeneration and repair. Part A: current challenges in BMP delivery," *Biotechnology Letters*, vol. 31, no. 12, pp. 1817–1824, 2009.

[45] D. H. R. Kempen, L. Lu, A. Heijink et al., "Effect of local sequential VEGF and BMP-2 delivery on ectopic and orthotopic bone regeneration," *Biomaterials*, vol. 30, no. 14, pp. 2816–2825, 2009.

[46] J. D. Boerckel, Y. M. Kolambkar, K. M. Dupont et al., "Effects of protein dose and delivery system on BMP-mediated bone regeneration," *Biomaterials*, vol. 32, no. 22, pp. 5241–5251, 2011.

[47] Q. Li, T. Hou, J. Zhao, and J. Xu, "Vascular endothelial growth factor release from alginate microspheres under simulated physiological compressive loading and the effect on human vascular endothelial cells," *Tissue Engineering A*, vol. 17, no. 13-14, pp. 1777–1785, 2011.

[48] A. Lochmann, H. Nitzsche, S. von Einem, E. Schwarz, and K. Mäder, "The influence of covalently linked and free polyethylene glycol on the structural and release properties of rhBMP-2 loaded microspheres," *Journal of Controlled Release*, vol. 147, no. 1, pp. 92–100, 2010.

[49] J. Patterson, R. Siew, S. W. Herring, A. S. P. Lin, R. Guldberg, and P. S. Stayton, "Hyaluronic acid hydrogels with controlled degradation properties for oriented bone regeneration," *Biomaterials*, vol. 31, no. 26, pp. 6772–6781, 2010.

[50] N. Bock, M. A. Woodruff, D. W. Hutmacher, and T. R. Dargaville, "Electrospraying, a reproducible method for production of polymeric microspheres for biomedical applications," *Polymers*, vol. 3, no. 1, pp. 131–149, 2011.

[51] A. Sukarto and B. G. Amsden, "Low melting point amphiphilic microspheres for delivery of bone morphogenetic protein-6 and transforming growth factor-$\beta 3$ in a hydrogel matrix," *Journal of Controlled Release*, vol. 158, no. 1, pp. 53–62, 2012.

[52] L. Solorio, C. Zwolinski, A. W. Lund, M. J. Farrell, and J. P. Stegemann, "Gelatin microspheres crosslinked with genipin for local delivery of growth factors," *Journal of Tissue Engineering and Regenerative Medicine*, vol. 4, no. 7, pp. 514–523, 2010.

[53] D. H. R. Kempen, L. Lu, T. E. Hefferan et al., "Retention of in vitro and in vivo BMP-2 bioactivities in sustained delivery vehicles for bone tissue engineering," *Biomaterials*, vol. 29, no. 22, pp. 3245–3252, 2008.

[54] S. Chen, A. Osaka, T. Ikoma et al., "Fabrication, microstructure, and BMP-2 delivery of novel biodegradable and biocompatible silicate-collagen hybrid fibril sheets," *Journal of Materials Chemistry*, vol. 21, no. 29, pp. 10942–10948, 2011.

[55] E. Martínez-Sanz, D. A. Ossipov, J. Hilborn, S. Larsson, K. B. Jonsson, and O. P. Varghese, "Bone reservoir: injectable hyaluronic acid hydrogel for minimal invasive bone augmentation," *Journal of Controlled Release*, vol. 152, no. 2, pp. 232–240, 2011.

[56] T. Jiang, Y. Khan, L. S. Nair, W. I. Abdel-Fattah, and C. T. Laurencin, "Functionalization of chitosan/poly(lactic acid-glycolic acid) sintered microsphere scaffolds via surface heparinization for bone tissue engineering," *Journal of Biomedical Materials Research A*, vol. 93, no. 3, pp. 1193–1208, 2010.

[57] G. J. Dawes, L. E. Fratila-Apachitei, B. S. Necula et al., "Effects of dexamethasone loaded plga microspheres on human fetal osteoblasts," *Journal of Biomaterials Applications*, vol. 27, no. 4, pp. 477–483, 2012.

[58] P. C. Bessa, E. R. Balmayor, H. S. Azevedo et al., "Silk fibroin microparticles as carriers for delivery of human recombinant BMPs. Physical characterization and drug release," *Journal of Tissue Engineering and Regenerative Medicine*, vol. 4, no. 5, pp. 349–355, 2010.

[59] M. R. Johnson, H. J. Lee, R. V. Bellamkonda, and R. E. Guldberg, "Sustained release of BMP-2 in a lipid-based microtube vehicle," *Acta Biomaterialia*, vol. 5, no. 1, pp. 23–28, 2009.

[60] T. M. Filion, A. Kutikov, and J. Song, "Chemically modified cellulose fibrous meshes for use as tissue engineering scaffolds," *Bioorganic and Medicinal Chemistry Letters*, vol. 21, no. 17, pp. 5067–5070, 2011.

[61] G. T. S. Kirby, L. J. White, C. V. Rahman et al., "PLGA-based microparticles for the sustained release of BMP-2," *Polymers*, vol. 3, no. 1, pp. 571–586, 2011.

[62] J. P. Schmitz and J. O. Hollinger, "The critical size defect as an experimental model for craniomandibulofacial nonunions," *Clinical Orthopaedics and Related Research*, vol. 205, no. 205, pp. 299–308, 1986.

[63] Y. Ji, G. P. Xu, Z. P. Zhang, J. J. Xia, J. L. Yan, and S. H. Pan, "BMP-2/PLGA delayed-release microspheres composite graft, selection of bone particulate diameters, and prevention of aseptic inflammation for bone tissue engineering," *Annals of Biomedical Engineering*, vol. 38, no. 3, pp. 632–639, 2010.

[64] G. Hulsart-Billström, Q. Hu, K. Bergman et al., "Calcium phosphates compounds in conjunction with hydrogel as carrier for BMP-2: a study on ectopic bone formation in rats," *Acta Biomaterialia*, vol. 7, no. 8, pp. 3042–3049, 2011.

[65] M. Larsen, W. F. Willems, M. Pelzer, P. F. Friedrich, M. J. Yaszemski, and A. T. Bishop, "Augmentation of surgical angiogenesis in vascularized bone allotransplants with host-derived A/V bundle implantation, fibroblast growth factor-2, and vascular endothelial growth factor administration," *Journal of Orthopaedic Research*, vol. 28, no. 8, pp. 1015–1021, 2010.

[66] S. N. Rath, G. Pryymachuk, O. A. Bleiziffer et al., "Hyaluronan-based heparin-incorporated hydrogels for generation of axially vascularized bioartificial bone tissues: in vitro and in vivo evaluation in a PLDLLA-TCP-PCL-composite system," *Journal of Materials Science*, vol. 22, no. 5, pp. 1279–1291, 2011.

[67] W. F. Willems, M. Larsen, G. Giusti, P. F. Friedrich, and A. T. Bishop, "Revascularization and bone remodeling of frozen allografts stimulated by intramedullary sustained delivery of FGF-2 and VEGF," *Journal of Orthopaedic Research*, vol. 29, no. 9, pp. 1431–1436, 2011.

[68] M. D. Schofer, P. P. Roessler, J. Schaefer et al., "Electrospun plla nanofiber scaffolds and their use in combination with bmp-2 for reconstruction of bone defects," *PLoS One*, vol. 6, no. 9, Article ID e25462, 2011.

[69] A. Suzuki, H. Terai, H. Toyoda et al., "A biodegradable delivery system for antibiotics and recombinant human bone morphogenetic protein-2: a potential treatment for infected bone defects," *Journal of Orthopaedic Research*, vol. 24, no. 3, pp. 327–332, 2006.

[70] S. Young, Z. S. Patel, J. D. Kretlow et al., "Dose effect of dual delivery of vascular endothelial growth factor and bone morphogenetic protein-2 on bone regeneration in a rat critical-size defect model," *Tissue Engineering A*, vol. 15, no. 9, pp. 2347–2362, 2009.

[71] Z. X. Wu, D. Liu, S. Y. Wan, G. Cui, Y. Zhang, and W. Lei, "Sustained-release rhBMP-2 increased bone mass and bone strength in an ovine model of postmenopausal osteoporosis," *Journal of Orthopaedic Science*, vol. 16, no. 1, pp. 99–104, 2011.

[72] M. Li, X. Liu, X. Liu, and B. Ge, "Calcium phosphate cement with BMP-2-loaded gelatin microspheres enhances bone healing in osteoporosis: a pilot study," *Clinical Orthopaedics and Related Research*, vol. 468, no. 7, pp. 1978–1985, 2010.

[73] Y. M. Kolambkar, J. D. Boerckel, K. M. Dupont et al., "Spatiotemporal delivery of bone morphogenetic protein enhances functional repair of segmental bone defects," *Bone*, vol. 49, no. 3, pp. 485–492, 2011.

[74] C. Kirker-Head, V. Karageorgiou, S. Hofmann et al., "BMP-silk composite matrices heal critically sized femoral defects," *Bone*, vol. 41, no. 2, pp. 247–255, 2007.

[75] K. V. Brown, B. Li, T. Guda, D. S. Perrien, S. A. Guelcher, and J. C. Wenke, "Improving bone formation in a rat femur segmental defect by controlling bone morphogenetic protein-2 release," *Tissue Engineering A*, vol. 17, no. 13-14, pp. 1735–1746, 2011.

[76] O. Jeon, S. J. Song, S. W. Kang, A. J. Putnam, and B. S. Kim, "Enhancement of ectopic bone formation by bone morphogenetic protein-2 released from a heparin-conjugated poly(l-lactic-co-glycolic acid) scaffold," *Biomaterials*, vol. 28, no. 17, pp. 2763–2771, 2007.

[77] O. Jeon, J. W. Rhie, I. K. Kwon, J. H. Kim, B. S. Kim, and S. H. Lee, "In vivo bone formation following transplantation of human adipose-derived stromal cells that are not differentiated osteogenically," *Tissue Engineering A*, vol. 14, no. 8, pp. 1285–1294, 2008.

[78] A. M. Henslee, P. P. Spicer, D. M. Yoon et al., "Biodegradable composite scaffolds incorporating an intramedullary rod and delivering bone morphogenetic protein-2 for stabilization and bone regeneration in segmental long bone defects," *Acta Biomaterialia*, vol. 7, no. 10, pp. 3627–3637, 2011.

[79] L. Luca, A. L. Rougemont, B. H. Walpoth et al., "Injectable rhBMP-2-loaded chitosan hydrogel composite: Osteoinduction at ectopic site and in segmental long bone defect," *Journal of Biomedical Materials Research A*, vol. 96, no. 1, pp. 66–74, 2011.

[80] J. C. Reichert, D. R. Epari, M. E. Wullschleger et al., "Establishment of a preclinical ovine model for tibial segmental bone defect repair by applying bone tissue engineering strategies," *Tissue Engineering B*, vol. 16, no. 1, pp. 93–104, 2010.

[81] M. J. Whitaker, R. A. Quirk, S. M. Howdle, and K. M. Shakesheff, "Growth factor release from tissue engineering scaffolds," *Journal of Pharmacy and Pharmacology*, vol. 53, no. 11, pp. 1427–1437, 2001.

[82] S. E. Sakiyama-Elbert and J. A. Hubbell, "Controlled release of nerve growth factor from a heparin-containing fibrin-based cell ingrowth matrix," *Journal of Controlled Release*, vol. 69, no. 1, pp. 149–158, 2000.

[83] S. W. Kang, J. S. Kim, K. S. Park et al., "Surface modification with fibrin/hyaluronic acid hydrogel on solid-free form-based scaffolds followed by BMP-2 loading to enhance bone regeneration," *Bone*, vol. 48, no. 2, pp. 298–306, 2011.

[84] M. Sokolsky-Papkov, K. Agashi, A. Olaye, K. Shakesheff, and A. J. Domb, "Polymer carriers for drug delivery in tissue engineering," *Advanced Drug Delivery Reviews*, vol. 59, no. 4-5, pp. 187–206, 2007.

[85] W. Chaisri, A. H. Ghassemi, W. E. Hennink, and S. Okonogi, "Enhanced gentamicin loading and release of PLGA and PLHMGA microspheres by varying the formulation parameters," *Colloids and Surfaces B*, vol. 84, no. 2, pp. 508–514, 2011.

[86] E. Saito, E. E. Liao, W. W. Hu, P. H. Krebsbach, and S. J. Hollister, "Effects of designed PLLA and 50:50 PLGA scaffold architectures on bone formation in vivo," *Journal of Tissue Engineering and Regenerative Medicine*. In press.

[87] J. M. Kanczler, J. Barry, P. Ginty, S. M. Howdle, K. M. Shakesheff, and R. O. C. Oreffo, "Supercritical carbon dioxide generated vascular endothelial growth factor encapsulated poly(dl-lactic acid) scaffolds induce angiogenesis in vitro," *Biochemical and Biophysical Research Communications*, vol. 352, no. 1, pp. 135–141, 2007.

[88] W. L. Murphy, M. C. Peters, D. H. Kohn, and D. J. Mooney, "Sustained release of vascular endothelial growth factor from mineralized poly(lactide-co-glycolide) scaffolds for tissue

engineering," *Biomaterials*, vol. 21, no. 24, pp. 2521–2527, 2000.

[89] X. B. Yang, M. J. Whitaker, W. Sebald et al., "Human osteoprogenitor bone formation using encapsulated bone morphogenetic protein 2 in porous polymer scaffolds," *Tissue Engineering*, vol. 10, no. 7-8, pp. 1037–1045, 2004.

[90] S. M. Howdle, M. S. Watson, M. J. Whitaker et al., "Supercritical fluid mixing: preparation of thermally sensitive polymer composites containing bioactive materials," *Chemical Communications*, no. 1, pp. 109–110, 2001.

[91] J. M. Kanczler, P. J. Ginty, L. White et al., "The effect of the delivery of vascular endothelial growth factor and bone morphogenic protein-2 to osteoprogenitor cell populations on bone formation," *Biomaterials*, vol. 31, no. 6, pp. 1242–1250, 2010.

[92] M. L. Macdonald, R. E. Samuel, N. J. Shah, R. F. Padera, Y. M. Beben, and P. T. Hammond, "Tissue integration of growth factor-eluting layer-by-layer polyelectrolyte multilayer coated implants," *Biomaterials*, vol. 32, no. 5, pp. 1446–1453, 2011.

[93] E. H. Sanders, R. Kloefkorn, G. L. Bowlin, D. G. Simpson, and G. E. Wnek, "Two-phase electrospinning from a single electrified jet: microencapsulation of aqueous reservoirs in poly(ethylene-co-vinyl acetate) fibers," *Macromolecules*, vol. 36, no. 11, pp. 3803–3805, 2003.

[94] H. Jiang, Y. Hu, Y. Li, P. Zhao, K. Zhu, and W. Chen, "A facile technique to prepare biodegradable coaxial electrospun nanofibers for controlled release of bioactive agents," *Journal of Controlled Release*, vol. 108, no. 2-3, pp. 237–243, 2005.

[95] I. C. Liao, S. Y. Chew, and K. W. Leong, "Aligned core-shell nanofibers delivering bioactive proteins," *Nanomedicine*, vol. 1, no. 4, pp. 465–471, 2006.

[96] M. Hadjiargyrou and J. B. Chiu, "Enhanced composite electrospun nanofiber scaffolds for use in drug delivery," *Expert Opinion on Drug Delivery*, vol. 5, no. 10, pp. 1093–1106, 2008.

[97] H. Nie, W. S. Beng, Y. C. Fu, and C. H. Wang, "Three-dimensional fibrous PLGA/HAp composite scaffold for BMP-2 delivery," *Biotechnology and Bioengineering*, vol. 99, no. 1, pp. 223–234, 2008.

[98] C. Li, C. Vepari, H. J. Jin, H. J. Kim, and D. L. Kaplan, "Electrospun silk-BMP-2 scaffolds for bone tissue engineering," *Biomaterials*, vol. 27, no. 16, pp. 3115–3124, 2006.

[99] Y. C. Fu, H. Nie, M. L. Ho, C. K. Wang, and C. H. Wang, "Optimized bone regeneration based on sustained release from three-dimensional fibrous PLGA/HAp composite scaffolds loaded with BMP-2," *Biotechnology and Bioengineering*, vol. 99, no. 4, pp. 996–1006, 2008.

[100] J. S. Choi, K. W. Leong, and H. S. Yoo, "In vivo wound healing of diabetic ulcers using electrospun nanofibers immobilized with human epidermal growth factor (EGF)," *Biomaterials*, vol. 29, no. 5, pp. 587–596, 2008.

[101] S. Y. Chew, J. Wen, E. K. F. Yim, and K. W. Leong, "Sustained release of proteins from electrospun biodegradable fibers," *Biomacromolecules*, vol. 6, no. 4, pp. 2017–2024, 2005.

[102] J. R. J. Paletta, K. Erffmeier, C. Theisen et al., "Influence of poly-(L-lactic acid) nanofiber functionalization on maximum load, Young's modulus, and strain of nanofiber scaffolds before and after cultivation of osteoblasts: an in vitro study," *TheScientificWorldJournal*, vol. 9, pp. 1382–1393, 2009.

[103] M. D. Schofer, S. Fuchs-Winkelmann, C. Gräbedünkel et al., "Influence of poly(L-lactic acid) nanofibers and BMP-2-containing poly(L-lactic acid) nanofibers on growth and osteogenic differentiation of human mesenchymal stem cells," *TheScientificWorldJournal*, vol. 8, pp. 1269–1279, 2008.

[104] L. Chen, L. Liu, C. Li, Y. Tan, and G. Zhang, "A new growth factor controlled drug release system to promote healing of bone fractures: nanospheres of recombinant human bone morphogenetic-2 and polylactic acid," *Journal of Nanoscience and Nanotechnology*, vol. 11, no. 4, pp. 3107–3114, 2011.

[105] F. Ungaro, M. Biondi, I. d'Angelo et al., "Microsphere-integrated collagen scaffolds for tissue engineering: effect of microsphere formulation and scaffold properties on protein release kinetics," *Journal of Controlled Release*, vol. 113, no. 2, pp. 128–136, 2006.

[106] J. Schnieders, U. Gbureck, E. Vorndran, M. Schossig, and T. Kissel, "The effect of porosity on drug release kinetics from vancomycin microsphere/calcium phosphate cement composites," *Journal of Biomedical Materials Research B*, vol. 99, no. 2, pp. 391–398, 2011.

[107] D. F. dos Santos, C. S. Bitencourt, G. M. Gelfuso et al., "Biodegradable microspheres containing leukotriene B4 and cell-free antigens from Histoplasma capsulatum activate murine bone marrow-derived macrophages," *European Journal of Pharmaceutical Sciences*, vol. 44, no. 5, pp. 580–588.

[108] L. de Laporte, A. des Rieux, H. M. Tuinstra et al., "Vascular endothelial growth factor and fibroblast growth factor 2 delivery from spinal cord bridges to enhance angiogenesis following injury," *Journal of Biomedical Materials Research A*, vol. 98, no. 3, pp. 372–382, 2011.

[109] H. Shen, X. Hu, F. Yang, J. Bei, and S. Wang, "An injectable scaffold: rhBMP-2-loaded poly(lactide-co-glycolide)/hydroxyapatite composite microspheres," *Acta Biomaterialia*, vol. 6, no. 2, pp. 455–465, 2010.

[110] W. L. Lee, C. Loei, E. Widjaja, and S. C. J. Loo, "Altering the drug release profiles of double-layered ternary-phase microparticles," *Journal of Controlled Release*, vol. 151, no. 3, pp. 229–238, 2011.

[111] T. Jiang, R. R. Petersen, G. Call, G. Ofek, J. Gao, and J. Q. Yao, "Development of chondroitin sulfate encapsulated PLGA microsphere delivery systems with controllable multiple burst releases for treating osteoarthritis," *Journal of Biomedical Materials Research B*, vol. 97, no. 2, pp. 355–363, 2011.

[112] C. Wischke and S. P. Schwendeman, "Principles of encapsulating hydrophobic drugs in PLA/PLGA microparticles," *International Journal of Pharmaceutics*, vol. 364, no. 2, pp. 298–327, 2008.

[113] K. Fu, Q. Xu, J. Czernuszka et al., "Prolonged osteogenesis from human mesenchymal stem cells implanted in immunodeficient mice by using coralline hydroxyapatite incorporating rhBMP2 microspheres," *Journal of Biomedical Materials Research A*, vol. 92, no. 4, pp. 1256–1264, 2010.

[114] D. H. R. Kempen, M. J. Yaszemski, A. Heijink et al., "Non-invasive monitoring of BMP-2 retention and bone formation in composites for bone tissue engineering using SPECT/CT and scintillation probes," *Journal of Controlled Release*, vol. 134, no. 3, pp. 169–176, 2009.

[115] R. Reyes, B. de la Riva, A. Delgado, A. Hernández, E. Sánchez, and C. Évora, "Effect of triple growth factor controlled delivery by a brushite-PLGA system on a bone defect," *Injury*, vol. 43, no. 3, pp. 334–342, 2012.

[116] J. Rui, M. Dadsetan, M. B. Runge et al., "Controlled release of vascular endothelial growth factor using poly-lactic-co-glycolic acid microspheres: in vitro characterization and application in polycaprolactone fumarate nerve conduits," *Acta Biomaterialia*, vol. 8, no. 2, pp. 511–518, 2012.

[117] W. J. King, M. W. Toepke, and W. L. Murphy, "Facile formation of dynamic hydrogel microspheres for triggered growth

factor delivery," *Acta Biomaterialia*, vol. 7, no. 3, pp. 975–985, 2011.
[118] Q. Sun, E. A. Silva, A. Wang et al., "Sustained release of multiple growth factors from injectable polymeric system as a novel therapeutic approach towards angiogenesis," *Pharmaceutical Research*, vol. 27, no. 2, pp. 264–271, 2010.
[119] D. J. Mooney, P. M. Kaufmann, K. Sano et al., "Localized delivery of epidermal growth factor improves the survival of transplanted hepatocytes," *Biotechnology and Bioengineering*, vol. 50, no. 4, pp. 422–429, 1996.
[120] N. H. Dormer, K. Busaidy, C. J. Berkland, and M. S. Detamore, "Osteochondral interface regeneration of rabbit mandibular condyle with bioactive signal gradients," *Journal of Oral and Maxillofacial Surgery*, vol. 69, no. 6, pp. e50–e57, 2011.
[121] X. Xu, A. K. Jha, R. L. Duncan, and X. Jia, "Heparin-decorated, hyaluronic acid-based hydrogel particles for the controlled release of bone morphogenetic protein 2," *Acta Biomaterialia*, vol. 7, no. 8, pp. 3050–3059, 2011.
[122] A. Hernández, E. Sánchez, I. Soriano, R. Reyes, A. Delgado, and C. Évora, "Material-related effects of BMP-2 delivery systems on bone regeneration," *Acta Biomaterialia*, vol. 8, no. 2, pp. 781–791, 2012.
[123] S. Freitas, H. P. Merkle, and B. Gander, "Microencapsulation by solvent extraction/evaporation: reviewing the state of the art of microsphere preparation process technology," *Journal of Controlled Release*, vol. 102, no. 2, pp. 313–332, 2005.
[124] M. Li, O. Rouaud, and D. Poncelet, "Microencapsulation by solvent evaporation: state of the art for process engineering approaches," *International Journal of Pharmaceutics*, vol. 363, no. 1-2, pp. 26–39, 2008.
[125] J. S. Park, J. H. Lee, H. S. Shin et al., "Biodegradable polymer microspheres for controlled drug release," *Tissue Engineering and Regenerative Medicine*, vol. 4, no. 3, pp. 347–359, 2007.
[126] S. Freiberg and X. X. Zhu, "Polymer microspheres for controlled drug release," *International Journal of Pharmaceutics*, vol. 282, no. 1-2, pp. 1–18, 2004.
[127] S. Cohen, T. Yoshioka, M. Lucarelli, L. H. Hwang, and R. Langer, "Controlled delivery systems for proteins based on poly(lactic/glycolic acid) microspheres," *Pharmaceutical Research*, vol. 8, no. 6, pp. 713–720, 1991.
[128] U. Bilati, E. Allémann, and E. Doelker, "Strategic approaches for overcoming peptide and protein instability within biodegradable nano- and microparticles," *European Journal of Pharmaceutics and Biopharmaceutics*, vol. 59, no. 3, pp. 375–388, 2005.
[129] P. J. Johnson, S. L. Skornia, S. E. Stabenfeldt, and R. K. Willits, "Maintaining bioactivity of NGF for controlled release from PLGA using PEG," *Journal of Biomedical Materials Research A*, vol. 86, no. 2, pp. 420–427, 2008.
[130] G. Crotts and T. G. Park, "Protein delivery from poly(lactic-co-glycolic acid) biodegradable microspheres: release kinetics and stability issues," *Journal of Microencapsulation*, vol. 15, no. 6, pp. 699–713, 1998.
[131] J. P. Bertram, M. F. Rauch, K. Chang, and E. B. Lavik, "Using polymer chemistry to modulate the delivery of neurotrophic factors from degradable microspheres: delivery of BDNF," *Pharmaceutical Research*, vol. 27, no. 1, pp. 82–91, 2010.
[132] T. Morita, Y. Sakamura, Y. Horikiri, T. Suzuki, and H. Yoshino, "Protein encapsulation into biodegradable microspheres by a novel S/O/W emulsion method using poly(ethylene glycol) as a protein micronization adjuvant," *Journal of Controlled Release*, vol. 69, no. 3, pp. 435–444, 2000.
[133] T. Suciati, D. Howard, J. Barry, N. M. Everitt, K. M. Shakesheff, and F. R. Rose, "Zonal release of proteins within tissue engineering scaffolds," *Journal of Materials Science*, vol. 17, no. 11, pp. 1049–1056, 2006.
[134] Z. Cao, X. Chen, J. Yao, L. Huang, and Z. Shao, "The preparation of regenerated silk fibroin microspheres," *Soft Matter*, vol. 3, no. 7, pp. 910–915, 2007.
[135] M. Ye, S. Kim, and K. Park, "Issues in long-term protein delivery using biodegradable microparticles," *Journal of Controlled Release*, vol. 146, no. 2, pp. 241–260, 2010.
[136] H. Valo, L. Peltonen, S. Vehviläinen et al., "Electrospray encapsulation of hydrophilic and hydrophobic drugs in poly(L-lactic acid) nanoparticles," *Small*, vol. 5, no. 15, pp. 1791–1798, 2009.
[137] B. Almería, T. M. Fahmy, and A. Gomez, "A multiplexed electrospray process for single-step synthesis of stabilized polymer particles for drug delivery," *Journal of Controlled Release*, vol. 154, no. 2, pp. 203–210, 2011.
[138] F. Wang, Z. Q. Li, K. Tamama, C. K. Sen, and J. J. Guan, "Fabrication and characterization of prosurvival growth factor releasing, anisotropic scaffolds for enhanced mesenchymal stem cell survival/growth and orientation," *Biomacromolecules*, vol. 10, no. 9, pp. 2609–2618, 2009.
[139] A. K. Ekaputra, G. D. Prestwich, S. M. Cool, and D. W. Hutmacher, "The three-dimensional vascularization of growth factor-releasing hybrid scaffold of poly (eopencaprolactone)/collagen fibers and hyaluronic acid hydrogel," *Biomaterials*, vol. 32, no. 32, pp. 8108–8117, 2011.
[140] G. B. Wei, Q. M. Jin, W. V. Giannobile, and P. X. Ma, "The enhancement of osteogenesis by nano-fibrous scaffolds incorporating rhBMP-7 nanospheres," *Biomaterials*, vol. 28, no. 12, pp. 2087–2096, 2007.
[141] B. Li, T. Yoshii, A. E. Hafeman, J. S. Nyman, J. C. Wenke, and S. A. Guelcher, "The effects of rhBMP-2 released from biodegradable polyurethane/microsphere composite scaffolds on new bone formation in rat femora," *Biomaterials*, vol. 30, no. 35, pp. 6768–6779, 2009.
[142] T. J. Cho, L. C. Gerstenfeld, and T. A. Einhorn, "Differential temporal expression of members of the transforming growth factor β superfamily during murine fracture healing," *Journal of Bone and Mineral Research*, vol. 17, no. 3, pp. 513–520, 2002.
[143] D. H. R. Kempen, M. C. Kruyt, L. Lu et al., "Effect of autologous bone marrow stromal cell seeding and bone morphogenetic protein-2 delivery on ectopic bone formation in a microsphere/poly(propylene fumarate) composite," *Tissue Engineering A*, vol. 15, no. 3, pp. 587–594, 2009.
[144] B. H. Woo, B. F. Fink, R. Page et al., "Enhancement of bone growth by sustained delivery of recombinant human bone morphogenetic protein-2 in a polymeric matrix," *Pharmaceutical Research*, vol. 18, no. 12, pp. 1747–1753, 2001.
[145] M. K. Nguyen and D. S. Lee, "Injectable biodegradable hydrogels," *Macromolecular Bioscience*, vol. 10, no. 6, pp. 563–579, 2010.
[146] J. Zhu, "Bioactive modification of poly(ethylene glycol) hydrogels for tissue engineering," *Biomaterials*, vol. 31, no. 17, pp. 4639–4656, 2010.
[147] W. Zhang, X. Wang, S. Wang et al., "The use of injectable sonication-induced silk hydrogel for VEGF165 and BMP-2 delivery for elevation of the maxillary sinus floor," *Biomaterials*, vol. 32, no. 35, pp. 9415–9424, 2011.
[148] L. Luca, A. L. Rougemont, B. H. Walpoth, R. Gurny, and O. Jordan, "The effects of carrier nature and pH on rhBMP-2-induced ectopic bone formation," *Journal of Controlled Release*, vol. 147, no. 1, pp. 38–44, 2010.

[149] B. Wen, M. Karl, D. Pendrys, D. Shafer, M. Freilich, and L. Kuhn, "An evaluation of BMP-2 delivery from scaffolds with miniaturized dental implants in a novel rat mandible model," *Journal of Biomedical Materials Research B*, vol. 97, no. 2, pp. 315–326, 2011.

[150] C. Hayashi, U. Hasegawa, Y. Saita et al., "Osteoblastic bone formation is induced by using nanogel-crosslinking hydrogel as novel scaffold for bone growth factor," *Journal of Cellular Physiology*, vol. 220, no. 1, pp. 1–7, 2009.

[151] M. P. Lutolf, J. L. Lauer-Fields, H. G. Schmoekel et al., "Synthetic matrix metalloproteinase-sensitive hydrogels for the conduction of tissue regeneration: engineering cell-invasion characteristics," *Proceedings of the National Academy of Sciences of the United States of America*, vol. 100, no. 9, pp. 5413–5418, 2003.

[152] S. Q. Liu, Q. Tian, L. Wang et al., "Injectable biodegradable polyethylene glycol/ RGD peptide hybrid hydrogels for in vitro chondrogenesis of human mesenchymal stern cellsa," *Macromolecular Rapid Communications*, vol. 31, no. 13, pp. 1148–1154, 2010.

[153] A. Zieris, K. Chwalek, S. Prokoph et al., "Dual independent delivery of pro-angiogenic growth factors from starPEG-heparin hydrogels," *Journal of Controlled Release*, vol. 156, no. 1, pp. 28–36, 2011.

[154] P. Hänseler, U. W. Jung, R. E. Jung et al., "Analysis of hydrolyzable polyethylene glycol hydrogels and deproteinized bone mineral as delivery systems for glycosylated and non-glycosylated bone morphogenetic protein-2," *Acta Biomaterialia*, vol. 8, no. 1, pp. 116–123, 2012.

[155] X. Hao, E. A. Silva, A. Månsson-Broberg et al., "Angiogenic effects of sequential release of VEGF-A165 and PDGF-BB with alginate hydrogels after myocardial infarction," *Cardiovascular Research*, vol. 75, no. 1, pp. 178–185, 2007.

[156] C. A. Simmons, E. Alsberg, S. Hsiong, W. J. Kim, and D. J. Mooney, "Dual growth factor delivery and controlled scaffold degradation enhance in vivo bone formation by transplanted bone marrow stromal cells," *Bone*, vol. 35, no. 2, pp. 562–569, 2004.

[157] G. Schmidmaier, B. Wildemann, T. Gäbelein et al., "Synergistic effect of IGF-I and TGF-β1 on fracture healing in rats: single versus combined application of IGF-I and TGF-β1," *Acta Orthopaedica Scandinavica*, vol. 74, no. 5, pp. 604–610, 2003.

[158] F. B. Basmanav, G. T. Kose, and V. Hasirci, "Sequential growth factor delivery from complexed microspheres for bone tissue engineering," *Biomaterials*, vol. 29, no. 31, pp. 4195–4204, 2008.

[159] A. Jaklenec, A. Hinckfuss, B. Bilgen, D. M. Ciombor, R. Aaron, and E. Mathiowitz, "Sequential release of bioactive IGF-I and TGF-β1 from PLGA microsphere-based scaffolds," *Biomaterials*, vol. 29, no. 10, pp. 1518–1525, 2008.

[160] D. Kaigler, P. H. Krebsbach, E. R. West, K. Horger, Y. C. Huang, and D. J. Mooney, "Endothelial cell modulation of bone marrow stromal cell osteogenic potential," *The FASEB Journal*, vol. 19, no. 6, pp. 665–667, 2005.

[161] J. M. Kanczler, P. J. Ginty, J. J. A. Barry et al., "The effect of mesenchymal populations and vascular endothelial growth factor delivered from biodegradable polymer scaffolds on bone formation," *Biomaterials*, vol. 29, no. 12, pp. 1892–1900, 2008.

[162] J. L. Cleland, E. T. Duenas, A. Park et al., "Development of poly-(D,L-lactide-coglycolide) microsphere formulations containing recombinant human vascular endothelial growth factor to promote local angiogenesis," *Journal of Controlled Release*, vol. 72, no. 1–3, pp. 13–24, 2001.

[163] T. W. King and C. W. Patrick, "Development and in vitro characterization of vascular endothelial growth factor (VEGF)-loaded poly(DL-lactic-co-glycolic acid)/poly(ethylene glycol) microspheres using a solid encapsulation/single emulsion/solvent extraction technique," *Journal of Biomedical Materials Research*, vol. 51, no. 3, pp. 383–390, 2000.

[164] Z. S. Patel, S. Young, Y. Tabata, J. A. Jansen, M. E. K. Wong, and A. G. Mikos, "Dual delivery of an angiogenic and an osteogenic growth factor for bone regeneration in a critical size defect model," *Bone*, vol. 43, no. 5, pp. 931–940, 2008.

[165] T. P. Richardson, M. C. Peters, A. B. Ennett, and D. J. Mooney, "Polymeric system for dual growth factor delivery," *Nature Biotechnology*, vol. 19, no. 11, pp. 1029–1034, 2001.

A Nanocellulose Polypyrrole Composite Based on Tunicate Cellulose

Dawei Zhang,[1,2,3] Qing Zhang,[1,2,3] Xin Gao,[1,2,3] and Guangzhe Piao[1,2,3]

[1] *Key Laboratory of Rubber-Plastics of Ministry of Education, Qingdao 266042, China*
[2] *Shandong Provincial Key Laboratory of Rubber-Plastics, Qingdao 266042, China*
[3] *School of Polymer Science and Engineering, Qingdao University of Science and Technology, Qingdao 266042, China*

Correspondence should be addressed to Guangzhe Piao; piao@qust.edu.cn

Academic Editor: Zhou Yang

The water-dispersed conductive polypyrrole (PPy) was prepared via the *in situ* oxidative chemical polymerization by using ammonium persulfate (APS) as oxidant and tunicate cellulose nanocrystals (T-CNs) as a dopant and template for tuning the morphologies of PPy nanoparticles. Highly flexible paper-like materials of PPy/T-CNs nanocomposites with high electrical conductivity values and good mechanical properties were prepared. The structure of nanocomposites of PPy/T-CNs was investigated by using Fourier transform infrared spectroscopy. Scanning electron microscopy and transmission electron microscopy analyses of the composites revealed that PPy consisted of nanoparticles about 2.5 nm in mean size to form a continuous coating covered on the T-CNs. The diameters of the PPy nanoparticles increased from 10 to 100 nm with the increasing pyrrole amount. Moreover, electrical properties of the obtained PPy/T-CNs films were studied using standard four-probe technique and the electrical conductivity could be as high as 10^{-3} S/cm.

1. Introduction

The demand for new technologies that require high performance materials has driven materials research towards the development of novel functional nanoscaled materials with superior properties. Conducting polymers are a promising class of materials that possess unique properties that allow them to be used in a wide variety of applications [1]. The conductive properties of polyacetylene resulted in the 2000 Nobel Prize in chemistry and various analogues of it have been investigated throughout the years including polyphenylene, polyaniline, polythiophene, and polypyrrole [2]. PPy, one of the most prominent types of conjugated polymers, is a promising conducting polymer in electronics and biological and medical areas due to its straightforward polymerization, environmental stability, and high electrical conductivity that can be controlled by changing the doping degree [3]. But the poor processability and inadequate mechanical properties limit its commercial applications. In order to overcome these problems, several processing methods have been studied.

The first approach toward PPy nanoparticles was reported by Bjorklund and Liedberg, who polymerized pyrrole in water in the presence of methyl cellulose. Particles of 100 nm to 200 nm size were observed in films spread from the reaction mixture [4]. Arms and Vincent prepared colloidally stable dispersions of PPy nanoparticles by aqueous dispersion polymerization in the presence of polyvinylpyrrolidone (PVP) or polyvinyl alcohol (PVA) as a stabilizer [5]. PPy spherical nanoparticle dispersions have been prepared by a dispersion or microemulsion polymerization [6, 7]. One of the motivations of these studies was the desire to overcome the difficult processing of conjugated polymers, most of which are insoluble unless substituted with appropriate side chains.

However, all these approaches need external stabilizers, and pure PPy nanoparticles with a clean surface cannot simply be obtained unless the heavy contamination of the stabilizer is carefully removed. Unfortunately, the PPy nanoparticles without the protection of stabilizers are likely to aggregate to some extent in a few minutes, leading to instability [8]. Deposition of conducting polymers on fiber surface of fabrics, such as xylan [9], cotton [10], bacteria cellulose (BC) [1], cellulose microcrystal (CMC) [11], and cellulose derivatives

[12–14], has been widely investigated in the last few years due to its importance in emerging technologies. These materials were generally obtained through *in situ* oxidative polymerization of pyrrole by introducing an oxidant agent such as ferric chloride ($FeCl_3$) or ammonium persulfate (APS) [9, 12, 15, 16]. For providing finely structured heterogenic mixtures of cellulose and conducting polymers that are able to form films, it is necessary to provide both polymers in sufficiently dispersed or disintegrated form. The conducting organic polymers are not able to form films when cast and dried from dispersion as their H-bond forming ability is low or even absent. In this case, cellulose provides the stabilizing and film-forming matrix and is also able to stick to a supporting substrate, such as glass plates [17]. Paper-like *Cladophora* cellulose/PPy composites with highly porous surface have been proposed for ion-exchange applications [18–21].

Another material having high specific area is T-CNs, mainly composed of nanocrystals. Tunicate cellulose (TC) exhibits an ultrafine fibrous network, highly crystalline structure and aspect ratio, purity monoclinic cellulose I_β allomorph, low density, and remarkable mechanical strength [22–27]. Although TC is not a starting material for the textile industry, it is a versatile biopolymer with a wide range of properties that can be used for several innovative applications, including biomedical devices and scaffolds for tissue engineering [28]. Due to the presence of some amorphous regions, T-CNs may have better film-forming properties when compared to *Cladophora* [29].

This work reports the production of a new functional material that exploits the inherent properties of TC with electrical conductivity of PPy. For this purpose, the T-CNs doped PPy had been prepared via the chemical oxidative polymerization of Py with T-CNs as both polymeric template and dopant, and using APS as oxidant agent in aqueous solution. The electrically conducting membranes composed of T-CNs were coated with PPy as new promising polymer composite that presents a successful combination of the inherent properties of each single component. These properties include high tensile strength, toughness, biocompatibility, high surface areas of the TC, and electronic and chemical properties of PPy. However, the conductivity of the composites is still not comparable to that associated to solvent-cast PPy films. However, tunicate cellulose, already used as reinforcing agents in polymeric nanocomposites, appears as outstanding candidate for the production of highly flexible full paper-like materials. The structure of nanocomposites of PPy/T-CNs was investigated at a microscopic level using Fourier transform infrared spectroscopy. The morphology of the nanocomposites was also investigated at macroscopic level by scanning electron and transmission electron microscopies. Moreover, electrical properties of the obtained PPy/T-CNs films were studied using standard four-probe technique.

2. Experimental

2.1. Materials. Tunicate Cellulose was extracted from *Styela clava*, collected from the Yellow Sea near Yantai, China. The pyrrole (Py) monomer used was a product of TCI Development Co. Ltd. (Shanghai, China) and was used without further purification. The sulfuric acid (H_2SO_4) was obtained from Yantai Sanhe Co. Ltd. (Yantai, China). The ammonium persulfate (APS) as oxidant was purchased from Tianjin Bodi Co. Ltd. (Tianjin, China). Water was distilled before use.

2.2. Preparation of T-CNs Suspension. Suspensions were prepared from tunicate cellulose (*Styela clava*, collected from the Yellow Sea). The cellulose mantles were separated from the rest of the organs and were cut into small pieces. 200 g of *Styela clava* was soaked in a 10 wt% KOH solution for 24 h followed by washing and then bleached at 60°C with 30 mL of NaClO in 1 L of acetic buffer for 24 h, and the resultant white mantles were washed till neutrality. The previous procedure was repeated 3 times. Then, the pieces were homogenized into millimeter-size fragments using a multiblender mill (IKA-A11, Germany) and mixed with 64 wt% H_2SO_4 and stirred at 50°C for 5 h to obtain T-CNs. Then, the acid was removed through centrifugation and prolonged dialysis with distilled water until the pH outside dialysis bag was neutral. The sample thus obtained was concentrated by osmotic compression using molecular weight cutoffs of 14000 and the nanocrystal aggregates were disrupted by sonication for about 10 minutes under ice-water bath. The sample was poured into the dialysis bag and concentrated by a 15 wt% poly(ethylene glycol) (Mw = 20000) solution. Then, the T-CNs were dispersed by sonication and the concentration of the suspension was measured gravimetrically before and after evaporation of the water. The concentration of the sample suspension was 1.5 wt%.

2.3. Preparation of PPy/T-CNs Nanocomposites. PPy was synthesized via an oxidative synthetic approach in this study. Py monomer of different amounts was dissolved in 50 mL T-CNs dispersion (1.5 wt%) at room temperature for 1 h. The solutions were then cooled down, which was followed by the addition of APS solution (dissolved in 50 mL distilled water) with APS/Py = 1.25 mol/mol, in drops. The polymerization was carried out by using magnetic stirring in ice-water bath. After 6 h, the black suspensions of PPy/T-CNs could be observed and the reaction mixtures remained suspended. Aqueous suspensions of PPy/T-CNs were then prepared by washing the reaction mixtures with distilled water and dispersed by sonication. Films were deposited onto polystyrene petri dishes by casting directly from the solution of PPy/T-CNs and drying at room temperature for 3 days.

2.4. FT-IR Measurement. Fourier transform infrared (FT-IR) spectra of the PPy/T-CNs nanocomposites, PPy, and T-CNs were recorded with a FT-IR spectrometer (VERTEX70, Bruker Corporation, Germany) in the range of 200–4000 cm^{-1}. The dried samples were ground into powder and then blended with KBr before pressing the mixture into ultra-thin pellets.

2.5. Scanning Electron Microscopy (SEM). Morphology of the samples was studied by a scanning electron microscope (JSM-7500FA, JEOL, Japan). The electron energy (3–8 keV) was used to facilitate the imaging of the composite samples and

FIGURE 1: Photograph of a flexible film of the PPy/T-CNs nanocomposite.

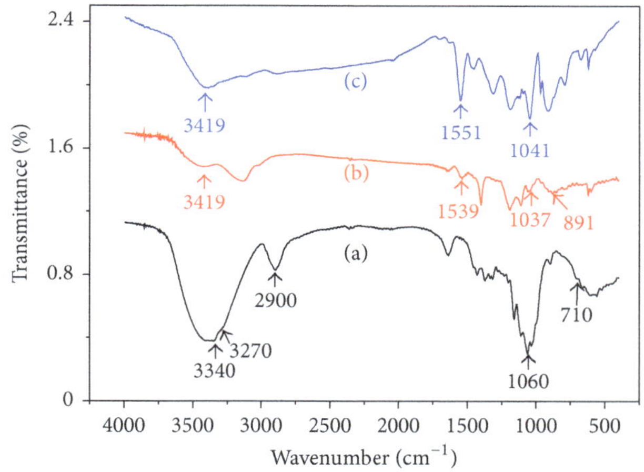

FIGURE 2: FT-IR spectra of TC (a), PPy (b) and the film of PPy/T-CNs nanocomposite (c).

neat T-CNs. All the samples were coated with a thin layer of gold to prevent charging and all the samples were not stained.

2.6. Transmission Electron Microscopy (TEM). The PPy/T-CNs nanocomposites suspensions were diluted by ethanol and then the as-prepared samples were deposited into glow-discharged carbon-coated TEM grids. They were observed after the specimen has been completely dried. Images were recorded using a TEM (JEOL, JEM-2100, Japan), operating at 200 kV with a CCD camera at different magnifications. All the samples were not stained.

2.7. Conductivity Measurements. The conductivities of the PPy/T-CNs nanocomposites films were measured by the four-probe technique at room temperature using a semiconductor device analyzer (RST-8, Guangzhou Four-Probe Technology Co., Ltd., China). The voltage U, which was scanned between −1 and +1 V, was applied with titanium needle probes and the resulting dc current I was measured. The conductivity σ was then calculated. Each sample was measured five times and the average value was taken.

3. Results and Discussion

3.1. Mechanical Properties. We prepared highly flexible paper-like materials from the PPy/T-CNs suspensions by casting it onto the polystyrene petri dishes (Figure 1). Previous work on PPy coated on other cellulose materials employed the addition of a steric stabilizer in order to maintain a stable suspension and an insulating matrix to produce the films. In the present work, no stabilizer was used to maintain the solution nor was another polymer matrix required to cast the film.

The new nanocomposites might find interesting uses in chemical sensors or paint formulations for antistatic applications, as well as in the development of electrically conductive nanocomposites with better mechanical properties due to the high mechanical strength of the CNs.

3.2. FT-IR Analysis. Fourier transform infrared spectroscopy in combination with attenuated total reflectance has been shown to be an extremely powerful tool for surface characterization of materials, particularly when the composite material is constituted by fibers coated with an external conducting polymer layer. This technique can also be applied to evaluate specific surface interactions between composite components.

The FT-IR spectra of T-CNs, PPy, and the film of PPy/T-CNs nanocomposite are shown in Figure 2. The peaks near 3340, 3270, 2900, 1060, and 710 cm^{-1} in Figure 2(a) were associated with cellulose. The characteristic broad band of cellulose O–H group appears at 3340 cm^{-1}. The band at 2900 cm^{-1} is attributed to the asymmetrically stretching vibration of C–H. The broad absorption band in the range of 1060 to 1030 cm^{-1} is attributed to the contribution of various functional groups, such as C–O and C–O–C. The peaks at 3270 and 710 cm^{-1} were attributed to the I_β phase of cellulose. Figure 2(b) was the FT-IR spectrum of PPy. The peaks at 3419 cm^{-1} could be attributed to N–H stretching vibrations. The peaks at 1539 and 1450 cm^{-1} could be attributed to C–N and C–C asymmetric and symmetric ring stretching, respectively. Additionally, the strong peaks near 1170 and 891 cm^{-1} present the doping state of PPy; the peak at 1037 cm^{-1} is attributed to C–H deformation and N–H deformation vibrations, and the broadband at 1300 cm^{-1} demonstrates the C–H and C–N in-plane deformation vibration, respectively. Moreover, these characteristic peaks of T-CNs and PPy were also found in that of the PPy/T-CNs nanocomposites. However the band at 1539 cm^{-1} for pure PPy is blue-shifted to 1551 cm^{-1} for composites and the bending modes for OH groups of T-CNs were weakened, which indicates the existance of PPy coating layer. The observed shift of the band may be caused by chemical bonding between –N–H in the Py ring and the –OH functional group of cellulose, although the –OH absorption band of the composites is too weak to confirm that OH groups of the cellulose are intermolecularly bonded to –N–H in the Py ring [2]. The results indicated that the *in situ* polymerization of Py onto the T-CNs still retains its chemical structures well.

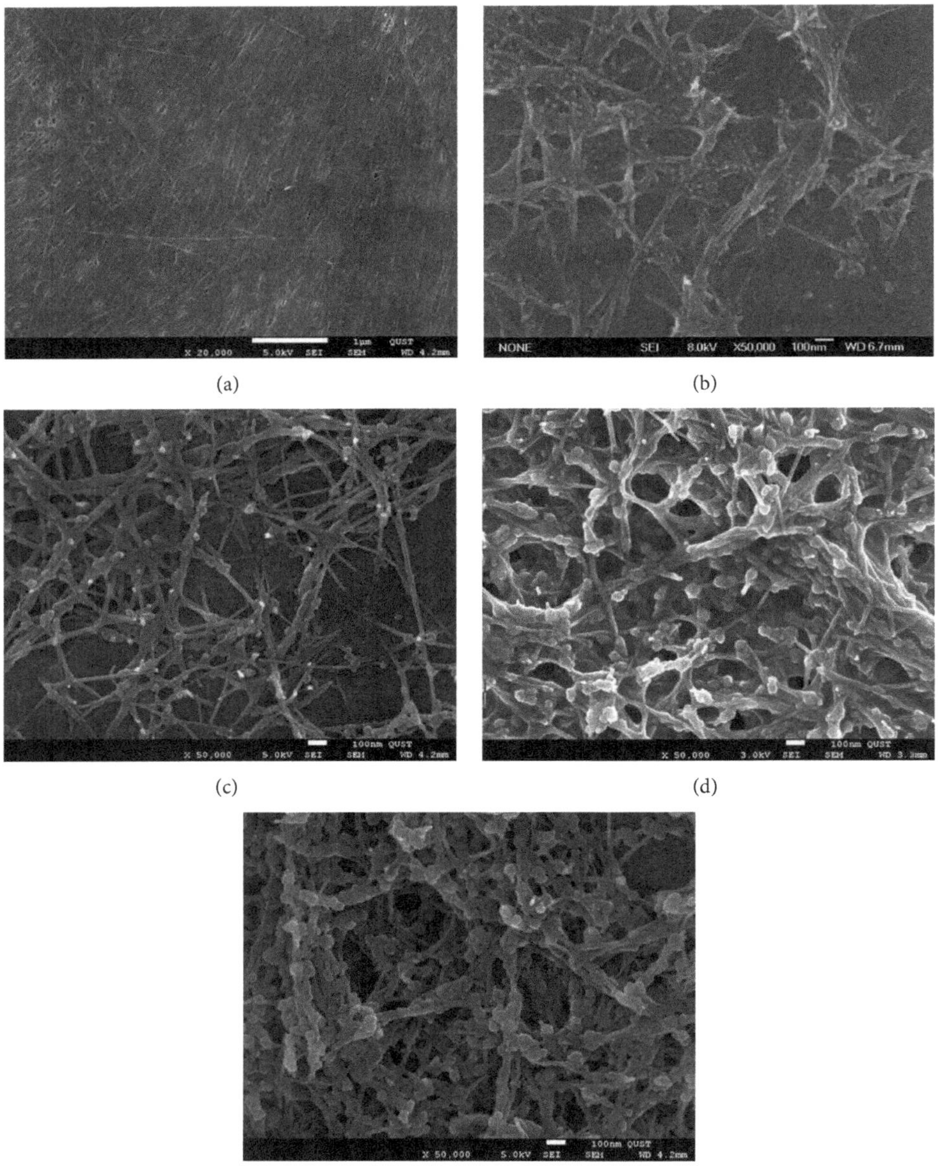

Figure 3: SEM images of T-CNs (a) and the PPy/T-CNs nanocomposites (b)–(e). The amount of Py in the nanocomposites was 0.5 (b), 1.0 (c), 2.0 (d), and 5.0 mmol (e), respectively.

3.3. SEM Analysis. Morphological analysis showed that the string-like T-CNs (Figure 3(a)) length ranges from 1 μm to 2 μm and width averages 25 nm, and the aspect ratio is about 40 to 80. Moreover, such high aspect ratio nanocrystals should create a huge entanglements network, and thus, it is responsible for such highly mechanical stiffness and highly flexible paper-like materials. The influence of the amount of Py on the structure and morphologies of the PPy/T-CNs nanocomposites was investigated by SEM (as shown in Figures 3(b)–3(e)). The size of PPy particles increased from 10 nm to 100 nm with the increasing Py. The adhesion of PPy particles to the cellulose surface could attribute to the formation of hydrogen bonds between cellulose hydroxyl groups and Py NH moieties. The formed PPy particles on cellulose surface had spherical shape and controlled size, which was probably due to the charge density on T-CNs that impeded, further, disordered growth [11]. And the increase of the T-CNs may have prevented the aggregation of PPy particles and acted as the separant and dispersant during the *in situ* polymerization of Py which eventually decreased the particles size.

3.4. TEM Analysis. In order to further prove the previous speculation, the TEM images of the PPy/T-CNs nanocomposites dispersed in ethanol were given in Figure 4. The

TABLE 1: Electrical conductivities of the PPy/T-CNs films.

Sample	Py, mmol	Resistivity, KΩ · cm	Conductivity, S/cm
1	0.5	4.9	2.1×10^{-4}
2	1.0	3.4	2.9×10^{-4}
3	2.0	1.5	6.7×10^{-4}
4	5.0	0.7	1.4×10^{-3}

morphologies of the composites became smaller with the increasing T-CNs. The width of the T-CNs was about 20 nm while the composites were 30 nm to 50 nm. A thickness layer of PPy about 30 nm was deposited on the T-CNs. It is found that on the surface of the nanocomposites absorbed a lot of uniform spherical nanoparticles and their diameters are about 2.5 nm. According to the previous work, they might be PPy nanoparticles [7, 30]. And we could also find on the surface of the covered layer adhered lots of spherical nanoparticles ranging from about 20 nm to 100 nm, which is consistent with the SEM results. This indicated that the T-CNs molecules acted as the templates as well as the dopant, as reported previously [31]. The micrographs revealed that PPy was constituted of particles that uniformly coated on the surface of T-CNs. The polymer coating adhered to the nanofiber and a continuous conducting pathway was formed which was responsible for the high electrical conductivity values.

3.5. Conductivity Measurements. Electrical conductivities of the samples were determined and the results were shown in Table 1. It was observed that the conductivity could be up to 1.4×10^{-3} S/cm without the addition of any external dopants. Additionally, T-CNs are not an electric conductor and the conductivity decreased with decreasing the amount of Py. This behavior is attributed to the formation of a continuous conducting PPy layer that homogeneously coated the surface of T-CNs. As expected, the amount of PPy incorporated on T-CNs increased with increasing monomer concentration and the conductivity also increased with the increase of the ratio of Py. Thus, we could conclude that T-CNs play a good role of dopant as well as template.

4. Conclusions

In conclusion, we have developed a renewable resource T-CNs and utilized it as a dopant for water-dispersed and electrically conductive nanocomposites by the *in situ* oxidative chemical polymerization of Py onto T-CNs. Highly flexible paper-like materials of PPy/T-CNs nanocomposites with electrical conductivities as high as 10^{-3} S/cm were also obtained. These films have interesting properties that may find important technological applications such as sensors, electronic devices, antistatic and anticorrosive nanocoatings, intelligent clothes, flexible electrodes, and tissue engineering scaffolds.

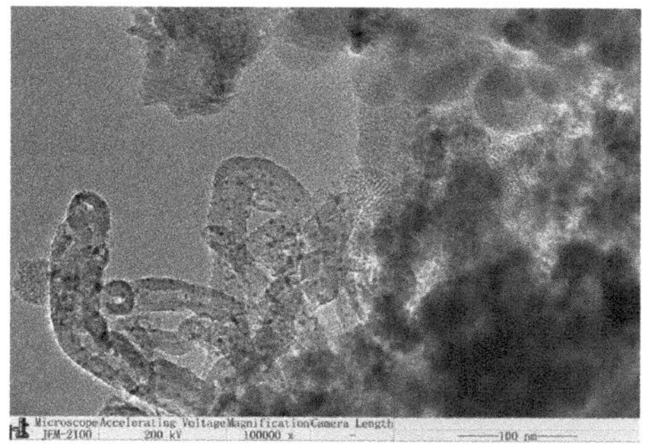

FIGURE 4: TEM image of the PPy/T-CNs nanocomposite. The amount of Py in the nanocomposite was 2.0 mmol.

Conflict of Interests

The authors declare that they have no conflict of interests.

Acknowledgments

The authors gratefully acknowledge Professor Tsunehisa Kimura and Dr. Fumiko Kimura (Kyoto University, Japan) for their helpful discussions and technical support. This work was partially supported by the Program for International S&T Cooperation Projects of China (2011DFA50430), National Natural Science Foundation of China (50773033), Science Foundation of Shandong Province (Y2007F01), and Doctoral Fund of QUST.

References

[1] A. Mihranyan, L. Nyholm, A. E. Garcia Bennett, and M. Strømme, "A novel high specific surface area conducting paper material composed of polypyrrole and Cladophora cellulose," *Journal of Physical Chemistry B*, vol. 112, no. 39, pp. 12249–12255, 2008.

[2] D. Müller, C. R. Rambo, C. R. Recouvreux, L. M. Porto, and G. M. O. Barra, "Chemical in situ polymerization of polypyrrole on bacterial cellulose nanofibers," *Synthetic Metals*, vol. 161, no. 1-2, pp. 106–111, 2011.

[3] J. Pecher and S. Mecking, "Nanoparticles of conjugated polymers," *Chemical Reviews*, vol. 110, no. 10, pp. 6260–6279, 2010.

[4] R. B. Bjorklund and B. Liedberg, "Electrically conducting composites of colloidal polypyrrole and methylcellulose," *Journal of the Chemical Society, Chemical Communications*, no. 16, pp. 1293–1295, 1986.

[5] S. P. Arms and B. Vincent, "Dispersions of electrically conducting polypyrrole particles in aqueous media," *Journal of the Chemical Society, Chemical Communications*, no. 4, pp. 288–290, 1987.

[6] A. Y. Men'shikova, B. M. Shabsel's, and T. G. Evseeva, "Synthesis of polypyrrole nanoparticles by dispersion polymerization," *Journal of Applied Chemistry*, vol. 76, no. 5, pp. 822–826.

[7] J. Jang, J. H. Oh, and G. D. Stucky, "Fabrication of ultrafine conducting polymer and graphite nanoparticles," *Angewandte Chemie*, vol. 41, no. 21, pp. 4016–4019, 2002.

[8] X.-G. Li, Z.-Z. Hou, M.-R. Huang, and M. G. Moloney, "Efficient synthesis of intrinsically conducting polypyrrole nanoparticles containing hydroxy sulfoaniline as key self-stabilized

units," *Journal of Physical Chemistry C*, vol. 113, no. 52, pp. 21586–21595, 2009.

[9] C. Sasso, N. Bruyant, D. Beneventi et al., "Polypyrrole (PPy) chemical synthesis with xylan in aqueous medium and production of highly conducting PPy/nanofibrillated cellulose films and coatings," *Cellulose*, vol. 18, no. 6, pp. 1455–1467, 2011.

[10] D. W. Zhang, L. H. Zhang, B. Z. Wang, and G. Z. Piao, "Nanocomposites of polyaniline and cellulose nanocrystals prepared in lyotropic chiral nematic liquid crystals," *Journal of Materials*, vol. 2013, Article ID 614507, 6 pages, 2013.

[11] C. Sasso, E. Zeno, M. Petit-Conil et al., "Highly conducting polypyrrole/cellulose nanocomposite films with enhanced mechanical properties," *Macromolecular Materials and Engineering*, vol. 295, no. 10, pp. 934–941, 2010.

[12] L. Dall'Acqua, C. Tonin, R. Peila, F. Ferrero, and M. Catellani, "Performances and properties of intrinsic conductive cellulose-polypyrrole textiles," *Synthetic Metals*, vol. 146, no. 2, pp. 213–221, 2004.

[13] L. Dall'Acqua, C. Tonin, A. Varesano, M. Canetti, W. Porzio, and M. Catellani, "Vapour phase polymerisation of pyrrole on cellulose-based textile substrates," *Synthetic Metals*, vol. 156, no. 5-6, pp. 379–386, 2006.

[14] D. Beneventi, S. Alila, S. Boufi, D. Chaussy, and P. Nortier, "Polymerization of pyrrole on cellulose fibres using a $FeCl_3$ impregnation—Pyrrole polymerization sequence," *Cellulose*, vol. 13, no. 6, pp. 725–734, 2006.

[15] J. Molina, A. I. del Río, J. Bonastre, and F. Cases, "Electrochemical polymerisation of aniline on conducting textiles of polyester covered with polypyrrole/AQSA," *European Polymer Journal*, vol. 45, no. 4, pp. 1302–1315, 2009.

[16] I. Cucchi, A. Boschi, C. Arosio, F. Bertini, G. Freddi, and M. Catellani, "Bio-based conductive composites: preparation and properties of polypyrrole (PPy)-coated silk fabrics," *Synthetic Metals*, vol. 159, no. 3-4, pp. 246–253, 2009.

[17] A. Rußler, K. Sakakibara, and T. Rosenau, "Cellulose as matrix component of conducting films," *Cellulose*, vol. 18, no. 4, pp. 937–944, 2011.

[18] G. Nyström, A. Razaq, M. Strømme, L. Nyholm, and A. Mihranyan, "Ultrafast all-polymer paper-based batteries," *Nano Letters*, vol. 9, no. 10, pp. 3635–3639, 2009.

[19] G. Nyström, A. Mihranyan, A. Razaq, T. Lindström, L. Nyholm, and M. Strømme, "A nanocellulose polypyrrole composite based on microfibrillated cellulose from wood," *Journal of Physical Chemistry B*, vol. 114, no. 12, pp. 4178–4182, 2010.

[20] A. Mihranyan, "Cellulose from cladophorales green algae: from environmental problem to high-tech composite materials," *Journal of Applied Polymer Science*, vol. 119, no. 4, pp. 2449–2460, 2011.

[21] A. Razaq, L. Nyholm, M. Strømme, and A. Mihranyan, "Paper-based energy-storage devices comprising carbon fiber-reinforced polypyrrole-cladophora nanocellulose composite electrodes," *Advanced Energy Materials*, vol. 2, no. 4, pp. 445–454, 2012.

[22] M. Wu, S. Kuga, and Y. Huang, "Quasi-one-dimensional arrangement of silver nanoparticles templated by cellulose microfibrils," *Langmuir*, vol. 24, no. 18, pp. 10494–10497, 2008.

[23] T. Saito, Y. Nishiyama, J.-L. Putaux, M. Vignon, and A. Isogai, "Homogeneous suspensions of individualized microfibrils from TEMPO-catalyzed oxidation of native cellulose," *Biomacromolecules*, vol. 7, no. 6, pp. 1687–1691, 2006.

[24] S. Berlioz, S. Molina-Boisseau, Y. Nishiyama, and L. Heux, "Gas-phase surface esterification of cellulose microfibrils and whiskers," *Biomacromolecules*, vol. 10, no. 8, pp. 2144–2151, 2009.

[25] S. Kimura and T. Itoh, "Cellulose synthesizing terminal complexes in the ascidians," *Cellulose*, vol. 11, pp. 377–383, 2004.

[26] M. M. S. Lima and R. Borsali, "Static and dynamic light scattering from polyelectrolyte microcrystal cellulose," *Langmuir*, vol. 18, no. 4, pp. 992–996, 2002.

[27] D. Li, Z. Liu, M. Al-Haik et al., "Magnetic alignment of cellulose nanowhiskers in an all-cellulose composite," *Polymer Bulletin*, vol. 65, no. 6, pp. 635–642, 2010.

[28] J. M. Dugan, J. E. Gough, and S. J. Eichhorn, "Directing the morphology and differentiation of skeletal muscle cells using oriented cellulose nanowhiskers," *Biomacromolecules*, vol. 11, no. 9, pp. 2498–2504, 2010.

[29] M. A. S. A. Samir, F. Alloin, and A. Dufresne, "Review of recent research into cellulosic whiskers, their properties and their application in nanocomposite field," *Biomacromolecules*, vol. 6, no. 2, pp. 612–626, 2005.

[30] J. Jang and H. O. Joon, "Novel crystalline supramolecular assemblies of amorphous polypyrrole nanoparticles through surfactant templating," *Chemical Communications*, no. 19, pp. 2200–2201, 2002.

[31] F. F. Bruno, R. Nagarajan, S. Roy, J. Kumar, and L. A. Samuelson, "Biomimetic Synthesis of Water Soluble Conducting Polypyrrole and Poly(3,4-ehtylenedioxythiophene)," *Journal of Macromolecular Science A*, vol. 40, no. 12, pp. 1327–1333, 2003.

10

Thermoresponsive and Redox Behaviors of Poly(*N*-isopropylacrylamide)-Based Block Copolymers Having TEMPO Groups as Their Side Chains

Toru Uemukai, Tomoya Hioki, and Manabu Ishifune

Department of Applied Chemistry, Faculty of Science and Engineering, Kinki University, 3-4-1 Kowakae, Higashi-Osaka, Osaka 577-8502, Japan

Correspondence should be addressed to Manabu Ishifune; ishifune@apch.kindai.ac.jp

Academic Editor: Haojun Liang

Thermoresponsive and redox-active block copolymers having 2,2,6,6-tetramethylpiperidine-1-oxyl (TEMPO) moieties have been synthesized by using the reversible addition-fragmentation chain transfer (RAFT) polymerization technique. *N*-Isopropylacrylamide (NIPAAm) and 2,2,6,6-tetramethylpiperidyl methacrylate (TEMPMA) monomers were copolymerized stepwise under RAFT polymerization conditions to afford the thermoresponsive block copolymers, PNIPAAm-*block*-PTEMPMA and PNIPAAm-*block*-PTEMPMA-*block*-PNIPAAm. Oxidation of tetramethylpiperidine groups in the copolymers successfully afforded the corresponding TEMPO-containing block copolymers. The resulting triblock copolymer was found to be thermoresponsive showing lower critical solution temperature (LCST) at 34°C in its aqueous solution. Redox behavior of the resulting copolymer was observed by cyclic voltammetry. The potential of anodic current peak changed below and above the LCST of the block copolymer. These results indicate that the phase transition of thermoresponsive polymer influences the redox potential of TEMPO moieties.

1. Introduction

Much attention has been paid to stimuli-responsive polymers in chemistry and materials science, because of their unique reversible change of physical properties in response to a chemical or physical stimulus [1–3]. Responsive behavior typically involves physical changes in polymer's properties. For instance, unique changes in solubility are often exhibited by the polymers that have lower critical solution temperature (LCST). While many synthetic polymers exhibit LCSTs, poly(*N*-isopropylacrylamide) (PNIPAAm) and its copolymers are the most popular materials with LCSTs. Drug delivery systems [4–6], chromatographic separations [7, 8], separation and purification of metal ions [9, 10], and molecular recognition [11] are among the most important fields in which thermoresponsive polymers are used. In previous study, we have found that the phase transition behavior of PNIPAAm-based polymer having pyrrolidine groups as side chains was influenced by pH values [12]. Their LCST's changes were influenced by deionization of pyrrolidine hydrogen chloride moieties. Moreover, the pyrrolidine side chains of the thermoresponsive copolymer promoted the aldol reaction between cyclohexanone and *p*-nitrobenzaldehyde in water, and the reaction proceeded most smoothly at its LCST. The thermoresponsive polymer in water aggregate to form a rather hydrophobic reaction field and the substrates are probably concentrated and promoted the aldol reaction in the resulting hydrophobic field. Thus, such a hydrophobic field formed by thermoresponsive polymers could be utilized for various organic reactions in aqueous media. Beside the pH dependence on phase transition, several examples of thermoresponsive polymers with redox responsiveness have been reported such as PNIPAAm-based polymers containing ferrocenyl pendant groups or 2,2,6,6-tetramethylpiperidine 1-oxyl (TEMPO) groups [13–16]. The polymers containing TEMPO moieties, which were prepared by the polymer reaction of poly(*N*-acryloxysuccinimide) with a mixture of 4-amino TEMPO and isopropylamine, were reported to show redox-sensitive LCST behavior [14]. These polymers are promising for the application to the TEMPO-mediated redox reactions in the hydrophobic field generated by the phase transition of the thermoresponsive polymers. For this purpose it is important to control the repeat units sequence so that the redox centers are effectively surrounded by the thermoresponsive polymer chains. However, most hitherto known thermoresponsive polymers having TEMPO moieties have been prepared in a random manner [14–16], and few thermoresponsive behaviors with redox responsiveness of the corresponding polymers having block structures have been reported. In this study, sequence-ordered thermoresponsive polymers having TEMPO pendant groups as mediatory centers were synthesized by using the reversible addition-fragmentation chain transfer (RAFT) polymerization technique, and thermoresponsivity and redox behaviors of the resulting polymers were investigated.

2. Experimental

2.1. General. IR spectra were obtained on an FT-IR-470 Plus spectrometer. ^{1}H NMR spectra were measured on a Varian OXFORD NMR300 (300 MHz) or JEOL JNM-AL400 (400 MHz) spectrometer, and the chemical shift values (δ) were expressed in ppm downfield from the internal TMS standard. The molecular weights of polymers were determined by using a gel permeation chromatography (GPC). The GPC analyses were carried out on a Hitachi L-6000 high-performance liquid chromatograph, L-3350 RI detector, and Shodex GPC KF-804L column. THF was used as the eluent with a flow rate of 1.0 mL/min, and molecular weight values were relative to the polystyrene standards (Shodex STANDARD SM-105 and SL-105). UV-visible spectra were recorded with Jasco V-630 spectrometer (Tokyo, Japan). ESR spectra were recorded by JEOL JES-PX1050. Cyclic voltammetry was measured by BAS CV-50.

2.2. Materials. 4-Cyano-4-[(dodecylsulfanylthiocarbonyl) sulfanyl]pentanoic acid was synthesized from 4,4′-azobis (4-cyanopentanoic acid), carbon disulfide (Wako Pure Chemical Industries, Ltd., Osaka, Japan) and 1-dodecanethiol (SIGMA-ALDRICH) [17]. *N*-Isopropylacrylamide (NIPAAm) (Wako Pure Chemical Industries, Ltd., Osaka, Japan) was purified by recrystallization from hexane-benzene (95 : 5). 2,2′-Azobisisobutyronitrile (AIBN) as a radical initiator was recrystallized from methanol below 40°C. Tetrahydrofuran (THF) (Wako Pure Chemical Industries, Ltd., Osaka, Japan) was distilled from Na-benzophenoneketyl under Ar atmosphere. 2,2,6,6-Tetramethyl-4-piperidyl methacrylate (TEMPMA) and the other materials were commercially available and used without further purification.

*2.3. Synthesis of Poly(TEMPMA-ran-NIPAAm) (**1**).* TEMPMA (0.26 g, 1.1 mmoL), NIPAAm (3.16 g, 27.9 mmoL), AIBN (0.1 moL% relative to monomers) as an initiator, and 4-cyano-4-[(dodecylsulfanylthiocarbonyl)sulfanyl]pentanoic acid (1.0 mol% relative to monomers) were dissolved in THF (4 mL). After oxygen was removed repeatedly by freeze-evacuation-thaw cycles, polymerization was carried out at 60°C for 15 h. After cooling in an ice-water bath, the polymer solution was poured into a large amount of diethyl ether, and the resulting precipitate was filtered and dried under reduced pressure at room temperature. The crude polymer was purified by reprecipitation from THF into diethyl ether. ^{1}H NMR ($CDCl_3$) δ 0.88–2.87 (m, NIPAAm unit, TEMPAMA unit), 3.87–4.12 (bs, $CH(CH_3)_2$), 5.03–5.19 (bs, O-CH); IR (KBr) 3299, 3077, 1726, 1653, 1545 cm^{-1}.

2.4. Synthesis of PNIPAAm-block-PTEMPMA-block-PNIPAAm

2.4.1. RAFT Reagent Terminated PNIPAAm. NIPAAm (6.40 g, 56.6 mmoL), AIBN (0.1 moL% relative to monomer) as an initiator, and 4-cyano-4-[(dodecylsulfanylthiocarbonyl)sulfanyl]pentanoic acid (1.0 moL% relative to monomer) as a chain transfer reagent (RAFT reagent) were dissolved in THF (6 mL). After oxygen was removed repeatedly by freeze-evacuation-thaw cycles, polymerization was carried out at 60°C for 15 h. After cooling in an ice-water bath, the polymer solution was poured into a large amount of diethyl ether, and the resulting precipitate was filtered and dried under reduced pressure at room temperature. The crude polymer was purified by reprecipitation from THF into diethyl ether. ^{1}H NMR ($CDCl_3$) δ 1.02–3.17 (m, CH_2-CH, $CH(CH_3)_2$), 3.85–4.18 (bs, $CH(CH_3)_2$); IR (KBr) 3293, 3079, 1650, 1550 cm^{-1}.

*2.4.2. PNIPAAm-block-PTEMPMA (**2**).* TEMPMA (1.03 g, 4.55 mmoL), AIBN (0.1 moL% relative to monomer) as an initiator, and RAFT reagent terminated PNIPAAm (1.0 moL% relative to monomer) as a macrochain transfer agent (macro-CTA) were dissolved in THF (4 mL). After oxygen was removed repeatedly by freeze-evacuation-thaw cycles, polymerization was carried out at 60°C for 60 h. After cooling in an ice-water bath, the polymer solution was poured into a large amount of diethyl ether, and the resulting precipitate was filtered and dried under reduced pressure at room temperature. The crude polymer was purified by reprecipitation from THF into diethyl ether. ^{1}H NMR ($CDCl_3$) δ 0.76–2.85 (m, NIPAAm unit, TEMPAMA unit), 3.86–4.19 (bs, $CH(CH_3)_2$), 4.96–5.18 (bs, O-CH); IR (KBr) 3299, 3074, 1730, 1652, 1559 cm^{-1}.

*2.4.3. PNIPAAm-block-PTEMPMA-block-PNIPAAm (**3**).* NIPAAm (1.66 g, 1.47 mmoL), AIBN (1.7 mg, 0.01 mmoL) as an initiator and PNIPAAm-*block*-PTEMPMA (**2**) (0.69 g) as a macro-CTA were dissolved in THF (4 mL). After oxygen was removed repeatedly by freeze-evacuation-thaw cycles, polymerization was carried out at 60°C for 80 h. After cooling in an ice-water bath, the polymer solution was poured into a large amount of diethyl ether, and the resulting precipitate was filtered and dried under reduced pressure at room temperature. The crude polymer was purified by reprecipitation from THF into diethyl ether. ^{1}H NMR ($CDCl_3$) δ 0.80–3.17 (m, NIPAAm unit, TEMPAMA unit), 3.90–4.17 (bs, $CH(CH_3)_2$), 4.98–5.18 (bs, O-CH); IR (KBr) 3298, 3074, 1726, 1649, 1546 cm^{-1}.

2.5. Oxidation of Tetramethylpiperidine Moieties in the Polymers. Polymer having TEMPMA moieties (**1**, **2**, or **3**) (0.80 g), $Na_2WO_4{\cdot}2H_2O$ (0.37 g, 1.1 mmoL), and 30% H_2O_2 (6 mL) was dissolved in H_2O (24 mL). This reaction mixture was stirred at 4°C for 24 h. Into the mixture was added K_2CO_3 until the solution was saturated. The aqueous solution was extracted with ethyl acetate until the organic layer was colorless. The combined organic layers were dried over $MgSO_4$ and concentrated under reduced pressure to give an orange to pale red powder.

Poly(TEMPOMA-ran-NIPAAm); IR (KBr) 3308, 3081, 1719, 1653, 1539 cm^{-1}.

PNIPAAm-block-PTEMPOMA; IR (KBr) 3298, 3080, 1726, 1649, 1546 cm^{-1}.

PNIPAAm-block-PTEMPOMA-block-PNIPAAm; IR (KBr) 3297, 3080, 1730, 1650, 1548 cm^{-1}.

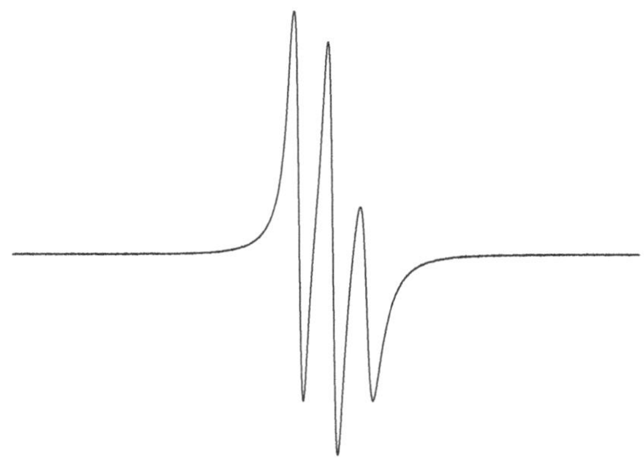

FIGURE 1: ESR spectrum of poly(TEMPOMA-*ran*-NIPAAm).

2.6. Measurement of Cloud Points. The cloud points of aqueous polymer solutions (1.0 w/v%) were determined by transmittance measurements using a UV-visible spectrometer equipped with a temperature controller. The transmittance of the polymer solutions was recorded as a function of temperature (15–45°C). The wavelength was fixed at 500 nm, and the temperature was raised at 1.0°C/min. The cloud points of the polymer solutions were defined as the temperatures when the transmittance was decreased to half of final transmittance from initial.

2.7. Electrochemical Testing of the Copolymers. Cyclic voltammograms were measured in 0.1 M KCl aqueous solution after bubbling with nitrogen gas (20 min) at various temperatures (5–55°C). As a working electrode and a counter electrode were used a GCE Glassy carbon (BAS 6.0 × 1.6 mm) and a platinum wire (BAS VC-3 cell 5 cm Pt), respectively, and a reference electrode consisted of Ag/AgCl (BAS RE-1B). Scan rate was 100 mV/s. The concentration of aqueous solution of poly(TEMPOMA-*ran*-NIPAAm) or PNIPAAm-*block*-PTEMPOMA was adjusted at 2.0 mM (based on TEMPO moieties). The concentration of aqueous solution of PNIPAAm-*block*-PTEMPOMA-*block*-PNIPAAm was 0.47 g/mL.

3. Results and Discussion

3.1. Synthesis and Characterization of Poly(TEMPOMA-ran-NIPAAm). Thermoresponsive polymers having 2,2,6,6-tetramethylpiperidine-1-oxyl (TEMPO) pendant groups as mediatory centers were synthesized by using the reversible addition-fragmentation chain transfer (RAFT) polymerization technique and oxidation of 2,2,6,6-tetramethyl-4-piperidyl methacrylate (TEMPMA) units. *N*-Isopropylacrylamide (NIPAAm) and TEMPMA were copolymerized in a random manner in the presence of 4-cyano-4-[(dodecylsulfanylthiocarbonyl)sulfanyl]pentanoic acid to afford poly(TEMPMA-*ran*-NIPAAm) (Scheme 1). The resulting random copolymer was analyzed by ^1H NMR spectroscopy; the ratio of TEMPMA/NIPAAm units was determined to be 6 : 94. The number average molecular weight (M_n) of the random copolymer was estimated by GPC based on polystyrene standard to be 4840 (M_w/M_n = 1.26). Oxidation of the TEMPMA units afforded poly(TEMPOMA-*ran*-NIPAAm). The resulting TEMPO-containing random copolymer was characterized by ESR spectroscopy, and its spectrum showed characteristic three-line signals of a nitroxyl group (Figure 1) [14, 15]. The random copolymer, poly(TEMPMA-*ran*-NIPAAm), showed thermoresponsivity, and its lower critical solution temperature (LCST) was observed at 31°C, which was slightly increased to 33°C after oxidation of tetramethylpiperidine moieties (Figure 2(a)). The transmittance of the oxidized random copolymer solution was observed to be decreased from 25°C, and at 0.1°C/min of heating rate the thermoresponsivity was still slow, but the LCST was observed at 32°C, which was almost unchanged in comparison with that before oxidation of the TEMPMA moieties.

3.2. Synthesis and Characterization of the Diblock and Triblock Copolymers. The diblock and triblock polymers were also synthesized by using the RAFT polymerization technique and oxidation of TEMPMA units. Firstly, the RAFT reagent terminated PNIPAAm was prepared, and, in the second step, PNIPAAm-*block*-PTEMPMA was synthesized by using the RAFT reagent terminated PNIPAAm as a macro-chain transfer agent (macro-CTA) (yield 58%). Finally, PNIPAAm-*block*-PTEMPMA-*block*-PNIPAAm was synthesized by using the resulting diblock polymer as a macro-CTA (yield 94%) (Scheme 2). The M_n of the resulting block polymers having TEMPO precursor as side chains were estimated by determining the molar ratio of the repeating units and the end group from the ratio of signal intensities of $-NHCH(CH_3)_2$ of NIPAAm units, the proton at 4-position of piperidine moieties, and $-S-CH_2-$ of the end group in the ^1H NMR spectra. The molecular weight increased stepwise with the progress of polymerization and the M_n values estimated from ^1H NMR were almost same as those estimated from GPC (Table 1). Oxidation of tetramethylpiperidine moieties in the resulting diblock and triblock copolymers

TABLE 1: Molecular weight data of the resulting copolymers.

	Content of TEMPMA units[a] (%)	M_n[a]	M_n[b]	M_w/M_n[b]
Poly(TEMPOMA-*ran*-NIPAAm) **1**	6	—	4840	1.26
RAFT reagent terminated PNIPAAm	—	6310	6600	1.20
PNIPAAm-*block*-PTEMPMA **2**	4	7750	6970	1.17
PNIPAAm-*block*-PTEMPMA-*block*-PNIPAAm **3**	1	15200	13940	1.35

[a]Determined by integrated intensity of ^{1}H NMR spectra. [b]Estimated by GPC based on polystyrene standard in THF.

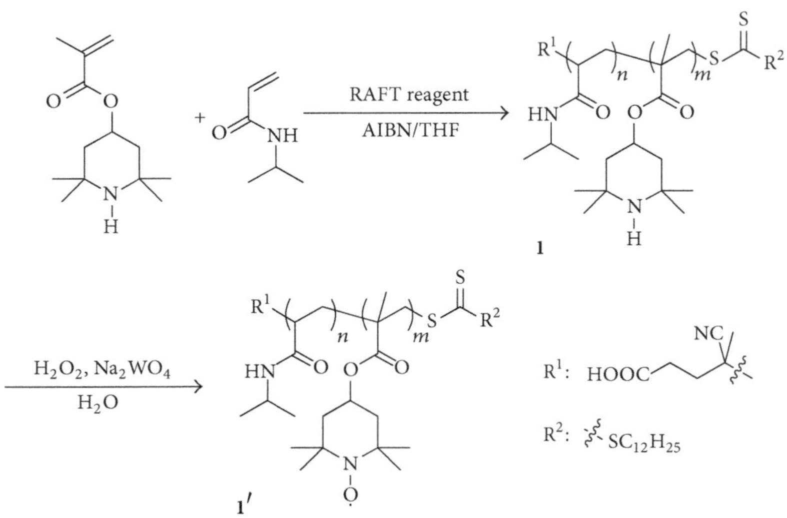

SCHEME 1: Synthesis of poly(TEMPOMA-*ran*-NIPAAm).

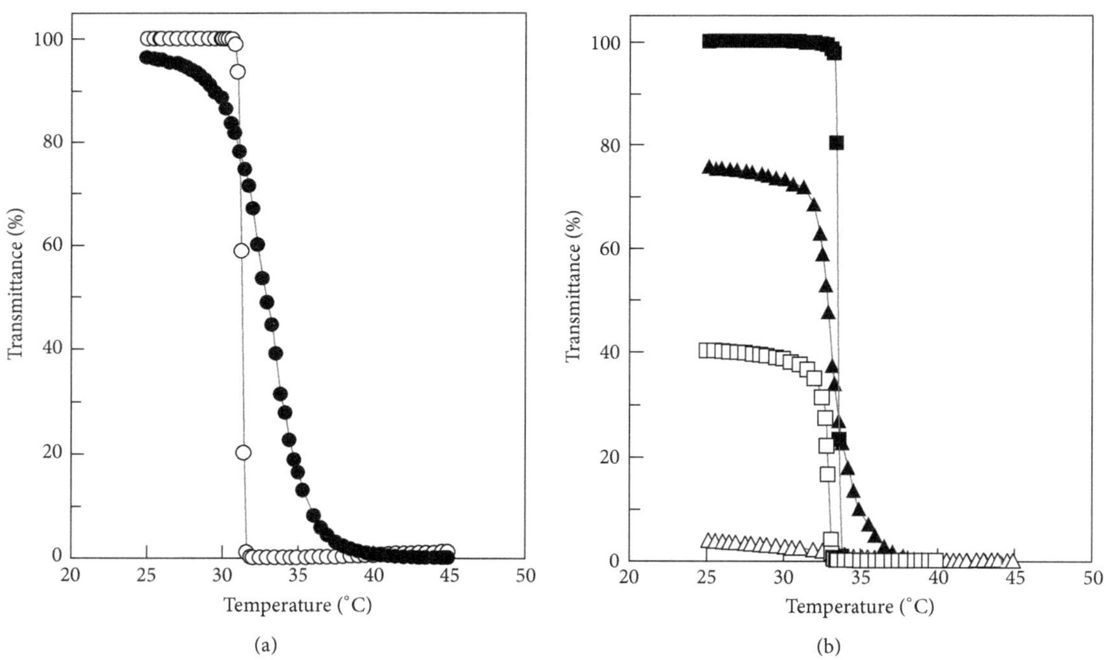

FIGURE 2: Temperature dependence of transmittance of 10 mg/mL aqueous solutions of (a) poly(TEMPMA-*ran*-NIPAAm) (**1**) (solid circle), poly(TEMPOMA-*ran*-NIPAAm) (**1′**) (open circle); (b) PNIPAAm-*block*-PTEMPMA (**2**) (solid triangle) PNIPAAm-*block*-PTEMPOMA (**2′**) (open triangle); PNIPAAm-*block*-PTEMPOMA-*block*-PNIPAAm (**3**) (solid square) and PNIPAAm-*block*-PTEMPOMA-*block*-PNIPAAm (**3′**) (open square).

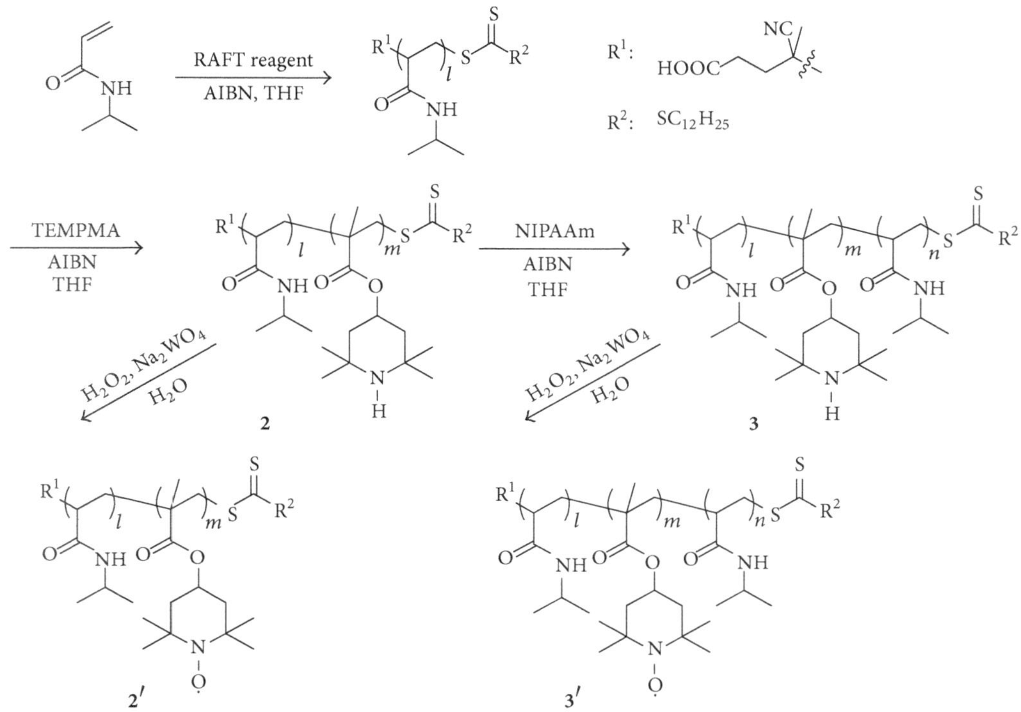

SCHEME 2: Synthesis of the block polymers consisting of NIPAAm and TEMPOMA units.

afforded PNIPAAm-*block*-PTEMPOMA and PNIPAAm-*block*-PTEMPOMA-*block*-PNIPAAm. The block polymers having TEMPO moieties as side chains were characterized by ESR spectroscopy, and their spectra showed characteristic three-line signals of nitroxyl groups. The diblock and triblock copolymers containing TEMPMA and NIPAAm groups showed thermoresponsivity, and their LCSTs were observed at 33°C and 34°C, respectively. After oxidation of tetramethylpiperidine moieties, the diblock copolymer became insoluble in water; on the other hand, the triblock copolymer was found to have LCST at 33°C (Figure 2(b)). The molecular weight of water-insoluble diblock copolymer was 6970 and contained 4% of TEMPO moiety, while that of the corresponding random copolymers containing 6% of TEMPO moiety was 4840 and soluble in water. Thus a continuous structure of TEMPO moieties probably leads to the insolubility in water, and the NIPAAm unit sequence seems to be too short to form a polymer micelle. On the other hand, TEM images of the triblock copolymer aggregates obtained from their aqueous solutions showed *ca.* 50 nm of spheres under LCST, and these spherical particles aggregated to form more than 100 nm of clusters over LCST. Moreover, the initial transmittance of the oxidized triblock copolymer solution was observed to be 40% (Figure 2(b)). These results indicate the triblock copolymers form a polymer micelle.

3.3. Thermoresponsive Behavior of the Resulting Copolymers. TEMPO derivatives are reduced with L-ascorbic acid to form the corresponding hydroxylamine [18]. The thermoresponsive behavior of the resulting random and block copolymers having TEMPO moieties was observed in 0.1 M aqueous solution of Na_2SO_4 in presence of L-ascorbic acid. The solution of the random copolymer was found to show LCST at 22°C in 0.1 M Na_2SO_4, and the LCST value increased to 33°C with increasing concentration of L-ascorbic acid because of the reduction of the TEMPO moieties (Figure 3(a)). The similar PNIPAAm-based polymer having TEMPO moieties has been reported to show almost the same LCST changes from 19°C to 33°C under reduction conditions with L-ascorbic acid [14]. On the other hand, the diblock polymer was slightly soluble in 0.1 M Na_2SO_4, and after 6.4 mM ascorbic acid was added in 0.1 M Na_2SO_4 the diblock polymer became water-soluble and showed LCST at 30°C (Figure 4(a)). The LCST of the diblock polymer slightly decreased to 28°C with increasing concentration of L-ascorbic acid. The solution of the triblock copolymer was found to show LCST at 27.2°C in 0.1 M Na_2SO_4, and the LCST was almost unchanged (28°C) with increasing concentration of L-ascorbic acid (Figure 5(a)). Reoxidation of the hydroxylamine-containing copolymer by $K_3[Fe(CN)_6]$ reverses the change of the LCST seen on reduction. The LCST of the reduced random copolymer was decreased with increasing concentration of $K_3[Fe(CN)_6]$ (Figure 3(b)). The reduced diblock copolymer became insoluble again when 4.8 mM $K_3[Fe(CN)_6]$ was added (Figure 4(b)), and the LCST of the reduced triblock copolymer was slightly decreased by adding $K_3[Fe(CN)_6]$ (Figure 5(b)). In Figures 4(b) and 5(b), initial transmittances of solutions of the random and triblock copolymers were decreased with increasing $K_3[Fe(CN)_6]$. It is due to the byproduct in the reaction of $K_3[Fe(CN)_6]$ with hydroxylamine. In both the reduction with L-ascorbic acid and the reoxidation with $K_3[Fe(CN)_6]$, the LCST's shifts

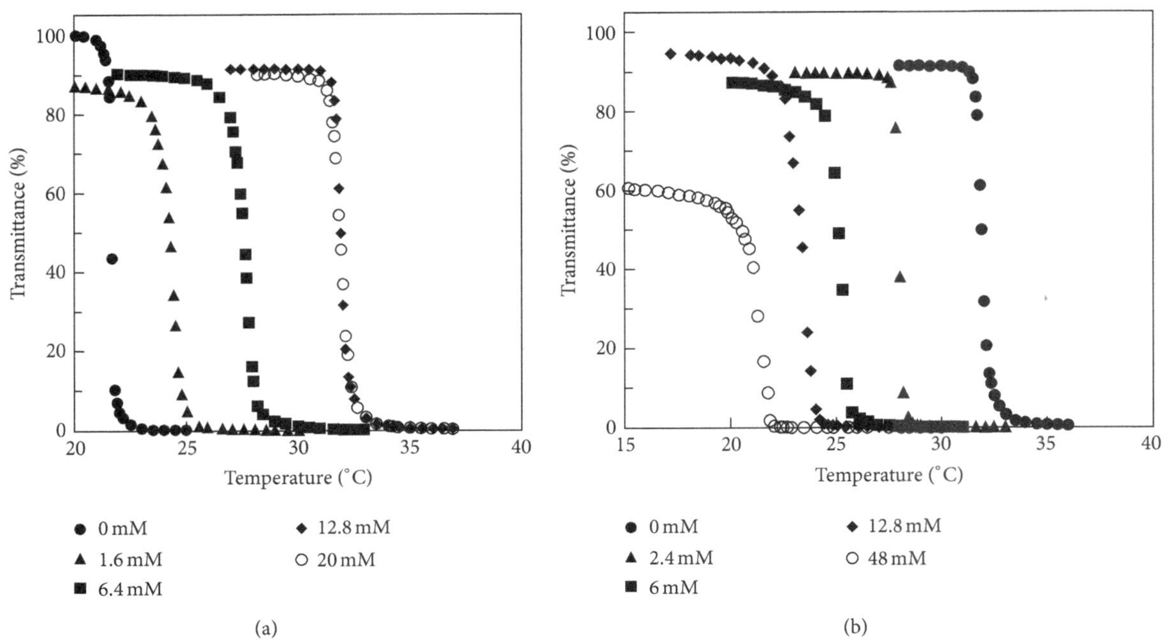

FIGURE 3: (a) Transmittance change of 10 mg/mL solutions of the poly(TEMPOMA-*ran*-NIPAAm) (**1′**) in 0.1 M Na_2SO_4aq in the presence of various amounts of L-ascorbic acid 0–20.0 mM and (b) transmittance change of the resulting solutions of the corresponding reduced copolymer in the presence of various amounts of $K_3[Fe(CN)_6]$.

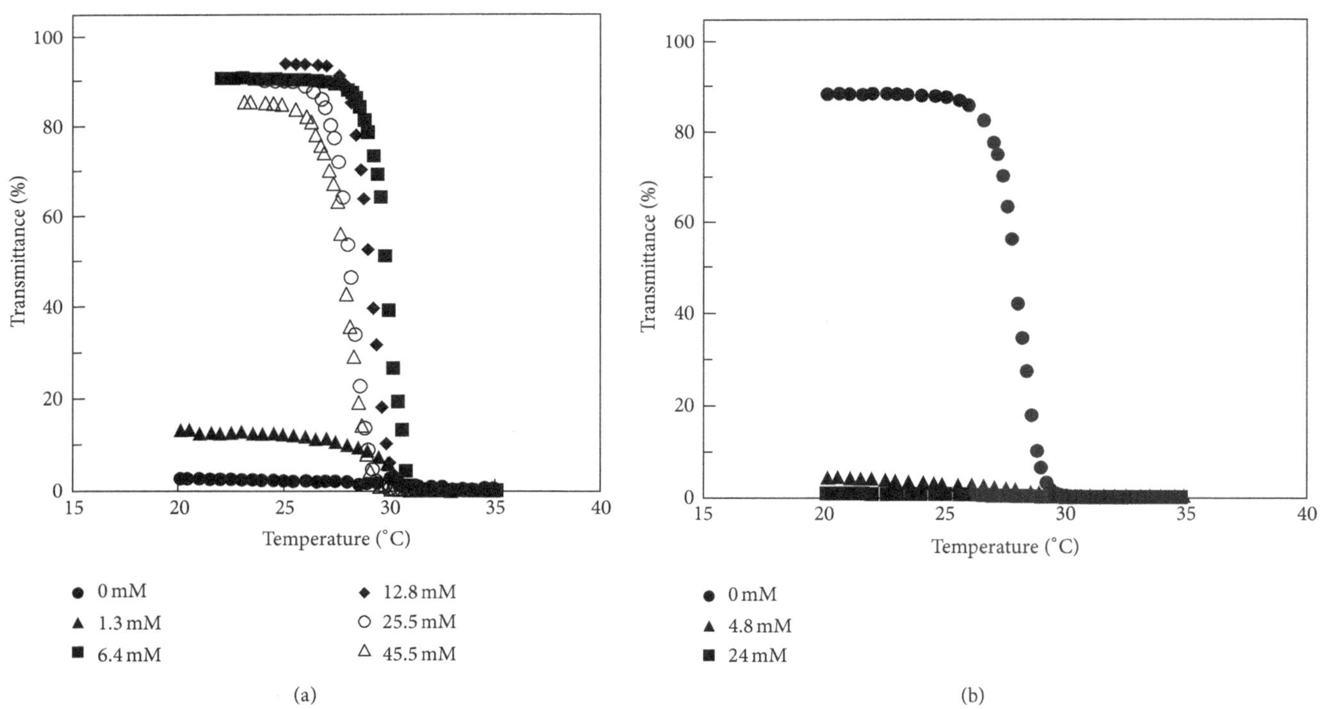

FIGURE 4: (a) Transmittance change of 10 mg/mL solutions of the PNIPAAm-*block*-PTEMPOMA (**2′**) in 0.1 M Na_2SO_4aq in the presence of various amounts of L-ascorbic acid 0–45.5 mM and (b) transmittance change of the resulting solutions of the corresponding reduced copolymer in the presence of various amounts of $K_3[Fe(CN)_6]$.

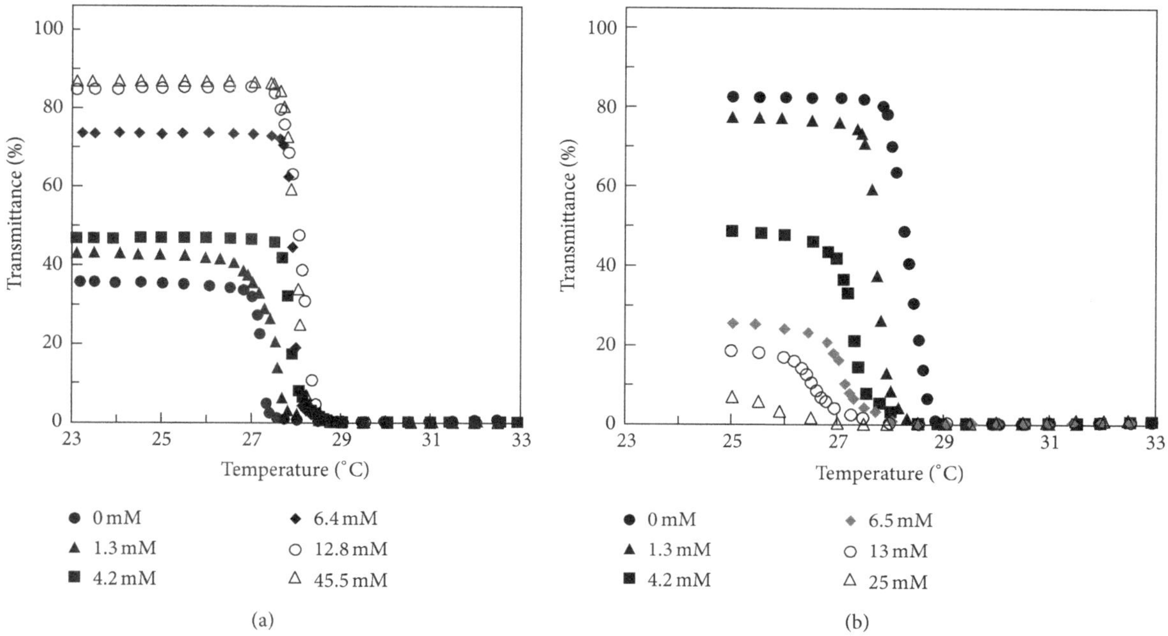

FIGURE 5: (a) Transmittance change of 10 mg/mL solutions of the PNIPAAm-*block*-PTEMPOMA-*block*-PNIPAAm (**3′**) in 0.1 M Na_2SO_4*aq* in the presence of various amounts of L-ascorbic acid 0–45.5 mM and (b) transmittance change of the resulting solutions of the corresponding reduced copolymer in the presence of various amounts of $K_3[Fe(CN)_6]$.

of the diblock and triblock copolymers were less than 2°C. The LCST of PNIPAAm decreases *ca.* 2°C by the addition of ascorbic acid or $K_3[Fe(CN)_6]$ [14]; therefore, the phase transition behavior of the block copolymers seems to be not affected by the redox reactions of the TEMPO moieties. Thus, the TEMPO-containing units sequence and PNIPAAm chains individually function, and thermoresponsivity of the PNIPAAm chains predominantly appears in the block copolymers [19].

3.4. Redox Behavior of the Resulting Copolymers at Various Temperatures. The resulting copolymers having TEMPO groups were analyzed by cyclic voltammetry in 0.1 M KCl aqueous solution at various temperatures (5–55°C). In the cyclic voltammogram of the random copolymer, poly(TEMPOMA-*ran*-NIPAAm), a pair of anodic and cathodic peaks was observed at around 0.66 V under its LCST (31°C) (Figure 6(a)), but no clear redox waves were observed over its LCST (Figure 6(b)). It was reported that TEMPO and the PNIPAAm having TEMPO groups have a redox potential of 0.63 and 0.66 V versus Ag/AgCl, respectively [14, 20]. The diblock copolymer, PNIPAAm-*block*-PTEMPOMA, showed no redox waves, due to low solubility in water. On the other hand, triblock copolymer, PNIPAAm-*block*-PTEMPOMA-*block*-PNIPAAm, showed a redox wave at 0.72-0.73 V at lower than its LCST (33°C) (Figure 7(a)). Moreover, the redox behavior of the TEMPO moieties in the triblock copolymer was found to be influenced by the phase transition of the PNIPAAm chains; namely, the anodic peaks were positively shifted to 0.81 V (Figure 7(b)). Below 33°C, the PNIPAAm chains are hydrated and TEMPO moieties are located in hydrophilic environment. On the other hand, the environment around TEMPO moieties becomes relatively hydrophobic during the phase transition of the PNIPAAm chains above LCST. The positive shift of the anodic peak is probably due to the change of hydrophilicity in the polymer field. Thus, the phase transition of the triblock copolymer having thermoresponsive repeat units was found to influence the redox potential of TEMPO moieties.

4. Conclusion

PNIPAAm-based thermoresponsive random and block copolymers having TEMPO moieties as mediatory centers were successfully synthesized by using the RAFT polymerization technique. Both the random copolymer and the triblock copolymer having TEMPO moieties were found to have LCST at 33°C, but the diblock copolymer was not soluble in water probably due to relatively hydrophobic sequence of TEMPO-containing unit at one side of the polymer chain. All copolymers containing TEMPO moieties exhibited redox sensitive LCST behavior. The observed LCSTs of the random copolymer in aqueous Na_2SO_4 increased from 22°C to 32°C by reduction with ascorbic acid and decreased in reverse by oxidation with $K_3[Fe(CN)_6]$. On the other hand, the LCST of the triblock copolymer was not so shifted under reduction and oxidation conditions. It is probably due to its block structure consisting of two relatively long PNIPAAm chains connected to the both sides of TEMPO-containing unit sequence. The TEM images of the triblock copolymer aggregates obtained from their aqueous solutions indicate the formation of a polymer micelle. Therefore the oxidation state

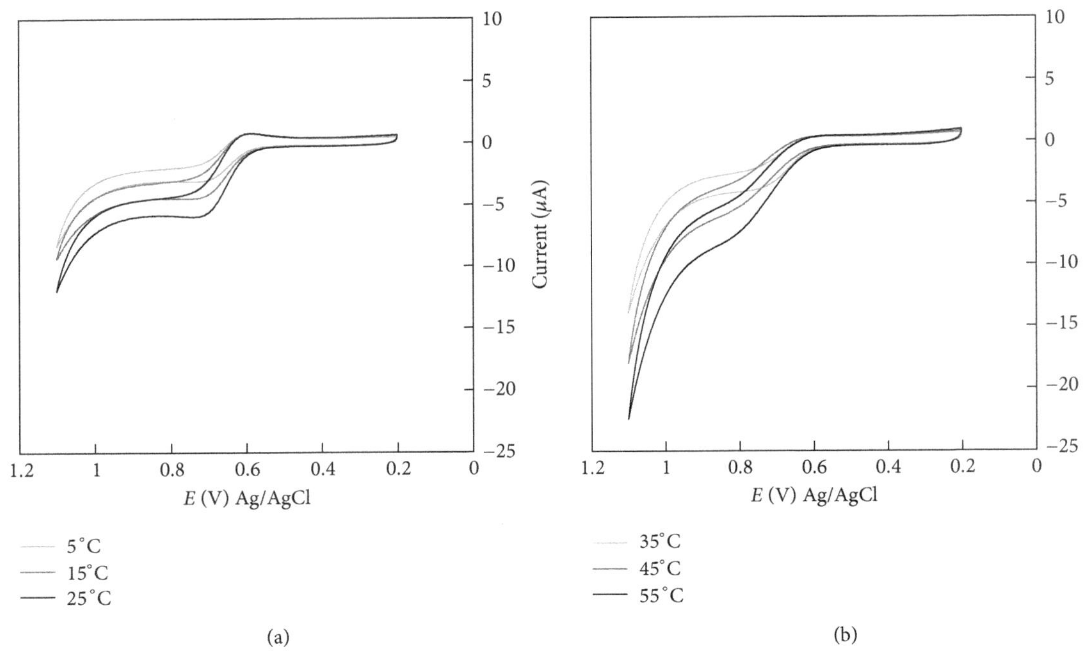

FIGURE 6: Cyclic voltammograms of poly(TEMPOMA-*ran*-NIPAAm) (**1′**) in 0.1 M KCl aqueous solution at (a) 5, 15, and 25°C, (b) 35, 45, and 55°C.

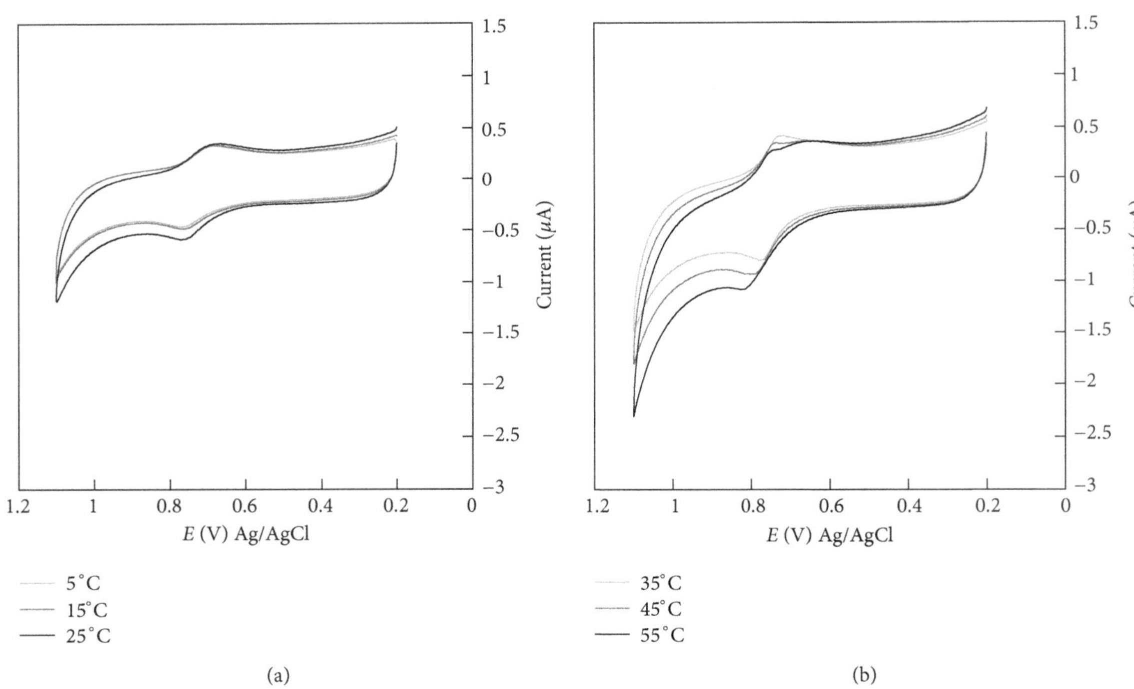

FIGURE 7: Cyclic voltammograms of PNIPAAm-*block*-PTEMPOMA-*block*-PNIPAAm (**3′**) in 0.1 M KCl aqueous solution at (a) 5, 15, and 25°C, (b) 35, 45, and 55°C.

of the TEMPO moieties within the hydrophobic core of the micelle does not influence apparent solubility of the triblock copolymer, but the observed LCST is determined only by the thermoresponsivity of the PNIPAAm chains. The redox behavior of the TEMPO moieties in the triblock copolymer was found to be influenced by the phase transition of the PNIPAAm chains. The anodic peaks of the TEMPO moieties were positively shifted above the LCST of the copolymers. This potential shift indicates that the TEMPO moieties are surrounded by the PNIPAAm chains and placed on relatively hydrophobic environment during the phase transition at LCST. We have also observed the concentration behavior of organic substrates into the hydrophobic field generated by PNIPAAm during the phase transition [12, 21]. These phenomena encourage us to perform the TEMPO-mediated oxidation of organic compounds in the thermoresponsive polymer field. Further study on the utilization of the thermoresponsive block copolymers containing TEMPO moieties to organic reactions and also electrochemical reactions in aqueous media is currently in progress [22].

References

[1] H. G. Börner, H. Kühnle, and J. Hentschel, "Making "smart polymers" smarter: modern concepts to regulate functions in polymer science," *Journal of Polymer Science Part A*, vol. 48, no. 1, pp. 1–14, 2010.

[2] I. Tokarev, M. Motornov, and S. Minko, "Molecular-engineered stimuli-responsive thin polymer film: a platform for the development of integrated multifunctional intelligent materials," *Journal of Materials Chemistry*, vol. 19, no. 38, pp. 6932–6948, 2009.

[3] M. W. Urban, "Stratification, stimuli-responsiveness, self-healing, and signaling in polymer networks," *Progress in Polymer Science*, vol. 34, no. 8, pp. 679–687, 2009.

[4] N. Rapoport, "Physical stimuli-responsive polymeric micelles for anti-cancer drug delivery," *Progress in Polymer Science*, vol. 32, no. 8-9, pp. 962–990, 2007.

[5] M. Miyajima, M. Yoshida, H. Sato, H. Omichi, R. Katakai, and W. I. Higuchi, "*In vitro* release of 9-β-d-arabinofuranosyladenine from thermo-responsive copoly(acryloyl-l-proline methyl ester/styrene) gels," *European Polymer Journal*, vol. 30, no. 7, pp. 827–831, 1994.

[6] A. Matsumoto, T. Kurata, D. Shiino, and K. Kataoka, "Swelling and shrinking kinetics of totally synthetic, glucose-responsive polymer gel bearing phenylborate derivative as a glucose-sensing moiety," *Macromolecules*, vol. 37, no. 4, pp. 1502–1510, 2004.

[7] T. Yakushiji, K. Sakai, A. Kikuchi, T. Aoyagi, Y. Sakurai, and T. Okano, "Effects of cross-linked structure on temperature-responsive hydrophobic interaction of poly(*N*-isopropylacrylamide) hydrogel-modified surfaces with steroids," *Analytical Chemistry*, vol. 71, no. 6, pp. 1125–1130, 1999.

[8] H. Feil, Y. H. Bae, J. Feijen, and S. W. Kim, "Molecular separation by thermosensitive hydrogel membranes," *Journal of Membrane Science*, vol. 64, no. 3, pp. 283–294, 1991.

[9] X.-J. Ju, S.-B. Zhang, M.-Y. Zhou, R. Xie, L. Yang, and L.-Y. Chu, "Novel heavy-metal adsorption material: ion-recognition P(NIPAM-*co*-BCAm) hydrogels for removal of lead(II) ions," *Journal of Hazardous Materials*, vol. 167, no. 1–3, pp. 114–118, 2009.

[10] M. J. Snowden, D. Thomas, and B. Vincent, "Use of colloidal microgels for the absorption of heavy metal and other ions from aqueous solution," *The Analyst*, vol. 118, no. 11, pp. 1367–1369, 1993.

[11] S. Gingter, E. Bezdushna, and H. Ritter, "Chiral recognition of macromolecules with cyclodextrins: pH- and thermosensitive copolymers from *N*-isopropylacrylamide and *N*-acryloyl-D/L-phenylalanine and their inclusion complexes with cyclodextrins," *Beilstein Journal of Organic Chemistry*, vol. 7, pp. 204–209, 2011.

[12] T. Uemukai and M. Ishifune, "Synthesis and catalytic activity of the thermoresponsive polymers having pyrrolidine side chains as base functionalities," *Journal of Applied Polymer Science*, vol. 129, no. 5, pp. 2554–2560, 2013.

[13] N. Kuramoto and Y. Shishido, "Property of thermo-sensitive and redox-active poly(*N*-cyclopropylacrylamide-co-vinylferrocene) and poly(*N*-isopropylacrylamide-co-vinylferrocene)," *Polymer*, vol. 39, no. 3, pp. 669–673, 1998.

[14] H. Fu, D. M. Policarpio, J. D. Batteas, and D. E. Bergbreiter, "Redox-controlled 'smart' polyacrylamide solubility," *Polymer Chemistry*, vol. 1, no. 5, pp. 631–633, 2010.

[15] R. Yu, H. Zhao, Z. Zhao et al., "A pH dependent thermosensitive copolymer drug carrier incorporating 4-amino-2,2,6,6-tetramethylpiperidin-1-oxyl (4-NH_2-TEMPO) residues for electron spin resonance (ESR) labeling," *Journal of Colloid and Interface Science*, vol. 362, no. 2, pp. 584–593, 2011.

[16] P. Schattling, F. D. Jochum, and P. Theato, "Multi-responsive copolymers: using thermo, light- and redox stimuli as three independent inputs towards polymeric information processing," *Chemical Communications*, vol. 47, no. 31, pp. 8859–8861, 2011.

[17] G. Moad, Y. K. Chong, A. Postma, E. Rizzardo, and S. H. Thang, "Advances in RAFT polymerization: the synthesis of polymers with defined end-groups," *Polymer*, vol. 46, no. 19, pp. 8458–8468, 2005.

[18] H. Yang, J.-Q. Chen, J. Li, Y. Lv, and S. Gao, "The effect of TEMPO in the hydroxylation of benzene to phenol on the $[(CH_3)_4N]_4PMo_{11}VO_{40}$/ascorbic acid/TEMPO/$O_2$ catalytic system: formation of ascorbic acid radicals through hydrogen exchange of ascorbic acid and TEMPO," *Applied Catalysis A*, vol. 415-416, pp. 22–28, 2012.

[19] K. Skrabania, W. Li, and A. Laschewsky, "Synthesis of double-hydrophilic BAB triblock copolymers via RAFT polymerisation and their thermoresponsive self-assembly in water," *Macromolecular Chemistry and Physics*, vol. 209, no. 13, pp. 1389–1403, 2008.

[20] H. Nishide, S. Iwasa, Y.-J. Pu, T. Suga, K. Nakahara, and M. Satoh, "Organic radical battery: nitroxide polymers as a cathode-active material," *Electrochimica Acta*, vol. 50, no. 2-3, pp. 827–831, 2004.

[21] M. Ishifune, R. Suzuki, M. Yamane, H. Tanabe, Y. Nakagawa, and K. Uchida, "Polymerization of acrylamide in aqueous solution of poly(*N*-isopropylacrylamide) at lower critical solution temperature," *Journal of Macromolecular Science Part A*, vol. 45, no. 7, pp. 523–528, 2008.

[22] T. Uemukai, T. Hioki, and M. Ishifune, "Preparation of thermo-responsive polymer-modified electrodes having a TEMPO moiety," *Electrochemistry*, vol. 81, no. 5, pp. 383–387, 2013.

11

Studies on the Transport of Aromatic Solvents through Filled Natural Rubber

Isaac O. Igwe and Obumneme E. Ezeani

Department of Polymer and Textile Engineering, Federal University of Technology, Owerri, P.M.B 1526, Owerri, Nigeria

Correspondence should be addressed to Isaac O. Igwe, izikigwe@yahoo.com

Academic Editor: Avraam I. Isayev

The transport of three aromatic solvents (benzene, toluene and xylene) through snail shell powder filled natural rubber was studied at 313, 333, and 353 K by conventional weight-gain experiments. The effects of snail shell powder content, particle size, nature of solvent, and temperature on the transport characteristics of natural rubber were determined. The estimated Arrhenius activation energies for the processes of sorption, diffusion, and permeation showed that the activation energies were highest in xylene at all the filler contents investigated. The calculated enthalpies, and entropies of sorption were all positive for the solvents investigated. Similarly, the change in the estimated free energies of sorption were all positive; an indication of the non-spontaneity of the solubility of snail shell powder filled natural rubber in the aromatic solvents at 313 k.

1. Introduction

Natural rubbers are widely used in the transportation, consumer, industrial, hygiene, and medical sectors where they provide decorative finish, high resilience, high tear strength, and so forth. Towards the end of the last century, great attention has been focused on the diffusion, sorption, and permeation of solvents into elastomers since these basic phenomena play vital roles in several important areas of engineering and industry. Thus, in the design and fabrication of barrier rubber materials for the transportation of liquids and gases, packaging of food stuffs, and so forth, it is of considerable importance to carry out transport studies so as to eliminate the diffusion of chemicals into such reinforced natural rubber products. This is necessary because the presence of these chemicals may affect the mechanical performance of the rubber material, degrade the product that the reinforced natural rubber is to protect, damage the interface between the rubber and another material, or even pollute the environment (e.g., diffusion leakage from storage vessels). Thus, it is necessary to understand the sorption, diffusion, and permeation processes in natural rubbers since this is critical to the development of reliable reinforced natural rubber products for use in the demanding environment.

Therefore, the transport behaviour of various organic solvents through different polymers and rubbers is of great technological importance since it plays a vital role in a variety of barrier applications [1–3]. The transports of solvents through polymer systems have received adequate attention in the scientific literature, and are reviewed below.

The solubility, diffusivity, and permeability of ethylbenzene in poly[1-trimethylsilyl-1-propyne] (PTMSP) were studied by Dixon-Garrett et al. [4]. They found that the diffusion coefficient of ethylbenzene in PTMSP decreased with increased concentration of ethylbenzene and decreased temperature of study.

The rate of absorption of benzene by open cell polyurethane foam of various sizes was studied by Sefton and Mann [5]. The rate of absorption was found to depend on the size of the foam. Experimental results obtained suggested that bulk flow was more rapid than pore diffusion, and it was concluded that pore diffusion was more important than bulk flow in foams.

Similarly, Schneider et al. [6] investigated the sorption and diffusion of a series of liquids through polyurethane membrane. The sorption experiments in n-heptane and incremental sorption experiments in n-heptane vapour were found to be Fickian. With increasing polarity of liquids, the swelling behaviour of the liquids was found to increase to a maximum level.

Aminabhavi and Aithal [2] studied the interactions of industrially important solvents with commercially available polyurethane. With the exception of trichloromethylene, sorption was found to increase with increase in temperature. This was attributed to the specific interaction of trichloroethylene with the hard segment part of polyurethane. The diffusivity and permeability values were found to be higher for trichloroethylene and 1,2-dichloroethylene when compared to other penetrants studied.

The diffusion characteristics of aromatic solvents into polypropylene, and polyethylene films have been reported in the literature [7, 8]. The diffusivity and permeability of the solvents into polypropylene films increased in the order of solvents, benzene, toluene, and xylene at the temperatures of investigation. The above properties were also found to be temperature dependent. Similarly, the permeability, solubility, and diffusivity of the aromatic solvents in polyethylene were found to increase with increase in temperature, and the calculated enthalpies of sorption for the solvents in polyethylene were positive while the calculated entropies of sorption were negative.

The study of diffusion, sorption, and permeation in polymer blend structures provides a valuable means for additional characterization of polymer blends. The diffusion behaviour in polymer blends was first reported by Cates and White [9–11] who investigated the water sorption characteristics of polyacrylonitrile (PAN)/cellulose, PAN/silk, and PAN/cellulose acetate blends. The diffusion behaviour of polyethylene-polystyrene (PE-PS) semi-interpenetrating polymer networks (IPNs) in toluene and chloroform was investigated by Hong and Duda [12]. In the blend of PE and nylon 6, it was found that the permeability to heptane, methyl salicylate, and methyl alcohol varied with blend ratio.

The diffusion characteristics of toluene, an aromatic solvent into natural rubber/linear low-density polyethylene blends were investigated by Obasi et al. [13]. The transport of toluene through most of the blend compositions was anomalous, although at 55°C, the transport of toluene through the 60/40 NR/LLDPE blend was Fickian, and at 35°C, pseudo-Fickian. The diffusion coefficient and permeability to toluene in 50/50 and 60/40 NR/LLDPE blends were found to increase with increase in sorption temperature.

The diffusivity in a given polymer system, whether it is rubbery polymers, glassy polymers, polymer blend, graft, or interpenetrating polymer networks, varies from one polymer system to another [14, 15]. Diffusivity depends on the free volume within the material and the segmental mobility of polymer chains, crosslinking of component polymer phases, size of penetrants, and so forth [16, 17].

The effect of fillers on the transport characteristics of polymers has been of immense interest to scientists, and reports on the effect of fillers on the diffusion and sorption processes exist [18–21]. Thus, Kwei and Kunins [22] found that the sorption of chloroform by an epoxy resin was lowered by about 70% when 5% filler was incorporated. Similarly, Boonstra and Dannenberg [23] presented the equilibrium swelling data of filled natural rubber and a number of synthetic rubbers in a variety of solvents. They observed that fillers like carbon black caused a reduction in swelling of the membranes, and which was commensurate with the volume loading of the filler. However, the effect they observed was not specific on a particular solvent or elastomer. The non-carbon black filler caused a reduction in rubber swelling, which was not dependent on the filler content.

Stickney and Mueller [24] studied the kinetics of swelling of carbon black-filled styrene-butadiene rubber vulcanizates in isooctane. It was found that, for rubber vulcanizates, diffusivity increased with the concentration of the penetrant.

Lawandy and Helaly [25] investigated the diffusion of chloroform in neoprene rubber vulcanizates containing different types of carbon black and reported that the penetration rate decreased with increase in the particle size of carbon black. Similarly, a decrease in penetration rate at high degree of equilibrium volume swelling was found. These results were attributed to the winkles formed at the surface of the rubber at high equilibrium swelling.

Ahmad et al. [26] investigated the swelling properties of filled natural rubber/linear low-density polyethylene blends in toluene for 24 h. It was found that the swelling index decreased with increase in filler loading, and this was attributed to increases in the cross-link density.

In the present study, we report the studies on the transport of aromatic solvents (benzene, toluene, and xylene) through snail shell powder filled natural rubber at 313, 333, and 353 K. This work is an extension of our investigations on the use of snail shell powder as a filler for natural rubber. The use of snail shell powder as a filler in compounding natural rubber was first reported by Igwe and Ejim [27].

The transport of liquids through snail shell powder filled natural rubber has not been reported in the scientific literature to our knowledge. The central objectives of the present study are as follows.

(i) Investigate the effects of snail shell powder filler content and particle size on the transport of benzene, toluene, and xylene through vulcanized natural rubber.

(ii) Investigate the effects of temperature on the transport of aromatic solvents through the filled natural rubber.

(iii) Determine the effects of solvent properties on the transport process in vulcanized natural rubber.

(iv) Determine the mechanism of sorption of the aromatic solvents investigated through snail shell powder filled natural rubber.

2. Experimental

2.1. Materials Used. The snail shell powder filled natural rubber vulcanizates used in this study were compounded at Dunlop Plc, Lagos, Nigeria. The snail shell was obtained locally, processed, and sieved to 0.075 and 0.30 μm mesh sizes, respectively. The following weights (g) of snail shell powder were used in compounding the natural rubber: 2, 5, 10, 15, and 20. The rubber vulcanizates are in form of rectangular sheets measuring 7.40 × 23.10 cm, with

TABLE 1: Properties of solvents.

Solvent	Density (g/cm^3)	Molecular mass volume (M_v), (g/mol)	Dipole moment (D)	Solubility parameter (Mpa)$^{1/2}$
Benzene	0.879	78.12	0	18.7
Toluene	0.866	92.14	0.36	18.3
Xylene	0.865	106.17	0.70	18.2

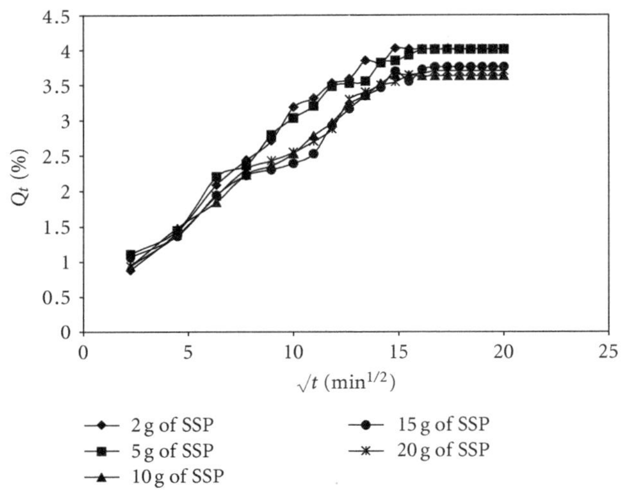

FIGURE 1: Plot of % Q_t versus $\sqrt{t}$ (min$^{1/2}$) for snail shell powder (particle size, 0.075 μm) filled natural rubber in benzene at different filler content at 313 K.

approximate thickness, 0.35 cm. The solvents used in this study are analytical-grade benzene, toluene, and xylene and were used without further purifications. The properties of the solvents used are presented in Table 1. The procedure for the sorption experiment was as described previously [7, 8], except that in this study, the following temperatures were investigated: 313, 333, and 353 K. An approximately 0.175 g of the snail shell powder filled natural rubber was used for the sorption experiment.

3. Results and Discussion

The uptake of the aromatic solvents (benzene, toluene, and xylene) by snail shell powder-filled natural rubber vulcanizates have been investigated using the conventional weight-gain method. The sorption behaviour of the rubber vulcanizates, prepared in two different particle sizes of snail shell powder, namely, 0.075, and 0.300 μm, were studied at the following temperatures 313, 333, and 353 K. The sorption data obtained for the filled natural rubber at the temperatures investigated were expressed as the molar percentage uptake (% Q_t) of the solvents per gram of the rubber vulcanizate. % Q_t was calculated using the following equation [28]:

$$\% Q_t = (\text{Mass of solvent absorbed/Molecular weight of solvent/Initial mass of the rubber vulcanizate}) \times 100. \quad (1)$$

The molar percentage uptake (%Q_t) at any particular temperature and snail shell powder particle size were plotted against the square root of time ($\sqrt{t}$) as illustrated in Figures 1, 2, 3, and 4. Each of these figures shows initial increases in the mass of the solvent sorbed, until maximum absorption was reached, at which time, the mass of the absorbed solvent remained constant, that is, equilibrium absorption was attained.

The higher initial solvent absorption rates in polymers have been explained in terms of rapid cavitations, which expose a greater surface area, thus enhancing solvent percolation [29]. On the other hand, at equilibrium, the solvent uptake is counter balanced by solvent extraction from the polymer. Obasi et al. [13] who investigated the transport of toluene through linear low-density polyethylene/natural rubber blends also found that the mass of toluene sorbed generally increased with time at the temperatures investigated until the maximum absorption was reached, at which time, the mass of toluene sorbed remained constant. Similar observation had been noted for other polymer systems [16, 28, 30, 31].

The figures show that solvent uptake for any particular solvent at the filler particle sizes used (0.075 and 0.30 μm) generally increased with increase in sorption temperature. This occurs because the diffusing solvent molecule is facilitated by the higher thermal energies at increased temperature. This results in a subsequent reduction in the viscosity of the solvents, and a further increase in the segmental mobility of the polymers. The increase in the mass of solvent sorbed by polymers with increase in temperature had been reported in the scientific literature. For example, Johnson and Thomas [30] who studied the effect of epoxidation on the transport behavior and mechanical properties of natural rubber reported that the mole percentage uptake of hexane in epoxidized rubber (ENR-50) showed increases with increase in temperature.

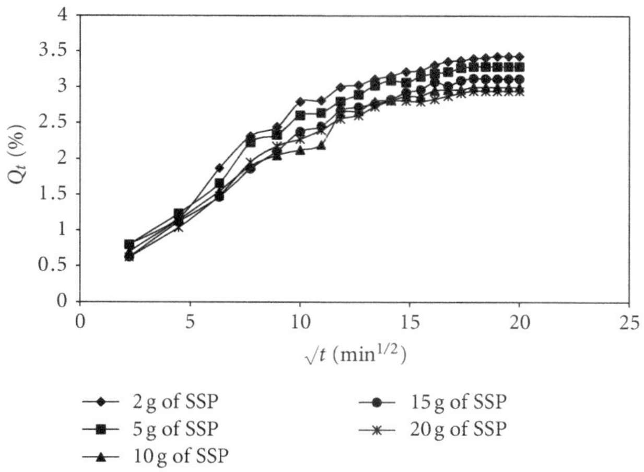

FIGURE 2: Plot of % Q_t versus $\sqrt{t}$ (min$^{1/2}$) for snail shell powder (particle size, 0.075 μm) filled natural rubber in xylene at different filler content at 333 K.

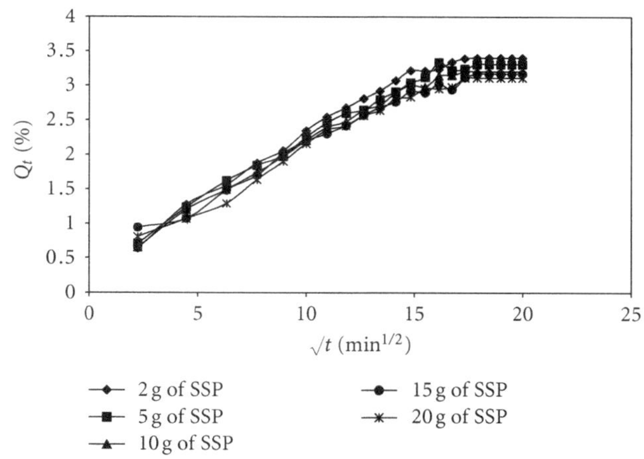

FIGURE 3: Plot of % Q_t versus $\sqrt{t}$ (min$^{1/2}$) for snail shell powder (particle size, 0.300 μm) filled natural rubber in toluene at different filler content at 313 K.

4. Analysis of the Sorption Data

4.1. The Diffusion Coefficient (D). The kinetic parameter, the diffusion coefficient (D), for the different filled natural rubber- (NR-) penetrant systems was calculated [2]:

$$D = \Pi\left(\frac{h\theta}{4Q_\infty}\right)^2, \quad (2)$$

where h is the initial sample thickness, θ is the slope of the linear portion of the sorption curve of the plot of % Q_t versus $\sqrt{t}$, and Q_∞ is the equilibrium absorption. The values of D for the different rubber systems are given in Table 2. Table 2 shows that D values generally increase with increase in the molecular mass of the solvent used, where the order in the molecular weight of the solvents is xylene > toluene > benzene. This order is also in the direction of increasing molecular mass volume, and dipole moment, and decreasing density and solubility parameter of the solvents as shown in Table 1. The higher diffusion coefficient observed for xylene could be due to the closeness of its solubility parameter (18.2 Mpa) to that of natural rubber (16.2 Mpa). The solubility parameter, which is a measure of intermolecular attraction, is a very important factor that affects solubility of polymers in solvents. Thus, polymers and solvents soften or dissolve if they have similar solubility parameter values. In other words, the present study has shown a dependence of diffusivity on the molecular mass of solvents. This is in contrast to the findings of some authors [30, 32] who reported an inverse dependence of D on the molecular weight of solvents. An increase in the diffusion coefficient of some polymers with increases in the molecular mass of penetrant molecules has been reported in the literature [7, 8]. It is important to note that diffusivity in a given polymer system, whether it is rubbery or glassy polymers, polymer blends, graft, or interpenetrating polymer networks as pointed out earlier varies from one polymer system to another.

Table 2 also shows that the D values generally decrease with increase in sorption temperature for any particular solvent and filler particle size considered. The temperature effect may be because the diffusing molecules are defacilitated by the higher thermal energies at the increased temperature. However, Johnson and Thomas [30] had reported an increase in the diffusion coefficient (D) with increase in sorption temperature.

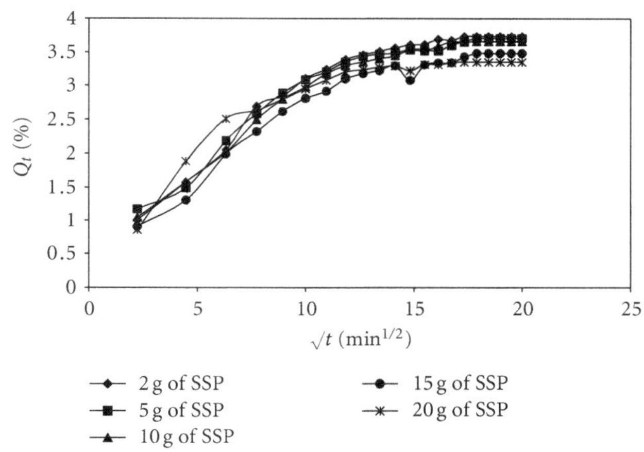

FIGURE 4: Plot of % Q_t versus $\sqrt{t}$ (min$^{1/2}$) for snail shell powder (particle size, 0.300 μm) filled natural rubber in toluene at different filler content at 353 K.

Table 2 also shows that there was no strong dependence of D on filler content at any temperature of investigation considered. However, Stickney and Muller [24] who studied the swelling of carbon black filled styrene-butadiene rubber vulcanizates in isooctane found that the diffusivity increased with the concentration of the penetrant. Similarly, Johnson and Thomas [30] in their studies found that D decreased with increase in the level of epoxidation in the rubber system investigated. The particle size of the snail shell powder used in compounding the natural rubber has no clear-cut effect on the D values obtained in the various solvents (Table 2). The table shows that at any particular filler content, solvent, and temperature investigated, the D value may increase or

Table 2: Values of diffusion coefficients (D) for snail shell powder filled natural rubber at different temperatures (filler particle sizes 0.075 and 0.300 μm).

Solvent	Filler content (g)	$D\times10^{-5}$ (cm^2/sec) 0.075 μm			$D\times10^{-5}$ (cm^2/sec) 0.300 μm		
		313 K	333 K	353 K	313 K	333 K	353 K
Benzene	2	4.38	3.53	3.23	4.84	3.71	3.05
	5	4.07	3.90	2.97	4.67	3.67	3.26
	10	4.20	3.39	2.65	4.38	3.94	3.54
	15	4.41	4.10	2.51	5.17	3.81	2.98
	20	4.38	3.33	2.57	4.07	3.47	2.86
Toluene	2	4.47	4.39	3.47	4.94	3.89	3.34
	5	4.84	4.33	3.34	4.74	3.83	3.06
	10	4.11	3.49	3.23	4.86	3.64	3.33
	15	4.47	4.17	3.41	4.41	3.80	3.24
	20	4.53	3.79	3.57	4.83	3.98	2.28
Xylene	2	4.59	3.86	4.26	4.71	4.29	3.74
	5	4.07	3.93	4.02	4.37	4.29	3.29
	10	4.37	4.24	3.87	4.70	4.29	3.38
	15	4.68	4.52	4.14	4.77	4.09	3.49
	20	4.45	4.12	3.92	4.48	4.65	3.24

decrease with filler particle size. Since diffusivity depends on the free volume within the polymer and the segmental mobility of polymer chains, the smaller-sized snail shell powder-filled vulcanizate was expected to sorb less solvent than the larger-sized snail shell powder filled rubber vulcanizate, but this was not borne out by the outcome of this study.

4.2. Sorption Coefficient (S). The sorption coefficient (S) was obtained from the plateau regions of the equilibrium sorption curves [33] and is presented in Table 3 for the vulcanizates. The table shows that the sorption coefficient decreased with increase in snail shell powder content in the rubber vulcanizate at any temperature, solvent, and filler particle size considered.

Mathai et al. [32] who studied the transport of substituted benzene through nitrile rubber/natural rubber blend membranes found that the sorption coefficient value decreased with increase in nitrile rubber content. Table 3 also shows that the sorption coefficient was highest in benzene, followed by toluene, and then xylene for any particular temperature and filler particle size considered. This finding contrasts the dependence of D on the solvents used in this study.

It is evident from Table 3 that for the solvents studied, their solubilities in filled natural rubber increased with increase in the sorption temperature from 40 to 80°C. This result is in agreement with the findings of Michaels et al. [34] who reported that the solubility of solvents in polymers increases with the increase in sorption temperature. Similarly, Johnson and Thomas [30] and Igwe et al. [7] in their sorption studies reported that the sorption coefficient increased with increase in sorption temperature. The sorption coefficient of the rubber vulcanizates is observed to decrease with increase in filler particle size. Sorption is a surface phenomenon, and it is an indication of the tendency of the penetrant to dissolve into the polymer.

4.3. Permeability Coefficient (P). The permeability or permeation coefficient (P) of a penetrant in a polymer membrane depends on the diffusivity as well as solubility or sorption of the penetrant in the polymer membrane. The permeability coefficient (P) of the aromatic solvents in the rubber vulcanizates was obtained using the following expression [35]:

$$P = D \cdot S, \quad (3)$$

where D is the diffusion coefficient and S is the sorption coefficient.

The values of P are given in Table 4. The permeability coefficients were generally observed to decrease with increase in sorption temperature for rubber vulcanizates of snail shell powder, particle size: 0.30 μm.

Generally, higher temperatures will tend to fluidize or make more flexible the polymer chains thereby facilitating the solvent permeability. The permeability coefficient in rubber vulcanizates of snail shell powder, particle size: 0.075 μm, did not show any definite relationship with the sorption temperature. Johnson and Thomas [30] who studied the transport of n-alkanes through epoxidized natural rubber found that the permeability coefficient increased with temperature. Also, Unnikrishnan et al. [36] who investigated the diffusion of aromatic hydrocarbons through filled natural rubber found that the permeability values of NR-penetrant systems followed the same trend as that of diffusivity.

The permeability was, however, found to be independent of the molecular mass, molecular mass volume, density, and dipole moment of the aromatic solvents. This is in contrast to the variation of diffusion coefficient on solvent properties

TABLE 3: Values of the sorption coefficient (S) for snail shell powder filled natural rubber at different temperatures (filler particle sizes 0.075 and 0.300 μm).

Solvents	Filler contents (g)	S (mol %) 0.075 μm			S (mol %) 0.300 μm		
		313 K	333 K	353 K	313 K	333 K	353 K
Benzene	2	4.01	4.59	4.53	4.02	4.26	4.34
	5	4.01	4.51	4.41	4.00	4.25	4.19
	10	3.76	4.10	4.07	3.91	4.26	4.12
	15	3.63	4.32	4.15	3.76	4.05	3.86
	20	3.70	3.89	3.82	3.61	3.82	3.83
Toluene	2	3.63	3.78	4.07	3.40	3.54	3.74
	5	3.43	3.61	3.84	3.31	3.52	3.71
	10	3.20	3.44	3.54	3.20	3.51	3.66
	15	3.26	3.37	3.62	3.17	3.34	3.48
	20	3.07	4.14	3.51	3.11	3.29	3.35
Xylene	2	3.37	3.43	3.50	3.08	3.23	3.31
	5	3.11	3.29	3.31	3.04	3.19	3.27
	10	2.95	3.10	3.20	3.00	3.02	3.19
	15	2.90	2.99	3.11	2.89	2.96	2.99
	20	2.84	2.94	2.96	2.95	2.78	2.95

TABLE 4: Values of the permeability coefficient (P) for snail shell powder filled natural rubber at different temperatures (filler particles sizes 0.075 and 0.30 μm).

Solvent	Filler content (g)	$P \times 10^{-5}$ (cm^2 s^{-1} mol %) 0.075 μm			$P \times 10^{-5}$ (cm^2 s^{-1} mol %) 0.300 μm		
		313 K	333 K	353 K	313 K	333 K	353 K
Benzene	2	17.56	16.20	14.63	19.46	15.80	13.24
	5	16.32	17.59	13.10	18.68	15.60	13.66
	10	15.79	13.90	10.79	17.13	16.78	14.58
	15	16.01	17.71	10.42	19.44	15.43	11.50
	20	16.21	12.85	9.82	14.69	13.26	10.95
Toluene	2	16.23	16.59	14.12	16.80	13.77	12.49
	5	16.60	15.63	12.83	15.63	13.48	11.35
	10	13.15	12.01	11.43	15.55	12.78	12.19
	15	14.57	14.05	12.34	13.98	12.69	11.28
	20	14.52	15.59	12.53	15.02	13.09	7.28
Xylene	2	15.47	13.24	14.91	14.51	13.86	12.58
	5	12.66	12.93	13.31	13.28	13.69	10.76
	10	12.89	13.14	12.38	14.10	12.87	10.76
	15	13.57	13.51	12.88	13.79	12.11	10.44
	20	12.64	12.11	11.60	13.22	12.93	9.56

obtained in this study. Igwe et al. [7] who studied the uptake of aromatic solvents into polypropylene films found that the permeability increased with increase in dipole movement, molecular mass volume, and molecular mass of the aromatic solvents.

4.4. Activation Energy of Sorption. To obtain the activation energy of sorption of the aromatic solvents in vulcanized natural rubber, the data for the diffusion coefficient (D) were treated with an Arrhenius-type expression [30]:

$$D = D_0 \exp\left(\frac{-E_D}{RT}\right), \tag{4}$$

where E_D is the activation energy of diffusion, which is a function of the intra- and interchain forces that must be overcome to create the space for a diffusional jump of

Table 5: Values of activation energy (E_D and E_P) of snail shell powder filled natural rubber (filler particle size, 0.075 μm).

Solvent	Filler content (g)	E_D (KJ/mol)	E_P (KJ/mol)
Benzene	2	-30.60×10^2	-18.15×10^2
	5	-30.92×10^2	-21.14×10^2
	10	-45.86×10^2	-37.69×10^2
	15	-55.29×10^2	-41.51×10^2
	20	-53.16×10^2	-49.86×10^2
Toluene	2	-24.79×10^2	-13.51×10^2
	5	-36.65×10^2	-25.39×10^2
	10	-91.59×10^2	-14.05×10^2
	15	-26.69×10^2	-16.37×10^2
	20	-23.96×10^2	-14.07×10^2
Xylene	2	-7.98×10^2	-4.22×10^2
	5	-1.35×10^2	-4.98×10^2
	10	-11.99×10^2	-3.88×10^2
	15	-12.09×10^2	-5.12×10^2
	20	-12.69×10^2	-8.53×10^2

Table 6: Values of activation energy (E_D and E_P) of snail shell powder filled natural rubber (filler particle size, 0.300 μm).

Solvent	Filler content (g)	E_D (KJ/mol)	E_P (KJ/mol)
Benzene	2	-46.13×10^2	-38.42×10^2
	5	-36.05×10^2	-31.23×10^2
	10	-21.22×10^2	-15.82×10^2
	15	-55.01×10^2	-52.17×10^2
	20	-35.07×10^2	-29.09×10^2
Toluene	2	-39.17×10^2	-29.73×10^2
	5	-43.58×10^2	-32.21×10^2
	10	-38.07×10^2	-24.55×10^2
	15	-30.69×10^2	-21.34×10^2
	20	-74.05×10^2	-71.26×10^2
Xylene	2	-22.87×10^2	-15.69×10^2
	5	-27.78×10^2	-20.42×10^2
	10	-32.58×10^2	-26.57×10^2
	15	-31.12×10^2	-20.14×10^2
	20	-30.83×10^2	-58.32×10^2

the penetrant molecule, D_0 is a pre-exponential factor, and RT has the conventional usual meaning. Plots of $\log D$ against $1/T$ for the aromatic solvents were made, and the calculated activation energies of diffusion (E_D) are shown in Tables 5 and 6. E_D values were all negative. Generally, the activation energy of diffusion (E_D) will be greater, the larger the size of the solvent molecule and vice versa. For rubber vulcanizates of snail shell powder, particle size, 0.075 μm, the E_D was highest in xylene as expected for all the snail shell powder contents investigated. However, for rubber vulcanizates of snail shell powder, particle size 0.30 μm, E_D was highest in xylene for the following snail shell powder contents: 2, 5, and 20 g. No definite order in the variation of E_D with the solvents studied or snail shell powder contents investigated was observed in this study. Johnson and Thomas [30] who investigated the effect of epoxidation on the transport behaviour of natural rubber found that the activation energy of diffusion increased with increase in the size of the penetrants. From Tables 5 and 6, it is evident that the particle size of the snail shell powder used in compounding the natural rubber vulcanizates influenced the E_D obtained in the various solvents used in this study. The tables equally showed that as the particle size of snail shell powder increased, the E_D values generally increased in xylene, and toluene. The later observation was also noted in the solvent benzene at snail shell powder contents, 2 and 5 g only. Thus, the activation energy of diffusion into snail shell powder filled natural rubber may depend to a large extent on the snail shell powder particle size.

4.5. Enthalpy of Absorption (ΔH_s) and Entropy of Absorption (ΔS_s). In order to determine ΔH_s and ΔS_s for the solvents in the vulcanized natural rubber, the equilibrium absorption constant (K_s) of the solvents was first determined using the following formula [36]:

$$K_s = \frac{\text{Number of moles of the solvent absorbed at equilibrium}}{\text{Mass of the natural rubber vulcanizate}}. \quad (5)$$

The data obtained were used in calculating the enthalpy of absorption (ΔH_s) and entropy of absorption (ΔS_s) in natural rubber vulcanizates. The values of K_s were substituted into Van't Hoff's equation [30]:

$$\operatorname{Log} K_s = \frac{\Delta S_s}{2.303R} - \frac{\Delta H_s}{2.303RT}, \quad (6)$$

and plots of Van't hoff's equations made were used in calculating ΔH_s and ΔS_s. The values of ΔH_s and ΔS_s obtained are given in Tables 7 and 8. For rubber vulcanizates of snail shell powder (0.075 μm), the calculated ΔH_s is generally positive and showed no relationship to the penetrant (solvent) size, the dipole moment, filler content, or filler particle size. ΔH_s is a composite parameter involving contributions from (1) Henry's law, which is needed for the formation of a site and the dissolution of the species into that site (the formation of the site involves an endodermic contribution), and (2) Langmuir's (hole-filling) sorption mechanism, in which case, the site already exists in the polymer matrix and sorption by hole filling gives exothermic heat of sorption.

In this study, the positive ΔH_s values obtained for the aromatic solvents suggest that sorption in this case is dominated by Henry's mode with endodermic contributions. Unnikrishnan and Thomas [16] in their study found that the ΔH_s obtained were negative in all cases, and increased regularly from benzene to mesitylene for a given sample. This trend could indicate an increased exothermicity of the sorption process with increase in molecular size of solvents.

The calculated ΔS_s for the solvents were all positive, and did not show any relationship to the snail shell powder content, particle size, penetrant (solvent) size, or solvent dipole moment.

TABLE 7: Values of enthalpy of absorption (ΔH_s), entropy of absorption (ΔS_s), and Gibbs free energy (ΔG_s) of snail shell powder filled natural rubber (filler particle size, 0.075 μm).

Solvent	Fillercontent (g)	ΔH_s (KJ/mol)	ΔS_s (J/mol)	ΔG_s (KJ/mol)
	2	28.74×10^2	18.85	28.68×10^2
	5	22.42×10^2	16.81	22.37×10^2
Benzene	10	18.93×10^2	15.11	18.89×10^2
	15	32.83×10^2	19.08	31.74×10^2
	20	7.81×10^2	11.37	7.77×10^2
	2	26.42×10^2	18.41	26.36×10^2
	5	25.72×10^2	17.75	25.67×10^2
Toluene	10	23.05×10^2	16.34	23.00×10^2
	15	24.05×10^2	16.80	24.00×10^2
	20	32.83×10^2	19.72	32.77×10^2
	2	8.77×10^2	13.38	8.73×10^2
	5	14.48×10^2	14.61	14.44×10^2
Xylene	10	18.09×10^2	15.20	18.05×10^2
	15	5.57×10^2	11.54	5.53×10^2
	20	9.82×10^2	12.33	9.78×10^2

TABLE 8: Values of enthalpy of absorption (ΔH_s), entropy of absorption (ΔS_s), and Gibbs free energy (ΔG_s) of snail shell powder filled natural rubber (filler particle size, 0.300 μm).

Solvent	Filler content (g)	ΔH_s (KJ/mol)	ΔS_s (J/mol)	ΔG_s (KJ/mol)
	2	18.04×10^2	15.31	17.99×10^2
	5	10.71×10^2	12.98	10.67×10^2
Benzene	10	12.71×10^2	13.49	12.67×10^2
	15	6.81×10^2	11.27	6.77×10^2
	20	14.09×10^2	13.17	14.05×10^2
	2	21.48×10^2	16.32	21.43×10^2
	5	26.08×10^2	17.59	26.03×10^2
Toluene	10	30.87×10^2	18.90	30.81×10^2
	15	21.49×10^2	15.77	21.44×10^2
	20	17.12×10^2	14.26	17.08×10^2
	2	16.63×10^2	15.17	16.59×10^2
	5	16.87×10^2	15.14	16.82×10^2
Xylene	10	13.98×10^2	14.01	13.94×10^2
	16	7.88×10^2	11.83	7.84×10^2
	20	15.41×10^2	13.76	15.36×10^2

4.6. Gibbs Free Energy of Sorption (ΔG_s). The change in ΔG_s for the aromatic solvents in filled natural rubber was obtained using the following expression:

$$\Delta G_s = \Delta H_s - T\Delta S_s, \tag{7}$$

where T is the temperature in Kelvin. The values of ΔH_s and ΔS_s earlier determined for the solvents were substituted into (7), and the calculated values of ΔG_s are also shown in Tables 7 and 8 for the solvents benzene, toluene, and xylene. The calculated ΔG_s values were all positive, and this was an indication of the nonspontaneity of the solubility of filled natural rubber in the aromatic solvents at 313 K. The calculated ΔG_s did not show any reasonable relationship to the snail shell powder content, particle size, solvent size, or dipole moment of solvent.

Igwe [8] who studied the uptake of aromatic solvents (benzene, toluene, and xylene) by polyethylene films found that the solvents showed similar ΔG_s values, which indicated that the change in ΔG_s of the aromatic solvents in polyethylene films might be independent of the solvent properties.

4.7. Transport Mechanism. The mechanism of diffusion of the aromatic solvents into filled natural rubber was analysed using the following empirical relation [37]:

$$\text{Log}\,\frac{Q_t}{Q_\infty} = \log k + n \log t, \tag{8}$$

where Q_t and Q_∞ are the mol % sorption at time t and equilibrium, respectively. k is a constant that depends on the structural characteristics of the rubber and gives information

Table 9: Values of n and k for snail shell powder filled natural rubber at different temperatures (filler particle size, 0.075 μm).

Solvent	Filler content (g)	n			k		
		313 K	333 K	353 K	40°C	60°C	80°C
Benzene	2	0.42	0.31	0.32	0.11	0.18	0.18
	5	0.35	0.38	0.32	0.15	0.16	0.18
	10	0.36	0.37	0.28	0.13	0.14	0.20
	15	0.33	0.39	0.27	0.16	0.18	0.23
	20	0.36	0.31	0.30	0.14	0.18	0.21
Toluene	2	0.40	0.38	0.32	0.10	0.12	0.17
	5	0.37	0.38	0.31	0.12	0.12	0.18
	10	0.34	0.33	0.33	0.15	0.16	0.17
	15	0.36	0.34	0.34	0.13	0.15	0.17
	20	0.35	0.35	0.35	0.14	0.14	0.15
Xylene	2	0.37	0.36	0.40	0.12	0.14	0.16
	5	0.33	0.35	0.38	0.13	0.14	0.16
	10	0.36	0.37	0.37	0.13	0.15	0.18
	15	0.17	0.40	0.38	0.14	0.14	0.17
	20	0.33	0.39	0.35	0.11	0.12	0.14

Table 10: Values of n and k of snail shell powder filled natural rubber at different temperatures (filler particle size, 0.300 μm).

Solvent	Filler content (g)	n			k		
		313 K	333 K	353 K	313 K	333 K	353 K
Benzene	2	0.39	0.36	0.30	0.11	0.14	0.20
	5	0.39	0.34	0.35	0.11	0.15	0.16
	10	0.36	0.36	0.29	0.13	0.14	0.17
	15	0.43	0.35	0.37	0.10	0.14	0.16
	20	0.36	0.36	0.34	0.14	0.14	0.17
Toluene	2	0.40	0.35	0.33	0.10	0.13	0.16
	5	0.38	0.39	0.30	0.12	0.13	0.20
	10	0.39	0.37	0.32	0.11	0.14	0.17
	15	0.33	0.36	0.34	0.14	0.14	0.16
	20	0.36	0.35	0.30	0.13	0.15	0.20
Xylene	2	0.39	0.42	0.36	0.10	0.10	0.14
	5	0.37	0.40	0.33	0.13	0.14	0.16
	10	0.42	0.38	0.35	0.10	0.12	0.15
	15	0.39	0.36	0.35	0.11	0.13	0.16
	20	0.38	0.38	0.34	0.11	0.12	0.16

about the interaction between the rubber and solvents. The value of $n = 0.5$ indicates a Fickian mode of transport while $n = 1$ indicates case II transport. The value of n between 0.5 and 1 indicates anomalous transport behaviour. The values of n and k for the rubber vulcanizates were obtained by regression analysis of $\log(Q_t/Q_\infty)$ versus $\log t$, and the results are given in Tables 9 and 10. From Tables 9 and 10, it can be seen that the values of n obtained for the rubber vulcanizates at different temperatures, and snail shell powder particle sizes are Fickian. The values of n did not show any relationship to the amount of snail shell powder incorporated into the natural rubber.

Mathew et al. [28] who studied the transport of substituted benzene through natural rubber/polystyrene (NR/PS) interpenetrating polymer network membranes found the mode of transport to be anomalous. For the prepared rubber vulcanizates of snail shell powder, the values of k were observed to increase with increase in sorption temperature for the solvents studied. The values of k obtained in this study did not show any relationship to snail shell powder content or solvent employed. The similarity in the values of k obtained in the various solvents for the natural rubber vulcanizates study may be an indication of the similarity of interaction between the rubber vulcanizates and aromatic solvents employed in this study. Kumnuantip and Sombatsompop [35] in their studies reported a general increase of k with increase in reclaimed natural rubber.

5. Conclusions

The transport of three aromatic solvents (benzene, toluene, and xylene) through snail shell powder filled natural rubber has been studied. The molar percentage solvent uptake (% Q_t) in the filled natural rubber was found to show initial increases in the mass of the solvents sorbed, until the maximum absorption was reached, at which time, the mass of the absorbed solvent remained constant. The diffusion and sorption coefficients obtained for the aromatic solvents in snail shell powder filled natural rubber increased with increase in the sorption temperature. The permeability coefficients (P) for the solvents were found to decrease with increase in the sorption temperature for rubber vulcanizates of snail shell powder, particle size: 0.300 μm. The rubber vulcanizates of snail shell powder, particle size: 0.075 μm, did not show any relationship with the sorption temperature. The calculated enthalpies of sorption (ΔH_s), entropies of sorption (ΔS_s), and Gibbs free energies of sorption (ΔG_s) were all positive while the Arrhenius activation energies (E_D) were all negative. The positive ΔG_s obtained in this study is an indication of nonspontaneity of the solubility of snail shell powder filled natural rubber in the aromatic solvents at 313 K. The mode of transport of the aromatic solvents into filled natural rubber has been found to be Fickian. The transport parameters presented in this study have not only provided additional characterization of the snail shell powder filled natural rubber but gave an insight into the behaviour of snail shell powder filled natural rubber in external liquid environment which is essential for their successful applications. The data obtained could be of importance in problem solving like designing a barrier material or tubes for transporting liquids.

References

[1] Y. M. Lee, D. Bourgeois, and G. Belfort, "Sorption, diffusion, and pervaporation of organics in polymer membranes," *Journal of Membrane Science*, vol. 44, no. 2-3, pp. 161–181, 1989.

[2] T. M. Aminabhavi and U. S. Aithal, "Molecular transport of oxygen and nitrogen through polymer films," *Journal of Macromolecular Science Part C*, vol. 31, no. 2-3, pp. 117–163, 1991.

[3] W. J. Koros, "Barrier polymers and structures; overview," in *Barrier Polymers and Structures*, W. J. Koros, Ed., vol. 423 of *ACS Symposium Series*, chapter 1, pp. 1–21, American Chemical Society, Washington, DC, USA, 1990.

[4] S. V. Dixon-Garrett, K. Nagai, and B. D. Freeman, "Sorption, diffusion, and permeation of ethylbenzene in poly(1-trimethylsilyl-1-propyne)," *Journal of Polymer Science Part B*, vol. 38, no. 8, pp. 1078–1089, 2000.

[5] M. V. Sefton and J. L. Mann, "Absorption of benzene by open-cell polyurethane foams," *Journal of Applied Polymer Science*, vol. 25, no. 5, pp. 829–839, 1980.

[6] N. S. Schneider, J. L. Illinger, and M. A. Cleaves, "Liquid sorption in a segmented polyurethane elastomer," *Polymer Engineering and Science*, vol. 26, no. 22, pp. 1547–1551, 1986.

[7] I. O. Igwe, C. M. Ewulonu, and I. Igboanugo, "Studies on the diffusion characteristics of some aromatic solvents into polypropylene film," *Journal of Applied Polymer Science*, vol. 102, no. 2, pp. 1985–1989, 2006.

[8] I. O. Igwe, "Uptake of aromatic solvents by polyethylene films," *Journal of Applied Polymer Science*, vol. 104, no. 6, pp. 3849–3854, 2007.

[9] M. D. Cates and J. H. White Jr., "Preparation and properties of fibers containing mixed polymers. I. Polyacrylonitrile-cellulose acetate fibers," *Journal of Polymer Science*, vol. 20, no. 94, pp. 155–180, 1956.

[10] M. D. Cates and J. H. White Jr., "Preparation and properties of fibres containing mixed polymers. II. Polyacrylonitrile-silk fibres," *Journal of Polymer Science*, vol. 21, pp. 181–195, 1956.

[11] M. D. Cates and J. H. White Jr., "Preparation and properties of fibers containing mixed polymers. III. Polyacrylonitrile-silk fibers," *Journal of Polymer Science*, vol. 21, no. 97, pp. 125–138, 1956.

[12] S. U. Hong and J. L. Duda, "Penetrant transport in polyethylene-polystyrene semi-interpenetrating polymer networks," *Journal of Applied Polymer Science*, vol. 65, no. 1, pp. 51–57, 1997.

[13] H. C. Obasi, O. Ogbobe, and I. O. Igwe, "Diffusion characteristics of toluene into natural rubber/linear low density polyethylene blends," *International Journal of Polymer Science*, vol. 2009, Article ID 140682, 6 pages, 2009.

[14] T. M. Aminabhavi, U. S. Aithal, and S. S. Shukla, "Molecular transport of organic liquids through polymer films," *Journal of Macromolecular Science*, vol. 29, no. 2-3, pp. 319–363, 1989.

[15] E. Sada, H. Kumazawa, P. Xu, and S. T. Wang, "Permeation of mixed gases in glassy polymer membranes," *Journal of Applied Polymer Science*, vol. 40, no. 7-8, pp. 1391–1399, 1990.

[16] G. Unnikrishnan and S. Thomas, "Interaction of crosslinked natural rubber with chlorinated hydrocarbons," *Polymer*, vol. 39, no. 17, pp. 3933–3938, 1998.

[17] M. J. Smith and N. A. Peppas, "Effect of the degree of crosslinking on penetrant transport in polystyrene," *Polymer*, vol. 26, no. 4, pp. 569–574, 1985.

[18] G. J. Kraus, "Swelling of filler-reinforced vulcanizates," *Journal of Applied Polymer Science*, vol. 7, no. 3, pp. 861–871, 1963.

[19] R. M. Barrer, J. A. Barrie, and M. G. Rogers, "Heterogeneous membranes diffusion in filled rubber," *Journal of Polymer Science Part A*, vol. 1, no. 8, pp. 2565–2586, 1963.

[20] F. de Candia, L. Gargani, and A. Renzulli, "Transport properties of filled elastomeric networks," *Journal of Applied Polymer Science*, vol. 41, no. 5-6, pp. 955–964, 1990.

[21] S. N. Lawandy and S. H. Botros, "Effect of type and amount of carbon black on the interaction between polychloroprene rubber and motor oil," *Journal of Applied Polymer Science*, vol. 42, no. 1, pp. 137–141, 1991.

[22] T. K. Kwei and C. A. Kunins, "Polymer filler interaction: vapour sorption studies," *Journal of Applied Polymer Science*, vol. 8, pp. 1483–1490, 1964.

[23] B. B. S. T. Boonstra and E. M. Dannenberg, "The equilibrium swelling data of filled natural rubber," *Rubber Age*, vol. 1, pp. 825–838, 1958.

[24] P. B. Stickney and W. Mueller, "The kinetics of swelling of carbon black filled styrene butadiene rubber vulcanizates in iso-octane," *Rubber Chemistry and Technology*, vol. 42, pp. 604–611, 1969.

[25] S. N. Lawandy and F. H. Helaly, "Diffusion of a volatile liquid in polychloroprene rubber," *Journal of Applied Polymer Science*, vol. 32, no. 6, pp. 5279–5286, 1989.

[26] A. Ahmad, D. H. Mohd, and I. Abdullah, "Mechanical properties of filled natural rubber/linear low density polyethylene blends," *Iranian Polymer Journal*, vol. 13, no. 3, pp. 173–178, 2004.

[27] I. O. Igwe and A. M. Ejim, "The effects of incorporating snail shell powder as filler on the properties of natural rubber," *Material Science and Application*, vol. 2, pp. 801–809, 2011.

[28] A. P. Mathew, S. Packirisamy, M. G. Kumaran, and S. Thomas, "Transport of styrene monomer through vulcanized natural rubber," *Polymer*, vol. 25, no. 35, pp. 4935–4942, 2002.

[29] A. B. Desai and G. L. Wilkes, "Solvent-induced crystallization of polyethylene terephthalate," *Journal of Polymer Science*, no. 46, pp. 291–319, 1974.

[30] T. Johnson and S. Thomas, "Effect of epoxidation on the transport behaviour and mechanical properties of natural rubber," *Polymer*, vol. 41, no. 20, pp. 7511–7522, 2000.

[31] H. Ismail and S. Suzaimah, "Styrene butadiene rubber/epoxidized natural rubber blends: dynamic properties, curing characteristics and swelling studies," *Polymer Testing*, vol. 19, no. 8, pp. 879–888, 2000.

[32] A. E. Mathai, R. P. Singh, and S. Thomas, "Transport of substituted benzenes through nitrile rubber/natural rubber blend membranes," *Journal of Membrane Science*, vol. 202, pp. 35–54, 2002.

[33] T. M. Aminabhavi and H. T. S. Phayde, "Molecular transport of alkanes through thermoplastic miscible blends of ethylene-propylene random copolymer and isostatic polypropylene," *Journal of Applied Polymer Science*, vol. 55, pp. 1335–1352, 1995.

[34] A. S. Michaels, W. Vieth, A. S. Hoffman, and H. A. Alcalay, "Structure-property relationships for liquid transport in modified polypropylene," *Journal of Applied Polymer Science*, vol. 13, pp. 577–598, 1969.

[35] C. Kumnuantip and N. Sombatsompop, "Dynamic mechanical properties and swelling behaviour of NR/reclaimed rubber blends," *Materials Letters*, vol. 57, no. 21, pp. 3167–3174, 2003.

[36] G. Unnikrishnan, S. Thomas, and S. Varghese, "Sorption and diffusion of aromatic hydrocarbons through filled natural rubber," *Polymer*, vol. 37, no. 13, pp. 2687–2693, 1996.

[37] L. M. Lucht and N. A. Peppas, "Transport of penetrants in the macromolecular structure of coals. V. Anomalous transport in pre-treated coal particles," *Journal of Applied Polymer Science*, vol. 33, no. 5, pp. 1557–1566, 1987.

12

MonoMax Suture: A New Long-Term Absorbable Monofilament Suture Made from Poly-4-Hydroxybutyrate

Erich K. Odermatt,[1] Lutz Funk,[2] Rainer Bargon,[1] David P. Martin,[3] Said Rizk,[3] and Simon F. Williams[3]

[1] *Aesculap AG, Am Aesculap-Platz, 78532 Tuttlingen, Germany*
[2] *B. Braun Surgical S.A., Carretera de Terrassa 121, 08191 Rubi, Spain*
[3] *Tepha Inc., 99 Hayden Avenue, Lexington, MA 02421, USA*

Correspondence should be addressed to Erich K. Odermatt, erich.odermatt@aesculap.de

Academic Editor: Kibret Mequanint

A long-term absorbable monofilament suture was developed using poly-4-hydroxybutyrate (P4HB) made from a biosynthetically produced homopolymer of the natural metabolite 4-hydroxybutyrate. The suture, called MonoMax, has prolonged strength retention. At 12 weeks, a size 3-0 MonoMax suture retains approximately 50% of its initial tensile strength *in vivo* and is substantially degraded in one year with minimal tissue reaction. In contrast, PDS II monofilament suture (Ethicon, Inc., Somerville, NJ) has no residual strength *in vivo* after 12 weeks. *In vivo*, the MonoMax suture is hydrolyzed primarily by bulk hydrolysis, and is then degraded via the Krebs cycle. MonoMax is substantially more compliant than other monofilament sutures, and incorporates an element of elasticity. Its tensile modulus of 0.48 GPa is approximately one-third of the value of the PDS II fiber providing an exceptionally flexible and pliable fiber with excellent knot strength and security. These features are further enhanced by the fiber's elasticity, which also improves knot security and may help prevent wound dehiscence. Because of its performance advantages, this suture may find clinical utility in applications where prolonged strength retention, and greater flexibility are required, particularly in procedures like abdominal wall closure where wound dehiscence is still a significant post-surgical complication.

1. Introduction

The development of synthetic absorbable sutures was one of the most important clinical advances in wound closure and suture technologies. When compared to catgut sutures, these synthetic alternatives typically have greater tensile strengths, more uniform degradation profiles, and a significantly reduced incidence of infection and tissue reaction. As a result, there has been widespread acceptance and use of synthetic absorbable sutures, and these products are now the most commonly used sutures in surgery and wound repair, where temporary wound support is needed.

In the early 1970s, suture manufacturers initially introduced synthetic absorbable sutures in a braided configuration made from polyglycolide (Dexon, Davis & Geck) or copolymers of glycolide and lactide (Vicryl, Ethicon), because monofilament fibers of these materials were too stiff and difficult to handle [1]. However, braided synthetic absorbable sutures have a tendency to tear through tissue, and can potentiate infection via the interstices of their braided structure. To address these drawbacks, suture manufacturers initially developed coatings to reduce tissue drag, but in the 1980s introduced new polymers for use in monofilament sutures as a way to reduce the tissue drag and infection rates associated with the use of braided sutures. As a result, there is a broad range of synthetic, absorbable braided and monofilament sutures available today.

The first monofilament absorbable synthetic suture, PDS, a homopolymer of *p*-dioxanone, was introduced in 1984 by Ethicon (Somerville, NJ, USA) and eliminated some of the concerns associated with the use of braided sutures [1]. An improved version, PDS II, was later introduced

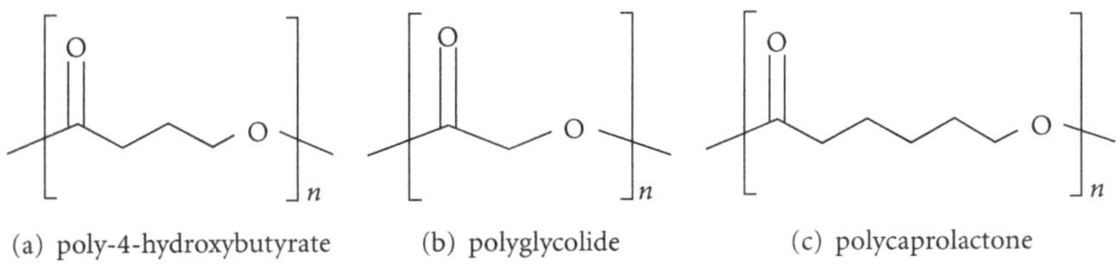

FIGURE 1: Chemical structures of (a) poly-4-hydroxybutyrate (P4HB), (b) polyglycolide, and (c) ε-polycaprolactone.

by Ethicon. Other absorbable monofilaments have since been introduced, but in general the existing monofilament absorbable sutures do not handle as well as braids and typically require more knotting throws. Moreover, PDS II degrades relatively rapidly with a loss of approximately 50% of its initial strength 4-5 weeks after implantation making it unsuitable for use, where longer strength retention is required. Thus, there is still room to improve the performance of absorbable monofilament sutures.

To address the current shortcomings of existing absorbable monofilament sutures, a major research effort was initiated to develop a P4HB suture (MonoMax, B.Braun), the most pliable synthetic absorbable monofilament suture in clinical use. The suture is derived from a biosynthetic polyester known as poly-4-hydroxybutyrate (P4HB, Tepha, Inc.). This new absorbable polymer promises to be useful in a variety of medical device applications, but is seeing its first clinical utility as a monofilament suture. In this paper, the tensile strength, breaking strength retention, knot strength, *in vitro* and *in vivo* degradation, and biocompatibility of the P4HB suture are described as well as the *in vivo* functionality of the MonoMax suture in an animal model for abdominal wall repair. Each of these parameters is critical in determining suture performance. Additional studies (not described herein) have also been undertaken to evaluate the clinical performance of the MonoMax suture in abdominal wall repair and demonstrate the superior performance of this new monofilament suture.

2. Experimental Method

2.1. Biosynthesis of Poly-4-Hydroxybutyrate. MonoMax Suture is derived from poly-4-hydroxybutyrate homopolymer, which is produced by Tepha, Inc. (Lexington, MA, USA) using a proprietary fermentation process [2]. Despite the polymer's biological method of production, the chemical structure of the polymer is very simple (Figure 1) and closely resembles the structure of existing synthetic absorbable biomaterials such as polyglycolide or poly-ε-caprolactone. The polymer belongs to a larger class of materials called polyhydroxyalkanoates (PHAs) that are produced in nature by numerous microorganisms [3–5]. In nature these polyesters are produced as storage granules inside cells and serve to regulate energy metabolism (see [2–5] and references therein). They are also of commercial interest because of their thermoplastic properties, and relative ease of production from sustainable resources. While there has been considerable effort to develop other PHA materials for medical applications [3, 4], P4HB is currently the only PHA cleared by the USFDA and EU regulatory agencies for use in clinical applications.

P4HB is not naturally produced by any wildtype microorganism, but can be produced at scale for medical applications through biotechnology. Several biosynthetic routes are known to produce P4HB [2, 6, 7]. Notably, the chemical synthesis of P4HB has been attempted, but it is generally considered impossible to produce the polymer with sufficiently high molecular weight necessary for thermoplastic applications [8]. Recent theoretical calculations have tried to elucidate the mechanistic reason for the difficulty in synthesizing P4HB [9]. Tepha, Inc. produces P4HB utilizing a genetically engineered *Escherichia coli* K12 microorganism that incorporates new biosynthetic pathways to produce the homopolymer. The polymer accumulates as distinct granules inside the cells during fermentation and can be extracted at the end of the process in a very pure form. Typically, yields of P4HB exceed 50 grams per liter of fermentation broth in less than 48 hours making large-scale production attractive. The ^{1}H-NMR spectrum of purified P4HB shows three characteristic peaks that are consistent with the structure for the polymer (^{1}H -NMR (400 MHz, CDCl3) $\delta = 1.94$ (*m*, 2H), 2.37 (*t*, 2H, $J = 7.4$ Hz), 4.10 (*t*, 2H, $J = 6.6$ Hz)). No extraneous peaks were found that would otherwise indicate the presence of soluble organic contaminants in the sample or other monomeric constituents in the polymer.

The weight average molecular weight (M_w) for the polymer was found to be 400–600 kDa as measured by gel permeation chromatography (GPC) in chloroform relative to polystyrene standards of narrow polydispersity. This M_w is well suited to melt spinning of the high-strength monofilament suture fibers. Endotoxin content of the purified polymer, measured by the Limulus amebocyte lysate (LAL) assay, is very low and is well below the requirements established for medical devices.

2.2. Preparation of the MonoMax Suture. Poly-4-hydroxybutyrate polymer was pelletized and melt extruded into monofilaments, quenched, and drawn in line with stretch ratios of 6 to 11x in a multistage drawing process in order to form the suture fiber and achieve the desired tensile and handling properties of the MonoMax suture.

2.3. Sterilization of the MonoMax Suture. Monofilament fibers of the MonoMax suture were cut to uniform lengths, packaged in ethylene oxide sterilization bags and exposed to a cold ethylene oxide (EtO) sterilization cycle. Sterility was confirmed by testing random samples by direct transfer to test media per USP standard test method ⟨25⟩. The EtO sterilization cycle was subsequently validated. EtO residuals (EtO, epichlorohydrine ECH, and ethylene glycol EG) were found to be below 2 ppm after the aeration [10].

2.4. Mechanical Properties of the MonoMax Suture. The tensile properties of sterile monofilament fibers from multiple lots were determined using a universal mechanical tester according to procedures described in the US Pharmacopeia (USP) standard for testing tensile properties of surgical sutures (USP ⟨881⟩) [11]. Suture diameter was determined in accordance with the USP standard by making three measurements at the two ends and middle of the sample, except for explanted samples, where diameter was determined by averaging measurements from both ends of the sample. This was done to prevent damaging the middle section of explanted samples prior to tensile testing. The tensile properties of the monofilament samples were determined using either an Instron 3535 or a Zwick Z005 universal testing machine equipped with pneumatic fiber grips using a gauge length of 200 mm, and a strain rate of 300 mm/minute. During testing, the location of the break was recorded. The specimen was rejected if the break occurred at the grips (i.e., outside the gauge length). Values for straight tensile strength, Young's modulus, elongation-to-break, and knot tensile strength were determined. The latter value was determined by following the procedure described in the European Pharmacopeia [12]. The bending stiffness of the P4HB suture was measured using a Frank 56585 bending stiffness testing device, applying a speed of determination of 6°/s (max. angle 30°; holding time 2 sec.), a measuring length of 1.0 mm, and a preforce of 1 mN.

2.5. In Vitro Degradation of the MonoMax Monofilament. Sterilized P4HB monofilament suture fibers of size 3–0 were incubated in Dulbeco's phosphate-buffered saline (pH 7.4; 37°C) to determine the stability of the monofilament under these conditions. The results were compared to control samples of the same fiber lot tested at the beginning of the study without any preliminary treatment in buffer solution. Sodium azide (0.05% wt/vol) was included in the buffer as a biocide. Samples were removed at $t = 1, 2, 3, 4, 6, 8, 12, 16$, and 20 weeks after incubating and tested for tensile strength. Additionally, the stability of the P4HB suture was determined in an accelerated model of *in vitro* degradation. The suture was incubated in 3 M hydrochloric acid at 37°C. Samples were removed for straight tensile testing and M_w determinations after 5, 24, and 48 hours incubation. To image the morphology of a fiber after accelerated *in vitro* degradation, the suture was incubated in a less concentrated 2 M aqueous HCl solution for 7 days, in order to capture the degradation process in more detail.

2.6. In Vivo Breaking Strength Retention of the MonoMax Monofilament Suture after Ethylene Oxide Sterilization. A subcutaneous implantation study in rabbits was used to compare the breaking strength retention of the MonoMax suture to PDS II suture (Ethicon, Inc.) as a control. Strength retention of the sutures was determined at 1, 2, 3, 4, 5, 6, 12, 16, and 20 weeks after the implantation date. A total of eighteen SPF albino rabbits (New Zealand White, Charles River Deutschland GmbH, weight 2-3 kgs) were arbitrarily assigned to the termination intervals, two rabbits per each time point. In preparation for the surgery, the fur was clipped from the area over the paravertebral muscles and the back scrubbed with 70% ethyl alcohol. The rabbits were anesthetized by an intramuscular injection of ketamine and xylazine applying a dosage of 40 mg/kg (ketamine) and 6 mg/kg (xylazine). Before the implantation, packages containing reels of ethylene oxide sterilized size 3–0 MonoMax suture were opened on a sterile field and strands of approximately 150 cm length were cut from the reels. Small coils were made from the sutures in order to implant them later into the subcutaneous tissue. The same procedure was repeated using 90 cm stands of PDS II as reference samples. After making 3 small lateral incisions 3 cm in length on both sides of the spinal column, small subcutaneous pockets were formed, and the suture coils were placed inside the pockets. MonoMax fibers were implanted on the left, while PDS II control samples were implanted on the right side of the spinal column. The skin was closed with atraumatic nonabsorbable suture material (Premilene, B. Braun) and finally sealed with a tissue adhesive (Histoacryl, B. Braun). At appropriate intervals, the designated rabbits were euthanized by an intravenous injection of 0.3 mL/kg T61. The skin of the rabbits was opened by a longitudinal incision along the spinal column, and extreme care was used during the dissection to avoid damaging the implants. The three MonoMax suture samples and three PDS II control suture samples from each animal were explanted, put in prelabeled containers and prepared for the tensile strength measurements. Straight pull tensile strengths and knot pull strengths of the explanted suture fibers were determined as described above according to USP ⟨881⟩ [11]. After tensile testing, the remaining fiber fragments were retained for M_w analysis by GPC (Polymer Standards Service GmbH, Mainz, Germany). The GPC measurements were conducted using chloroform as eluent and calibrating against polystyrene standards.

2.7. In Vitro Knotting Studies. To evaluate the knot security of the suture, the following *in vitro* test procedure was developed to simulate the real-life behavior of a surgeon performing a knot. For this test, the suture was tied around a plastic tube with a diameter of 9 cm, and the two ends of the strand were knotted tightly with a surgical knot (2 = 1) to form a loop according to Tera and Aberg [13]. After making the knot, the suture loop was cut on the side opposite the position of the knot. This results in a suture strand with a 2 = 1 knot called surgical knot in the middle (see Figure 2).

Both fiber ends were clamped in a tensile testing machine and the knotted strand was pulled apart with a traverse speed of 100 mm/min. If the suture breaks at the knot for 10 samples, the knot security factor (KSF) is defined as KSF = 0. If the knot slips and unties when applying a force for one or more samples out of ten, then another ten samples are prepared and tested as above but with an additional

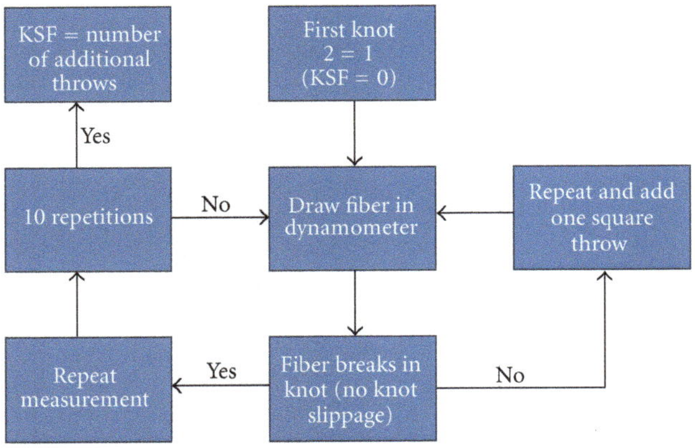

Figure 2: Schematic drawing of the knot security factor procedure.

square knot. The knot security factor (KSF) increases by 1 for each additional throw needed on top of the first surgical knot (2 = 1). The KSF is then defined as the number of additional throws needed to prevent all ten knotted samples from untying when an external force is applied. A knot security factor (KSF) of 2–4 proves to be excellent for a monofilament.

2.8. Biocompatibility Testing. The testing described in this section was performed to establish the biocompatibility of the ethylene oxide sterilized MonoMax suture in accordance with the International Standard ISO-10993 "Biological Evaluation of Medical Devices Part-1: Evaluation and Testing" [14]. Biocompatibility testing was performed at NAMSA (Ohio) and consisted of the tests for: cytotoxicity, irritation and sensitization, systemic toxicity, genotoxicity, carcinogenicity and reproductive toxicity, interactions with blood, and the local effects after implantation.

2.9. Histological Evaluation of MonoMax Suture Degradation. To examine the mass absorption process and the tissue reaction involved in the degradation process, ethylene oxide and gamma radiation sterilized MonoMax sutures were implanted in the paravertebral muscles of white SPF-Wistar rats (Dr. Schreiter, Freytox GmbH, Osteroda, Germany). A total of 56 rats were implanted with size 3–0 MonoMax sutures using a mattress stitch so that 2 cm of material was placed in the muscle. To distinguish between the different test objects, the EtO sterilized sutures were implanted to the left of the vertebral column, whereas the gamma sterilized sutures were placed on the right side. Prior to the implantation in the muscle, the Wistar rats were anaesthetized using 80 mg ketamine and 12 mg xylazine per kg of bodyweight, clipped free of fur on the back, and the skin was disinfected with 70% ethanol. The skin was opened with a 4 cm median incision along the vertebral column and the subcutaneous tissue as well as the paravertebral muscle were prepared. After implantation, the skin incision was closed by a subcutaneous suture and sealed using a tissue adhesive (Histoacryl, BBraun). The animals were distributed to termination dates according to the scheme in Table 1.

Table 1: Implantation scheme of size 3–0 P4HB sutures in the paravertebral muscle of SPF-Wistar rats.

	Implantation scheme MonoMax USP 3–0	
Group	Time of termination	Number of animals
1	4 weeks p.op.	5
2	8 weeks p.op.	5
3	12 weeks p.op.	5
4	16 weeks p.op.	5
5	24 weeks p.op.	6
6	32 weeks p.op.	6
7	37 weeks p.op.	6
8	42 weeks p.op.	6
9	50 weeks p.op.	6
10	64 weeks p.op.	6

p.op.: post operation.

At the appropriate termination dates, the animals were euthanized by CO_2 inhalation, the paravertebral muscle tissue removed entirely, labeled precisely, fixed in 4% neutral buffered formalin (NBF) and histologically processed (embedded, sectioned and stained in hematoxylin and eosin) for macroscopic, and microscopic evaluation by a qualified pathologist (Dr. Weber, Labor für Veterinärmedizinische Diagnostik, Bad Salzungen, Germany).

2.10. Surgical Functionality. The performance and surgical functionality of the MonoMax suture was evaluated in a rabbit model of soft tissue approximation. This study was performed at the animal research facilities of the Friedrich-Schiller-University in Jena, Germany by Dr. Ulrich Stock. Full-thickness abdominal incisions were repaired with either MonoMax suture or PDS II sutures (Ethicon, Inc.) as a comparative control. Eighteen (18) healthy adult New Zealand white rabbits (weighing 2.5–4 kg) were used in the evaluation. Following intramuscular induction and maintenance using intravenous anesthesia, the abdominal implant sites were clipped free of fur. Under sterile conditions, an abdominal midline skin incision of 4 cm length was made, and the skin spread using a self-retainer. Two paramedian muscular and peritoneal incisions (3 cm length) were made and closed either with running 3/0 PDS II or running 3/0 MonoMax sutures (see Figure 9). After thorough control for hemostasis, the skin was closed using absorbable Vicryl sutures. Special care was taken to ensure that the distance from the skin closure to the study suture implant site was at least 2 cm. Each rabbit was implanted with one test and one control suture. Animals were maintained for periods of 4, 8, and 12 weeks (6 rabbits per time point) and observed daily to ensure proper healing of the implant sites. At the end of the appropriate time points, the animals were weighed and sacrificed by an injectable barbiturate and tissue sections containing the implanted sutures were excised from the animals. Gross observations were made for hernias as indicators of weakening sutures and adhesions.

TABLE 2: Comparison of tensile properties of size 1 absorbable suture materials.

Suture	Chemical composition	Diameter x_{ave} (mm)	s.d.	LPTS (N)	s.d.	KPTS (N)	s.d.	Bending stiffness (mN)	s.d.	Young's modulus (GPa)	s.d.
MonoMax	Poly-4-hydroxybutyrate	0,517	0,007	118	6,1	74	3,1	363	7	0,485	0,01
PDS II	Poly-p-dioxanone	0,532	0,008	109	4,9	63	5,2	753	72	1,370	0,03
Monocryl	Polyglecaprone 25	0,531	0,002	183	7,3	92	5,9	532	21	0,725	0,03
Biosyn	Glycomer 631	0,538	0,005	176	24,9	89	10,5	679	64	0,838	0,03
Maxon	Polyglyconate	0,550	0,006	177	8,2	118	9,3	794	*41*	0,941	0,06

TABLE 3: Knot security factor (KSF) and knot pull tensile strength determination for MonoMax sutures size 1 to 3–0.

Size	KSF	Ave. KPTS (N)	s.d.
1	3	45,4	3,4
0	3	42,4	5,0
2–0	2	28,4	3,2
3–0	2	20,9	2,2

3. Results

3.1. Mechanical Properties. The tensile properties of the MonoMax suture size 1 are compared with those of PDS II, Monocryl, Biosyn [15], and Maxon [16] sutures in Table 2.

As is evident from Table 2, the size 1 MonoMax suture is stronger than the other homopolymer suture, namely, PDS II. However, these homopolymer sutures are not as strong as the copolymer sutures, which are made of an ABA block copolymer type structure with two glycolide hard segments. It should be noted that the MonoMax suture has the smallest diameter of all the sutures tested. This was done to keep MonoMax sizing close to the USP specification for diameter. In addition to offering a high tensile strength, the MonoMax suture is very flexible. It has the lowest Young's modulus value of the current commercial monofilament sutures and its bending stiffness is about half that of the other sutures listed in Table 2. The bending stiffness of Monocryl is the lowest of the competitive sutures, but is still 1.5 fold that of the MonoMax suture. The lower values of modulus and bending stiffness are important properties of the MonoMax suture that provide the suture with good handling for surgical suturing and knot tying, and improved flexibility and pliability compared to current commercial offerings. Besides flexibility, a smooth surface is also important for knotting to allow for knot run down and repositioning during surgical usage. The combination of the smooth surface with the flexibility of the MonoMax suture provides a suture with low tissue drag that minimizes trauma to the sutured tissues. The smooth, nonporous surface of monofilaments like the MonoMax suture can also reduce the chance of infection due to the capillary effect that can be caused by the interstices present in braided, multifilament sutures.

3.2. Knot Security of the MonoMax Monofilament. Due to their smooth surface, some surgeons have raised concerns about the knot security of monofilament sutures relative to braided sutures. Therefore, many surgeons traditionally use up to 6 throws to tie a secure knot when suturing with a monofilament. We performed knot security tests with size 1 MonoMax material as described in Section 2.7, tying a surgical knot 2 = 1 [13]. A knot security factor was determined by accounting for every additional square knot needed to secure the knot. Due to the soft and interlocking surface of the MonoMax suture, a total of only 4-5 throws (2 = 1 = 1 = 1; 2 = 1 = 1 = 1 = 1) was enough to tie a secure knot, corresponding to a knot security factor of 2 to 3 (see Table 3).

The tensile strength of the fiber after performing the KSF test is significantly lower than the knot pull tensile strength measured for the respective sizes of the MonoMax suture according to the US Pharmacopeia ⟨881⟩ [11]. The reason for this behavior lies in the application of multiple knots (KSF method) instead of making only one single knot which leads to additional frictional forces lowering the strength of the fiber.

3.3. In Vitro Strength Retention of MonoMax Monofilament. The breaking strength retention of an absorbable suture is defined as:

$$\text{BSR}(\%) = \frac{\text{actual strength}}{\text{initial strength}} * 100\%. \quad (1)$$

Figure 3 shows the *in vitro* knot pull tensile strength (KPTS) retention of size 3–0 MonoMax and PDS II suture in Dulbeco's phosphate-buffered saline (pH 7.4; 37°C) for a period of 20 weeks.

The size 3–0 MonoMax monofilament shows high-strength retention during the first 12 weeks of the incubation when the overall strength remains nearly unchanged. This period is followed by a decrease of breaking strength retention to 20% to 30% of the suture's initial strength until week 20, which is attributed to a slow hydrolysis of the polymer. In contrast, PDS II suture shows a more steady decrease of its KPTS, retaining an average value of 51% of its initial strength after 8 weeks and losing almost all of its tensile strength after 12 weeks, when only 7% of the strength is left.

When comparing the absolute knot pull tensile strength of the size 3–0 suture materials, the initial KPTS of MonoMax suture (36.2 N) is significantly higher than for PDS II

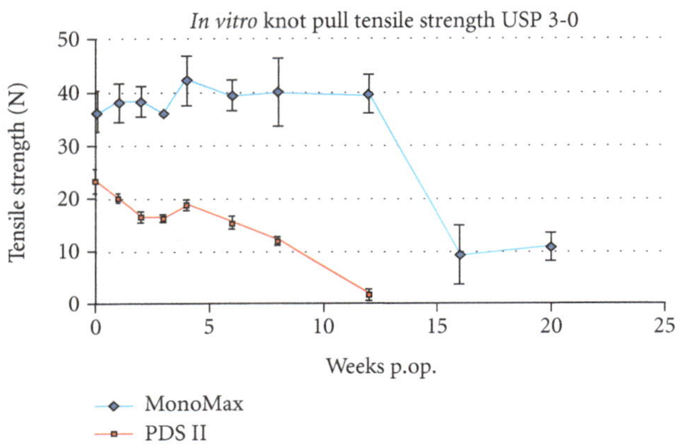

FIGURE 3: Knot pull tensile strength degradation profile of a P4HB monofilament fiber of size 3–0 incubated in Dulbecco's phosphate buffered saline (pH 7.4; 37°C).

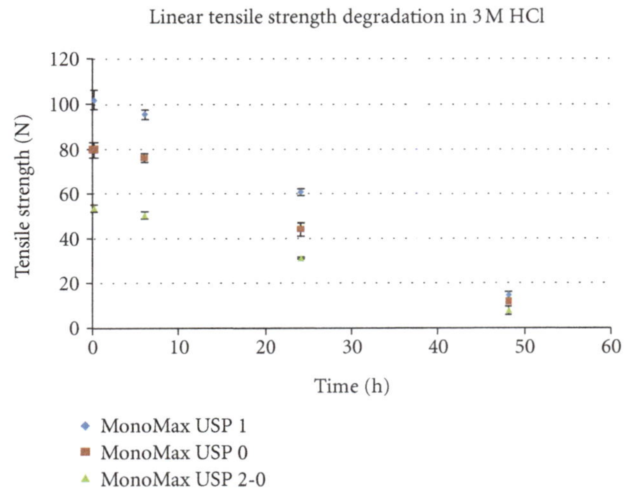

FIGURE 4: Accelerated *in vitro* degradation profile of a P4HB monofilament fiber size USP 1, USP 0, and USP 2–0 incubated in hydrochloric acid (3 M; 37°C).

(23.3 N), which only has about 64% of the initial strength of MonoMax suture (see Table 4).

Under the highly acidic conditions of the accelerated *in vitro* testing (3 M HCl, 37°C), the MonoMax suture undergoes a much more rapid loss of strength (see Figure 4).

These extreme conditions are meant as a rapid bench test for comparing different lots of material, rather than being representative of physiologic conditions. Within 48 hours, the linear breaking strength of MonoMax size 3–0 decreases to about 15% of its initial value under the afore-mentioned conditions. Figures 5(a) and 5(b) show a scanning electron micrograph (SEM) of the MonoMax suture before and after a similar accelerated *in vitro* degradation in 2 M HCl at 37°C solution for 7 days showing an advanced state of degradation caused by the hydrolytic digestion of the suture material. Figures 5(c) and 5(d) show SEM images of the MonoMax suture after 12 weeks *in vivo*. During this time period it is evident that the MonoMax suture is degrading by surface erosion as well as bulk hydrolysis that is shown through M_w analysis.

3.4. In Vivo Strength Retention of the MonoMax Monofilament. The *in vivo* knot pull tensile strength (KPTS) retention of size 3–0 MonoMax suture in a rabbit model is shown in Figure 6 and in more details in Table 5 and is compared with PDS II suture of the same size as a control.

As can be seen in Figure 6 and Table 5, the MonoMax suture retains its strength significantly longer *in vivo* than the PDS II suture. At 8 to 12 weeks, the size 3–0 MonoMax suture still retains approximately 50% of its initial knot pull tensile strength, whereas PDS II retains 50% after 4 to 6 weeks. Furthermore, PDS II does not provide any tensile strength to the sutured tissue after 12 weeks p.op. The MonoMax monofilament, however, still has about 20% to 30% of its initial strength at 20 weeks after implantation. As mentioned before in the *in vitro* testing (phosphate buffer), the absolute breaking strength values of MonoMax suture after 20 weeks correspond to 46% of the initial breaking strength of PDS II because of the superior initial breaking strength of the MonoMax suture. As shown in Figure 7, the molecular weight of the MonoMax monofilament gradually decreases *in vivo* from an initial value of M_w 286,000 Da, and the suture is substantially degraded within one year (see Section 3.6).

3.5. Biocompatibility of MonoMax Sutures. The biocompatibility of the P4HB suture was found to be excellent. Test results are shown in Table 6. The suture passed all tests with no evidence of toxicity.

The MonoMax suture is derived biosynthetically from the natural metabolite 4-hydroxybutyrate. This monomer is a normal constituent of the mammalian blood serum that has been extensively studied and is known to occur within a variety of tissues, including brain, heart, kidney, liver, lung, muscle, and brown fat [17]. The concentration of the monomer in these tissues ranges from 2.3 to 37.4 nmoles/g of tissue (or 0.24 to 3.8 mg/kg of tissue, based on a formula weight of 104 g/mol for the monomer). The body quickly metabolizes 4-hydroxybutyrate via the Krebs cycle to carbon dioxide and water and its half-life is just 35 minutes [18]. Two metabolic precursors to the 4-hydroxy-butyrate monomer, namely 1,4-butanediol and gamma-butyrolactone, have both been thoroughly studied in two NIH sponsored National Toxicology Program (NTP) studies that evaluated their toxicology, genotoxicity, and carcinogenesis [19, 20]. Both precursors were found to lack toxic and carcinogenic potential, and the NTP concluded that 4-hydroxybutyrate also lacks toxic and carcinogenic potential because of the rapid and complete conversion of the precursors to this metabolite. Prior to this study, however, the biocompatibility of the poly-4-hydroxybutyrate homopolymer had not been fully evaluated *in vivo*. The results of the studies on the polymer are reported in Table 6.

3.6. Histology. The tissue surrounding the implanted MonoMax monofilament and PDS II control sutures was analyzed

TABLE 4: Results of the *in vitro* degradation of the knot pull tensile strength of USP 3–0 MonoMax and PDS II sutures in phosphate buffer at pH 7.4.

	MonoMax USP 3–0					PDS II USP 3–0				
Weeks	x_{ave} [N]	s.d.	x_{min} [N]	x_{max} [N]	BSR [%]	x_{ave} [N]	s.d.	x_{min} [N]	x_{max} [N]	BSR [%]
0	36,2	3,81	29,4	42,1	100	23,3	2,33	20,5	27,2	100
1	38,1	3,55	32,1	42,7	105	20,1	0,95	18,5	21,7	86
2	38,4	2,86	33,3	43,1	106	16,5	0,92	15,4	18,2	71
3	36,2	4,13	28,2	41,3	100	15,9	0,64	15,2	17,2	68
4	42,2	4,61	32,6	48,0	116	18,7	0,90	17,2	20,1	80
6	39,5	2,84	33,7	44,1	109	14,8	1,10	13,6	16,5	63
8	40,0	6,37	25,4	51,2	110	11,9	0,85	10,7	13,3	51
12	39,7	3,49	32,4	41,6	104	1,7	0,38	1,1	2,2	7
16	9,2	5,55	3,2	16,7	25	n.d.	n.d.	n.d.	n.d.	n.d.
20	10,7	2,57	8,3	16,2	29	n.d.	n.d.	n.d.	n.d.	n.d.

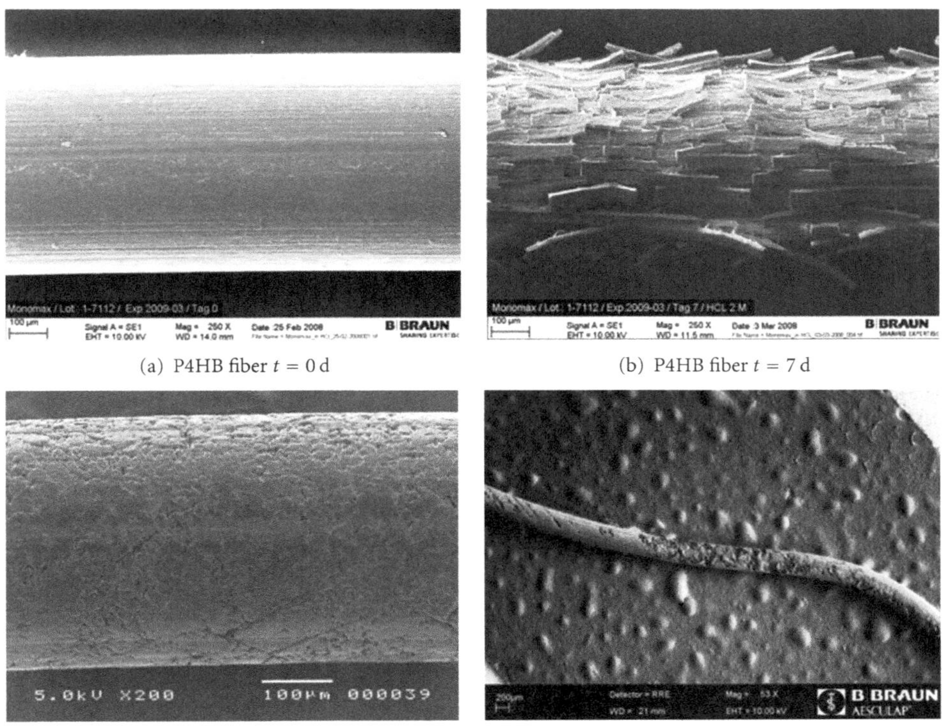

(a) P4HB fiber $t = 0$ d

(b) P4HB fiber $t = 7$ d

(c) P4HB fiber 12 weeks p.op.

(d) P4HB fiber 12 weeks p.op.

FIGURE 5: SEM photomicrograph showing the surface of a P4HB monofilament fibers: (a) P4HB fiber size 1 before incubation in 2 M HCl, (b) after incubation in 2 M HCl for 7 d, (c) and (d) P4HB monofilament size 3–0 after being implanted subcutaneously in rabbits for 12 weeks.

for a tissue reaction at 4, 8, 12, 16, 24, 32, 37, 42, 50, and 64 weeks. Formalin-fixed tissue samples from each test animal were sectioned and assessed by a board certified pathologist for the following: inflammation, fibrosis, hemorrhage, necrosis, degeneration, foreign debris, and relative size of the involved area. Histopathological evaluation showed that there were no significant indications of a local toxic effect for the MonoMax suture. Ethylene oxide and gamma-sterilized MonoMax fiber samples showed similar histological behavior at the examined termination intervals. After 8 weeks p.op., the tissue reaction was characterized primarily by the presence of mononuclear cells (macrophages, lymphocytes, and plasma cells) involved in the digestion of the capsule present around the fiber. Almost no granulocytic reaction was present at that time point and the suture fiber appeared intact (see Figure 8(a)). Histological signs of digestion of the fiber by mononuclear cells (macrophages) appear at 24 weeks. At this time, degradation is apparent at the surface of the fiber. The 32 weeks p.op. results show decomposition of the fiber structure and phagocytosis by macrophages,

TABLE 5: Results of the knot pull tensile strength (KPTS) *in vivo* degradation of USP 3–0 MonoMax and PDS II sutures subcutaneously implanted in the subcutis of rabbits.

	MonoMax USP 3/0					PDS II USP 3/0				
Weeks p.op.	x_{ave} [N]	s.d.	x_{min} [N]	x_{max} [N]	BSR [%]	x_{ave} [N]	s.d.	x_{min}	x_{max}	BSR
0	36,2	3,81	29,4	42,1	100	23,3	2,3	20,5	27,2	100
1	37,0	2,17	34,5	41,6	102	22,6	1,1	20,8	24,2	97
2	34,5	3,05	29,4	39,3	95	19,1	4,3	8,3	22,4	82
3	36,9	5,83	23,3	43,6	102	16,6	1,3	14,7	18,8	71
4	30,0	7,80	20,6	44,8	83	15,8	1,5	13,9	18,8	68
6	20,8	8,78	3,5	34,3	57	11,5	1,3	8,9	12,8	49
8	22,1	4,94	15,0	30,2	61	7,1	1,9	2,8	8,8	30
12	22,1	6,31	11,8	33,3	61	n.d.	n.d.	n.d.	n.d.	n.d.
16	9,2	5,55	3,0	16,7	25	n.d.	n.d.	n.d.	n.d.	n.d.
20	10,7	2,57	7,8	16,2	29	n.d.	n.d.	n.d.	n.d.	n.d.

TABLE 6: Biocompatibility Testing Performed on the P4HB Suture Fiber.

Tests For	Method	Results	Remark
Cytotoxicity	MEM elution	Passed	No evidence of cell lysis or toxicity
Cytotoxicity	Agarose overlay	Passed	No evidence of cell lysis or toxicity
Irritation and sensitization	Intracutaneous reactivity with saline and sesame oil extracts	Passed	No evidence of significant irritation
Irritation and sensitization	Maximization sensitization with saline and sesame oil extracts	Passed	No evidence of delayed dermal sensitization
Systemic toxicity	Acute systemic toxicity with saline and sesame oil extracts	Passed	No evidence of systemic toxicity
Systemic toxicity	Material-mediated pyrogenicity	Passed	Nonpyrogenic
Genotoxicity, carcinogenicity, and reproductive toxicity	Bacterial reverse mutation study with saline and ethanol extracts	Passed	Nonmutagenic
Genotoxicity, carcinogenicity, and reproductive toxicity	Chromosomal aberration in CHO cells	Passed	Not genotoxic
Genotoxicity, carcinogenicity, and reproductive toxicity	Mouse micronucleus study with saline and sesame oil extracts	Passed	Not genotoxic. No evidence of cellular toxicity.
Local effects after implantation	Subchronic implantation	Passed	Nonirritant
Local effects after implantation	Chronic implantation	Passed	Nonirritant
Interaction with blood	*In vitro* hemolysis with saline extract	Passed	Nonhemolytic

with the macrophages penetrating the whole fiber cross-section. The fiber material appears amorphous and swollen (see Figure 8(c)). The histology images after 50 weeks p.op. indicate a complete loss of the initial fiber structure. The voids are filled with a fine vacuolar, granular material, and cellular reaction was present at the implantation site. No granulocytic reaction was detectable and all histological samples showed low tissue reaction intensity. After 64 weeks p.op. (see Figure 8(d)), no fiber could be detected in two of the four animals. The samples recovered from the other two animals had similar histological findings like those of the 50 week p.op. group exhibiting loss of fiber structure and a vacuolar, granular material replacing the suture.

Summarizing the results, the MonoMax monofilament size 3–0 shows a good biocompatibility as well as a very low tissue reaction throughout the whole study. After 24 weeks, the MonoMax suture undergoes a progressive phagocytosis with the participation of macrophages resulting in a complete loss of the fiber structure after 50 weeks p.op., when the fiber is replaced by a fine vacuolar, granular material. Finally, the MonoMax monofilament is essentially absorbed after 64 weeks.

3.7. Surgical Functionality of MonoMax Monofilament. The MonoMax suture was evaluated in a rabbit model of abdominal soft tissue approximation with PDS II as a control (see Figure 9).

All the animals ($n = 17$) recovered uneventfully from the surgeries, and appeared normal throughout the remaining course of the study (12 weeks). There was no indication of hernia or adhesions in the test animals. Macroscopically the closed sites showed no abnormalities, and wound healing was not disturbed. Histological examination of the tissue at the surgical incision lines showed progressive healing with no evidence of wound dehiscence at any time during the study.

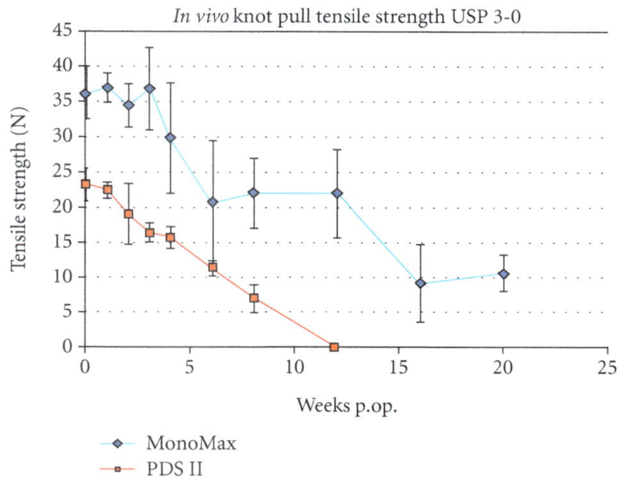

FIGURE 6: *In Vivo* Knot Pull Tensile Strength (KPTS) degradation profile of a P4HB monofilament and PDS II suture size 3–0 implanted subcutaneously in White New Zealand rabbits up to 20 weeks.

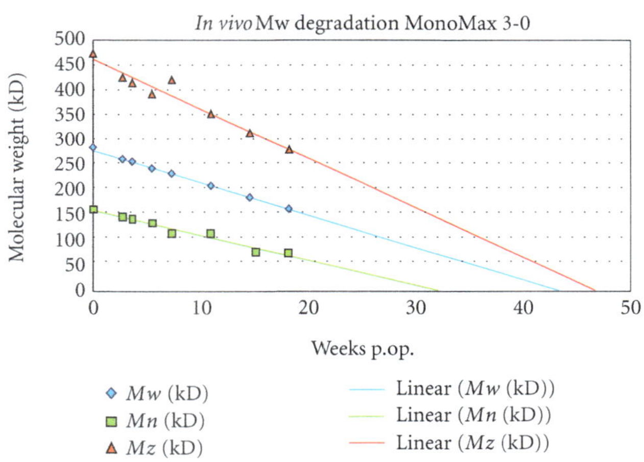

FIGURE 7: *In Vivo* molecular weight degradation of a P4HB monofilament fiber size 3–0 implanted subcutaneously in White New Zealand rabbits up to 20 weeks.

4. Discussion

MonoMax suture fiber was prepared by melt extrusion of poly-4-hydroxybutyrate homopolymer for preclinical evaluation as a new absorbable monofilament suture. The properties of the MonoMax suture were compared to PDS II, another monofilament suture material, in order to identify the possible clinical benefits of this new biomaterial.

As shown in Table 2, the MonoMax suture has comparable tensile and knot strength to the PDS II suture material however, the MonoMax suture is significantly more pliable and flexible than the PDS II suture or any of the other absorbable monofilaments tested. These properties can be quantified by measuring the value of the Young's modulus and the bending force of the respective sutures. As far as the Young's modulus is concerned, the value of MonoMax suture is about one-third the value of the PDS II suture fiber. Even compared to the hard segment-soft segment type sutures like Monocryl and Biosyn, the value of MonoMax monofilament ranges between one-half to one-third the values of these sutures. The lower Young's modulus contributes to the very good knot security of the MonoMax monofilament by providing a more flexible suture with a softer surface resulting in a superior interlocking of the fiber that results in a very good knot safety factor. Consequently, a lower number of knot throws is needed to make a secure knot, resulting in less irritating mass and potentially less foreign body reaction. Furthermore the MonoMax fiber's intrinsic elasticity allows the surgeon to apply more tension to the knot during tying.

The low bending stiffness value of the MonoMax fiber makes the suture extremely pliable and flexible for surgical manipulation, which is advantageous when the surgeon uses suture techniques that include close radii. The MonoMax suture material shows little memory and its fiber ends are less irritating than for a stiffer suture fiber. This is of some importance in situations when for instance the surgeon cuts the fiber a short distance from the knot and has to cover the fiber ends in the surrounding tissues.

The strength retention of the MonoMax suture was evaluated both *in vitro* and *in vivo*. *In vitro*, the MonoMax monofilament was found to be less susceptible to hydrolysis than currently available suture materials, giving more safety to slow-healing tissues. The MonoMax suture is in fact fairly resistant to neutral conditions and maintains strength beyond 12 weeks. Accelerated degradation can be achieved under more severe acidic conditions. Figure 4 shows the accelerated degradation profile of a size 3–0 MonoMax suture exposed to 3 M hydrochloric acid for 48 hours. In 3 M HCl, the suture loses approx. 50% of its initial tensile strength within one day and is 90% degraded after 2 days. Figure 6 shows the *in vivo* strength retention of a size 3–0 MonoMax monofilament implanted in the subcutaneous tissue of rabbits. The MonoMax fiber undergoes a relatively slow degradation process *in vivo*. As the measurements from the implanted size 3–0 suture demonstrate, the MonoMax monofilament fiber retains about 50% of its initial tensile strength after 3 months, which is a significantly slower degradation profile than for other commercially available absorbable monofilaments (Maxon 50% BSR after 28 days, PDS II after 35 days). Therefore, the MonoMax fiber may be an excellent choice for surgical indications involving the apposition of slow healing tissues such as fascia or tendons when an application of a nonabsorbable suture is not appropriate.

The biocompatibility of the MonoMax suture was demonstrated in a variety of *in vitro* and *in vivo* biocompatibility testing. Histological examination of the MonoMax monofilament size 3–0 fiber implanted in the dorsal muscle of rats up to 64 weeks revealed again the very good biocompatibility of the MonoMax suture without any adverse effects. First signs of phagocytosis of the suture were visible after 16 weeks, and at 32 weeks, the fiber showed a surface degradation with macrophages eventually penetrating the whole fiber diameter. After 50 weeks, the fiber was replaced

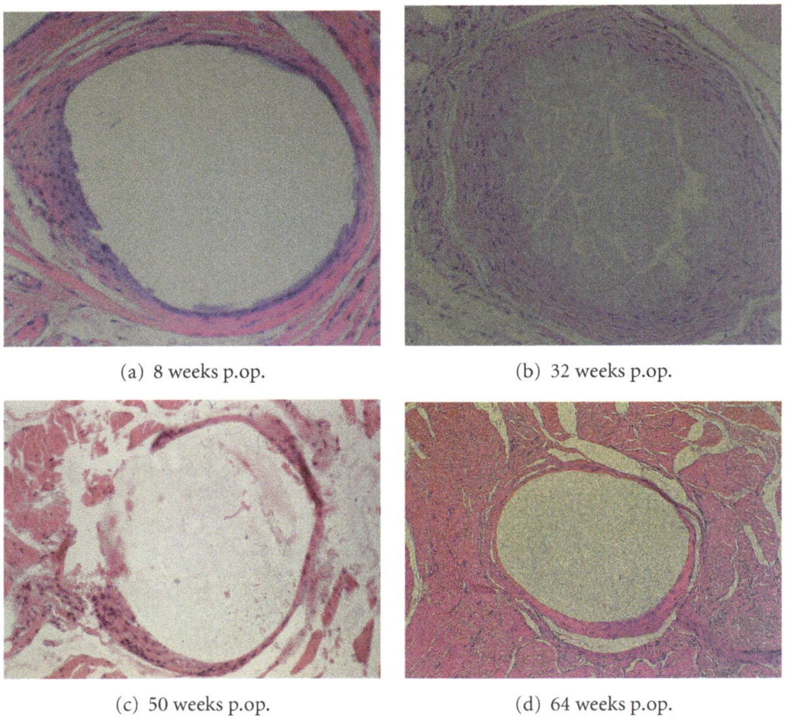

(a) 8 weeks p.op.

(b) 32 weeks p.op.

(c) 50 weeks p.op.

(d) 64 weeks p.op.

FIGURE 8: Histology images of EtO sterilized P4HB monofilament size 3–0 implanted in the long. dorsi j. of male Wistar rats.

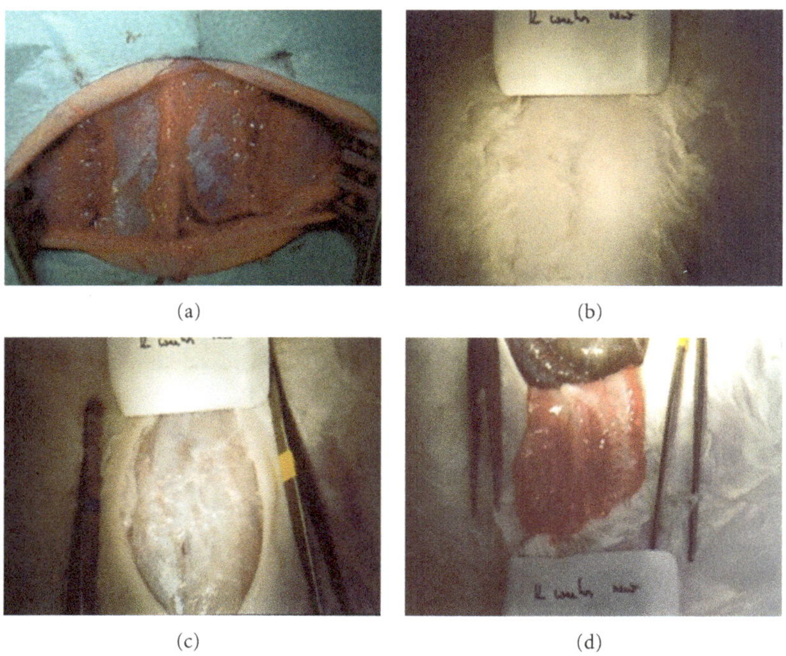

(a)

(b)

(c)

(d)

FIGURE 9: Implantation of P4HB and PDS II sutures in a rabbit model to evaluate surgical functionality. Full thickness abdominal incisions were closed with PDS II (left) or poly-4-hydroxybutyrate (right) sutures. (a) at implantation. (b) skin incision line at 12 weeks. (c) laparotomy suture line (skin side) at 12 weeks. (d) laparotomy suture line (intestinal side) at 12 weeks.

by fine granular material, leading to an essential mass absorption after 64 weeks.

Finally, the surgical functionality of the suture was demonstrated in a rabbit model for abdominal wound closure. The MonoMax suture performed well and no instances of wound dehiscence, hernias, or adhesions were found. There was also no evidence of tissue irritation or infections.

5. Conclusion

MonoMax suture, a new monofilament suture based on the poly-4-hydroxybutyrate homopolymer, has been developed that combines a high tensile strength with the highest pliability and the lowest Young's modulus of any commercially available monofilament absorbable suture. In addition, this suture offers a number of clinical benefits to meet unmet needs, including prolonged strength retention, low tissue drag, and knot security.

The biocompatibility of the MonoMax suture was evaluated for cytotoxicity, irritation, sensitization, acute systemic toxicity, pyrogenicity, genotoxicity, subchronic systemic toxicity, and chronic toxicity, and tissue reaction was assessed by intramuscular implantation. All tests indicated that the MonoMax suture presents an excellent biocompatibility and physiologically is well integrated in the tissues. Absorption of a size 3–0 suture was found to be substantially complete at about 64 weeks.

A surgical functionality study was undertaken to evaluate the performance of the MonoMax suture in a rabbit model of abdominal soft tissue approximation. Full thickness abdominal incisions were successfully and satisfactorily repaired with the MonoMax suture without causing tissue irritation or infections. The properties of this suture material were demonstrated in a human clinical trial for the abdominal wall closure (ISSAAC Trial) [21] and the suture was found to be safe for abdominal wall closure after midline laparotomy [22].

Considering the excellent properties of the MonoMax suture, it could be especially useful as a suture material for slowly healing tissues. We expect this new suture to offer surgeons an alternative to faster degrading absorbable sutures or permanent nonabsorbable materials when long-term wound support is desired.

Acknowledgment

Tepha, Inc. would like to acknowledge NIH SBIR Grant no. 2 R44 EB003644-02 for partial support of this work.

References

[1] M. K. Hirko, P. H. Lin, H. P. Greisler, and C. C. Chu, "Biological properties of suture materials," in *Wound Closure Materials*, C. C. Chu, J. A. Von Fraunhofer, and H. P. Greisler, Eds., pp. 237–287, CRC Press, Boca Raton, Fla, USA, 1996.

[2] D. P. Martin and S. F. Williams, "Medical applications of poly-4-hydroxybutyrate: A strong flexible absorbable biomaterial," *Biochemical Engineering Journal*, vol. 16, no. 2, pp. 97–105, 2003.

[3] Y. Doi, *Microbial Polyesters*, VCH, New York, NY, USA, 1990.

[4] S. F. Williams and D. P. Martin, "Applications of PHAs in Medicine and Pharmacy," in *Biopolymers*, Y. Doi and A. Steinbüchel, Eds., pp. 91–127, Wiley-VCH, 2002.

[5] Q. Wu, Y. Wang, and G. Q. Chen, "Medical application of microbial biopolyesters polyhydroxyalkanoates," *Artificial Cells, Blood Substitutes, and Biotechnology*, vol. 37, no. 1, pp. 1–12, 2009.

[6] G. W. Huisman, F. Skraly, D. P. Martin, and O. P. Peoples, "Biological systems for manufacture of polyhydroxyalkanoate polymers containing 4-hydroxyacids," US Patent no. 6,316,262.

[7] D. E. Dennis and H. E. Valentin, "Methods of making polyhydroxyalkanoates comprising 4-hydroxybutyrate monomer units," US Patent no. 6,117,658.

[8] Y. Hori, A. Yamaguchi, and T. Hagiwara, "Chemical synthesis of high molecular weight poly(3-hydroxybutyrate-co-4-hydroxybutyrate)," *Polymer*, vol. 36, no. 24, pp. 4703–4705, 1995.

[9] K. N. Houk, A. Jabbari, H. K. Hall, and C. Alemán, "Why δ-valerolactone polymerizes and γ-butyrolactone does not," *Journal of Organic Chemistry*, vol. 73, no. 7, pp. 2674–2678, 2008.

[10] "Biological evaluation of medical devices - Part 7: Ethylene oxide sterilization residuals," ISO 10993-7:2008.

[11] The United States Pharmacopeia, The National Formulation USP 32-NF 27 (2010), chapter 881 Tensile Strength: 380, United States Pharmacopeial Convention INC. Rockville, Md, USA.

[12] *European Pharmacopeia*, The European Directorate for the Quality of Medicines & HealthCare, Strasbourg, France, 6th edition, 2010.

[13] H. Tera and C. Aberg, "Tensile strengths of twelve types of knot employed in surgery, using different suture materials," *Acta Chirurgica Scandinavica*, vol. 142, no. 1, pp. 1–7, 1976.

[14] "Biological evaluation of medical devices - Part 1: Evaluation and testing," ISO 10993-1:2003.

[15] R. S. Bezwada, D. D. Jamiolkowski, I. Y. Lee et al., "Monocryl suture, a new ultra-pliable absorbable monofilament suture," *Biomaterials*, vol. 16, no. 15, pp. 1141–1148, 1995.

[16] G. T. Rodeheaver, K. A. Beltran, C. W. Green et al., "Bio-mechanical and clinical performance of a new synthetic monofilament absorbable suture," *Journal of Long-Term Effects of Medical Implants*, vol. 6, no. 3-4, pp. 181–198, 1996.

[17] T. Nelson, E. Kaufman, J. Kline, and L. Sokoloff, "The extraneural distribution of γ-hydroxybutyrate," *Journal of Neurochemistry*, vol. 37, no. 5, pp. 1345–1348, 1981.

[18] S. L. Sendelbeck and C. L. Girdis, "Disposition of a 14C-labeled bioerodible polyorthoester and its hydrolysis products, 4-hydroxybutyrate and cis,trans-1,4-bis(hydroxymethyl)cyclohexane, in rats," *Drug Metabolism and Disposition*, vol. 13, no. 3, pp. 291–295, 1985.

[19] National Toxicology Program (NTP), "NTP Summary report on the metabolism, disposition and toxicity of 1,4-butanediol," Toxicity Report 54, NIH Publication no. 96-3932, 1996.

[20] National Toxicology Program (NTP), "Toxicology and carcinogenesis studies of gamma-butyrolactone (CAS No. 96-48-0) in F344/N rats and B6C3F1 mice (gavage studies)," Technical Report 406, NIH Publication no. 92-3137, 1992.

[21] L. Fischer, P. Baumann, J. Hüsing et al., "A historically controlled, single-arm, multi-centre, prospective trial to evaluate the safety and efficacy of MonoMax suture material for abdominal wall closure after primary midline laparotomy.

ISSAAC-Trial [NCT005725079]," *BMC Surgery*, vol. 8, article 12, 2008.

[22] M. Albertsmeier, C. M. Seiler, L. Fischer et al., "Evaluation of the safety and efficacy of MonoMax suture material for abdominal wall closure after primary midline laparotomy—a controlled prospective multicenter trial: ISSAAC [NCT005725079]," *Langenbeck's Archives of Surgery*, vol. 397, pp. 363–371, 2012.

13

Facile Solventless Synthesis of a Nylon-6,6/Silver Nanoparticles Composite and Its XPS Study

Raúl A. Morales-Luckie,[1] Víctor Sánchez-Mendieta,[2] Oscar Olea-Mejia,[1] Alfredo R. Vilchis-Nestor,[1] Gustavo López-Téllez,[1] Víctor Varela-Guerrero,[1] L. Huerta,[3] and Jesús Arenas-Alatorre[4]

[1] *Centro Conjunto de Investigación en Química Sustentable UAEM-UNAM (CCIQS), Facultad de Química, Universidad Autónoma del Estado de México, Km. 14.5 de la carretera Toluca-Atlacomulco, 50200 San Cayetano, MEX, Mexico*
[2] *Facultad de Química, Universidad Autónoma del Estado de México, Paseo Colón y Paseo Tollocan, 50120 Toluca, MEX, Mexico*
[3] *Instituto de Investigaciones en Materiales, Universidad Nacional Autónoma de México, Apartado Postal 70-360, 04510 México, DF, Mexico*
[4] *Instituto de Física, Universidad Nacional Autónoma de México, Apartado Postal 20-364, 01000 México, DF, Mexico*

Correspondence should be addressed to Víctor Sánchez-Mendieta; vsanchezm@uaemex.mx

Academic Editor: Carmina Menchaca-Campos

Silver nanoparticles were synthesized and supported on thin nylon membranes by means of a simple method of impregnation and chemical reduction of Ag ions at ambient conditions. Particles of less than 10 nm were obtained using this methodology, in which the nylon fibers behave as constrained nanoreactors. Pores on nylon fibres along with oxygen and nitrogen from amide moieties in nylon provide effective sites for *in situ* reduction of silver ions and for the formation and stabilization of Ag nanoparticles. Transmission electron microscopy (TEM) analysis showed that silver nanoparticles are well dispersed throughout the nylon fibers. Furthermore, an interaction between nitrogen of amides moieties of nylon-6,6 and silver nanoparticles has been found by X-ray photoelectron spectroscopy (XPS).

1. Introduction

Research on the synthesis of mesoporous materials containing nanoparticles represents a fast-developing area of nanoscience and nanotechnology. This interest is stimulated by several possible application areas of these materials including catalytic [1], magnetic [2], and optoelectronic [3, 4]. Metal nanoparticles dispersed in polymeric matrixes have recently been the subject of intense study aiming to develop nanocomposite films [5–8]. General approaches for the synthesis and support of nanoparticles inside porous materials include impregnation [9] and deposition-precipitation [10]. A drawback of this nanocomposites is the difficulty to disperse nanoparticles in most systems [11]; hence one potential advantage of such metal/polymer systems is that the size and distribution of dispersed metal nanoparticles can be readily controlled based on the properties of the host polymer [12, 13]. Silver particles with a narrow size distribution have been produced upon reversible chemical transformation between metallic and oxide states in a titania matrix [14] and in a mesoporous silica which was grafted with hydrophobic $-Si(CH_3)_3$ groups at the pore surface [15].

Nylon is an electron-rich and polar synthetic polymer (polyamide) usually made from the monomers adipoyl chloride and hexamethylene diamine to form a linear molecular chain (Figure 1). Synthetic nylon membranes have a porous structure [16] and are composed of microfibrils that are interconnected forming a three-dimensional network. Such morphological features provide a unique reaction vessel for synthesizing and supporting metal nanoparticles, allowing enhanced access of guest molecules to catalytic centres, compared with nonporous films.

In this work, a facile synthesis of silver nanoparticles of less than 10 nm in diameter with a narrow size distribution,

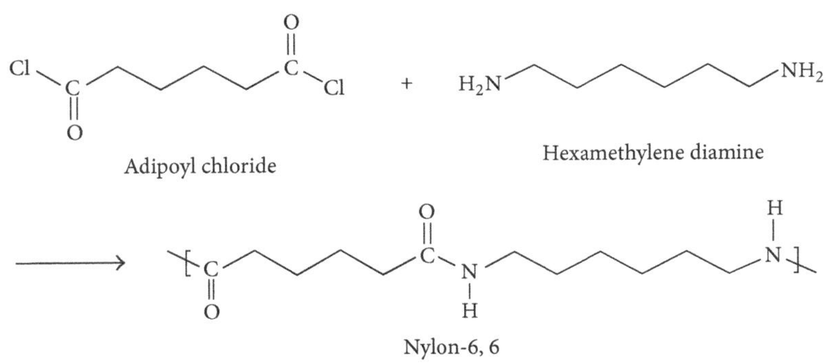

FIGURE 1: Characteristic synthesis of nylon-6,6 fibers.

using porous nylon fibers as unique nanoreactors to generate a nylon-6,6/silver nanoparticles composite, is reported.

2. Experimental

All chemical reagents were analytical grade and were used without further purification.

2.1. Preparation of the Nylon-6,6/Ag Nanoparticles Composite. In a typical experiment, nylon-6,6 membranes (Millipore Co., cat. number: GNWP02500; 150 μm thick and 0.22 μm pores average diameter) are impregnated with silver ions by immersing the membranes in $AgNO_3$ (0.1 mM, 0.5 mM, 1 mM, 1.5 mM, and 10 mM; Aldrich) aqueous solution for 1 min, followed by rinsing with ethanol for 30 s. Then, nylon membranes are immersed in aqueous $NaBH_4$ (2 mM; Aldrich), for reduction of silver ions, for 30 min and subsequently rinsed with deionized water for 1 min. Finally, the specimen is vacuum-dried overnight at room temperature.

2.2. Characterization. UV-visible absorption spectra were recorded, using a nylon-6,6 membrane as the reference, on a Cary 5000 UV-Vis-NIR scanning spectrophotometer. Scanning electron microscopy (SEM) observations were performed on a Philips XL30 electron scanning microscope. Transmission electron microscopy (TEM) observations were carried out on a JEOL JEM-2010F instrument with a point resolution of 1.9 Å and equipped with high-angle annular dark-field or Z-contrast detector. The nylon fibers with Ag nanoparticles were cut into pieces, dispersed in pure water, transferred onto a SiO-coated copper grid with disperser, dried in a vacuum overnight, and observed by TEM.

The measurements of lattice-fringe spacing and angles were made using digital image analysis of reciprocal space parameters, according to Akamatsu et al. [17]. Using this method the precision is 0.0001 nm for lattice spacing and 0.1° for lattice plane angles. This analysis was carried out with the Digital Micrograph software, Oxford, UK.

X-ray photoelectron spectroscopy (XPS) was performed using an UHV system of VG Microtech ESCA3000 Multilab, with an Al K_α X-ray source (1486.6 eV) and CLAM4MCD analyser. The surface of the sample was etched for 10 s with 3 kV Ar^+ at 0.16 $\mu A/mm^2$. The XPS spectrum was obtained at 55° from the normal surface in the constant pass energy mode; the pass energies used were 50 eV and 20 eV for survey and high resolution, respectively. The peak positions were referenced to the Shirley background Ag $3d_{5/2}$ photopeak at 368.21 eV and C 1s hydrocarbon groups in a binding energy of 285.00 eV. XPS spectra were fitted with the program SDP v 4.1 [18]. For the process of deconvolution the uncertainty in the binding energy was estimated to be 5% (i.e., ±0.05 eV).

3. Results and Discussion

The membranes used in this work are composed of nylon-6,6 fibers in the micrometer range, as shown in Figure 2(a). The fibers surface is rough (Figure 2(b)) and has several pores of less than 100 nm in diameter (inset Figure 2(b)).

These pores may allow guest molecules to enter. Thus, when a nylon membrane is immersed in aqueous $AgNO_3$, Ag^+ ions are readily impregnated into the nylon membrane fibers through the pores. Most of the incorporated Ag^+ ions are bound to nylon macromolecules probably via electrostatic (i.e., ion-dipole) interactions, because the electron-rich oxygen and nitrogen atoms of polar amides groups are expected to interact with positive metal ions. The posterior rinse with ethanol (ca. 30 s) removes Ag^+ ions that were not anchored to nylon fibres. Figure 3 shows the pure nylon membrane and nylon membranes having Ag nanoparticles, using 0.5 mM, 1 mM, 1.5 mM, and 10 mM $AgNO_3$ solutions, respectively. As the concentration of nanoparticles increases, the membranes exhibit different colours from light yellow to yellow and finally yellow-brownish.

Energy-dispersive X-ray spectroscopy (EDX) spectrum in Figure 4, which was collected from the nylon-6,6/silver nanoparticles composite, clearly shows the presence of Ag signal peak, indicating the effectiveness of the electrostatic assembly of the particles on the surface of the nylon fibers. The yellowish colour of Ag-nylon composite membrane (Figure 3) also suggests the presence of Ag nanoparticles on the polymer surface.

UV-Vis spectra of the nylon-6,6/silver nanoparticles composites are shown in Figure 5. Because the nylon-6,6 membrane was used as the reference, the absorption spectra can be considered as the absorption of Ag species. Using 0.5 mM $AgNO_3$ a narrow symmetrical absorption band is

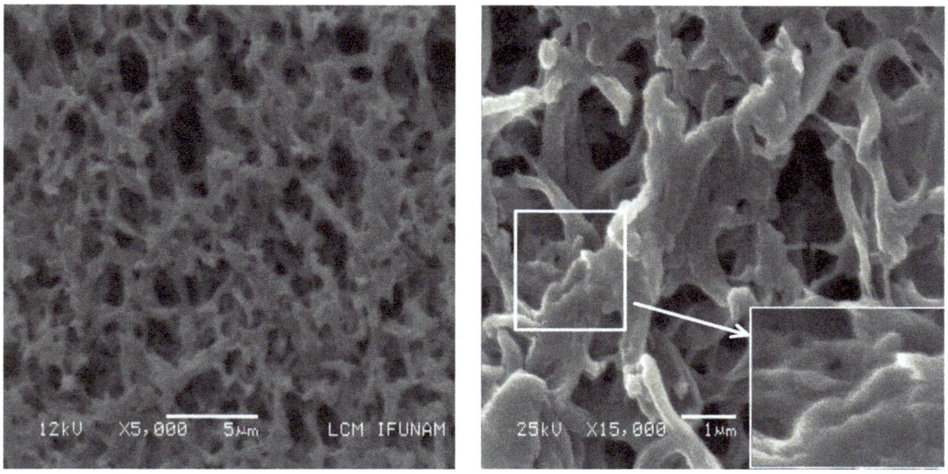

(a) (b)

FIGURE 2: SEM image of the nylon-6,6 membrane.

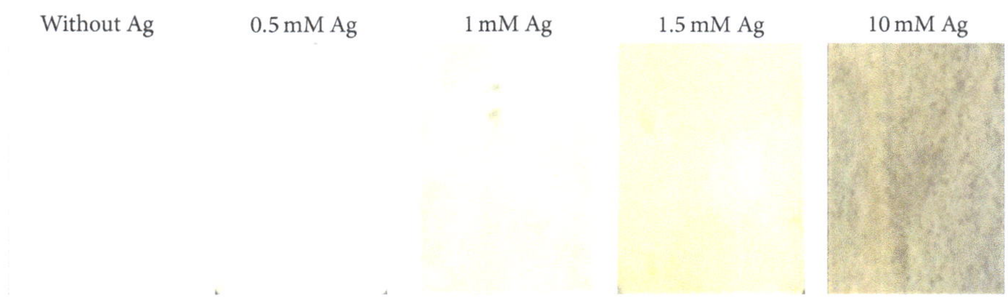

FIGURE 3: Nylon-6,6 without and with Ag nanoparticles formed at different $AgNO_3$ concentrations.

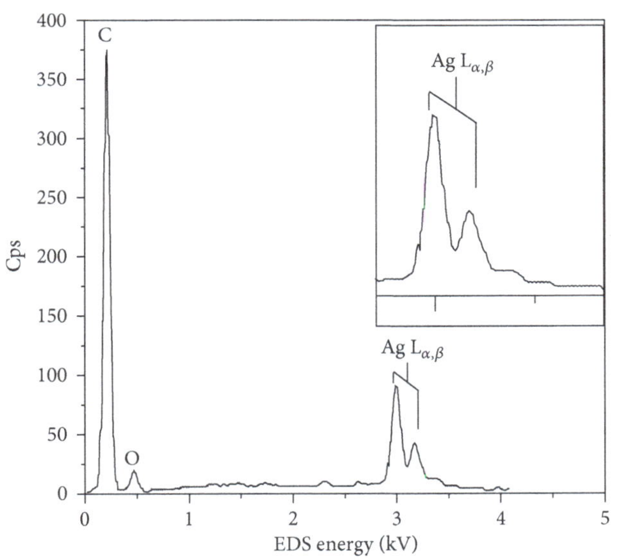

FIGURE 4: EDX spectrum of nylon-6,6 membrane coated with Ag nanoparticles.

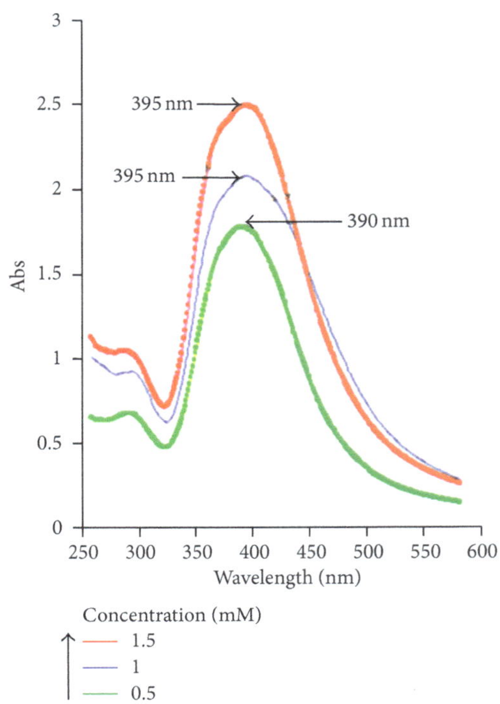

FIGURE 5: UV-Vis spectra of silver nanoparticles in nylon-6,6 membranes. Nanoparticles were prepared using aqueous $AgNO_3$ at 0.5 mM, 1.0 mM, and 1.5 mM, respectively.

located at 390 nm (Figure 5(a)); this is attributable to the surface plasmon resonance of silver nanoparticles in agreement with the observed yellow colour showed in Figure 3 [19]. No absorption was observed at wavelengths longer than 450 nm. These observations imply that Ag nanoparticles were formed. The surface plasmon peak underwent a shift to

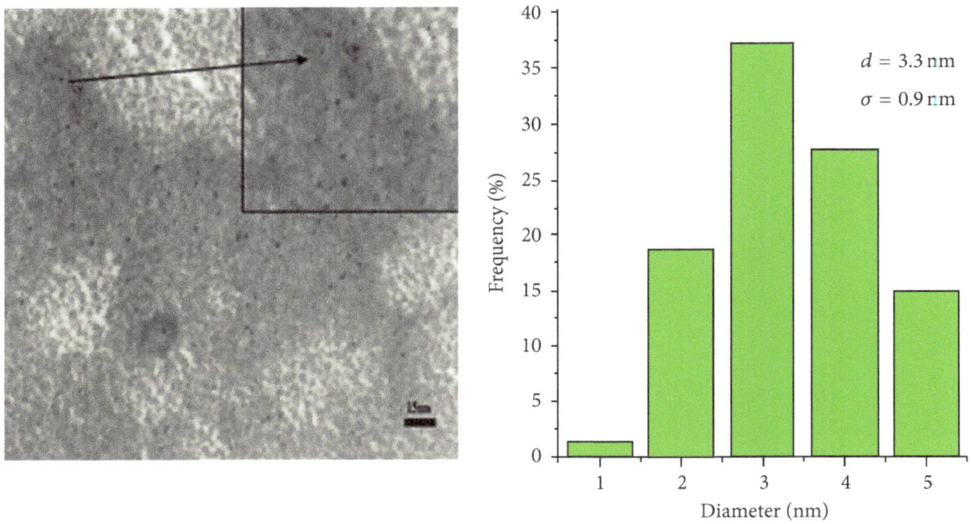

FIGURE 6: TEM micrograph and size distribution histogram of Ag nanoparticles in nylon-6,6 membranes ($AgNO_3$ 0.05 mM).

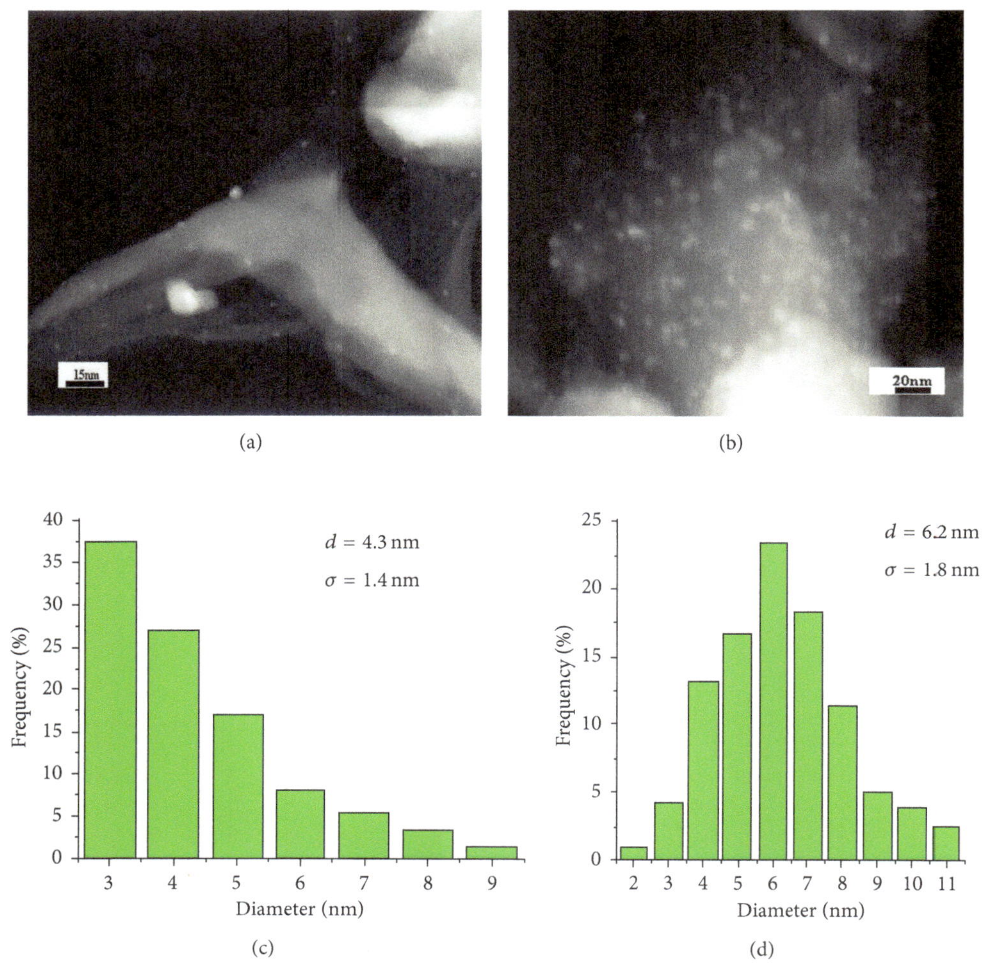

FIGURE 7: HAADF images and size distribution of Ag nanoparticles obtained from (a), (c) $AgNO_3$ 1.0 mM; (b), (d) $AgNO_3$ 1.5 mM.

395 nm and was slightly broadened when 1 mM $AgNO_3$ was employed (Figure 5(b)). When 1.5 mM $AgNO_3$ was used, the surface plasmon absorption becomes broader (Figure 5(c)). It is noted that the absorption intensity of the plasmon band increases as the concentration of aqueous $AgNO_3$ augments, also in agreement with the membrane intensity changes in colour, which can be attributed to the increase in the Ag nanoparticles concentration.

These assumptions are confirmed by TEM observations. As shown in Figure 6 monodisperse Ag nanoparticles were obtained at $AgNO_3$ 0.5 mM. Their mean diameter (d) was measured as 3.3 nm with a standard deviation (σ) of 0.9 nm.

Figure 7 shows Z-contrast images of nylon fibres with Ag particles. Figures 7(a) and 7(c) correspond to a concentration of 1.0 mM of $AgNO_3$ resulting in nanoparticles with an average size of 4.3 nm with a standard deviation of 1.4 nm. Using a solution of 1.5 mM $AgNO_3$ (Figures 7(b) and 7(d)), the average particle size slightly increases to 6.2 nm with a standard deviation of 1.8 nm. Therefore, it is possible to control the size and size distribution by adjusting the concentration of metal ions in solution.

Because of the porous structure of nylon fibre and the strong interactions between Ag^+ ions and the carbonyl and amide groups of nylon macromolecule, Ag^+ ions were uniformly and tightly anchored to the nylon fibres [20]. Such interactions would lower the mobility of Ag^+ ions, enhance the formation of silver nuclei, limit the formation of several morphologies, and prevent the growth of larger particles [21]. This is particularly true at low Ag^+ ion concentrations and can explain the formation of monodisperse Ag nanoparticles (after $NaBH_4$ reduction) under such conditions. At higher $AgNO_3$ concentrations, larger amounts of Ag^+ ions are embedded on nylon membranes, leading to large and widely distributed particles after reduction [22]. In these kinds of systems, it is expected that the carbonyl and amide groups may also play an important role in stabilization of metal nanoparticles in addition to the porous structure of nylon fibres.

The HRTEM image in Figure 8 shows Ag nanoparticles between 2 and 3 nm in diameter. This image possesses atomic resolution; therefore, several crystalline planes are distinguishable and the interplanar distances can be measured. The interplanar distances shown in the micrograph were measured from the Fourier transforms (FFT) of these nanoparticles (Figure 8, bottom), where the corresponding crystalline planes are specified. The interplanar distances and their corresponding crystalline planes match the ones of metallic Ag (FCC) phase. The measured interplanar distance is 2.36 Å and corresponds to the (111) plane [23].

In order to examine the chemical composition of the nylon-6,6/Ag nanoparticles fibre composite, as well as the possible interaction of silver metal with the nylon moieties, after formation of Ag nanoparticles, X-ray photoelectronic spectroscopy (XPS) was used. Figure 9 shows the XPS survey spectra obtained after Ar^+ etching for 10 s.

Figure 9 shows the increase in the Ag 3d signal corresponding to the different concentrations of $AgNO_3$ added in the solution reaction; a spectrum of nylon-6,6 is used as reference. Clearly an increase of the concentration of

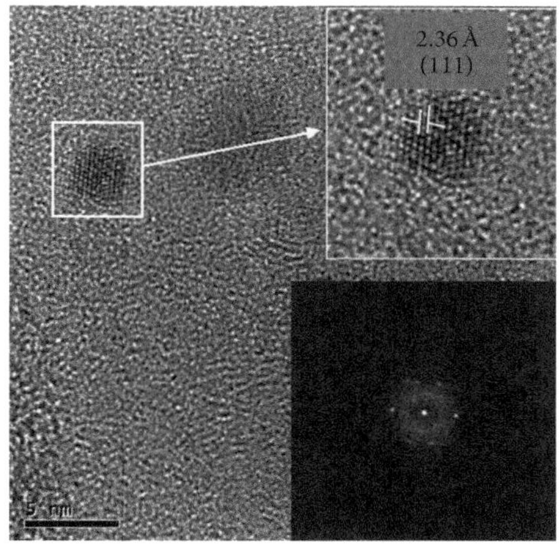

FIGURE 8: HRTEM image of Ag nanoparticles (top) and their corresponding Fourier transforms (FFT) (bottom) pattern.

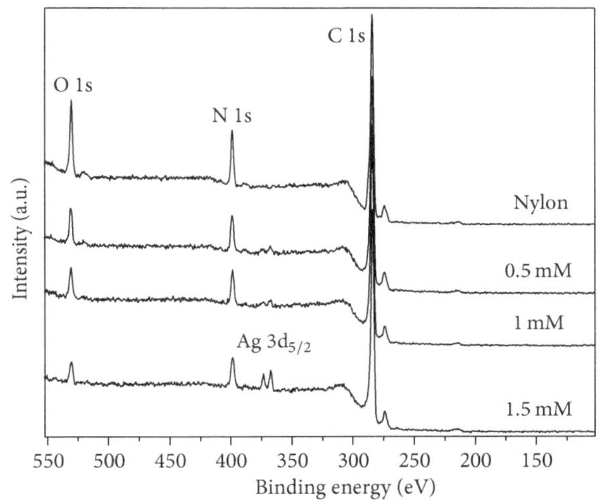

FIGURE 9: XPS spectra of nylon-6,6 and three different nylon-6,6/Ag composites.

the metal precursor salt favours the formation of the nylon/Ag composite.

In order to obtain more information about the chemical state of the Ag nanoparticles present in the composite a curve fit of the various signals was made and is shown in Figure 10.

Table 1 shows the binding energies of the XPS deconvolution of Ag 3d, C 1s, O 1s, and N 1s core levels samples with the Ag metallic, AgO, and nylon-6,6 as references. The peak energy position in the deconvolution and number of peaks were based in data reported by Beamson and Brigs [24] and Hoflund and Hazos [25] and calibrated with metallic Ag foil as reference.

From Table 1, a positive chemical shift in N 1s core level and negative chemical shift in Ag $3d_{5/2}$ of Ag(0) of nanoparticles, with respect to N 1s and Ag $3d_{5/2}$ of nylon and Ag metallic, are observed. From the data in Table 1 the general rule based on the electronegativity can explain both the positive chemical shift of the N 1s core level and the negative chemical shift of the Ag $3d_{5/2}$ core level. This effect can be seen

TABLE 1: The binding energy position by XPS of nylon-6,6- and Ag-containing samples core levels Ag 3d, C 1s, O 1s, and N 1s.

Sample	N 1s	O 1s		$Ag\ 3d_{5/2}$		*C 1s			
		O=C	O–Ag	Ag	AgO (eV)	1	2	3	4
References*	398.69	530.33		368.21					
0.5 mM	398.91	530.33	531.30						
1.0 mM	398.80	530.33	531.30	368.20	367.31	283.84	284.14	285.00	286.81
1.5 mM	398.75	530.33	531.30						

*References: nylon and metallic Ag.

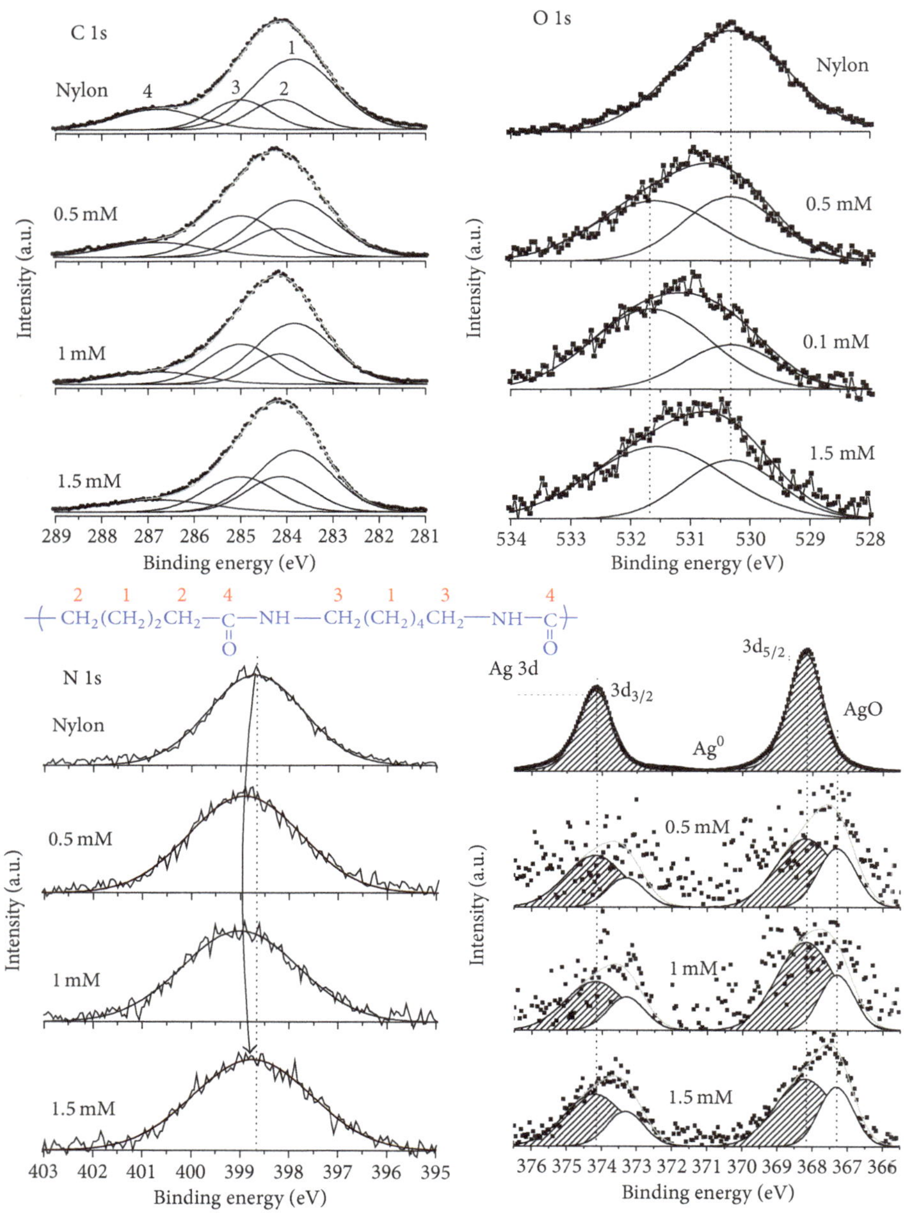

FIGURE 10: The XPS deconvolution for C 1s, O 1s, N 1s, and Ag 3d core levels.

in Figure 10 for N 1s. For the composite samples analysed, all C 1s, O 1s, and Ag $3d_{5/2}$ of AgO core levels are in the same positions as exhibited in Table 1 and Figure 10. O 1s (530.33 eV) corresponds to O=C and Ag–O at 367.31 eV; C 1s corresponds to (1) C–CH_2 at 283.84 eV, (2) CH_2–CH_2 at 284.14 eV, (3) CH_2–N at 285 eV, and (4) C=O at 286.81 eV. It can be also observed that for N 1s the maximum shifting (Δ) occurs for the 0.5 mM simple with $\Delta = 0.22$ eV; however, for 0.1 mM sample, $\Delta = 0.11$, it contains the largest amount of Ag nanoparticles, taking in account the amount of AgO. For the 1.5 mM sample, Ag 3d peaks can be deconvoluted satisfactorily obtaining thus all oxidation states even in the samples with less Ag concentration. In addition, in Figure 10, the data obtained from the curve fitting show that there is no change in the binding energy of the orbitals of C 1s despite the use of different $AgNO_3$ concentrations. There is, however, a 0.31 eV shift in the N 1s peak corresponding to the sample with 1.5 mM $AgNO_3$. This small shift in binding energy suggests there is an interaction of the nitrogen atoms in nylon during the stabilization of the silver nanoparticles. A similar chemical shift for N 1s binding energy in XPS spectra has been observed for the interactions of PVP with Pt nanoparticles and nitrocellulose with Ru [26, 27]. Hence, pores of nylon fibers, where the nitrogen atoms from the amide groups are found, not only interact with the metal ions presumably through iondipole forces, but once the reduction reaction occurs, they also stabilize the Ag nanoparticles. Analysis of the O 1s and Ag $3d_{5/2}$ peaks shows the presence of AgO, which was expected due to the high reactivity of the silver nanoparticles at the sizes obtained. Nevertheless, the fact that there is still a signal corresponding to Ag^0 suggests that the oxidation takes place only on the surface of the nanoparticles, while metallic silver remains at their core; again this is possible due to the stabilizing effect of the amide groups present in the nylon fibers. This nylon-6,6/Ag nanoparticles composite will be probed in dyes removal from aqueous solutions and in applications regarding its antibacterial properties.

4. Conclusions

It was demonstrated that using an aqueous Ag ion impregnation of nylon fibres followed by a reduction with $NaBH_4$, a composite of Ag nanoparticles attached to the polymer can be formed. The whole process is carried out at ambient conditions. SEM, HRTEM, and XPS studies confirmed the presence of such Ag nanoparticles in the fibres, with an average size of 3.3 nm. This very simple and versatile synthetic route could be applied to obtain other composites made of metal nanoparticles and natural or synthetic polymer fibres. Moreover, an interaction between nitrogen of amides moieties of nylon-6,6 and silver nanoparticles has been found by X-ray photoelectron spectroscopy.

Acknowledgments

Financial support of this work was provided by Universidad Autónoma del Estado de México (UAEM) (Project no. 3246/2012U). The authors are indebted to Fernando Ureña (ININ), Samuel Tehuacanero (IFUNAM) for their assistance in SEM and image digitalisation, respectively.

References

[1] J. Y. Ying, C. P. Mehert, and M. S. Wong, "Synthesis and applications of supramolecular-templated mesoporous materials," *Angewandte Chemie*, vol. 38, no. 1-2, pp. 56–77, 1999.

[2] M. S. Morey, J. D. Bryan, S. Schartz, and G. D. Stuchy, "Pore surface functionalization of MCM-48 mesoporous silica with tungsten and molybdenum metal centers: perspectives on catalytic peroxide activation," *Chemistry of Materials*, vol. 12, no. 11, pp. 3435–3444, 2000.

[3] J. Liu, X. Feng, G. E. Fryxell, L. Wang, A. Y. Kim, and M. Gong, "Hybrid mesoporous materials with functionalized monolayers," *Advanced Materials*, vol. 10, no. 2, pp. 161–165, 1998.

[4] J. J. Schneider, N. Czap, J. Hagen et al., "Metallorganic routes to nanoscale iron and titanium oxide particles encapsulated in mesoporous alumina: formation, physical properties, and chemical reactivity," *Chemistry*, vol. 6, no. 23, pp. 4305–4321, 2000.

[5] J. S. Jung, W. S. Chae, R. A. McIntyre, C. T. Selp, J. B. Wiley, and C. J. O'Connor, "Preparation and characterization of Ni nanoparticles in an MCM mesoporous material," *Materials Research Bulletin*, vol. 34, no. 9, pp. 1353–1360, 1999.

[6] K. Moller and T. Bein, "Inclusion chemistry in periodic mesoporous hosts," *Chemistry of Materials*, vol. 10, no. 10, pp. 2950–2963, 1998.

[7] H. Shi, L. Zhang, and W. Cal, "Composition modulation of optical absorption in Ag_xAu_{1-x} alloy nanocrystals *in situ* formed within pores of mesoporous silica," *Journal of Applied Physics*, vol. 87, no. 3, p. 1572, 2000.

[8] K. W. Powers and L. L. Hench, "The effect of pore size on metal cluster formation in silica sol gel monoliths," *Ceramic Transactions*, vol. 101, p. 253, 2000.

[9] Y. Plyuto, J. M. Berquer, C. Jaquiod, and C. Ricolleau, "Ag nanoparticles synthesised in template-structured mesoporous silica films on a glass substrate," *Chemical Communications*, no. 17, pp. 1653–1654, 1999.

[10] W. H. Zhang, J. L. Shi, L. Z. Wang, and D. S. Yan, "Preparation and characterization of ZnO clusters inside mesoporous silica," *Chemistry of Materials*, vol. 12, no. 5, pp. 1408–1413, 2000.

[11] B. Lebeau, C. E. Fowler, S. Mann, C. Farcet, B. Charleux, and C. Sanchez, "Synthesis of hierarchically ordered dye-functionalised mesoporous silica with macroporous architecture by dual templating," *Journal of Materials Chemistry*, vol. 10, no. 9, pp. 2105–2108, 2000.

[12] L. Zhang, T. Sun, and J. Y. Ying, "Oxidation catalysis over functionalized metalloporphyrins fixated within ultralarge-pore transition metal-doped silicate supports," *Chemical Communications*, no. 12, pp. 1103–1104, 1999.

[13] V. L. Colvin, M. C. Schlamp, and A. P. Alivisatos, "Light-emitting diodes made from cadmium selenide nanocrystals and a semiconducting polymer," *Nature*, vol. 370, no. 6488, pp. 354–357, 1994.

[14] B. O. Dabbousi, M. G. Bawendi, O. Onitsuka, and M. F. Rubner, "Electroluminescence from CdSe quantum-dot/polymer composites," *Applied Physics Letters*, vol. 99, p. 13834, 1995.

[15] R. Lamber, S. Wetjen, G. Schulz-Ekloff, and A. Baalmann, "Metal clusters in plasma polymer matrices: gold clusters," *Journal of Physical Chemistry*, vol. 99, no. 38, pp. 13834–13838, 1995.

[16] L. L. Beecroft and C. K. Ober, "Nanocomposite materials for optical applications," *Chemistry of Materials*, vol. 9, no. 6, pp. 1302–1317, 1997.

[17] K. Akamatsu, N. Tsuboi, Y. Hatakenaka, and S. Deki, "*In situ* spectroscopic and microscopic study on dispersion of Ag nanoparticles in polymer thin films," *Journal of Physical Chemistry B*, vol. 104, no. 44, pp. 10168–10173, 2000.

[18] SDP v4. 1 (32 bit) Copyright 2004, XPS International, LLC, Compiled January 17, 2004.

[19] J. He, I. Ichinose, T. Kunitake, and A. Nakao, "Reversible conversion of nanoparticles of metallic silver and silver oxide in ultrathin TiO_2 films: a chemical transformation in nano-space," *Chemical Communications*, no. 17, pp. 1910–1911, 2002.

[20] S. Besson, T. Gacoin, C. Ricolleau, and J. Boilot, "Silver nanoparticle growth in 3D-hexagonal mesoporous silica films," *Chemical Communications*, vol. 9, no. 3, pp. 360–361, 2003.

[21] T. M. Fahmy, P. M. Fong, A. Goyal, and W. M. Saltzman, "Targeted for drug delivery," *Nanotoday*, vol. 8, no. 8, pp. 18–26, 2005.

[22] J. He, I. Ichinose, T. Kunitake, A. Nakao, Y. Shiraishi, and N. Toshima, "Facile fabrication of Ag-Pd bimetallic nanoparticles in ultrathin TiO_2-gel films: nanoparticle morphology and catalytic activity," *Journal of the American Chemical Society*, vol. 125, no. 36, pp. 11034–11040, 2003.

[23] W. J. de Ruijter, R. Sharma, M. R. McCartney, and D. J. Smith, "Measurement of lattice-fringe vectors from digital HREM images: experimental precision," *Ultramicroscopy*, vol. 57, no. 4, pp. 409–422, 1995.

[24] G. Beamson and D. Brigs, *High Resolution XPS of Organic Polymers*, John Wiley & Sons, Chichester, UK, 1992.

[25] G. B. Hoflund and Z. F. Hazos, "Surface characterization study of Ag, AgO, and Ag_2O using x-ray photoelectron spectroscopy and electron energy-loss spectroscopy," *Physical Review B*, vol. 62, no. 16, pp. 11126–11133, 2000.

[26] L. Qiu, F. Liu, L. Zhao, W. Yang, and J. Yao, "Evidence of a unique electron donor—acceptor property for platinum nanoparticles as studied by XPS," *Langmuir*, vol. 22, no. 10, pp. 4480–4482, 2006.

[27] V. K. Kaushik, "XPS core level spectra and Auger parameters for some silver compounds," *Journal of Electron Spectroscopy and Related Phenomena*, vol. 56, no. 3, pp. 273–277, 1991.

Optical Properties of the Self-Assembling Polymeric Colloidal Systems

Alexandra Mocanu, Edina Rusen, and Aurel Diacon

Department of Polymer Science, University Politehnica of Bucharest, 149 Calea Victoriei, 010072 Bucharest, Romania

Correspondence should be addressed to Edina Rusen; edina rusen@yahoo.com

Academic Editor: Haojun Liang

In the last decade, optical materials have gained much interest due to the high number of possible applications involving path or intensity control and filtering of light. The continuous emerging technology in the field of electrooptical devices or medical applications allowed the development of new innovative cost effective processes to obtain optical materials suited for future applications such as hybrid/polymeric solar cells, lasers, polymeric optical fibers, and chemo- and biosensing devices. Considering the above, the aim of this review is to present recent studies in the field of photonic crystals involving the use of polymeric materials.

1. Photonic Crystals

Nanomaterials with well-defined 3D repetitive arrays have found application in photonic devices, sensors, selective membranes, and even microchips.

Compared to a classic crystalline array, colloidal spherical particles replace the molecules, atoms, or ions in hexagonal close packed (HCP) structure or cubic close packed (CCP) structures [1–3] (Figure 1).

Colloidal particles are able, through self-assembly processes, to be arranged into 3D structures leading to colloidal crystals. The resulting materials revealed interesting properties, presenting similar optical characteristics with synthetic opal and photonic crystals (Figure 2). One of the most facile routes for obtaining photonic crystals is based on the employment of polymer colloids with self-assembling properties [3–5].

A photonic crystal is a periodic nanostructured material that influences the propagation of electromagnetic waves in a similar manner as a semiconductor does for electrons [3].

Although photonic crystals have been studied since 1887, with one of the pioneers being Rayleigh [6], the term has been introduced only in 1987, after the papers of two physicist working independently Yablonovitch and John [7, 8].

In order to understand the behavior of photons inside a photonic crystal it is possible to compare it with the motion of electrons and vacancies in a semiconductor material as mentioned above. A proper example of photonic crystals occurring in nature is opals, which contain a natural periodic microstructure of silica microspheres of 15–900 nm, responsible for their iridescent color [9]. This structure allows the propagation of the photons through the crystal, while the interaction between the silica spheres and light makes possible the formation of allowed and forbidden energy bands, also known as band gaps or stop bands [8]. Thus, the propagation of the electromagnetic waves is forbidden for specific wavelengths.

Therefore, a photonic crystal is a powerful tool for light modulation in certain directions of space. Exploiting these properties, photonic crystals have been used in many optical applications, such as the use in sensors and monitors [10, 11], bioassays [12, 13], colour displays [14, 15], solar energy [16], and lasers [17, 18]. For all applications involving PCs one of the key parameters that have to be controlled is the band-gap position. The stop-band value depends primarily on three factors [19–22]: the refractive index contrast between the dielectric components (solid and air usually), the lattice constant (the spacing between the spheres), and the filling

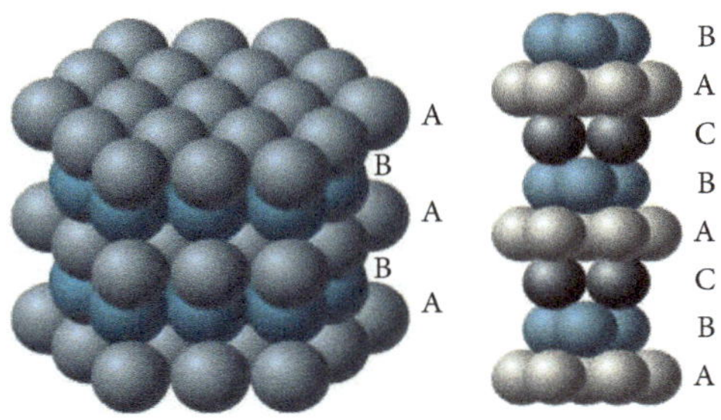

FIGURE 1: Three-dimensional arrays: (a) HCP, (b) CCP [3].

factor (volume of the colloids related to the volume of surrounding media). In addition, one of the most challenging applications consists in fabricating photonic crystals with 3D band gap around 1.55 μm representing the wavelength used in optical fibers [23].

2. Polymeric Photonic Crystals

In order to obtain colloidal crystals, the most common methods consist in surfactant-free emulsion polymerization [5, 24], precipitation polymerization [25, 26], dispersion polymerization [27], and seeded emulsion polymerization [28]. The absence of surfactants leads to polymer latex of high purity, and at the same time, the polymerization parameters (temperature, initiator and monomer concentrations, stirring, etc.) are facile to control. The final monodisperse polymer particles represent a class of materials that can be easily functionalized on the surface, therefore, increasing the number of possible applications [1].

As it was specified in the previous section, colloidal crystals represent HCP or CCP structures, consisting of colloidal microspheres ranging from 100 nm to 1 μm. Depending on the arrangement parameters the photons are forbidden to propagate inside colloidal crystals in certain directions, for specific wavelengths. If the lattice potential is strong enough, the gap might extend to all possible directions, resulting in a complete band gap [29]. This property allows the control of the light that interacts with the colloidal crystal.

Taking this into account, it is easy to realize that one of the most important optical properties of the colloidal crystals consists in light diffraction according to Bragg's law.

2.1. Bragg's Law. In order to better understand the phenomenon that occurs inside a polymer photonic crystal, it is necessary to analyze the path of an incident beam that travels through the crystal planes.

The diagram below [30] (Figure 3) explains the pathway the beam B, which represents the extra distance traveled by beam A. Perpendicular lines, (l) lines, were drawn to both rays. Following this method two right triangles were obtained [30].

The angle, marked θ, also called Braggs angle can be calculated from these triangles as $\sin\theta = s/d$. Therefore, the extra distance is $2s$ long or $2d \sin\theta$ long. This value is equal with a number of wavelengths (n) for a strong beam (Bragg's law)

$$n\lambda = 2d \sin 0. \quad (1)$$

When the Bragg equation is satisfied, the angle formed by the diffracted beam with the crystal plane is equal to the one between by the incident beam and the same crystal plane. In this situation, diffraction is similar to a reflection from a mirror (Braggs mirror effect), where the angle of incidence is identical to the angle of reflection [30].

In the case of molecular, atomic, or ionic crystals, the interplanar spacing, d, is in the range of Å, thus having the capacity of diffracting X-rays. In the case of colloidal crystals, d is in the range of 100 nm, which implies the diffraction of light with wavelengths of the same order of magnitude (400–800 nm). Therefore, the colloidal crystal will diffract light with wavelengths corresponding to violet-blue-green and transmit it in the complementary radiation (orange-red) of visible light.

2.2. Bragg's Law on Colloidal Crystal. The examination of a colloidal dispersion (Figure 4) during and after solvent evaporation allows the observation of iridescence in different areas of the visible spectrum due to diffraction of the light in accordance with Bragg's law [31].

The optical microscopy shows that the distance between the defaults due to water evaporation is only 150–200 μm, which proves the high quality of the film, respectively, of the obtained opal [32]. The crystallization process reveals permanent changes in the color of the colloidal dispersion. In the first step, the dispersion is white due to Mie scattering, and d is larger than polymer particles [33]. In the second step, the colloidal dispersion changes its colors as a consequence of reducing the interplanar distance.

Figure 5(a) shows a well-ordered close packed structure with a triangular arrangement that can be attributed to planes (111) of a CCP system [9, 31, 34]. Moreover the (fast fourier transform) FFT of the image (Figure 5(b)) displays sharp peaks confirming the existence of long-range crystalline order.

According to Bragg's law, the change of the angle between the incident radiation and photonic crystal leads to notable changes of the reflected wavelength, respectively, of the observed color (Figure 6) [35, 36].

UV-VIS spectra of polymer photonic crystals indicate in many cases sharp peaks, which confirm the existence of highly ordered structures, due to Braggs reflection of periodic arrangement of the polymer particles [37–40].

Figure 7(b) represents the reflection spectra of polymer colloidal crystals with different size particles (178 nm, 211 nm, 235 nm, and 270 nm) corresponding to photograph (Figure 7(a)) [41].

Moreover, these materials show structural color changes (Figure 8) under mechanical stress. In this case, the photonic crystal film, deposited on a plastic plate, changed from red to green as result of the modification of the interplanar distance, which is thus a change in the wavelength of the diffracted light [37].

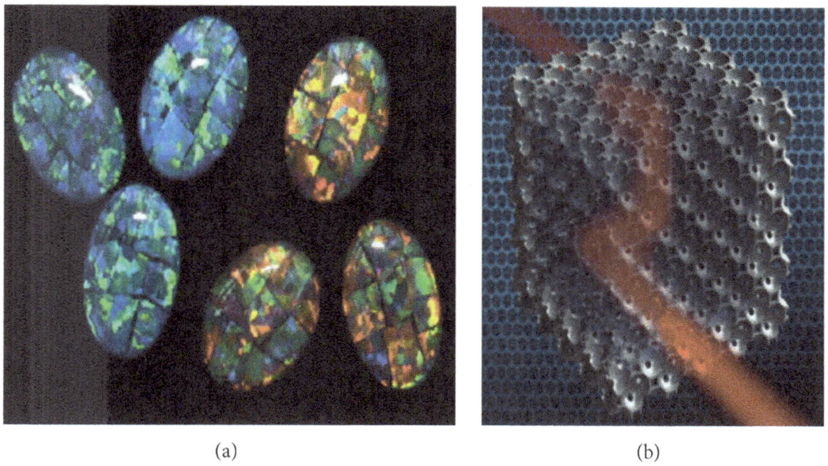

FIGURE 2: (a) Synthetic opal; (b) structure of a photonic crystal [4].

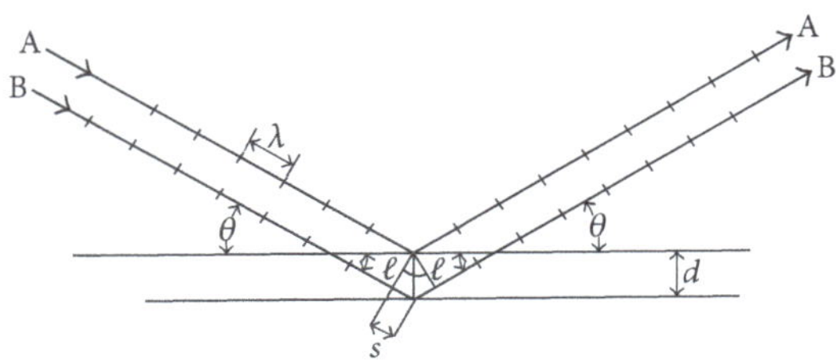

FIGURE 3: The diffraction of the light by crystal planes [30].

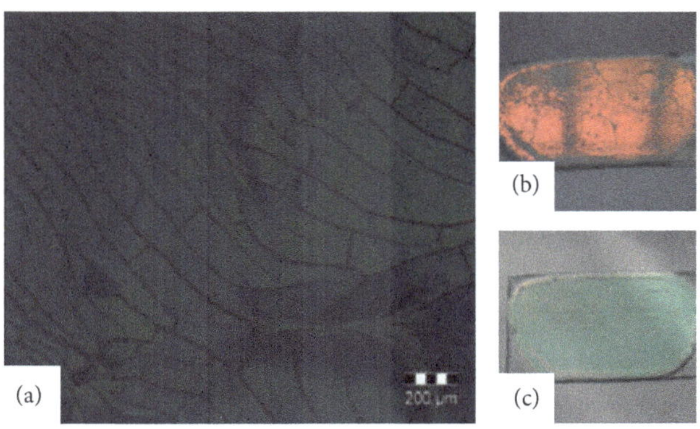

FIGURE 4: Light diffraction: iridescence of polymer colloidal spheres: (a) optical microscopy; (b) photo recorded at an angle of 45° following thermal treatment; (c) photo recorded at an angle of 90° following thermal treatment [31].

Although, until now, polymer photonic crystals were not expected to exhibit full stop bands, they were easy to prepare, leading to materials that allow the control of the light that interacts with the crystalline colloidal arrays [3, 42].

3. Applications of the Polymeric Photonic Crystals

In the last years, polymeric photonic crystals found application in optical devices [43–45] and sensors [46] or as templates for other interesting porous-media applications taking advantage of their optical properties, composition, or morphology [26, 47, 48].

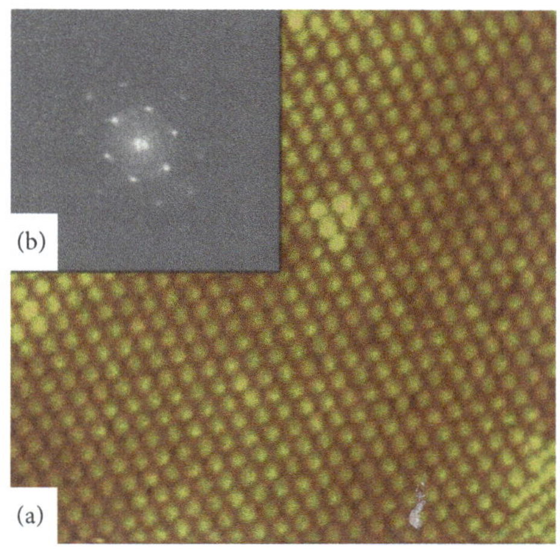

FIGURE 5: (a) AFM image of the top surface of the opal and (b) FFT of the image [31].

3.1. Templates for Porous-Media Applications. Polymer colloidal crystals without complete band gaps can still be used as templates to create new structures with complete band gaps [49].

The first step in obtaining inverse opal is to arrange by self-assembly polymer colloidal spheres into periodic arrays (Figure 9). The voids between the polymer particles can then be filled with different solutions that play the role of matrix. After this procedure, the polymer spheres are removed by dissolution or calcination leaving air spaces in the matrix. The resulting system is a photonic crystal with different optical properties (incremental shift and broadening with respect to those in spectra of parent opals) (Figure 10).

Waterhouse et al. [9] obtained inverse opal structures by depositing inorganic particles such as silica, titania, and ceria respectively, by a sol-gel method in the voids of

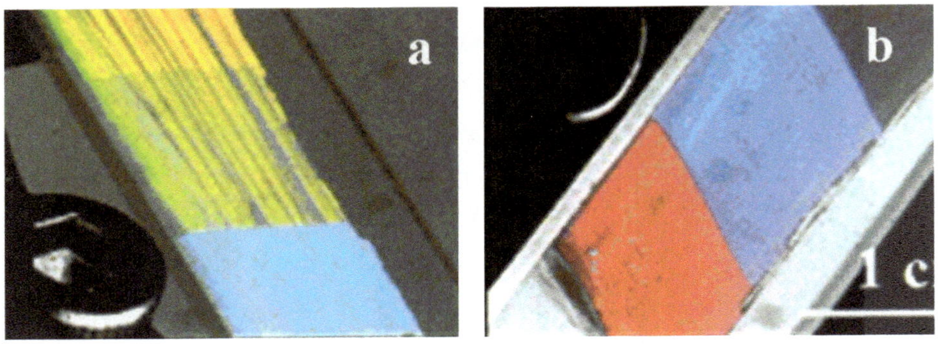

(a) (b)

FIGURE 6: Reflected colors of colloidal dispersions registered at different angles [36].

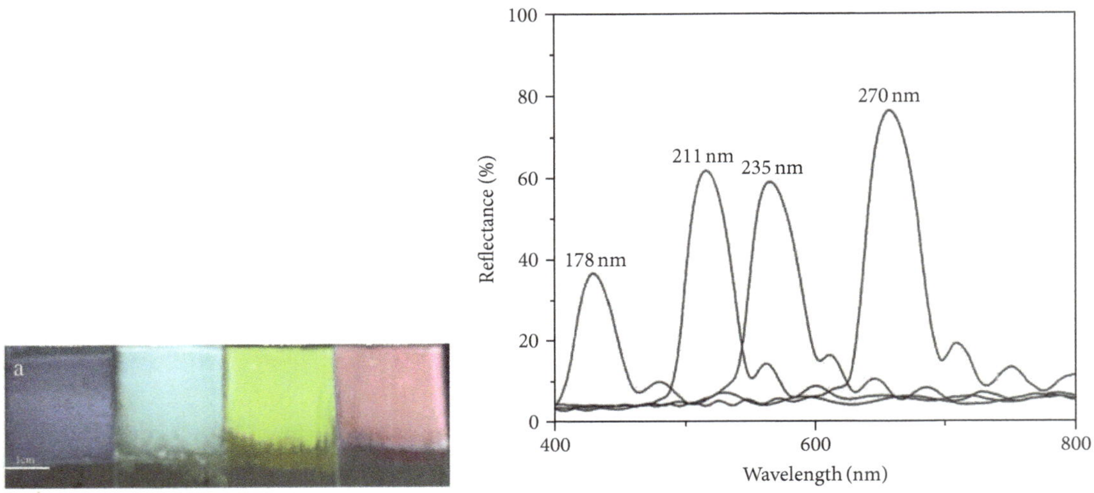

(a) (b)

FIGURE 7: Polymer colloidal dispersions: (a) photograph taken at normal incidence; (b) reflection spectra [41].

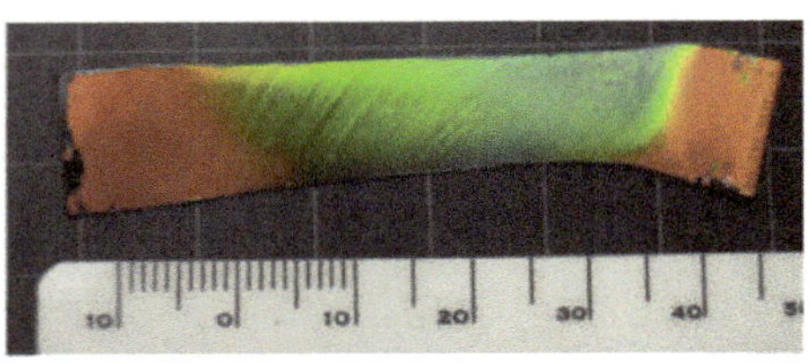

FIGURE 8: Structural color change in a crystal film after mechanical stress [37].

a polymeric photonic film. The poly(methyl methacrylate) polymer self-assembled spheres (Figure 11(b)) were removed by calcination at 550°C. The process resulted in obtaining a macroporous matrix with CCP ordered air spheres separated by inorganic nanoparticles according to the layout presented below (Figure 11(a)). The macroporous materials formed were characterized by SEM analysis in comparison with the polymer poly(methyl methacrylate) templates in order to put into evidence the inverse opal structures highlighted in Figure 11(c) [9].

Macroporous 3D structures were obtained also by a dip-coating procedure explained by Moon et al. [50]. In this study,

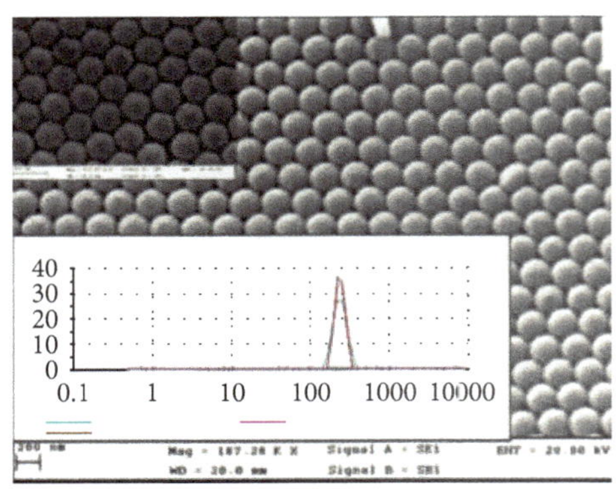

FIGURE 9: Polymer colloidal crystal obtained by soap free emulsion polymerization and DLS spectrum [65].

they obtained porous cylinders with interconnected air voids using commercial cylindrical glass fiber of 60 and 100 μm, respectively.

The fiber was dipped into the colloidal dispersion. The colloidal poly(styrene) particles self-assembled in a CCP

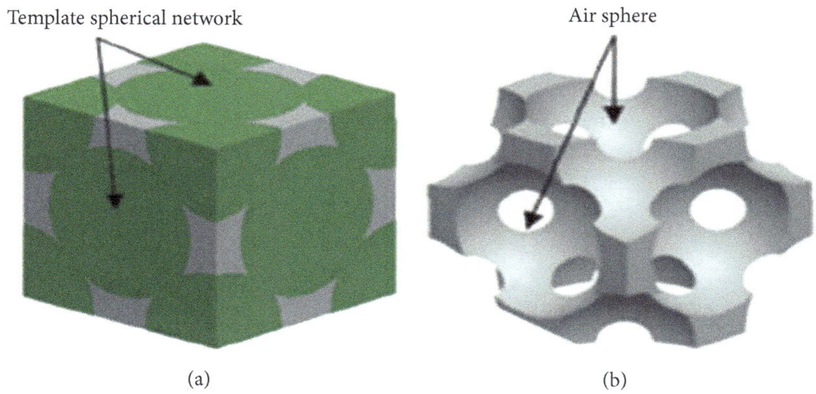

FIGURE 10: Unit cell of an infiltrated opal (a); unit cell of an inverse opal (b) [66].

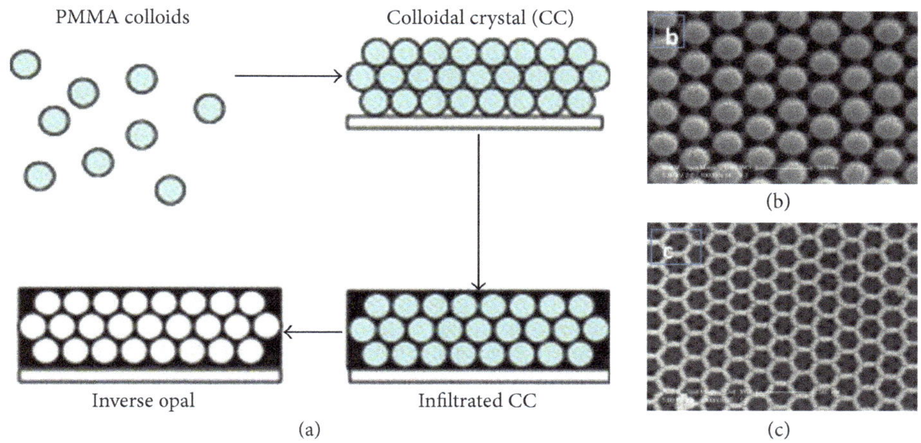

FIGURE 11: Fabrication scheme of an inverse opal (a); poly(methyl methacrylate) colloidal spheres (b); TiO_2 inverse opal (c) [9].

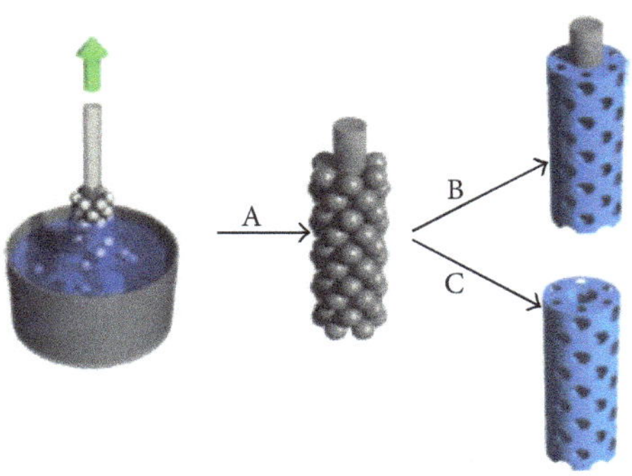

FIGURE 12: Schematic illustration of the self-assembly of colloidal spheres on a cylindrical glass fiber (A) and the inverted structure with glass fiber core (B), respectively, with hollow core (C) [50].

structure on the cylindrical glass substrate during the evaporation of the solvent. Then, a solution of polyurethane was infiltrated between the poly(styrene) arranged spheres and polymerized. The macroporous cylindrical structure was obtained on a glass fiber support by dissolving the poly(styrene) spheres. By removing the inner core an inverted porous structure with hollow core was obtained (Figures 12 and 13).

Moreover, the authors demonstrated that these structures have a controllable pore size by using colloidal particles with different dimensions. Therefore, these porous fibers are suitable for separation of membranes or catalyst support [50].

Kalinina et al. [51] proposed a new method in order to obtain materials with optically sensitive matrix starting from monodisperse core-shell colloidal particles. In the first step of this study, the particles that possess a hard core of poly(methyl methacrylate) and a relatively soft shell of poly(butyl methacrylate) are self-assembled in a crystalline array. The film is annealed at a temperature that is above the glass transition temperature T_g of the polymeric shell and below the T_g of the polymeric core. At this temperature, the shell fills the air spaces between the colloidal particles, leaving a matrix with intact cores and totally different optical properties towards the initial polymer film. The incorporation of a fluorescent dye during the polymerization process of the shell forming polymer resulted in a substantial reduction in optical contrast between the particles and the matrix [51].

The type of the matrix often influences the specific applications of inverse opal structures. A polymer matrix formed by poly(methyl methacrylate), polystyrene, polyurethane, or electroconducting polymers is suitable for optical devices. For example, polypyrrole, polythiophene, or polyaniline films

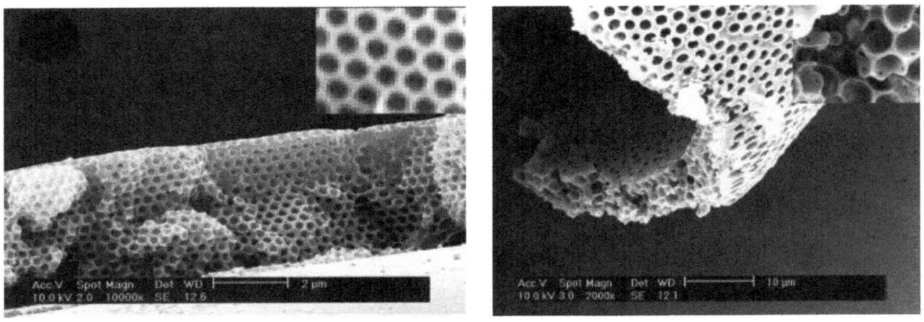

(a) (b)

FIGURE 13: Cross-section of polyurethane fiber with glass core (a) and with hollow core (b), respectively, [50].

with inverse opal structure are assembled on ITO conductive glass used in solar cells [52]. It is worth mentioning the fact that an inorganic matrix formed by semiconductors (like Ge, TiO_2, InP, and CdSe) or other metal compounds [53] found application in realization of infrared active optical materials due to their high refractive index [54, 55], while polymer inverse opals can be used in medicine as scaffolds for tissue engineering or cell culture [56] (Figure 14).

3.2. Sensors. The refractive index is a result of the interplanar distance and the type of packing of the polymer/inorganic spheres according to Bragg's law. Practically, the stop band is directly proportional with the interplanar distance, d. In other words, any variation of d should determine a shift in the diffraction peak of the reflected light. This property of the crystalline 3D arrays makes these materials suitable for optical sensors [1].

Polymer photonic crystals are used in optical sensors, coupled to a transduction device that identifies the optical diffraction shifts that occur once the external medium changes.

Optical sensors based on colloidal crystals (sometimes embedded in hydrogels) are known for their simplicity and low cost and efficiency. In the last years, remarkable results to the preparation of photonic crystal sensors that are highly sensitive to a series of pH [57], temperature [58], solvent swelling [59], ionic strength, ions [60], and biomolecules [61] have been obtained.

In addition, colloidal crystals were also used to develop new sensors for the detection of endocrine-disrupting chemicals like bisphenol A (BPA) at low costs. Guo et al. [62] explain the concept of photonic crystals-based sensors using combined photonic crystal technology and molecular imprinting techniques. First, poly(methyl methacrylate) spheres were modified with nanocavities derived from the BPA-imprinting method. The modification of the colloidal spheres by these techniques affords nanocavities distributed in the polymeric crystalline array which allow the photonic crystal sensor to identify BPA specifically. Using these procedures the changes in diffraction intensity, which are related to BPA concentrations, can be observed (Figure 15), even at very low concentrations of BPA ranging from 1 ng/mL to 1 μg/mL.

In a recent study, Hu et al. [63] developed an optical sensor sensitive for anions, starting from the properties of an ionic liquid based monomer and photonic crystals.

Therefore, a photopolymerization was conducted using an imidazolium-based ionic liquid monomer (*1-(2-acryloyloxyhexyl)-3-methylimidazolium bromide*) with methyl methacrylate, a cross-linker and initiator in a solvent mixture (methanol/chloroform), and infiltrated in a photonic crystal consisting of silica colloidal particles. After removal of the inorganic silica particles, a porous inverse opal film was obtained (Figure 16(a)).

The new photonic ionic liquid films were immersed in anionic aqueous solutions. The anions bonded onto the film surface, reducing its hydrophilicity. As a consequence, the volume of the film changed (shrunk) leading to strong modifications of the resulting photonic structures stop bands (Figure 16(b)).

In Figure 16(c), the color changes can also be observed with naked eye, representing a much easier and in-expensive method for identifying anions.

Moreover, it was also possible to identify the polarity of different solvents using these photonic ionic liquid films, due to the interactions between the terminal groups of the ionic liquid and solvent [63].

Although the use of ionic liquids in optical sensors is limited due to time-consuming or difficult synthetic procedures, this new approach based on photonic colloidal crystals afforded a system for rapid, naked-eye detection of anions.

Considering that there are about 10^8 ionic liquids [63] this procedure might allow the development of more ionic liquid monomers for advanced applications in photonic crystals.

3.3. Optical Devices. Recently, the propagation mechanism of light inside 3D polymer colloidal crystals gained much interest, leading to an important progress in photonic applications such as microlasers, waveguides, or photonic crystal fibers [45].

In other words, Furumi et al. [18] fabricated a colloidal crystal film using 200 nm polystyrene colloidal particles, followed by the thermal polymerization of dimethylsiloxane in the air voids of the 3D structure. Next step consisted in

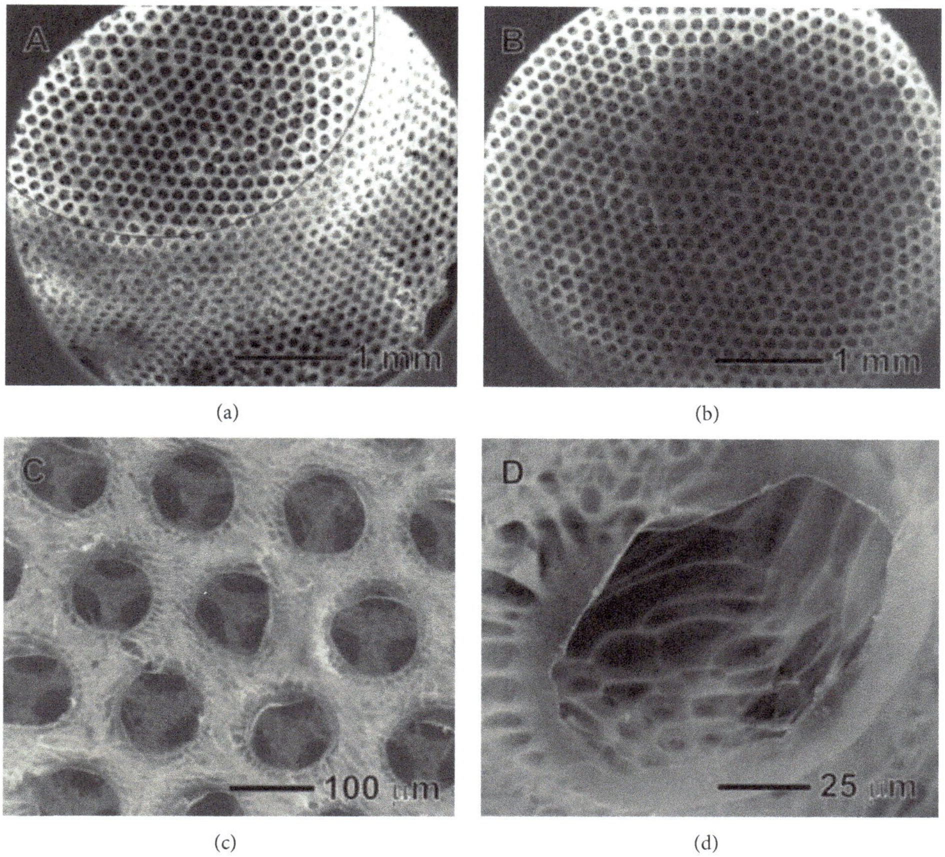

FIGURE 14: SEM images of the chitosan inverse opal scaffold: (a) side view, (b) top view, (c) a magnified view of the top surface, and (d) a magnified view of the side wall in a pore. The black line in (a) represents the edge of the scaffold [56].

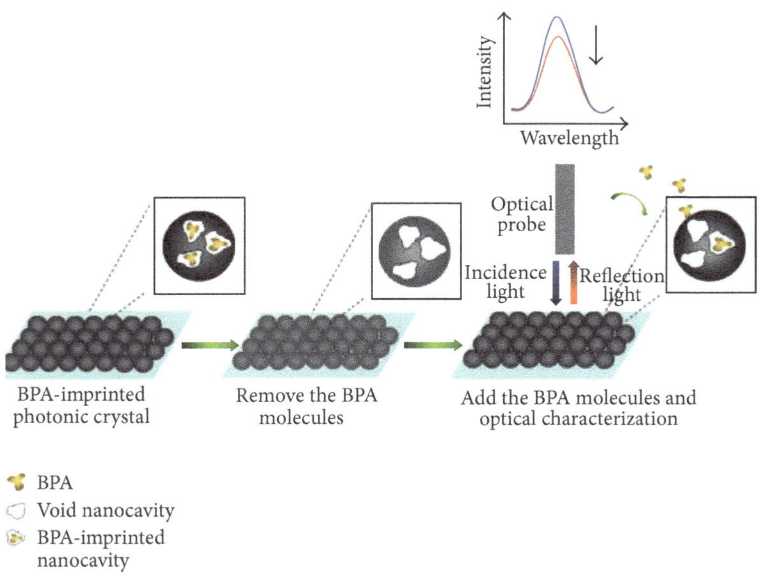

FIGURE 15: Experimental procedure for the detection of BPA using a photonic crystal sensor [62].

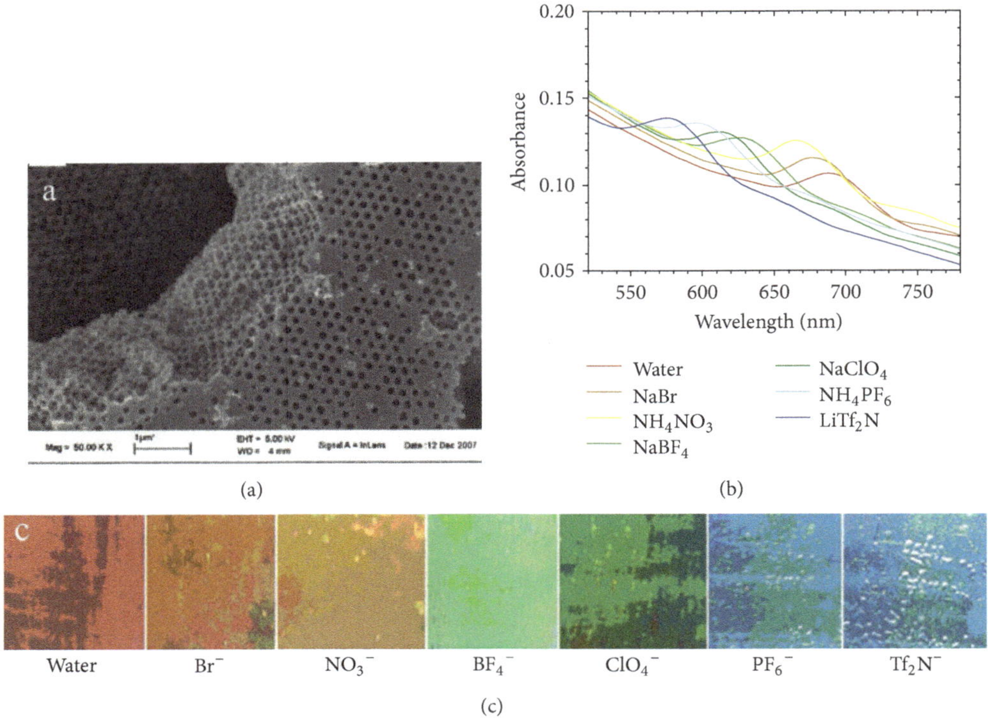

Figure 16: SEM image of the inverseopal photonic ionic liquid film (a). Stop-band shifts of the film upon soaking in various aqueous anionic solutions (b). Color changes of the film after anions aqueous solutions treatment (c) [63].

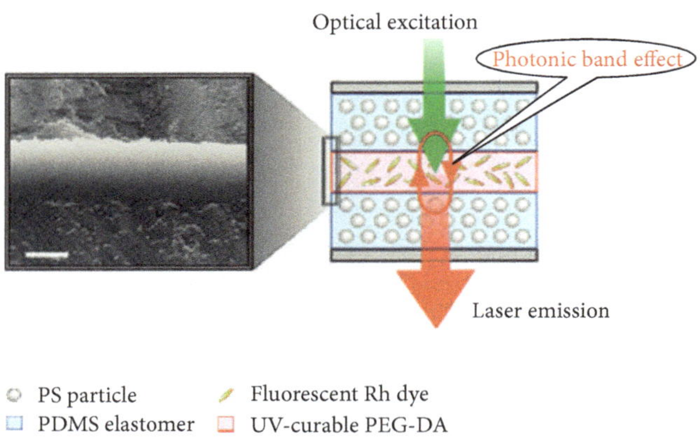

Figure 17: A schematic illustration of the colloidal crystal laser (CCL) proposed by Furumi et al. A light-emitting planar defect of Rhodamine (Rh)/PEG-DA is sandwiched between a pair of CC films of PS particles stabilized in PDMS elastomer. The left-hand picture represents a cross-sectional scanning electron microscopy (SEM) image of the magnified light-emitting planar defect between the PS/PDMS CC films. The white scale bar represents 2 μm [18].

introducing a light emitting planar defect based on fluorescent dye (rhodamine 640) dispersed in a poly(ethylene glycol) diacrylate film in order to obtain a laser effect, resulting in sandwich-type structure (Figure 17).

Therefore, the plastic film through excitation with a green wavelength affords a red laser emission as suggested in Figure 18.

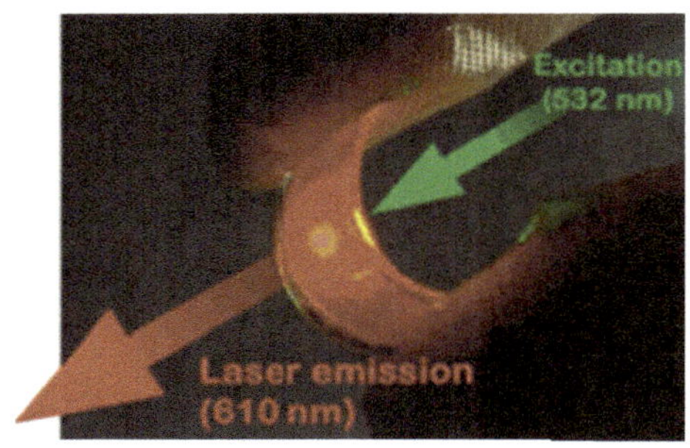

Figure 18: Laser effect of the plastic film [18].

Photonic crystal microfibers were obtained by Míguez et al. [64] using an inverted silicon colloidal crystal. First, in rectangular microchannels, colloidal particles of silica ranging from 250 nm to 900 nm were self-assembled in FCC structure. After a coating process of the silica spheres in order to control the connectivity between the particles, the microchannels are infiltrated with silicon by a chemical vapor deposition method. The silicon fills the interstitial spaces from the silica colloidal structure. In order to obtain the 3D silicon inverse structure the silica particles are removed, and the resulted microfibers are investigated by SEM and optical microscopy (Figure 19).

In Figure 19(a), the presented microfibers have several micrometers in length. An interesting optical behavior of the microfibers, which due to the various degrees of silicon infiltration exhibit different modulations of the dielectric

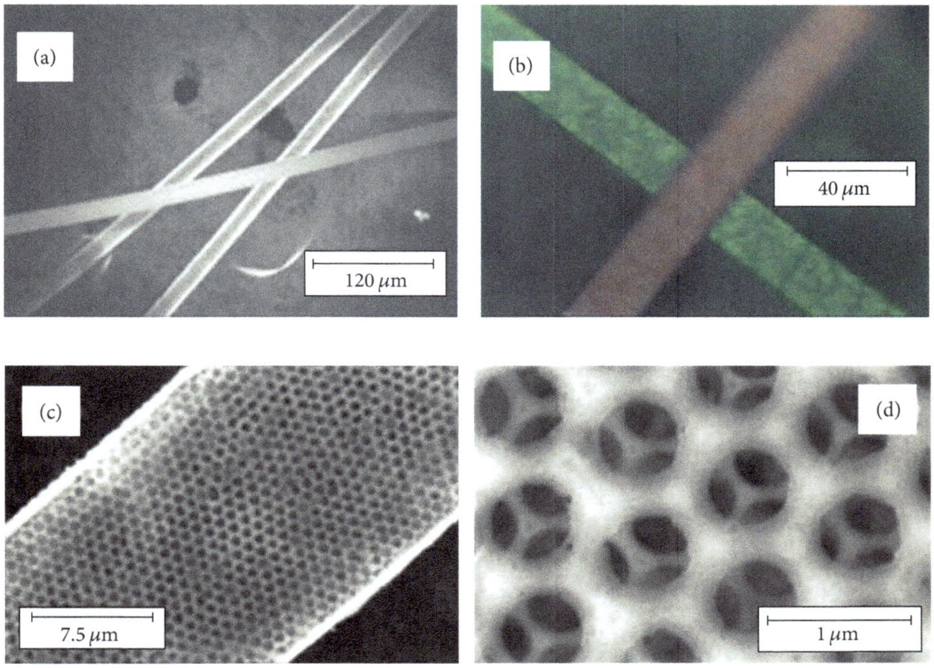

FIGURE 19: Images of inverse silicon microfibers. (a) SEM image of a bunch of fibers collected using a sticky carbon tape; (b) optical picture of two fibers presenting a different degree of infiltration, thus presenting different colors; ((c), (d)) details of the bottom surface of a free standing inverted colloidal crystal fiber showing the high degree of connectivity between the spherical cavities [64].

constant in the structure, thus resulting in distinct reflected colors, is illustrated in Figure 19(b) [64].

Compared to inorganic fibers, this method provided a simple synthesis procedure for obtaining microfibers with higher flexibility and controllable optical properties. As the authors suggested [64], the optical properties of the synthesized photonic crystal microfibers could make them suitable in future for optical components in microcomputers replacing the conventional semiconductor nanowires.

In conclusion, the aim of this review has been to briefly present the concept of photonic crystals and the advantages of obtaining these structures using polymeric materials. The property of self-assembling remains one of the key characteristics that motivate the use of these materials. Control of the self-assembly process as well as of the morphology attained allows the elaboration of new materials capable of interacting with light and able to be incorporated into novel devices.

The doping with various dyes, the interaction with inorganic particles in hybrid systems innovative synthesis techniques for specific applications (molecularly imprinted polymers, synthesis in the presence of ionic liquids) constitute new topics for novel and innovative applications based on polymeric colloidal particles and photonic crystals.

Acknowledgment

The authors would like to acknowledge thank for the financial support provided by the National Authority for Scientific Research from the Ministry of Education, Research and Youth of Romania through the PN-II-PT-PCCA-2011-3.2-0042-RPETUM Project.

References

[1] Y. Xia, B. Gates, Y. Yin, and Y. Lu, “Monodispersed colloidal spheres: old materials with new applications,” *Advanced Materials*, vol. 12, no. 10, pp. 693–713, 2000.

[2] Y. Chen, W. T. Ford, N. F. Materer, and D. Teeters, “Conversion of colloidal crystals to polymer nets: turning latex particles inside out,” *Chemistry of Materials*, vol. 13, no. 8, pp. 2697–2704, 2001.

[3] J. Zhang, Z. Sun, and B. Yang, “Self-assembly of photonic crystals from polymer colloids,” *Current Opinion in Colloid and Interface Science*, vol. 14, no. 2, pp. 103–114, 2009.

[4] S. A. Rinne, F. García-Santamaría, and P. V. Braun, “Embedded cavities and waveguides in three-dimensional silicon photonic crystals,” *Nature Photonics*, vol. 2, no. 1, pp. 52–56, 2008.

[5] A. Herzog Cardoso, C. A. P. Leite, M. E. D. Zaniquelli, and F. Galembeck, “Easy polymer latex self-assembly and colloidal crystal formation: the case of poly[styrene-co-(2-hydroxyethyl methacrylate)],” *Colloids and Surfaces A*, vol. 144, no. 1–3, pp. 207–217, 1998.

[6] L. Rayleigh, “XXVI. On the remarkable phenomenon of crystalline reflexion described by Prof. Stokes,” *Philosophical Magazine Series 5*, vol. 26, no. 160, pp. 256–265, 1888.

[7] E. Yablonovitch, “Inhibited spontaneous emission in solid-state physics and electronics,” *Physical Review Letters*, vol. 58, no. 20, pp. 2059–2062, 1987.

[8] S. John, “Strong localization of photons in certain disordered dielectric superlattices,” *Physical Review Letters*, vol. 58, no. 23, pp. 2486–2489, 1987.

[9] G. I. N. Waterhouse and M. R. Waterland, "Opal and inverse opal photonic crystals: fabrication and characterization," *Polyhedron*, vol. 26, no. 2, pp. 356–368, 2007.

[10] X. Zhao, Y. Cao, F. Ito et al., "Colloidal crystal beads as supports for biomolecular screening," *Angewandte Chemie*, vol. 45, no. 41, pp. 6835–6838, 2006.

[11] E. Rusen, A. Mocanu, A. Diacon, and B. Marculescu, "Fluorescence enhancement of rhodamine B in the presence of photonic crystal heterostructures," *Journal of Physical Chemistry C*, vol. 115, no. 30, pp. 14947–14953, 2011.

[12] Y. Zhao, X. Zhao, and Z. Gu, "Photonic crystals in bioassays," *Advanced Functional Materials*, vol. 20, no. 18, pp. 2970–2988, 2010.

[13] A. Mocanu, B. Marculescu, R. Somoghi, F. Miculescu, C. Boscornea, and I. C. Stancu, "Fluorescence enhancement for the complex PAMAM-BSA in the presence of photonic crystal heterostructures," *Colloids and Surfaces A*, vol. 392, no. 1, pp. 288–293, 2011.

[14] A. C. Arsenault, D. P. Puzzo, A. Ghoussoub, I. Manners, and G. A. Ozin, "Development of photonic crystal composites for display applications," *Journal of the Society for Information Display*, vol. 15, no. 12, pp. 1095–1098, 2007.

[15] A. C. Arsenault, D. P. Puzzo, I. Manners, and G. A. Ozin, "Photonic-crystal full-colour displays," *Nature Photonics*, vol. 1, no. 8, pp. 468–472, 2007.

[16] Z. Ye, J. M. Park, K. Constant, T. G. Kim, and K. M. Ho, "Photonic crystal: energy-related applications," *Journal of Photonics for Energy*, vol. 2, no. 1, Article ID 021012, 2012.

[17] J. R. Lawrence, Y. Ying, P. Jiang, and S. H. Foulger, "Dynamic tuning of organic lasers with colloidal crystals," *Advanced Materials*, vol. 18, no. 3, pp. 300–303, 2006.

[18] S. Furumi, H. Fudouzi, H. T. Miyazaki, and Y. Sakka, "Flexible polymer colloidal-crystal lasers with a light-emitting planar defect," *Advanced Materials*, vol. 19, no. 16, pp. 2067–2072, 2007.

[19] J. Ge and Y. Yin, "Responsive photonic crystals," *Angewandte Chemie*, vol. 50, no. 7, pp. 1492–1522, 2011.

[20] C. I. Aguirre, E. Reguera, and A. Stein, "Tunable colors in opals and inverse opal photonic crystals," *Advanced Functional Materials*, vol. 20, no. 16, pp. 2565–2578, 2010.

[21] S. Furumi, H. Fudouzi, and T. Sawada, "Self-organized colloidal crystals for photonics and laser applications," *Laser and Photonics Reviews*, vol. 4, no. 2, pp. 205–220, 2010.

[22] C. T. Chan and J. A. Yeh, "Tunable photonic crystal based on capillary attraction and repulsion," *Optics Express*, vol. 18, no. 20, pp. 20894–20899, 2010.

[23] S. S. Mishra and V. K. Singh, "Comparative study of fundamental properties of honey comb photonic crystal fiber at 1.55 μm wavelength," *Journal of Microwaves, Optoelectronics and Electromagnetic Applications*, vol. 10, no. 2, pp. 343–354, 2011.

[24] Y. Chen and S. Sajjadi, "Particle formation and growth in ab initio emulsifier-free emulsion polymerisation under monomer-starved conditions," *Polymer*, vol. 50, no. 2, pp. 357–365, 2009.

[25] D. Suzuki, J. G. McGrath, H. Kawaguchi, and L. A. Lyon, "Colloidal crystals of thermosensitive, core/shell hybrid microgels," *Journal of Physical Chemistry C*, vol. 111, no. 15, pp. 5667–5672, 2007.

[26] X. Xu and S. A. Asher, "Synthesis and utilization of monodisperse hollow polymeric particles in photonic crystals," *Journal of the American Chemical Society*, vol. 126, no. 25, pp. 7940–7945, 2004.

[27] S. M. Klein, V. N. Manoharan, D. J. Pine, and F. F. Lange, "Preparation of monodisperse PMMA microspheres in nonpolar solvents by dispersion polymerization with a macromonomeric stabilizer," *Colloid and Polymer Science*, vol. 282, no. 1, pp. 7–13, 2003.

[28] J. W. Kim, J. H. Ryu, and K. D. Suh, "Monodisperse micron-sized macroporous poly(styrene-co-divinylbenzene) particles by seeded polymerization," *Colloid and Polymer Science*, vol. 279, no. 2, pp. 146–152, 2001.

[29] X. Wu, A. Yamilov, X. Liu et al., "Ultraviolet photonic crystal laser," *Applied Physics Letters*, vol. 85, no. 17, pp. 3657–3659, 2004.

[30] E. A. Wood, *Crystals and Light: An Introduction to Optical Crystallography*, Dover Publications, 1977.

[31] E. Rusen, A. Mocanu, C. Corobea, and B. Marculescu, "Obtaining of monodisperse particles through soap-free polymerization in the presence of C60," *Colloids and Surfaces A*, vol. 355, no. 1–3, pp. 23–28, 2010.

[32] H. Fudouzi, "Novel coating method for artificial opal films and its process analysis," *Colloids and Surfaces A*, vol. 311, no. 1–3, pp. 11–15, 2007.

[33] H. Fudouzi, "Fabricating high-quality opal films with uniform structure over a large area," *Journal of Colloid and Interface Science*, vol. 275, no. 1, pp. 277–283, 2004.

[34] X. He, Y. Thomann, R. J. Leyrer, and J. Rieger, "Iridescent colors from films made of polymeric core-shell particles," *Polymer Bulletin*, vol. 57, no. 5, pp. 785–796, 2006.

[35] M. Allard, E. H. Sargent, E. Kumacheva, and O. Kalinina, "Characterization of internal order of colloidal crystals by optical diffraction," *Optical and Quantum Electronics*, vol. 34, no. 1–3, pp. 27–36, 2002.

[36] P. Jiang, G. N. Ostojic, R. Narat, D. M. Mittleman, and V. L. Colvin, "The fabrication and bandgap engineering of photonic multilayers," *Advanced Materials*, vol. 13, no. 6, pp. 389–393, 2001.

[37] H. Fudouzi, "Optical properties caused by periodical array structure with colloidal particles and their applications," *Advanced Powder Technology*, vol. 20, no. 5, pp. 502–508, 2009.

[38] S. Reculusa, P. Massé, and S. Ravaine, "Three-dimensional colloidal crystals with a well-defined architecture," *Journal of Colloid and Interface Science*, vol. 279, no. 2, pp. 471–478, 2004.

[39] H. Zhou, S. Xu, Z. Sun, X. Du, and J. Xie, "Rapid determination of colloidal crystal's structure by reflection spectrum," *Colloids and Surfaces A*, vol. 375, no. 1–3, pp. 50–54, 2011.

[40] A. Emoto, E. Uchida, and T. Fukuda, "Fabrication and optical properties of binary colloidal crystal monolayers consisting of micro- and nano-polystyrene spheres," *Colloids and Surfaces A*, vol. 396, pp. 189–194, 2012.

[41] Z. Z. Gu, H. Chen, S. Zhang, L. Sun, Z. Xie, and Y. Ge, "Rapid synthesis of monodisperse polymer spheres for self-assembled photonic crystals," *Colloids and Surfaces A*, vol. 302, no. 1–3, pp. 312–319, 2007.

[42] E. Yablonovitch, T. J. Gmitter, and K. M. Leung, "Photonic band structure: the face-centered-cubic case employing nonspherical atoms," *Physical Review Letters*, vol. 67, no. 17, pp. 2295–2298, 1991.

[43] W. H. Wong, K. K. Liu, K. S. Chan, and E. Y. B. Pun, "Polymer devices for photonic applications," *Journal of Crystal Growth*, vol. 288, no. 1, pp. 100–104, 2006.

[44] P. S. J. Russell, "Photonic-crystal fibers," *Journal of Lightwave Technology*, vol. 24, no. 12, pp. 4729–4749, 2006.

[45] K. Hansen and H. Imam, "Photonic crystal fiber," *Optik & Photonik*, vol. 5, no. 2, pp. 37–41, 2010.

[46] Z. Z. Gu, H. Xu, P. Wu, C. Zhu, and A. Elbaz, "Photonic crystal for gas sensing," *Journal of Materials Chemistry C*, 2013.

[47] Z. Cai, Y. J. Liu, J. Teng, and X. Lu, "Fabrication of large domain crack-free colloidal crystal heterostructures with superposition bandgaps using hydrophobic polystyrene spheres," *ACS Applied Materials & Interfaces*, vol. 4, no. 10, pp. 5562–5569, 2012.

[48] H. Huang, J. Chen, Y. Yu, Z. Shi, H. Möhwald, and G. Zhang, "Controlled gradient colloidal photonic crystals and their optical properties," *Colloids and Surfaces A*, vol. 428, pp. 9–17, 2013.

[49] Y. Li, Z. Sun, J. Zhang et al., "Polystyrene@TiO_2 core-shell microsphere colloidal crystals and nonspherical macro-porous materials," *Journal of Colloid and Interface Science*, vol. 325, no. 2, pp. 567–572, 2008.

[50] J. H. Moon, G. R. Yi, and S. M. Yang, "Fabrication of hollow colloidal crystal cylinders and their inverted polymeric replicas," *Journal of Colloid and Interface Science*, vol. 287, no. 1, pp. 173–177, 2005.

[51] O. Kalinina and E. Kumacheva, "Polymeric nanocomposite material with a periodic structure," *Chemistry of Materials*, vol. 13, no. 1, pp. 35–38, 2001.

[52] G. Cao and C. J. Brinker, *Annual Review of Nano Research*, World Scientific, 2006.

[53] M. Lanata, M. Cherchi, A. Zappettini, S. M. Pietralunga, and M. Martinelli, "Titania inverse opals for infrared optical applications," *Optical Materials*, vol. 17, no. 1-2, pp. 11–14, 2001.

[54] C. Lü, C. Guan, Y. Liu, Y. Cheng, and B. Yang, "PbS/polymer nanocomposite optical materials with high refractive index," *Chemistry of Materials*, vol. 17, no. 9, pp. 2448–2454, 2005.

[55] T. Huang, Q. Zhao, J. Xiao, and L. Qi, "Controllable self-assembly of pbs nanostars into ordered structures: close-packed arrays and patterned arrays," *ACS Nano*, vol. 4, no. 8, pp. 4707–4716, 2010.

[56] S. W. Choi, J. Xie, and Y. Xia, "Chitosan-based inverse opals: three-dimensional scaffolds with uniform pore structures for cell culture," *Advanced Materials*, vol. 21, no. 29, pp. 2997–3001, 2009.

[57] K. Lee and S. A. Asher, "Photonic crystal chemical sensors: pH and ionic strength," *Journal of the American Chemical Society*, vol. 122, no. 39, pp. 9534–9537, 2000.

[58] J. Ballato and A. James, "A ceramic photonic crystal temperature sensor," *Journal of the American Ceramic Society*, vol. 82, no. 8, pp. 2273–2275, 1999.

[59] H. Fudouzi and Y. Xia, "Colloidal crystals with tunable colors and their use as photonic papers," *Langmuir*, vol. 19, no. 23, pp. 9653–9660, 2003.

[60] H. C. Kolb, M. G. Finn, and K. B. Sharpless, "Click chemistry: diverse chemical function from a few good reactions," *Angewandte Chemie*, vol. 40, no. 11, pp. 2004–2021, 2001.

[61] S. Mandal, J. M. Goddard, and D. Erickson, "A multiplexed optofluidic biomolecular sensor for low mass detection," *Lab on a Chip*, vol. 9, no. 20, pp. 2924–2932, 2009.

[62] C. Guo, C. Zhou, N. Sai et al., "Detection of bisphenol A using an opal photonic crystal sensor," *Sensors and Actuators B*, vol. 166-167, pp. 17–23, 2012.

[63] X. Hu, J. Huang, W. Zhang, M. Li, C. Tao, and G. Li, "Photonic ionic liquids polymer for naked-eye detection of anions," *Advanced Materials*, vol. 20, no. 21, pp. 4074–4078, 2008.

[64] H. Míguez, S. M. Yang, N. Tétreault, and G. A. Ozin, "Oriented free-standing three-dimensional silicon inverted colloidal photonic crystal microfibers," *Advanced Materials*, vol. 14, no. 24, pp. 1805–1808, 2002.

[65] E. Rusen, A. Mocanu, and B. Marculescu, "Obtaining of monodisperse particles through soap-free and seeded polymerization, respectively, through polymerization in the presence of C 60," *Colloid and Polymer Science*, vol. 288, no. 7, pp. 769–776, 2010.

[66] D. Gaillot, T. Yamashita, and C. J. Summers, "Photonic band gaps in highly conformal inverse-opal based photonic crystals," *Physical Review B*, vol. 72, no. 20, Article ID 205109, 2005.

Mechanical Behavior of a Series of Copolyester Blends near the Glass Transition: Monotonic and Load-Hold Behavior in Compression

Gurucharan Chandrasekaran and Rebecca B. Dupaix

Mechanical and Aerospace Engineering Department, The Ohio State University, Columbus, OH 43210, USA

Correspondence should be addressed to Rebecca B. Dupaix, dupaix.1@osu.edu

Academic Editor: Wen Fu Lee

Monotonic loading tests were conducted on five commercial blends of poly(ethylene terephthalate) (PET) and poly(1,4-cyclohexylenedimethylene terephthalate) (PCT) at temperatures of 90°C and 100°C and strain rates of 0.1/s, 0.05/s, and 0.005/s in uniaxial and plane strain compression. On comparing the mechanical behavior of the five materials, it was found that the behavior of the low-PCT content materials was different from the high-PCT content materials only at conditions that favored strain-induced crystallization, particularly in plane strain compression. Load-hold tests were also conducted on three of the blends with similar results to the monotonic tests. Material differences were only pronounced at certain conditions, and in these cases the low-PCT content materials showed increased strain hardening after the hold period while the high-PCT content material did not. Therefore, it was found that the addition of a hold period was not exclusively required to observe differences in the crystallizable materials over the noncrystallizing blends. The increased strain hardening likely associated with crystallization in PET was only observed when the following conditions were met: (i) strain rates of 0.1/s and above, (ii) temperatures of 90°C–100°C, (iii) plane strain compression, and (iv) after a certain level of deformation.

1. Introduction

Poly(ethylene terephthalate) (PET) is a thermoplastic polymer that is extensively used in applications such as drawn fibers, bottles, and stretched films. PET has the ability to crystallize when deformed at temperatures slightly above the glass transition (T_g) by a phenomenon called stress- or strain-induced crystallization. When it undergoes strain-induced crystallization, the density, stiffness, resistance to gas permeability, and dimensional stability of the material increase without affecting its optical transparency, because of the very small (nanoscale) size of the crystallites. As a result, PET is frequently processed in the range of 10 to 20°C above T_g. The crystallization process and the mechanical behavior of PET above T_g are highly dependent on temperature, strain rate, and the mode of deformation. It is difficult to experimentally isolate these dependencies and determine their effects, particularly for complex 3 D processes, and the need exists for reliable material models that can be used in finite element simulations in this temperature regime. Any potential model must capture the high temperature, strain rate, and strain state-dependent behavior of PET. In addition to these dependencies, the material model should also incorporate the effect of strain-induced crystallization to be able to effectively simulate a manufacturing process that involves PET.

Previous work has largely involved conducting experiments at simple monotonic load conditions. Also, the issue of occurrence of strain-induced crystallization and its effect on the strain-hardening behavior of PET at large strains continues to be unclear. One previous study attempted to isolate effects of strain-induced crystallization by comparing PET3.5 (3.5% PCT) and PETG (31% PCT) [1]. This work looks to build on that previous research and address these open issues by (i) conducting monotonic tests to carry out a comparative study of the mechanical behavior of five different PET-PCT blends and (ii) conducting tests with nonmonotonic load histories to gain a better understanding of the factors that influence the occurrence and effects of crystallization.

2. Background

2.1. Experimental Background. Various experimentalists have studied the effects of strain-induced crystallization on the behavior of PET through mechanical and thermal characterization. Mechanical characterization has included uniaxial tension, biaxial tension, uniaxial compression, and plane strain compression tests. Thermal characterization has included the use of differential scanning calorimetry (DSC) to estimate the crystallinity content as a function of temperature, strain rate, strain state, and final strain. Wide-angle X-ray diffraction (WAXD) measurements are also used to determine the nature of the evolved crystallinity and the development of crystal texture with temperature, strain rate, strain state, and final strain.

Zaroulis and Boyce [2] conducted compression tests and DSC analysis on initially amorphous PET to determine the contribution of molecular orientation versus strain-induced crystallization to the significant strain-hardening behavior of PET observed at large strains. These tests were conducted up to T_g over a wide range of strain rates. They inferred from the results that strain hardening was caused solely by orientation in uniaxial compression and by both orientation and crystallization in plane strain compression. Llana and Boyce [3] added to this experimental work by performing mechanical tests and microstructural analyses on PET above T_g. They observed from the mechanical tests, that the stress-strain behavior of PET mainly consisted of four distinct features: (1) a relatively high initial stiffness, (2) a rollover to flow rather than a distinctive yield stress, (3) a moderate strain hardening with increasing strain, and (4) dramatic strain hardening at large strains. From the DSC results they found that the crystallinity content in PET increases with decreasing temperature and increasing strain rate. Using WAXD results, they also found the evolution of the crystallographic texture in both the strain states to be consistent with the molecular orientation of the respective strain state. Therefore, these tests established the fact that both the mechanical behavior and the structural evolution of PET depend on temperature, strain rate, and strain state above the glass transition temperature.

Complicating the ability to directly correlate mechanical experimental conditions with postexperimentally measured crystallinity is the practical difficulty of quenching a sample before crystallization occurs due to the brief annealing period at the end of deformation. As a result, it is difficult to determine whether crystallization occurs during or after deformation. Blundell et al. [4] used in situ experiments in which X-ray patterns were recorded during the deformation process to gain a better understanding of the occurrence and effects of crystallization during the large strain deformation of PET. By conducting tensile tests over a wide range of rates (0.05–12/s), they found that there are three regimes of PET deformation. (1) At draw rates faster than the rate of chain retraction, crystallization is delayed until the end of draw. (2) At draw rates that are slower than the chain reptation rate, no crystallization occurs. (3) For draw rates between the two above-mentioned rates, crystallization can start before the end of draw. Other important conclusions drawn from their work were that both orientation and crystallization occur only within a certain range of temperature and draw ratio and the presence of oriented noncrystalline polymer segments is essential for crystallization to occur.

Work by Dupaix and Boyce [1] looked at comparing the stress-strain behavior of PET with PETG, which is a noncrystallizing copolymer of PET. The comparison in behavior was done across a wide range of temperatures (25–110°C) and strain rates (.005–1.0/s) in both uniaxial and plane strain compression. Their results clearly indicated that at most of the conditions, the mechanical behavior of PET and PETG was similar and that they exhibit similar trends with respect to temperature, strain rate, and strain state, even though PET is a crystallizing polymer and PETG is not. This led to the suggestion that the dramatic hardening that occurs at large compressive strains in PET is mainly caused by increased molecular orientation and not necessarily by strain-induced crystallization.

2.2. Constitutive Modeling Background. Due to its wide industrial use, various research groups have tried to model the behavior of PET under deformation near the glass transition temperature. Several of the key models are summarized here.

Buckley et al. [5] created a physically based three-dimensional model to capture the behavior of amorphous polymers near the glass transition. This model was formulated on the idea that the total stress on deformation would be the sum of the individual stress components arising from bond stretching and conformational change. For the bond distortion part, the model used linear elasticity and Eyring viscous flow; for the conformation part, the model used the Edwards-Vilgis entropy function.

Boyce et al. [6] proposed a constitutive model for PET that captured its behavior above T_g across a wide range of strain rates in uniaxial and plane strain compression. In this model the total resistance offered by the material to deformation is broken down into (1) the resistance that exists between the polymer molecules due to secondary bonding, and (2) the network resistance that occurs due to the underlying network of polymer chains. Crystallization, which was considered to provide an additional strain-hardening mechanism, was incorporated through increasing the shear resistance after a critical network stretch was reached.

Another model was proposed by Vigny et al. [7] that incorporates several of the same features: thermally activated viscoplastic flow, the eight-chain model for orientation and hardening at large strains, and elements that stiffen with the evolution of crystallinity.

Dupaix and Boyce [8, 9] built on the Boyce model [6] to capture the strong dependence of the stress-strain behavior of the noncrystallizing PETG on temperature, strain rate, and strain state, from below to above T_g and up to very large strains. A physical parameter which is independent of temperature and strain rate, called the orientation parameter, was used as the governing variable for molecular relaxation and especially for its cessation in that work.

Work by Ahzi and coworkers [10, 11] has greatly advanced the modeling of strain-induced crystallization in PET through a series of recent papers. Their earlier model built on the Boyce model [6], adding a separate branch to account for the evolving semicrystalline phase in both an upper bound and lower bound configuration [10]. Later, they extended this to a self-consistent model [11]. In this work, they used a model by Flory [12] to predict the onset of crystallization as a function of temperature and the stretch in the material.

Dupaix and Krishnan [13] formulated a hybrid model to account for the behavior of PET and PETG under both simple and complex (load-hold) conditions. This hybrid model was the model of Dupaix and Boyce modified to include strain-induced crystallization following the upper bound model of Ahzi. In addition, criteria to delay the onset of strain-induced crystallization were included. The criteria restricted strain-induced crystallization to occur only (1) after a particular level of deformation has been reached and (2) after the strain rate has slowed below a specific rate. However, no load-hold experimental data was available to compare with the simulation results obtained from that work.

3. Experimental Methods

3.1. Material. The five different blends of poly(ethylene terephthalate) (PET) and poly(1,4-cyclohexylenedimethylene terephthalate) (PCT) used in these experiments were supplied by Eastman Chemical Co. as 10 cm × 10 cm × 3.2 mm injection molded plaques. The composition of the five blends is PET00 (0% PCT), PET1.5 (1.5% PCT), PET3.5 (3.5% PCT), PET12 (12.5% PCT), and PETG (31% PCT). The materials are identical to those used in Chen et al. [14]. The glass transition temperature of these materials is approximately 77°C and there is very little variation in glass transition temperature among the materials. In addition the materials as received were optically clear and amorphous. While some small initial anisotropy arising from the fabrication of the plaques is unavoidable, we expect this to be fairly small and we avoided taking samples from areas such as near the edges or the runners.

For uniaxial compression tests, cylindrical specimens with a diameter of 12.3 mm and a height to diameter ratio of 0.2 were used. For plane strain compression tests, rectangular samples with a width and length of 10 mm and a height of 3.2 mm were used. Prior to testing, the specimens were stored in a desiccant chamber.

3.2. Experimental Setup and Method. An Instron 5869 electromechanical test frame was used to carry out the experiments. A 50 KN load cell recorded force data while an LVDT measured the distance between the compression platens during testing. The crosshead speed was controlled using a personal computer with Windows XP and Instron Bluehill software. The speed was constantly decreased throughout the course of the experiment in order to maintain a constant true strain rate in compression.

Temperature was controlled using an environmental chamber that was heated to the required temperature for one hour prior to any testing to obtain a uniform temperature distribution within the test chamber. Teflon sheets were placed between the compression platens and the specimen and also between the specimen and the channel die (in the case of plane strain) with a light amount of WD-40 lubricant between the Teflon and the steel fixtures in order to minimize the effects of friction. Care was taken not to expose the specimen to the lubricant. Each test sample was heated in the chamber for 20 minutes to ensure uniform sample temperature. Care was taken to ensure that all the test samples were subjected to the same thermal history.

4. Experimental Results and Discussions

4.1. Monotonic Tests. Monotonic tests were conducted at test temperatures of 90°C and 100°C and at strain rates of 0.1/s, 0.05/s, and 0.005/s in both uniaxial compression and plane strain compression for all five PET-PCT blends.

In compression, it is observed that all five materials exhibit the four distinct features of the stress strain curve that have been observed by previous researchers for PET3.5 and PETG above the glass transition temperature. These features include (1) a relatively high initial stiffness, (2) a rollover to flow rather than a distinct yield stress, (3) moderate strain hardening with increasing strain, and (4) dramatic strain hardening at large strains. Each of these features was seen to depend on both strain rate and temperature for all five materials.

Figure 1 shows a representative plot of uniaxial compression stress-strain data of the five materials. This plot is for 90°C and a strain rate of 0.1/s, with data at other temperatures and strain rates being very similar. Namely, in uniaxial compression, there is no significant difference in the stress-strain behavior of these materials, other than the fact that PET12 tends to be the softest of the five materials. Otherwise, any measured difference among the samples was similar to the difference measured between two samples of the same material.

Figure 2 shows the strain rate dependence of PET00 at a temperature of 90°C in uniaxial compression. Both the initial modulus and flow stress increase with increasing strain rate for all five materials. The subsequent strain hardening that occurs with increasing strain increases as the strain rate increases. The strain level at which the dramatic hardening occurs shifts to lower strains with increasing strain rate.

Figure 3 shows the temperature dependence of PET00 at a strain rate of 0.1/s in uniaxial compression. It was observed that the initial modulus, flow stress, and the subsequent strain hardening decrease with increasing temperature. Also, the dramatic strain hardening at large strains shifts to higher strains at higher temperatures. Though Figures 2 and 3 show only the temperature and strain rate dependence of PET00, the behavior of all five materials showed similar temperature and strain rate dependence.

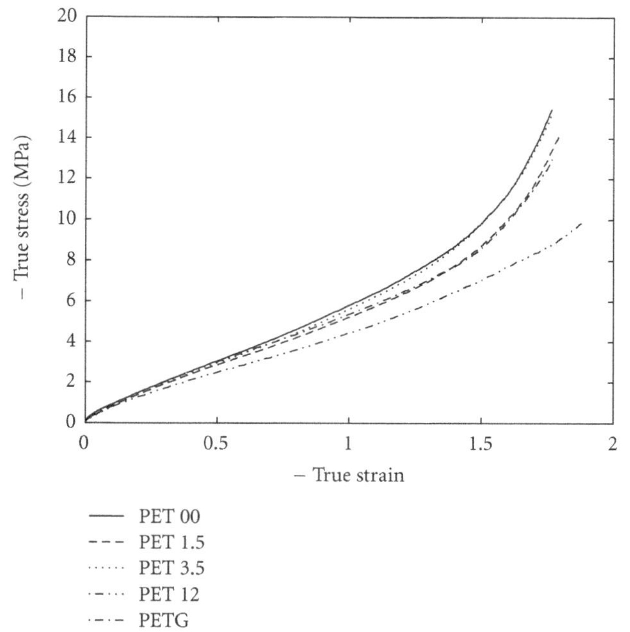

FIGURE 1: Material comparison, uniaxial compression at 90°C and 0.1/s.

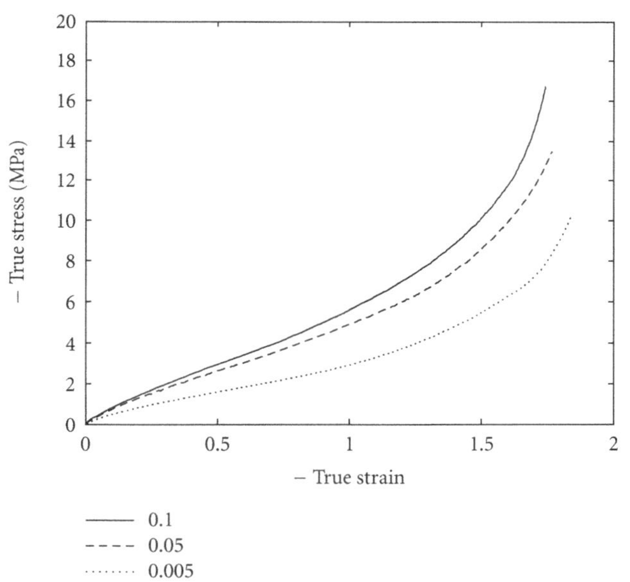

FIGURE 2: Strain rate comparison, PET00, and uniaxial compression at 90°C.

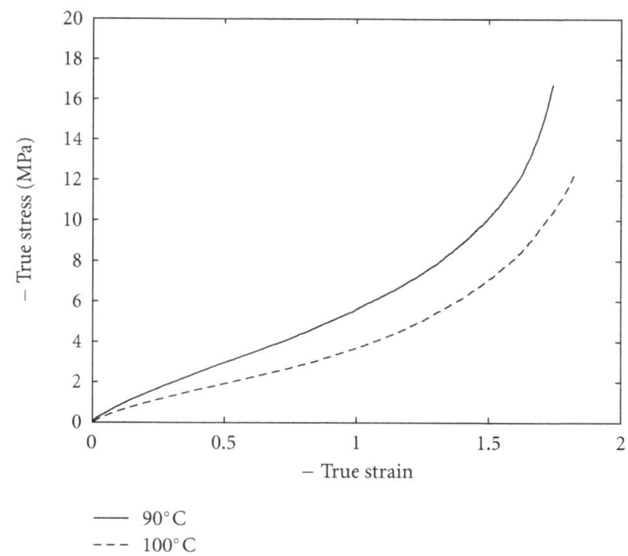

FIGURE 3: Temperature comparison, PET00, and uniaxial compression at 0.1/s.

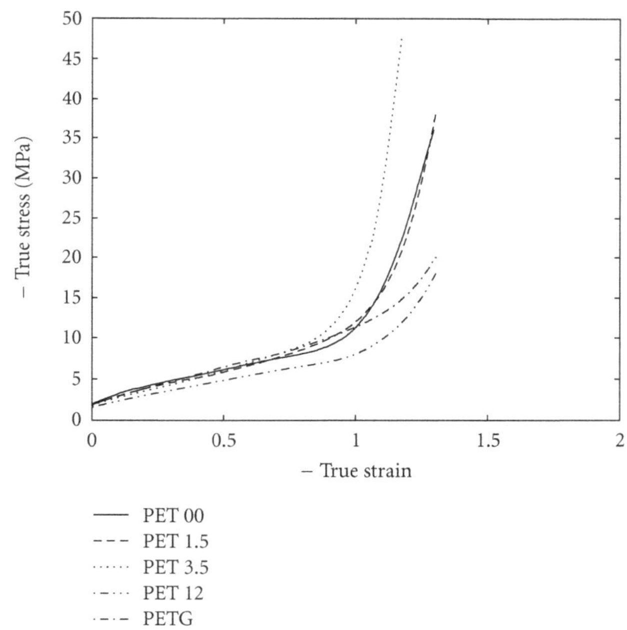

FIGURE 4: Material comparison, plane strain compression at 90°C and 0.1/s.

In Figures 4–6, the stress strain behavior of the five materials in plane strain compression is compared. From these curves it was observed that for a strain rate of 0.1/s and temperatures of 90°C and 100°C, PET00, PET1.5, and PET3.5 exhibited dramatic strain hardening at higher strains while PET12 and PETG did not. This could be attributed to the occurrence of strain-induced crystallization or a highly ordered mesophase in the three materials with low-PCT content. However, at the experimental condition of 90°C and 0.005/s (Figure 5), none of the materials exhibited any strain hardening at higher strains. This could be explained by the fact that at a slower rate of 0.005/s, molecular relaxation occurs more quickly, preventing the development of significant molecular orientation and also hindering strain-induced crystallization.

Figure 7 shows the strain state dependence of the mechanical behavior of PET00 above the glass transition temperature (90°C and 0.1/s). It was seen in these figures that the initial modulus, flow stress, and the subsequent strain hardening are much higher in plane strain compression

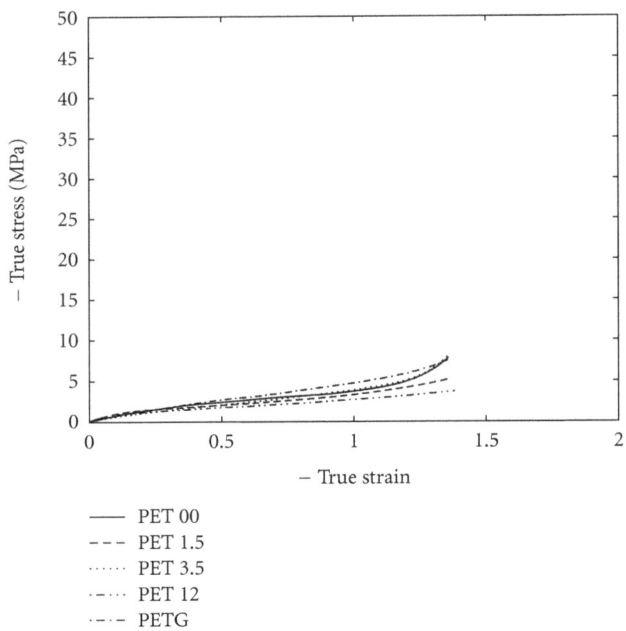

FIGURE 5: Material comparison, plane strain compression at 90°C and 0.005/s.

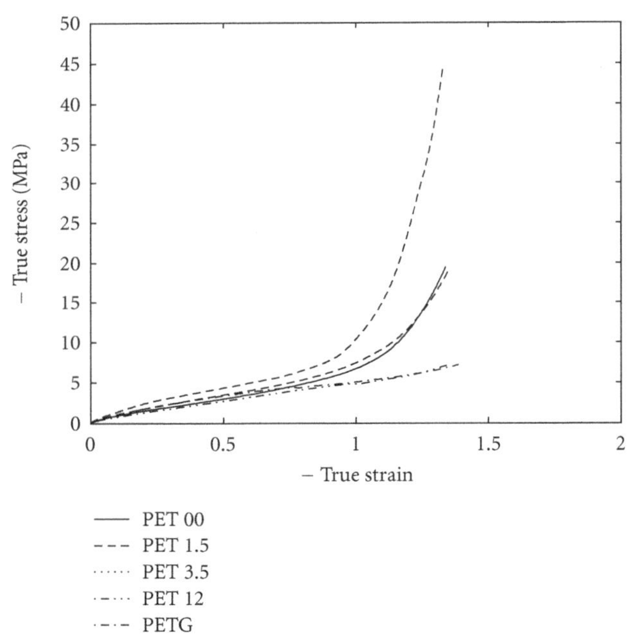

FIGURE 6: Material comparison, plane strain compression at 100°C and 0.1/s.

when compared to uniaxial compression for the same temperature and strain rate. The dramatic hardening that occurs at high strains also occurs at a lower strain level in plane strain compression than in uniaxial compression. The same trend was seen with the other materials. We also observed that the stress-strain curves exhibit the same dependence on strain rate and temperature in plane strain compression as in uniaxial compression (data not shown).

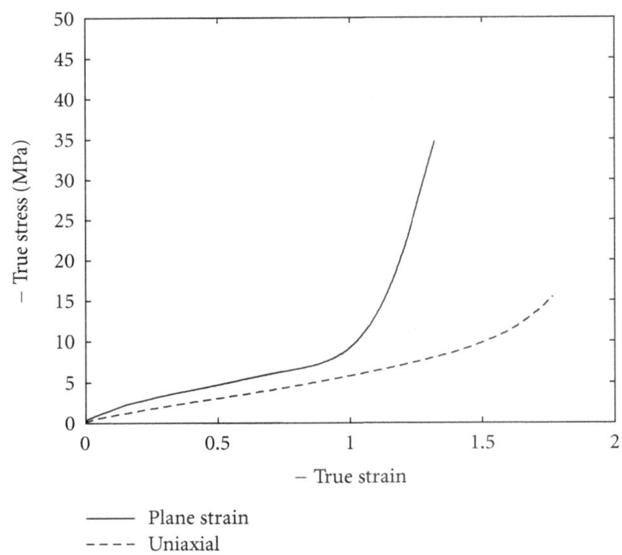

FIGURE 7: Strain state comparison, PET00, and 90°C and 0.1/s.

To conclude, PET00, PET1.5, and PET12 exhibit the same dependence on strain rate, temperature, and strain state as PET3.5 and PETG (studied in detail in previous research by Dupaix and Boyce [1]). In plane strain compression, it was seen that PET00 and PET1.5 behaved in a similar manner to PET3.5 while PET 12 behaved more like PETG. This was expected as PET00, PET1.5, and PET3.5 have lower PCT content while PET12 and PETG have relatively higher PCT content. It should also be noted that Dupaix and Boyce [1] found very little difference between PET and PETG in uniaxial compression and only saw significant differences in plane strain compression. Therefore, the finding in this work that the five materials behave similarly in uniaxial compression and exhibit differences only in plane strain compression is consistent with those results.

4.2. Load-Hold Tests. Crystallization is favored in PET when it is deformed at strain rates that are fast enough for molecular orientation to develop. However, at higher strain rates, the polymer chains have limited time to crystallize during the deformation. Therefore, it is hypothesized that deforming PET at a high strain rate and subsequently holding the material at a constant strain level would provide adequate molecular orientation during deformation and would also allow enough time for crystallization to occur during the hold period. This would also mean that PET and PETG should exhibit different behavior in load-hold tests.

In uniaxial compression, load-hold tests were conducted at temperatures of 90°C and 100°C and at strain rates of 0.1/s and 0.005/s. At each experimental condition, the deformation was paused and held at true strains of either −1.4 or −1.7. Plane strain compression tests were also conducted at the same strain rates and temperatures, but with the hold taking place at a true strain of −1.0, which provides the same chain stretch as holding a sample at −1.4 in uniaxial compression (chain stretch refers to the stretch along the diagonal of a unit cube aligned with the principal axes

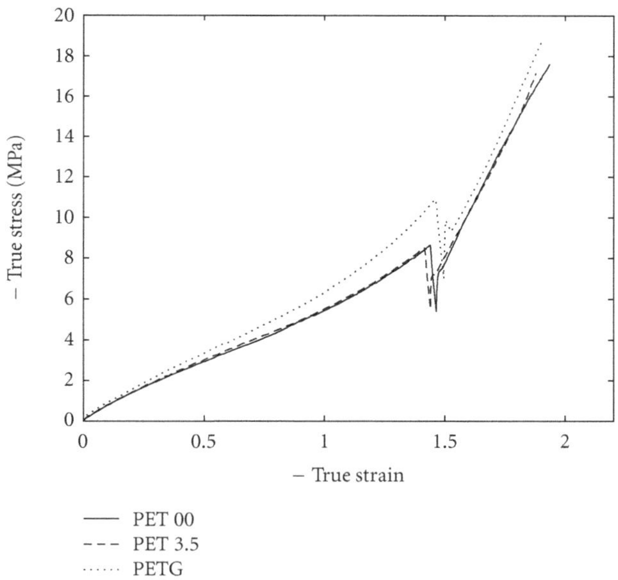

FIGURE 8: Material comparison, uniaxial compression at 90°C and 0.1/s, and hold at −1.4.

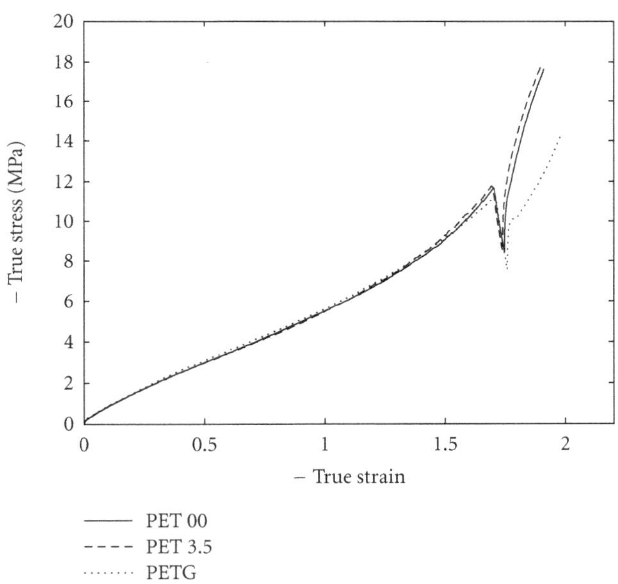

FIGURE 9: Material comparison, uniaxial compression at 90°C and 0.1/s, and hold at −1.7.

of deformation [15] and is used here as an effective stretch to better compare the amount of deformation a polymer sample would experience in two different deformation modes). Middleton et al. [16] observed that strain-induced crystallization occurs very rapidly after deformation (<1 sec). In this work, a hold of 20 seconds was chosen, which should be more than sufficient to allow crystallization to initiate.

Figure 8 shows a representative plot of uniaxial stress-strain behavior of the three materials at 90°C and a strain rate of 0.1/s, with a hold at a true strain of −1.4. Qualitatively similar results were observed at the other strain rates and

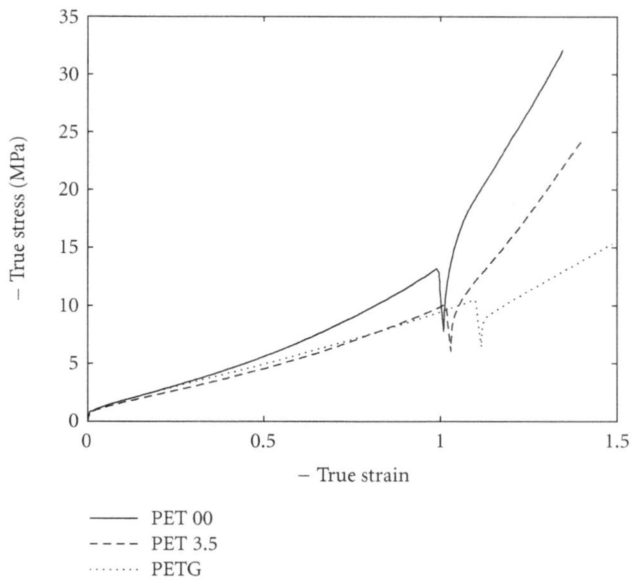

FIGURE 10: Material comparison, plane strain compression at 90°C and 0.1/s, and hold at −1.0.

temperature. Figure 9 shows the one exception in uniaxial compression. This figure shows data, also at 90°C and 0.1/s, but with the hold delayed until −1.7. In this one case, PET00 and PET3.5 show slightly increased hardening after the hold compared to PETG. Yet the difference does not seem significant enough to make a strong conclusion about the contribution of strain-induced crystallization.

In contrast, Figure 10 shows the stress strain behavior of the three materials in plane strain compression at 90°C and 0.1/s with a hold occurring at a true strain of −1.0. From these curves it was observed that at this temperature and strain rate, PET00 and PET3.5 exhibited dramatic strain hardening at higher strains while PETG did not. This could be attributed to the occurrence of strain-induced crystallization in PET00 and PET3.5 under these favorable conditions. This effect was still present but less pronounced at the other temperature and rates that were tested.

The experimental results indicate that even with the addition of a hold period, most of the test conditions did not show significant material differences, with the exception of one or two specific test conditions. The differences were most evident in plane strain compression, a result that was already observed in the monotonic results.

The conclusions we can draw are somewhat limited, in the absence of real-time data on crystallinity. However, whether crystallization begins during deformation or only upon holding/annealing, it appears that in uniaxial compression, the ability of the material to crystallize has little effect on the stress-strain curve, with or without a hold period. This is further supported by the strong similarity in strain-hardening behavior across all of the copolyesters tested. In plane strain compression, the ability of the material to crystallize has much more of an effect on the large strain stiffening, a result that is not appreciably changed with the addition of a hold period, though the hold period seems to cause the strain hardening to increase even further at particular conditions. Since the hold period has only a small effect, it probably implies that crystallization had already begun during the deformation. In uniaxial compression, strain-induced crystallization does not significantly affect the stress-strain behavior, but in plane strain compression it appears to, likely because of the single flow direction in plane strain compression that leads to a high degree of crystal alignment with that flow direction [3].

5. Crystallization Onset Criteria

We propose here some possible crystallization onset criteria based on these experimental results.

Amount of Deformation. Previous work has established that the polymer chains attain a highly oriented state after a significant level of deformation is attained which helps crystallization to occur. To better define the strain level at which crystallization occurs, tests were conducted with holds at true strains of −1.0, −0.9, −0.85, and −0.7. A temperature of 90°C, strain rate of −0.1/s, and plane strain compression were used, since these conditions showed the most strain hardening in the earlier tests. From Figure 11, it can be seen that the additional strain hardening occurs for holds at −0.9 and −1.0 and not for holds at −0.85 and −0.7. Therefore, it can be inferred that strain-induced crystallization first occurs for these experimental conditions at a true strain of −0.9.

To correlate the observed strain level with a more physical representation of deformation level, we use the chain stretch as an equivalent strain measure:

$$\lambda_{ch} = \frac{1}{\sqrt{3}}\sqrt{\lambda_1^2 + \lambda_2^2 + \lambda_3^2}, \tag{1}$$

where λ_1, λ_2, and λ_3 are the principal stretches.

The chain stretch corresponding to the true strain value of $\varepsilon = -0.9$ is determined to be $\lambda_{ch} = 1.55$. Therefore, strain-induced crystallization is taken to occur when the chain stretch reaches 1.55.

Condition 1: Crystallize if $\lambda_{ch} > 1.55$.

Mode of Deformation. Experimental results clearly showed that the additional strain hardening observed during large strain deformation of PET was predominantly present in plane strain compression (see Figures 8 and 10). This can be attributed to the occurrence of strain-induced crystallization which is favored by the unidirectional alignment of the polymer chains along the flow direction in plane strain compression. Therefore, this observation that crystallization occurs only in plane strain deformation should be incorporated into the constitutive model as an initiation condition. This is done using the fact that the principal stretches are different for the two different modes of deformation. In uniaxial compression, two principal stretches are greater than one (with the third principal stretch less than one). In plane strain compression, however, only one principal stretch is greater than one. Therefore, a criterion that prevents crystallization when two principal stretches are large will achieve

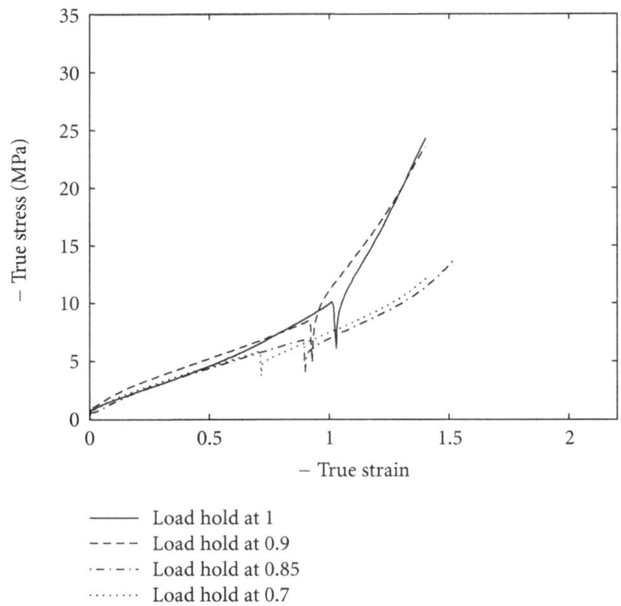

FIGURE 11: Stress-strain behavior of PET3.5 at 90°C and 0.1/s, plane strain compression with different hold strains.

our goal of only allowing crystallization during plane strain compression. Here, we arbitrarily choose 1.3 to prevent crystallization during uniaxial compression simulations.

Condition 2: Do not crystallize if both $\lambda_1 > 1.3$, $\lambda_2 > 1.3$.

Strain Rate. Strain rate clearly influences the occurrence of crystallization. Dupaix and Krishnan [13] proposed that PET should undergo a certain amount of deformation and the deformation should slow to a very low rate ($\dot{\gamma}_A < 0.009$/s) for strain-induced crystallization to occur during the deformation of PET. These conditions were based on the experimental investigations performed by Mahendrasingam et al. [17–19] and Middleton et al. [16].

Here, we observed that crystallization-caused strain hardening did not appear during testing at a strain rate of 0.005/s (lower than 0.009/s). As a result, if we want to restrict crystallization to only occur during a hold period, the critical strain rate must be lower than this; here we will use $\dot{\gamma}_A < 0.0009$/s as the criterion.

Condition 3: Crystallize if $\dot{\gamma}_A < 0.0009$/s.

Alternate Criteria to Consider. On comparing the monotonic test results with the load-hold test results, it was seen that the experimental conditions at which strain hardening occurs in PET00 and PET3.5 are the same in the monotonic tests and load-hold tests. Strain hardening was observed in PET00 and PET3.5 at temperatures of 90°C and 100°C and the fast strain rate of 0.1/s in the strain state of plane strain compression. This suggests that the onset of strain-induced crystallization during deformation does not necessarily depend upon the occurrence of a hold, but in all cases crystallization only significantly altered the stress-stiffening behavior at the fastest strain rate tested.

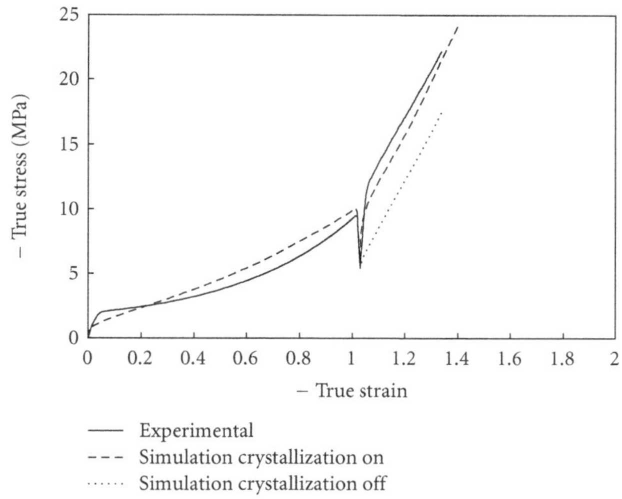

FIGURE 12: Comparison with experiment, PET3.5, 90°C, 0.1/s, plane strain compression.

Based on this evidence, the conditions required to cause strain-induced crystallization during deformation in PET may be more appropriately (1) fast strain rate (0.1/s and above), (2) deformation beyond a certain level ($\lambda_{ch} > 1.55$), and (3) mode of deformation that leads to a highly unidirectional molecular alignment (plane strain compression). An alternate approach would be to prescribe the chain stretch (λ_{ch}) to change according to rate (and possibly also temperature), similar to the approach used in previous models of strain-induced crystallization in PET [5, 6]. We leave these comparisons to future work and report here the results using our first proposed set of crystallization criteria.

6. Simulation Results and Discussions

To illustrate the effects of using these onset criteria in a constitutive model, the Dupaix-Krishnan [12] model was used with the proposed set of crystallization criteria.

Figure 12 shows a comparison of the experimental and simulated stress-strain curves (both with and without the crystallization part) of PET3.5 at a temperature of 90°C, a strain rate of −0.1/s in plane strain compression, and a hold period at a true strain of −1.0. The drop in the stress level at the start of the hold condition and the subsequent strain stiffening that occurs due to crystallization is captured very well when the crystallization part is included.

7. Conclusion and Future Work

7.1. Conclusions. Five PET-PCT blends (PET00, PET1.5, PET3.5, PET 12, and PETG) were tested over a range of temperatures and strain rates in uniaxial and plane strain compression. Results showed that under conditions favorable to strain-induced crystallization, the low-PCT content materials (PET00, PET1.5, and PET3.5) exhibited strain hardening at high strains while the high-PCT content materials (PET12 and PETG) did not.

Load-hold tests were conducted to verify the hypothesis that the hold condition was one of the criteria that initiated strain-induced crystallization during deformation in PET. The results showed that under experimental conditions that favored crystallization, PET3.5 exhibited dramatic strain hardening while PETG did not. Three significant observations were made based on these results and they were subsequently used to prescribe the values of the crystallization initiation criteria including the chain stretch, the drop in strain rate at the hold condition, and the combination of principal stretches used to identify the strain state.

However, on comparing the monotonic test results with the load-hold test results, it was seen that the experimental conditions at which strain hardening occurs in PET00 and PET3.5 are the same in the monotonic tests and the load-hold tests. This led to the conclusion that the onset of strain-induced crystallization during deformation may not depend upon the presence of a hold. The essential criteria for the occurrence of strain-induced crystallization were identified to be (1) fast strain rate (0.1/s and above), (2) deformation beyond a certain level ($\lambda_{ch} > 1.55$), (3) temperature of 90°C–100°C, and (4) mode of deformation that leads to a highly unidirectional molecular alignment (plane strain compression).

References

[1] R. B. Dupaix and M. C. Boyce, "Finite strain behavior of poly(ethylene terephthalate) (PET) and poly(ethylene terephthalate)-glycol (PETG)," *Polymer*, vol. 46, no. 13, pp. 4827–4838, 2005.

[2] J. S. Zaroulis and M. C. Boyce, "Temperature, strain rate, and strain state dependence of the evolution in mechanical behaviour and structure of poly(ethylene terephthalate) with finite strain deformation," *Polymer*, vol. 38, no. 6, pp. 1303–1315, 1997.

[3] P. G. Llana and M. C. Boyce, "Finite strain behavior of poly(ethylene terephthalate) above the glass transition temperature," *Polymer*, vol. 40, no. 24, pp. 6729–6751, 1999.

[4] D. J. Blundell, A. Mahendrasingam, C. Martin et al., "Orientation prior to crystallisation during drawing of poly(ethylene terephthalate)," *Polymer*, vol. 41, no. 21, pp. 7793–7802, 2000.

[5] C. P. Buckley, D. C. Jones, and D. P. Jones, "Hot-drawing of poly(ethylene terephthalate) under biaxial stress: application of a three-dimensional glass-rubber constitutive model," *Polymer*, vol. 37, no. 12, pp. 2403–2414, 1996.

[6] M. C. Boyce, S. Socrate, and P. G. Llana, "Constitutive model for the finite deformation stress-strain behavior of poly(ethylene terephthalate) above the glass transition," *Polymer*, vol. 41, no. 6, pp. 2183–2201, 2000.

[7] M. Vigny, A. Aubert, J. M. Hiver, M. Aboulfaraj, and C. G'Sell, "Constitutive viscoplastic behavior of amorphous PET during plane-strain tensile stretching," *Polymer Engineering and Science*, vol. 39, no. 12, pp. 2366–2376, 1999.

[8] R. B. Dupaix and M. C. Boyce, "Constitutive modeling of the finite strain behavior of amorphous polymers in and above the glass transition," *Mechanics of Materials*, vol. 39, no. 1, pp. 39–52, 2007.

[9] R. B. Dupaix, *Temperature and rate dependant finite strain behavior of poly (ethylene terephthalate) and poly (ethylene terephthalate)-glycol above the glass transition temperature*, Ph.D. thesis, MIT, 2003.

[10] S. Ahzi, A. Makradi, R. V. Gregory, and D. D. Edie, "Modeling of deformation behavior and strain-induced crystallization in poly(ethylene terephthalate) above the glass transition temperature," *Mechanics of Materials*, vol. 35, no. 12, pp. 1139–1148, 2003.
[11] A. Makradi, S. Ahzi, R. V. Gregory, and D. D. Edie, "A two-phase self-consistent model for the deformation and phase transformation behavior of polymers above the glass transition temperature: application to PET," *International Journal of Plasticity*, vol. 21, no. 4, pp. 741–758, 2005.
[12] P. J. Flory, "Thermodynamics of crystallization in high polymers. I. Crystallization induced by stretching," *Journal of Chemical Physics*, vol. 15, no. 6, pp. 397–408, 1947.
[13] R. B. Dupaix and D. Krishnan, "A constitutive model for strain-induced crystallization in poly(ethylene terephthalate) (PET) during finite strain load-hold simulations," *Journal of Engineering Materials and Technology*, vol. 128, no. 1, pp. 28–33, 2006.
[14] L. P. Chen, A. F. Yee, and E. J. Moskala, "The molecular basis for the relationship between the secondary relaxation and mechanical properties of a series of polyester copolymer glasses," *Macromolecules*, vol. 32, no. 18, pp. 5944–5955, 1999.
[15] E. M. Arruda and M. C. Boyce, "A three-dimensional constitutive model for the large stretch behavior of rubber elastic materials," *Journal of the Mechanics and Physics of Solids*, vol. 41, no. 2, pp. 389–412, 1993.
[16] A. C. Middleton, R. A. Duckett, I. M. Ward, A. Mahendrasingam, and C. Martin, "Real-time FTIR and WAXS studies of drawing behavior of poly(ethylene terephthalate) films," *Journal of Applied Polymer Science*, vol. 79, no. 10, pp. 1825–1837, 2001.
[17] A. Mahendrasingam, C. Martin, W. Fuller et al., "Effect of draw ratio and temperature on the strain-induced crystallization of poly (ethylene terephthalate) at fast draw rates," *Polymer*, vol. 40, no. 20, pp. 5553–5565, 1999.
[18] A. Mahendrasingam, D. J. Blundell, C. Martin et al., "Influence of temperature and chain orientation on the crystallization of poly(ethylene terephthalate) during fast drawing," *Polymer*, vol. 41, no. 21, pp. 7803–7814, 2000.
[19] A. Mahendrasingam, C. Martin, W. Fuller et al., "Observation of a transient structure prior to strain-induced crystallization in poly(ethylene terephthalate)," *Polymer*, vol. 41, no. 3, pp. 1217–1221, 2000.

Viscoelastic Properties of Mineral-Filled Poly(lactic acid) Composites

Adriana Gregorova,[1] Michal Machovsky,[2] and Rupert Wimmer[3]

[1] *Institute for Chemistry and Technology of Materials, Graz University of Technology, Stremayrgasse 9, 8010 Graz, Austria*
[2] *Polymer Centre, Faculty of Technology, Tomas Bata University in Zlin, 76272 Zlin, Czech Republic*
[3] *Institute for Wood Technology and Renewable Materials, University of Natural Resources and Life Sciences, Vienna, Konrad Lorenz-Strasse 24, 3430 Tulln, Austria*

Correspondence should be addressed to Rupert Wimmer, rupert.wimmer@boku.ac.at

Academic Editor: Kibret Mequanint

Poly(lactic acid) was filled with 20 wt% of the three mineral fillers Mica, Zeolite, and Vansil, differing in the particle shape and surface area. Viscoelastic properties of unfilled and filled composites were investigated via dynamic mechanical analysis, while filler and fracture surface morphology of the composites was analysed through scanning electron microscopy. Results demonstrate the relationships between viscoelastic damping behaviour of filled PLA composites and the filler distribution in the PLA matrix. Both damping reduction and scanning electron microscope analysis revealed that Zeolite was better distributed in the poly(lactic acid) matrix than the other used fillers Mica and Vansil. The interfacial filler/matrix adhesion has again proved to be the key factor determining thermal and mechanical properties of reinforced composite material.

1. Introduction

Annual global production of polymers produced from petroleum has arrived at 150 million tons, with an expected growth of 5% p.a. [1]. Formation of fossil fuels has taken millions of years to form and due to extensive utilization their supply is now limited. In addition, the majority of the produced short-term goods made from plastic end up in landfills for indefinite periods of time. From this it is evident that the development of competitive renewable and biodegradable plastic materials is inevitable [2]. Poly(lactic acid), PLA, is at present one of the most promising biodegradable thermoplastics, also due to its commercial availability. Successful applications are seen in packaging, textile industry, in the biomedical field, and also in fiber-reinforced composite manufacturing. To enhance competitiveness of PLA-based applications, the costs need to be optimized by using affordable filling materials. Due to its nucleating and reinforcing function mineral (inorganic) fillers such as talc, mica potassium titanate, calcium carbonate, calcium sulfate anhydrite, and hydroxyapatite are commonly used [3–8]. The influence of mineral fillers on thermal and mechanical properties of polymer composites depends on a number of factors, among filler concentration, filler particle size and shape, filler/polymer adhesion, filler dispersion in the polymer matrix, and the applied composite processing [8–11].

Dynamic mechanical analysis (DMA) is known as a technique that combines both mechanical and rheological characterization of polymer samples as a function of temperature, time, and frequency of the applied oscillating deformation force. Determination of relaxation phenomena enables us to evaluate glass transition, stress relaxation, dynamic fragility parameter, miscibility of the polymers, interfacial compatibility of the individual composite components, and filler effectiveness (in the case of filled composites) [12–16]. It is known that a good distribution of fillers in polymer matrices is essential for high performance of the composites [17–20]. We hypothesize that the state of filler dispersion in polymer matrices influences the viscoelastic properties. To address this hypothesis dynamic mechanical analysis was used as a method to predict filler distribution in polymer matrices. We assessed the state of dispersion in a PLA matrix of the mineral fillers Mica, Zeolite, and Vansil. The mineral fillers differed chemically as well as in their physical properties. However, the average particle size of 5 μm was constant across the filler types. Interfacial compatibility between PLA and mineral filler was improved using 4,4′-methylenediphenyl diisocyanate (PLA-g-MDI; [21]). Viscoelastic properties were analyzed by DMA, while morphology was investigated with scanning electron microscopy.

2. Experimental

2.1. Materials. PLA 7000D pellets obtained from NatureWorks LLC (Minnetonka, MN, USA) were used as matrix material. PLA had an average molecular weight (Mw) of 140 000 g mol^{-1}, with a polydispersity index of 2.06 and a density of 1.22 g cm^{-3}. Glass transition temperature and melting temperature of the native PLA resins were 60.5°C and 151.1°C, respectively. PLA-g-MDI polymer matrix with 1 wt% of MDI was prepared according to Li and Yang [22]. Chemical formulas along with physical properties of the mineral fillers are shown in Table 1; the structural nature is seen in Figure 1.

MDI-modified PLA film composites were filled with 20 wt% of the mineral fillers Mica, Zeolite, or Vansil. The fixed concentration of 20 wt% was chosen as it was as a mechanical optimum, exhibiting also minimal brittleness. Composites were prepared through solution casting in chloroform followed by thermoforming (see Figure 2) [13].

2.2. Dynamic Mechanical Analysis (DMA). The viscoelastic properties of PLA and PLA/mineral filler films, the storage modulus (E'), loss modulus (E''), and damping factor ($\tan\delta = E''/E'$) were both measured on a Netzsch DMA 242 C operated in tension mode at 1 Hz frequency, within a temperature range between −20°C and 120°C, and a heating rate of 3°C/min. Strips were cut from the films at the size of $10 \times 6 \times 0.18\ mm^3$.

From the values of storage moduli obtained at the glassy and rubbery regions, at a sweep frequency of 1 Hz, the filler effectiveness (Factor C) in polymer matrix can be evaluated [23].

$$\text{Factor } C = \frac{\left(E'_g/E'_r\right)\text{composite}}{\left(E'_g/E'_r\right)\text{PLA}}, \quad (1)$$

where, E'_g and E'_r are storage moduli determined at the glassy and rubbery regions, respectively.

The state of filler dispersion in the polymer matrix was determined by calculating the damping reduction (DR) [24], which is a ratio of the determined damping factors ($\tan\delta$) of the neat and filled PLA-g-MDI composite types:

$$\text{DR}(\%) = \frac{(\tan\delta)_{\text{neat PLA-g-MDI}} - (\tan\delta)_{\text{filled PLA-g-MDI}}}{(\tan\delta)_{\text{neat PLA-g-MDI}}} \times 100. \quad (2)$$

2.3. Scanning Electron Microscopy. Surface morphology of the mineral fillers and the fracture surface morphology of the composites were investigated by scanning electron microscopy (TESCAN VEGA/LMU). Sample surfaces were sputtered with Au/Pd. The microscope was operated under high vacuum at an acceleration voltage of 5 kV.

3. Results and Discussion

3.1. Tensile Fracture Surface Analysis. SEM micrographs showed the different filler particle shapes in the composites (Figure 3). Mica was mostly platy, the zeolites had a blocky structure, and Vansil created fibrous aggregates.

Zeolite and Vansil were uniformly distributed in the PLA-g-MDI matrices indicating the efficiency in mixing the filler within the polymer matrix. In contrast, Mica was less distributed as it deposited on the bottom of the films. This deposition is connected to the applied solution casting method. The fracture surfaces of PLA-g-MDI filled with Mica and Vansil (Figures 3(a) and 3(c)) have shown voids between filler particles and the polymer matrix, which was taken as an evidence for poor adhesion. In opposite, SEM microscopy of PLA-g-MDI matrix filled with Zeolite particles indicated much better adhesion. Even with the Zeolite forming small agglomerates to some extent, fracture surfaces of this composite did not show voids or any sign of irregular distribution of the filler.

3.2. Viscoelastic Properties. Viscoelastic properties of composites (E', E'', and $\tan\delta$) were detected through DMA to detect any changes in the sample in terms of mechanical, rheological, and thermal trends. Figures 4, 5, and 6 show storage moduli, loss moduli, and damping factors ($\tan\delta$) as related to temperature.

All mineral filler types incorporated in the PLA-g-MDI matrices have raised the storage modulus of composites. The surfaces of these particles hold a number of hydroxyl groups that may have created agglomerates or caused poor dispersion in the hydrophobic PLA-g-MDI matrix. The differences in the obtained storage modulus values can be explained by the expressed chemical and physical differences among the filler types, that are, surface area, particle shape, hardness, and pH level [17, 18]. All these factors are crucial for the resulting mechanical and thermal properties of the obtained composites. Filler effectiveness (Factor C, formula 1) indicating the composite reinforcing capacity can be seen in Table 2.

By definition, the unfilled polymer matrix film has a factor C equal to 1. Factor C lower than 1 indicates a mechanical stiffening effect as well as a thermal stability contribution of the used filler. According to the obtained values mineral filler contributed to the PLA-g-MDI stiffness in the order Mica > Zeolite > Vansil.

Maximum loss moduli (E'') of all PLA-g-MDI/mineral filler composites were higher than those obtained from unfilled PLA-g-MDI (Figure 5) indicating that the fillers contributed towards a greater viscous response due to the occurred friction between filler and polymer matrix. The

TABLE 1: Physical properties of the used mineral fillers.

Inorganic filler	Chemical formula	Density (g/cm^3)	Mohs hardness	Particle shape	Surface area BET (m^2/g)	pH
Mica SG[a]	$^{d}X_2Y_{4-6}Z_8O_{20}(OH,F)_4$	2.9	2.5	Platy	7	11
Natural Zeolite[b]	$(Ca,K_2,Na_2,Mg)_4Al_8Si_{40}O_{96}\cdot 24\ H_2O)$	2.3	2	Blocky	13	7
Vansil[c]	$CaSiO_3$	2.9	4.5	Acicular	4	9

[a]Aspanger Bergbau und Mineralwerke.
[b]IFA, Tulln, Austria.
[c]Vansil, C. H. Erbslöh.
[d]X is K, Na, Ca, Ba, Rb or Cs; Y is Al, Mg, Fe, Mn Cr, Ti or Li; Z is Si, Al, Fe^{3+} or Ti.

(a) (b) (c)

FIGURE 1: SEM micrographs of mineral fillers (a) Mica (monoclinic, tendency towards pseudohexagonal crystals), (b) Zeolite (silica, and alumina tetrahedrons), (c) Vansil (wollastonite; triclinic, with lamellar, compact, and fibrous aggregates).

values of glass transition temperature (T_g), determined at the loss modulus peak temperature, reflect differences in the polymer chain segment mobility. The Mica filler has caused a T_g shift of about 4°C towards higher temperatures, which is connected to a more restricted mobility of the PLA chains.

Damping ($\tan\delta$) is highly sensitive to structural transformations of the polymers. The incorporation of fillers in polymer matrices usually broadens the $\tan\delta$ peak and changes its intensity depending on the actual damping mechanisms in composite. Composites with mineral filler in contrast to neat polymer show broader $\tan\delta$ peaks (Figure 6). The broadening of damping peak may be connected with the polymer chain arrangement that took place during the solution casting procedure. Mica and Vansil in contrast to Zeolite increased in a PLA-g-MDI matrix, an expressed composite damping. Additional damping in the particle-filled polymer composites may have occurred due to particle-particle friction, particle-polymer friction, induced thermal stresses near the interface, and the changed polymer morphology [12, 19, 20]. Table 3 shows the damping characteristics of PLA-g-MDI composites. The negative values of the damping reduction (PLA-g-MDI/Mica and PLA-g-MDI/Vansil) indicated the increase in damping. In accordance to the SEM analyses it is suggested that the increase in damping has occurred due to the breakdown of filler agglomerates. The lower $\tan\delta$ intensity of the PLA-g-MDI/Zeolite composite indicates a better distribution of Zeolite prevalent in the PLA-g-MDI matrix.

4. Conclusions

Effect of 20 wt% of the mineral fillers Mica, Zeolite, and Vansil on the state of dispersion in PLA-g-MDI polymer matrix by the determination of viscoelastic properties as well as SEM microscopy was investigated. All used mineral fillers increased mechanical stiffness and have also changed the resulting molecular mobility of the PLA-g-MDI composites. The increase of storage modulus was accompanied with an increase of the loss modulus. However, the $\tan\delta$ intensity was different, depending on the present damping mechanism. With the obtained scanning electron microscope analyses it can be concluded that Zeolite was best dispersed in the PLA-g-MDI matrix, compared to Mica and Vansil. Also, the values of damping reduction (DR) showed differences among the

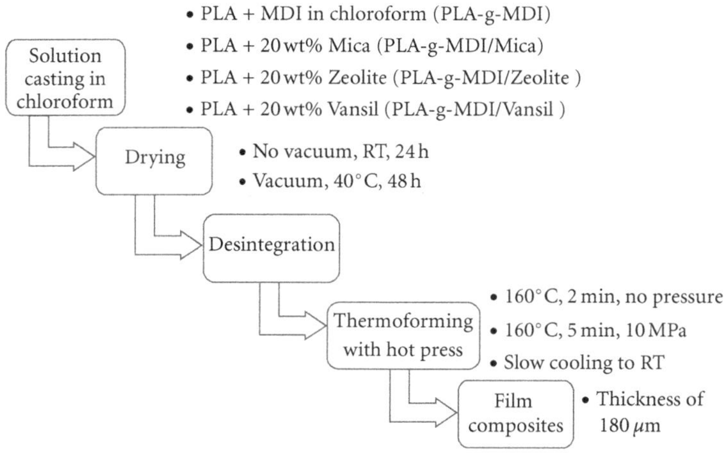

FIGURE 2: Preparation of mineral-filled PLA composites, compared with PLA modified with 4,4′-methylenediphenyl diisocyanate (MDI).

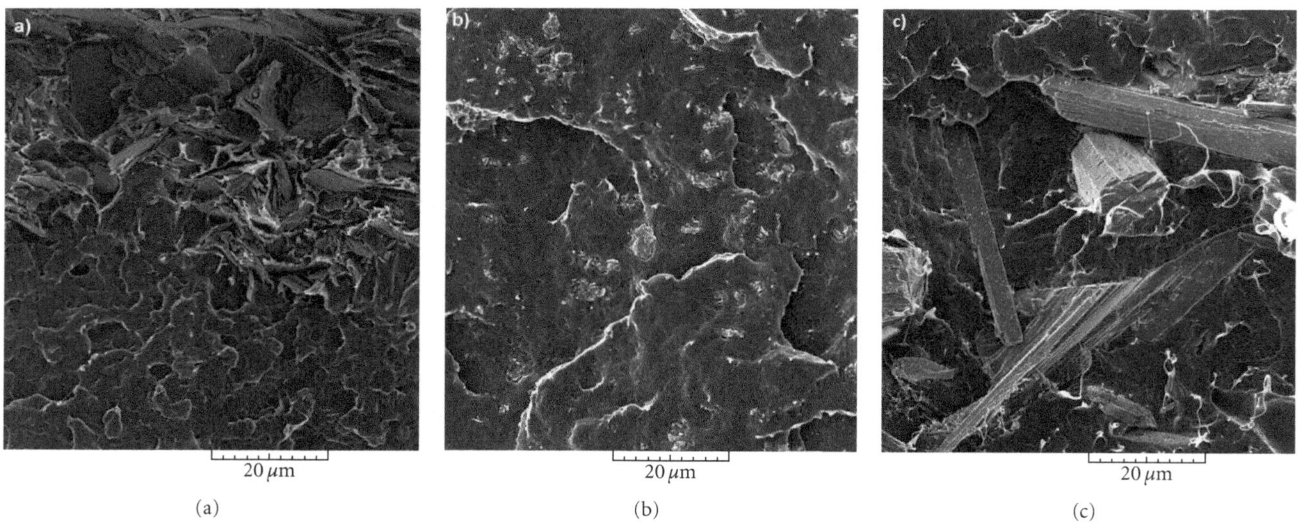

FIGURE 3: SEM micrographs of PLA-g-MDI composites filled with (a) Mica, (b) Zeolite, (c) Vansil.

TABLE 2: Storage moduli and factor C of PLA-g-MDI composites.

Sample	E' at 20°C (MPa)	E' at 80°C (MPa)	Factor C
PLA-g-MDI	3348	613	1
PLA-g-MDI/Mica	6041	1703	0.65
PLA-g-MDI/Zeolite	5010	1219	0.75
PLA-g-MDI/Vansil	4343	914	0.87

TABLE 3: Damping characteristics of PLA-g-MDI composites.

Sample	tan δ intensity at peak	tan δ temperature at peak (°C)	Damping reduction (%)
PLA-g-MDI	0,233	77.3	—
PLA-g-MDI/Mica	0,258	80.8	−10.5
PLA-g-MDI/Zeolite	0,207	77.3	+11.1
PLA-g-MDI/Vansil	0,250	75.8	−7.2

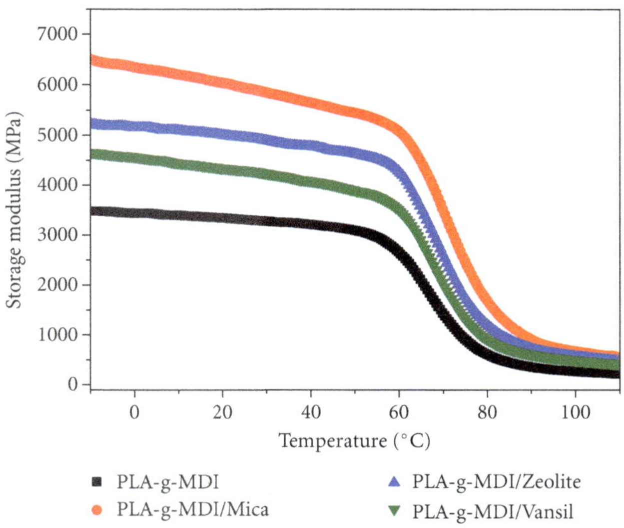

FIGURE 4: Storage moduli of PLA-g-MDI and PLA/mineral composites as a function of temperature. Frequency sweep was set at 1 Hz.

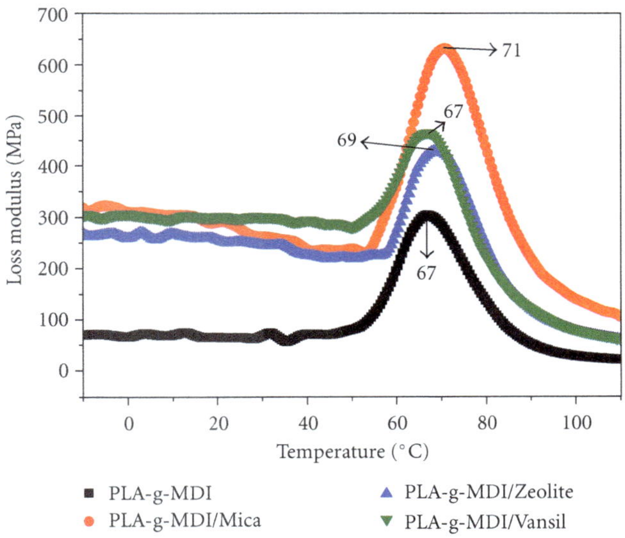

FIGURE 5: Loss moduli of PLA-g-MDI and PLA/mineral filler composites as a function of temperature. Frequency sweep was set at 1 Hz; peak temperatures are outlined.

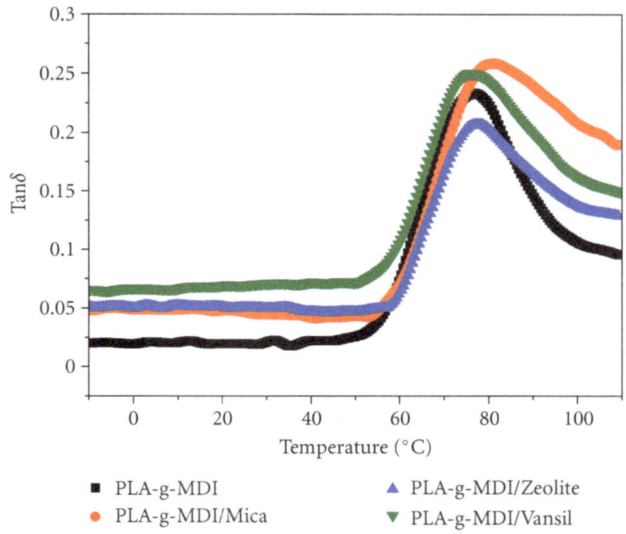

FIGURE 6: Damping factors tan δ of PLA-g-MDI and PLA/mineral filler composites as a function of temperature. Frequency sweep was set at 1 Hz.

used mineral filler types. The fact that filler-filler friction due to broken agglomerates has contributed to the higher damping intensity can be used as an indicator for the filler distribution status. On the base of coherent observations of the filler distribution by both, the damping reduction and the SEM analysis, we conclude that DMA is a quick and reliable method to predict the functional morphology of PLA-based composites.

Acknowledgments

This work was supported by the Austrian Science Fund, FWF, through project L319-B16. The authors also acknowledge support by the Austrian Agency for International Cooperation in Education and Research (OeAD-GmbH) (project no. WTZ 03/2011) and by the Ministry of Education, Youth, and Sports of the Czech Republic (Grant no. ME061102).

References

[1] E. T. H. Vink, K. R. Rábago, D. A. Glassner, and P. R. Gruber, "Applications of life cycle assessment to NatureWorks polylactide (PLA) production," *Polymer Degradation and Stability*, vol. 80, no. 3, pp. 403–419, 2003.

[2] G. Scott and D. M. Wiles, "Degradable hydrocarbon polymers in waste and litter control," in *Degradable Polymers*, G. Scott, Ed., Principles and Applications, pp. 449–479, Kluwer Academic, Dordrecht, The Netherlands, 2002.

[3] J. J. Kolstad, "Crystallization kinetics of poly(L-lactide-co-meso-lactide)," *Journal of Applied Polymer Science*, vol. 62, no. 7, pp. 1079–1091, 1996.

[4] H. Urayama, C. Ma, and Y. Kimura, "Mechanical and thermal properties of poly(L-lactide) incorporating various inorganic fillers with particle and whisker shapes," *Macromolecular Materials and Engineering*, vol. 288, no. 7, pp. 562–568, 2003.

[5] H.-S. Kim, B. H. Park, J. H. Choi, and J.-S. Yoon, "Mechanical properties and thermal stability of poly(L-lactide)/calcium carbonate composites," *Journal of Applied Polymer Science*, vol. 109, no. 5, pp. 3087–3092, 2008.

[6] M. J. Sobkowicz, J. L. Feaver, and J. R. Dorgan, "Clean and green bioplastic composites: comparison of calcium sulfate and carbon nanospheres in polylactide composites," *Clean*, vol. 36, no. 8, pp. 706–713, 2008.

[7] N. Ignjatovic and D. Uskokovic, "Synthesis and application of hydroxyapatite/polylactide composite biomaterial," *Applied Surface Science*, vol. 238, no. 1-4, pp. 314–319, 2004.

[8] A. Gregorova, V. Sedlarik, M. Pastorek, H. Jachandra, and F. Stelzer, "Effect of compatibilizing agent on the properties of highly crystalline composites based on poly(lactic acid)

and wood flour and/or mica," *Journal of Polymers and the Environment*, vol. 19, no. 2, pp. 372–381, 2011.

[9] J. Jancar, "Mineral fillers in thermoplastics," in *Raw Materials and Processing*, vol. 139 of *Advances in Polymer Science*, pp. 1–65, Springer, Berlin, Germany, 1999.

[10] C. Dupuy and P. Bussi, "New dispersion process for submicronic fillers in thermoplastics," *Macromolecular Symposia*, vol. 169, no. 1, pp. 103–107, 2001.

[11] T. Huuhilo, O. Martikka, S. Butylina, and T. Kärki, "Mineral fillers for wood-plastic composites," *Wood Material Science and Engineering*, vol. 5, no. 1, pp. 34–40, 2010.

[12] A. Gregorova, M. Hrabalova, R. Wimmer, B. Saake, and C. Altaner, "Poly(lactide acid) composites reinforced with fibers obtained from different tissue types of *Picea sitchensis*," *Journal of Applied Polymer Science*, vol. 114, no. 5, pp. 2616–2623, 2009.

[13] A. Gregorova, M. Hrabalova, R. Kovalcik, and R. Wimmer, "Surface modification of spruce wood flour and effects on the dynamic fragility of PLA/wood composites," *Polymer Engineering and Science*, vol. 51, no. 1, pp. 143–150, 2011.

[14] A. Gregorova, R. Wimmer, M. Hrabalova, M. Koller, T. Ters, and N. Mundigler, "Effect of surface modification of beech wood flour on mechanical and thermal properties of poly (3-hydroxybutyrate)/wood flour composites," *Holzforschung*, vol. 63, no. 5, pp. 565–570, 2009.

[15] M. Hrabalova, A. Gregorova, R. Wimmer, V. Sedlarik, M. Machovsky, and N. Mundigler, "Effect of wood flour loading and thermal annealing on viscoelastic properties of poly(lactic acid) composite films," *Journal of Applied Polymer Science*, vol. 118, no. 3, pp. 1534–1540, 2010.

[16] J. F. Mano, "Study of the segmental dynamics in semi-crystalline poly(lactic acid) using mechanical spectroscopies," *Macromolecular Bioscience*, vol. 5, no. 4, pp. 337–343, 2005.

[17] I. Spinu and G. B. McKenna, "Physical aging of nylon 66," *Polymer Engineering & Science*, vol. 34, no. 24, pp. 1808–1814, 1994.

[18] L. H. Sperling, *Introduction to Physical Polymer Science*, John Wiley & Sons, Toronto, Canada, 4th edition, 2006, Edited by L.H. Sperling.

[19] S. Farzaneh and A. Tcharkhtchi, "Viscoelastic properties of polypropylene reinforced with mica in T_{α} and $T_{\alpha c}$ transition zones," *International Journal of Polymer Science*, vol. 2011, Article ID 427095, 5 pages, 2011.

[20] Y. Li, K. Venkateshan, and S. S. Xiuzhi, "Mechanical and thermal properties, morphology and relaxation characteristics of poly(lactic acid) and soy flour/wood flour blends," *Polymer International*, vol. 59, no. 8, pp. 1099–1109, 2010.

[21] N. C. Bleach, S. N. Nazhat, K. E. Tanner, M. Kellomäki, and P. Törmälä, "Effect of filler content on mechanical and dynamic mechanical properties of particulate biphasic calcium phosphate—polylactide composites," *Biomaterials*, vol. 23, no. 7, pp. 1579–1585, 2002.

[22] B. H. Li and M.-C. Yang, "Improvement of thermal and mechanical properties of poly(L-lactic acid) with 4,4-methylene diphenyl diisocyanate," *Polymers for Advanced Technologies*, vol. 17, no. 6, pp. 439–443, 2006.

[23] A. Gregorova and R. Wimmer, "Filler-matrix compatibility of poly(lactic acid) based composites," in *Polylactic Acid: Synthesis, Properties and Applications*, V. Piemonte, Ed., chapter 5, pp. 97–119, Nova Science, New York, NY, USA, 2012.

[24] S. Pattanawanidchai, P. Saeoui, and C. Sirisinha, "Influence of precipitated silica on dynamic mechanical properties and resistance to oil and thermal aging in CPE/NR blends," *Journal of Applied Polymer Science*, vol. 96, no. 6, pp. 2218–2224, 2005.

17

Controlled Morphology and Mechanical Characterisation of Electrospun Cellulose Acetate Fibre Webs

B. Ghorani, S. J. Russell, and P. Goswami

Nonwovens Research Group, Centre for Technical Textiles, University of Leeds, Leeds LS2 9JT, UK

Correspondence should be addressed to B. Ghorani; behrooz.ghorani@gmail.com

Academic Editor: Wen-Fu Lee

The purpose was to interpret the varying morphology of electrospun cellulose acetate (CA) fibres produced from single and binary solvent systems based on solubility parameters to identify processing conditions for the production of defect-free CA fibrous webs by electrospinning. The Hildebrand solubility parameter (δ) and the radius of the sphere in the Hansen space ($D_{(s-p)}$) of acetone, acetic acid, water, *N,N*-dimethylacetamide (DMAc), methanol, and chloroform were examined and discussed for the electrospinning of CA. The Hildebrand solubility parameter (δ) of acetone and DMAc were found to be within an appropriate range for the dissolution of CA. The suitability of the binary solvent system of acetone: DMAc (2 : 1) for the continuous electrospinning of defect-free CA fibres was confirmed. Electrospun webs exhibited improved tensile strength and modulus after heat and alkali treatment (deacetylation) of the as-spun material, and no major fibre morphological degradation occurred during the deacetylation process.

1. Introduction

Cellulose acetate (CA) fibres and membranes have biomedical applications in ultrafiltration and reverse osmosis [1, 2] biosensors and affinity membranes. In webs produced by electrospinning fibre, morphology is an important parameter affecting product performance. The morphology of electrospun fibres is influenced by both the physical and electrical properties of the spinning solution, which is, to a large extent, determined by type of solvent [3]. The selection of solvents for electrospinning and their influence on fibre morphology has been one of the main areas of research in recent years [4]. Wannatong et al. [5] showed that both the density of electrospun webs and the presence of beads can be influenced by the choice of solvent. Uyar and Besenbacher [6] indicated that differences in the dryness of the collected fibres could be attributed to the density and boiling point of the solvent. The collected fibre dryness increased as the density and boiling point of the solvent decreased [5]. Similar work on solvent selection was reported by Jarusuwannapoom et al. [7].

Solvent type can also influence the fibre morphology of the as-spun web. Yang et al. [8] investigated the effects of dichloromethane (MC), ethanol, and *N,N*-dimethylformamide (DMF) as single and binary solvent systems on the formation of ultrathin uniform poly(vinyl pyrrolidone) (PVP) fibres during electrospinning and noted substantial differences in the collected fibre morphologies ranging from smooth to helical.

Van der Schueren et al. [9] showed that substantial reduction of the cross-sectional area of electrospun fibres could be achieved by selecting a binary rather than a single solvent system. The use of binary solvent systems has been utilised by various researchers to optimise the electrospinning process, for example, achieving a high productivity at the same time as a finer fibre diameter.

Numerous single and mixed solvent systems have been utilised for electrospinning CA [2, 8–12] with one group reporting the development of a ternary system for the preparation of protein A/G functionalised CA membranes [13, 14]. Son et al. [10] showed that when acetone was used as a solvent for electrospinning of CA, the tip of the needle was easily blocked because of rapid solvent evaporation indicating that acetone is not an appropriate solvent for continuous electrospinning using the prevailing experimental conditions. However, it was found that ultrafine CA fibres could be obtained using acetone as a solvent when the CA concentration exceeded 17% (w/w). Electrospinning continued until the tip of the needle was blocked by evaporation of the acetone.

Ultrafine CA fibres were successfully produced *via* electrospinning of CA in a mixed solvent of acetone and water (water content 10–15% (w/w)). Ultrafine cellulose fibres could then be regenerated from homogenous deacetylation of CA webs in KOH/ethanol solution. Electrospinning of a CA solution in acetone and water under acidic conditions produced larger fibres, whereas when the solution was basic, much finer fibres were produced [10]. Han et al. [11] reported that uniform CA fibres could be electrospun at CA concentrations above 17% (w/w) in a mixed solvent of acetic acid : water (75 : 25). The average diameter of these fibres was 180 nm. Tungprapa et al. [12] showed that smooth fibres were obtained from 16% (w/w) CA solutions in 1 : 1, 2 : 1, and 3 : 1 acetone : DMAc, 14–20% (w/w) CA solutions in 2 : 1 acetone : DMAc and 8%–12% (w/w) CA solutions in 4 : 1 methylene chloride (DCM) : methanol. This work also concluded that the most versatile solvent system for electrospinning of CA is a binary system of acetone : *N,N*-dimethylacetamide (DMAc) (2 : 1) with 12.5%–20% (w/w) CA concentration [12].

Previous studies on the electrospinning of CA have focused on solvent selection [12, 13], but the identification of suitable solvents has been based on trial and error experimentation, results for similar polymer systems or solubility models limited by the physico-chemical database [14]. The purpose of this research was to interpret the varying morphology of the CA electrospun cellulose fibres in single and binary solvent systems based on the solubility parameters. This enabled suitable solvent systems and optimum processing conditions to be identified as well as the production of defect-free cellulose acetate (CA) fibrous webs by electrospinning. Previously, researchers have investigated changes in tensile properties as a result of deacetylation [13, 15]. Ma et al. [13] demonstrated the feasibility of applying cellulose fibre membrane prepared by electrospinning as affinity membranes. The influence of heat treatment (post fibre formation) on the mechanical properties of the electrospun CA fibre webs was explored as means of maximising their strength [13]. A similar approach was attempted in the present study based on a previously reported method [13, 15].

2. Material and Methods

2.1. Materials. Cellulose acetate (Acetyl content, 39.8% (w/w), M_w 30,000 Da), acetone, acetic acid, *N,N*-dimethylacetamide, methanol, chloroform, and sodium hydroxide, was purchased from Sigma-Aldrich (UK) and used without any further purification.

2.2. Sample Preparation. To prepare spinning solutions, single and binary solvent systems were evaluated as follows: (1) Single-acetone, DMAc, acetic acid, chloroform, and methanol; (2) Binary-acetone : distilled water, acetic acid : distilled water, acetic acid : acetone, acetic acid : DMAc, DMAc : acetone, methanol : distilled water, acetone : methanol, and chloroform : methanol. Solutions were prepared by constant stirring at room temperature (25±5°C). The CA concentration in the spinning solutions was fixed at a concentration of 15% (w/v). This concentration was selected based on previous studies to establish the most suitable solvent systems [12, 15]. The suitability of each system was assessed in terms of the freedom from precipitation during preparation of the spinning solution and the spinnability of the resulting polymer solution. Spinnability in this context was assessed in terms of freedom from needle blockages, consistency of fibre morphology, and freedom from bead and spindle defects in the web. Clear spinning solutions were obtained after 24 h.

2.3. Electrospinning. Electrospinning of the CA polymer solutions was performed in horizontal alignment at an applied voltage of 20–25 kV and a tip-to-collector distance of 100 mm. The syringe pump feed rate was 0.002 to 0.02 mL min^{-1}. A copper electrode supplying the current and voltage was attached to the syringe needle at its base. Electrospinning was performed inside a fume cupboard and was carried out for 60 min at room temperature. Webs were collected on a square aluminium foil. The single and binary solvent systems were prepared at a fixed CA concentration of 15% (w/v) based on the findings of previous studies [12, 15]. Preliminary work indicated that no fibres could be successfully spun with a CA concentration of less than 13% (w/v). The presence of hydroxyl and acetyl groups in CA gives rise to strong intra- and intermolecular interactions, predominately H-bonding. Intermolecular H-bonding results in bridging the adjacent CA structural entities [16]. With concentrations less than 13% (w/v), it is proposed that the interactions between adjacent CA molecules (intermolecular H-bonding) are not strong enough for fibre formation that is, the critical chain entanglement concentration for fibre formation was not reached. Precipitation of CA occurred in the binary solvent systems of acetone : water (4 : 1), methanol : water (4 : 1), and methanol : acetone (3 : 2, 4 : 1) even at a low CA concentration of 10% (w/v). A summary of the solvent systems and electrospinning conditions utilised is given in Table 1. The effects of each solvent on the electrospinnability of the CA spinning solutions and the morphological appearance of the as-spun CA fibres were investigated to determine the optimum solvent system for electrospinning of CA.

2.4. Scanning Electron Microscopy (SEM). Prior to SEM imaging (*Philips XL30*-ESEM), all the samples were cut into small sizes (stubs 12.7 mm diameter) and were sputter coated (*EMscope Biorad SC500*) with a gold and platinum mixture in a vacuum. SEM images were taken at different magnifications: high (scale bar =10 μm), medium (scale bar =25 and 50 μm), and low (scale bar =100 μm) for all samples. Fibre morphology was determined directly from multiple SEM images of the same sample to determine a representative view.

2.5. Attenuated Total Reflectance Fourier-Transform Infra-Red (ATR-FTIR) Spectroscopy. The *Perkin-Elmer Spectrum BX* spectrophotometer with diamond ATR attachment was used in this study (ATR-FTIR) to measure the pre- and post-neutralised fibrous webs. Scanning was conducted from 4000 to 600 cm^{-1} with a resolution of 4 cm^{-1} and a scanning interval of 2 cm^{-1} with 64 repetitious scans averaged per sample.

2.6. Tensile Strength Testing. Tensile testing of electrospun webs was conducted using an *Instron 5544* single column

TABLE 1: Experimental conditions: combinations of solvent systems, polymer concentration, and spinning solution flow rate and voltage.

Solvent	Ratio (v/v)	Polymer concentration % (w/v)	Flow rate ($mL\,min^{-1}$)	Voltage (kV)	As-spun fibres
Acetone	—	15	0.02	20–25	Few ribbon-like fibres
DMAc	—	15	0.02	20–25	Only beads
Acetic acid	—	15	0.02	20–25	No production
Chloroform	—	15	0.02	20–25	Only beads
Methanol	—	15	0.02	20–25	Only beads
Acetone : water	17 : 3	15	0.02	20–25	No production
Acetone : water	4 : 1	10	—	—	Precipitation
Acetic acid : water	3 : 1	15	0.004	20–25	Large beads
Acetic acid : acetone	3 : 1	15	0.004	20–25	Few fibres and large beads
Acetic acid : DMAc	3 : 1	15	0.004	20–25	Large beads
DMAc : acetone	2 : 1	15	0. 004	20–25	Large beads
Methanol : water	4 : 1	10	—	—	Precipitation
Methanol : acetone	3 : 2 and 4 : 1	10	—	—	Precipitation
Acetone : methanol	3 : 2 and 4 : 1	15	0.004	20–25	Fibres and small beads
Chloroform : methanol	4 : 1	15	0.004	20–25	Some ribbon-like fibres
Acetone : DMAc	2 : 1	15	0.004	20–25	Fibres and small beads
		20	0.004	20–25	Fibres and small beads
		25	0.002	20–25	Fibres free of beads

system with the *Merlin* materials testing software. Samples of 5 mm × 20 mm were measured using a gauge length of 20 mm. The rate of extension was 10 mm min^{-1}. Each measurement was replicated three times per sample.

3. Results and Discussion

The solubility characteristics of the polymer are of fundamental importance in the process of selecting the suitable solvent systems for electrospinning. Table 1 summarises the solvent systems that were selected based on previous studies [10–12, 15] and evaluation of the solubility behaviour following the Hansen theory of solubility [14, 17]. The total solubility parameter (δ_t) is a numerical value that demonstrates the relative solvency behaviour of a specific solvent. The concept that solubility is related to the internal energy of solvents and solutes was first introduced by Hildebrand as the square root of the cohesive energy density (CED) of the material. The Hansen parameters extended the concept that the total cohesive term and thus the total solubility parameters (δ_t) of the total Hildebrand value may be divided into: dispersion component (δ_d), polar component (δ_p) and hydrogen bonding component (δ_h). The equation is as follow [17]:

$$\delta_t = \sqrt{\delta_d^2 + \delta_p^2 + \delta_h^2}. \quad (1)$$

The SI unit of all the Hansen parameters is $MPa^{1/2}$. Values of δ_d, δ_p, and δ_h at room temperature for a variety of CA solvents are given in Table 2 [14, 18]. According to Hansen, an approximately spherical area of solubility may be constructed in a three-dimensional coordinate system of δ_d, δ_p, and δ_h. The radius of that sphere is called the interaction radius (R), and the value of R for CA is 12.40 $MPa^{1/2}$ [18]. A polymer is probably soluble in a solvent if the distance between the solvent and the centre of the polymer solubility sphere ($D_{(s-p)}$) is less than the radius of interaction for the polymer ($D_{(s-p)} < R$) [14, 17, 18]. Accordingly, acetone, DMAc, acetic acid, chloroform, and partly methanol would be expected to dissolve CA (Table 2). The Hansen sphere ($D_{(s-p)}$) of water and methanol are 32.47 and 12.35 $MPa^{1/2}$, respectively, which is not within an appropriate range for the dissolution of CA, particularly for water (Table 2). Therefore, the preparation of CA solutions in the binary solvent systems of acetone : water (4 : 1), methanol : water (4 : 1), and methanol : acetone (3 : 2 and 4 : 1) were hampered due to precipitation of polymer during the stirring process (Table 1).

Table 3 summarises the boiling points, viscosity, surface tension, and the Hildebrand and the Flory-Huggins parameters for each solvent type [15, 19, 20]. The Hildebrand solubility parameter (δ) indicates the degree of interaction between materials and can be a good predictor of solubility, particularly for nonpolar polymer materials [15]. The Hildebrand solubility parameter (δ) is derived from the cohesive energy density of the solvent, which in turn is derived from the heat of vaporisation [21]. CA is soluble in solvents when the Hildebrand solubility parameters (δ) lie between 9.5 and 12.5 $(cal\,cm^{-3})^{1/2}$ [15].

The Hildebrand solubility parameters of acetic acid and chloroform are 13.1 and 9.21 $(cal\,cm^{-3})^{1/2}$, respectively, which is not within an appropriate range for the dissolution of CA (Table 3). However, Table 2 shows the Hansen sphere ($D_{(s-p)}$) of acetic acid and chloroform are 2.68 and 8.87 $MPa^{1/2}$, respectively, which is less than the radius of interaction (R)

TABLE 2: Total solubility parameters (δ_t) of various solvents and cellulose acetate [14, 18].

Liquid	Solubility parameter ($MPa^{1/2}$)					
	δ_t	δ_d	δ_p	δ_h	$^*D_{(s-p)}$	R
Acetone (AC)	19.93	15.50	10.40	7.00	5.39	
Dimethylacetamide (DMAc)	22.77	16.80	11.50	10.20	5.88	
Acetic acid	21.36	14.5	8.0	13.5	2.68	
Methanol	29.60	15.10	12.30	22.30	12.35	
Chloroform	19	17.80	3.10	5.70	8.87	
Water	47.83	15.60	16.0	42.3	32.47	
Celluloce acetate (CA)	19.89	14.90	7.10	11.10	—	12.40

$^*D_{(s-p)} = [4(\delta_{dS} - \delta_{dP})^2 + (\delta_{pS} - \delta_{pP})^2 + (\delta_{hS} - \delta_{hP})^2]^{1/2}$.
δ_{xs} = The Hansen component parameter for the solvent.
δ_{xp} = The Hansen component parameter for the polymer.

of CA and would be expected to dissolve the CA. This may be attributed to the fact that the predictions using the Hildebrand solubility parameters are made with the absence of any specific interactions, especially hydrogen bonds. [21]. Therefore, all experiments were conducted using acetone, acetic acid, DMAc, chloroform, and methanol. Other important considerations are the boiling temperature (°C), viscosity (η), and surface tension (γ) which were also taken into account (Table 3) in the selection of an appropriate solvent because they also affect the prospect of continuous electrospinning of ultrafine and uniform CA fibres. Additionally, binary solvents were studied since they could also potentially provide solvent systems with properties that were suitable for continuous, blockage-free electrospinning.

A summary of the spinning performance of each polymer solution is given in Table 1, which summarises the formation of continuous fibres or beads as a function of processing conditions. The solubility of CA in the selected solvent does not necessarily mean that fibres could be continuously produced during electrospinning; in many cases, beads were produced, and, in others, continuous spinning was interrupted by the volatility of the solvent in the polymer solution. In the preparation of the spinning solutions, different polymer concentrations were examined; however, except for the binary solvent systems containing acetone : DMAc, none of the solutions successfully produced substantially bead-free fibres (Table 1). Fibrous webs could not be obtained using a CA concentration of less than 13% (w/v). Similarly, spinnability issues were encountered due to high viscosity at polymer concentrations greater than 18%–20% (w/v) using single solvent systems. Specifically, continuous electrospinning of CA from solutions containing acetone, methanol, and chloroform alone was impeded due to their volatility and low boiling point (Table 3); beads were obtained on the screen collector for single solvent systems of chloroform and methanol (Figures 1(a) and 1(b)) [10, 15]. Acetic acid as a solvent did not enable production of a fibrous web, which is in agreement with the reported literature [11]. Liu and Hsieh [15] first reported that CA solutions in acetic acid could not be continuously electrospun, which may be attributable to the relatively low dielectric constant of acetic acid resulting in unstable electrospinning conditions [9].

Satisfactory spinning conditions could not be established for acetone : water in ratios of 17 : 3 (85 : 15) under the given experimental conditions [10, 22]. The solubility of CA in acetone : water is affected by the relative concentration of solvents such that the spinning solution could not be satisfactorily produced in a ratio of 4 : 1 (80 : 20) due to the precipitation of CA (Table 1). Reducing the concentration to 10% (w/v) did not improve the results. The poor performance of this particular solvent system may be explained in terms of the Hildebrand solubility (δ) and the Flory-Huggins (χ) parameters of water (Table 3). The Flory-Huggins parameter (χ) characterises the polymer-solvent interaction, wherein a smaller value of χ indicates a more thermodynamically compatible interaction between the solvent and the polymer [15]. Therefore, among the six solvents examined, DMAc was expected to be the most thermodynamically compatible solvent for CA, followed by chloroform, acetic acid, acetone, water, and methanol. The Hildebrand solubility and the Hansen sphere ($D_{(s-p)}$) for water are also 23.5 $(cal/cm^{-3})^{1/2}$ [19] and 32.47 $MPa^{1/2}$, respectively (Table 2), which is not within an appropriate range for the dissolution of CA. Therefore, it was observed that there was greater precipitation of CA as the amount of water increased in the binary solution of acetone and water. Incorporation of water also decreased the overall evaporation rate during electrospinning [23]. In the present study, satisfactory spinning conditions could not be established with 3 : 1 acetic acid : water as a solvent either, using the given experimental conditions because continuous electrospinning was hampered (Figures 1(c) and 1(d)).

Similarly, a satisfactory spinning solution could not be established for methanol : water (4 : 1). The poor performance of this binary solvent system is attributed to the high Hildebrand solubility (δ) and Hansen sphere ($D_{(s-p)}$) for water and methanol which is not within an appropriate range for the dissolution of CA (Tables 2 and 3). The Flory-Huggins (χ) parameter of methanol (1.87) is also the highest amongst the solvents used for this study followed by water (1.4) (Table 3) suggesting these two solvents are the least thermodynamically compatible for dissolving CA. The precipitation of cellulose acetate in the binary solvent systems of acetone : water (4 : 1), methanol : water (4 : 1), and methanol : acetone (4 : 1, 3 : 2) therefore prevented spinning trials using these particular solvent systems.

Tubular, collapsed ribbon, tape, branched, and various other morphological forms are frequently observed in electrospun materials depending on solvent properties, spinning solution rheology and process conditions [24]. Ribbon-like fibres were observed in the webs produced from acetone with a relatively large ribbon width of 28 μm (range of 10.28–61.13 μm) (Figure 2(a)). When a highly volatile solvent system is used for electrospinning, the tubular fibre skin of the jet collapses due to a rapid vaporisation of the solvent from within the fibre [24, 25]. Koombhongse et al. [24] reported that irregular shaped beads can be observed since the skin on the bead also collapses due to the rapid removal of solvent from the interior. The rapid rate of evaporation of acetone

TABLE 3: Physical properties of solvents [15, 19, 20, 31].

Solvent	Boiling temperature (°C)	Viscosity (η) cP at 20°C	The Flory-Huggins parameter (χ) at 25°C	Hildebrand Solubility parameter (δ) $(cal/cm^3)^{1/2}$	Surface tension (γ) (dyne cm^{-1}) at 20°C
Water	100	1	1.4	23.5	72.8
Acetone	56.29	0.36	0.48	9.77	23.32
DMAc	164	2.14	0.28	11.1	32.43
Acetic acid	117.9	1.31	0.41	13.1	27.6
Chloroform	61.2	0.56	0.34	9.21	27.50
Methanol	65	0.59	1.87	14.3	22

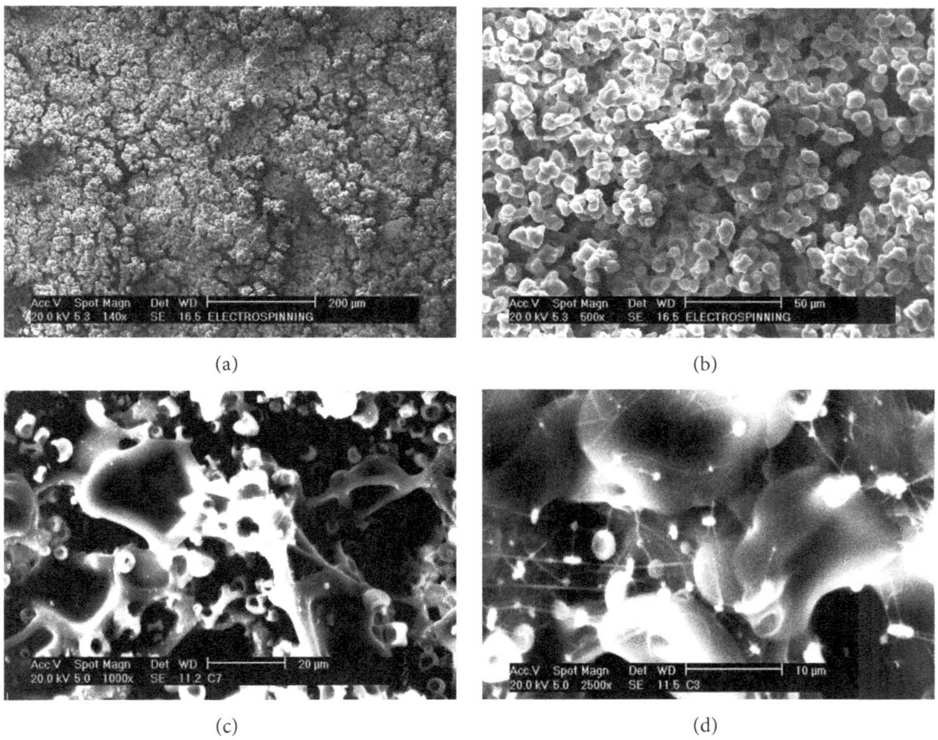

FIGURE 1: SEM micrographs showing the effect of: (a) chloroform as solvent on resulting CA electrospun fibre morphology, Mag 140x (b) Methanol, Mag 500x, and ((c)-(d)) Acetic acid : water (3 : 1) Mag 1000x–2500x.

therefore leads to ribbon-like fibres being observed in the as-spun web (Figure 2(a)). Moreover, the low boiling point of acetone (lowest amongst of the solvent used in this study) gives rise to needle tip blockage, which hampers continuous electrospinning (Table 3).

It was not possible to spin fibres from a CA solution in DMAc under the prevailing experimental conditions. Strong hydrogen bonding between CA and DMAc [26] and the high boiling point of the solvent (Table 3) resulted in poor evaporation of DMAc during electrospinning. The evidence of insufficient solvent evaporation may be observed in Figure 2(b). Similarly, a high incidence of bead formation was observed in the webs produced from the binary solvent system of acetic acid : DMAc (3 : 1) (Figure 2(c)). This could also be attributed to the high boiling point of the solvents and the strong hydrogen bonding between CA and DMAc which results in insufficient solvent evaporation [27].

CA solutions with concentration of 15% (w/v) in the binary solvent system of acetic acid : acetone (3 : 1) was also examined. Acetic acid : acetone (3 : 1) mixtures supported continuous electrospinning and web formation. The resultant fibres, however, contained many large beads on their surface (Figure 2(d)). The average fibre diameter was found to be 12.06 μm (range of 11.22–16.78 μm). Liu and Hsieh [15] indicated that CA solutions produced from acetic acid : DMAc and acetic acid : acetone mixtures could be continuously electrospun, however fibres with a diameter of 1-2 μm containing large beads were formed.

For the remaining solvents, acetone : methanol (3 : 2 and 4 : 1), chloroform : methanol (4 : 1), and acetone : DMAc (2 : 1)

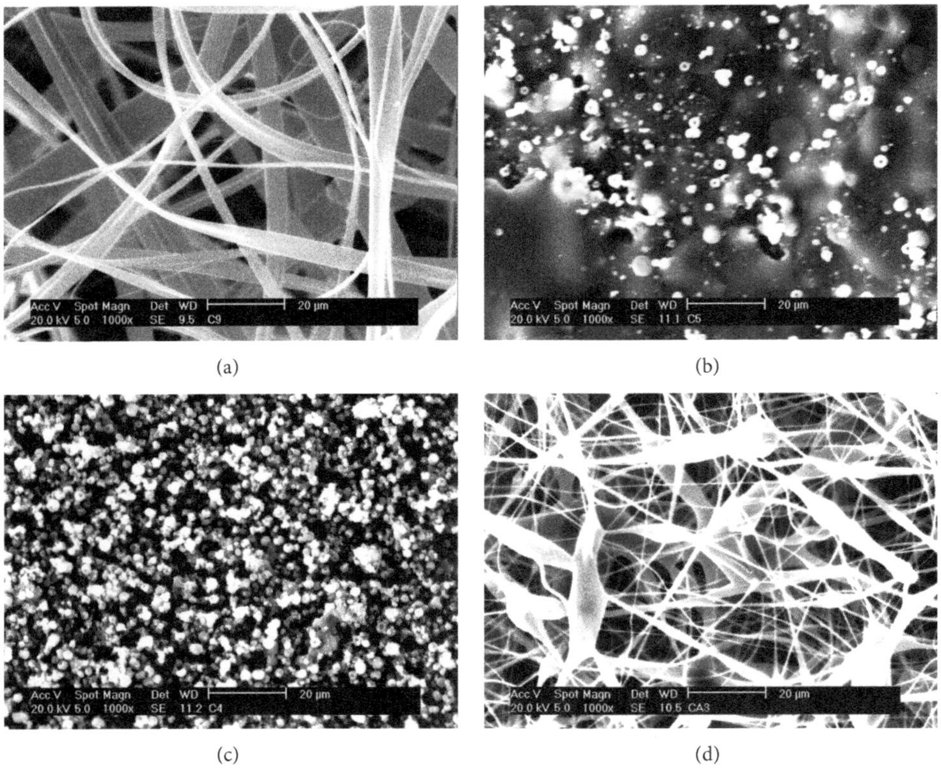

FIGURE 2: SEM micrographs showing the effect of individual and binary solvent systems on resulting CA electrospun fibre morphology. Mag 1000x–3000x (a) acetone; (b) DMAc; (c) acetic acid : DMAc (3 : 1), and (d) acetic acid : acetone (3 : 1).

fibrous webs were satisfactorily produced on a continuous basis for up to one hour confirming their suitability for continuous web production.

The solubility of CA in a binary solvent of methanol and acetone is affected by the ratio of methanol and acetone in the solution; spinning solutions could not be satisfactorily produced when the ratios were 3 : 2 and 4 : 1 methanol : acetone due to precipitation of CA. However, by increasing the amount of acetone in this binary solution (acetone : methanol 3 : 2 and 4 : 1), clear spinning solutions were produced; spinning from this solution resulted in fibrous webs containing small beads with fibre diameters ranging between 436 nm and 1.01 μm. Typical web structures produced from each solvent system are shown in Figures 3(a) and 3(b). The mean fibre diameter, fibre morphology, and the prevalence of bead formation were found to depend upon the solvent system. The mean fibre diameter of the webs produced from the 4 : 1 acetone : methanol binary solution was 4.35 ± 0.5 μm, and from the 3 : 2 acetone : methanol binary solution, it was 1.99 ± 0.5 μm. Morphologically, smooth fibres were observed in all webs.

Fibres produced from the binary solution of chloroform : methanol (4 : 1) were irregular in morphology and ribbon-like fibre structures were present as indicated in Figures 3(c) and 3(d). The presence of the ribbon-like fibres could be attributable to the rapid evaporation of chloroform from the polymer surfaces. The mean fibre diameter of the CA fibres produced from this spinning solution was 1 ± 0.5 μm. It has previously been reported that increasing the methanol content in solvent systems can contribute to the formation of both beaded and smooth fibres [28].

Table 3 demonstrates that DMAc and acetone possess the two extremes of solvent properties. To achieve intermediate surface tensions and viscosities, CA was dissolved in a binary solvent system of acetone : DMAc in different ratios. Electrospinning of CA in DMAc : acetone (2 : 1) at any concentration generated beads rather than fibres (Figure 4(a)), similar to the use of DMAc alone. Liu and Hsieh [15] also showed that DMAc : acetone (2 : 1) is not a suitable solvent for the electrospinning of CA due to its high surface tension. However, increasing the fraction of acetone to DMAc (2 : 1) improved spinning stability and produced substantiality smaller beads. Fibres of 140 nm–582 nm diameter (mean = 264 ± 0.5 nm) were produced as shown in Figure 4(b) at a polymer concentration of CA 15% (w/v).

The morphology of the electrospun CA fibres produced using the binary solution of acetone : DMAc (2 : 1) was found to vary with the concentration of the CA in the solution. The SEM images of fibres produced from 15% (w/v) and 20% (w/v) CA concentrations are shown in Figure 5. Spherical bead-like particles were observed for CA electrospun fibres produced from 15% (w/v) CA in acetone : DMAc (2 : 1) solution (Figures 5(a) and 5(b)). The morphology of electrospun webs were improved with an increase in the CA concentration to 20% (w/v) (Figures 5(c) and 5(d)). The size and number of beads decreased and the fibre diameter

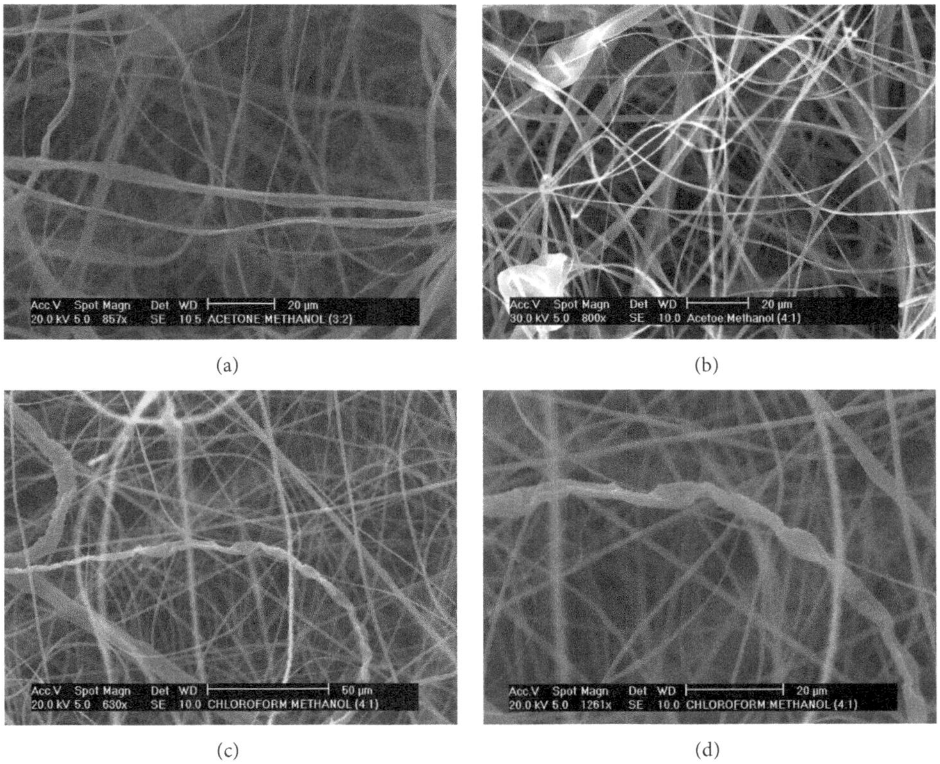

FIGURE 3: SEM micrographs showing the effect of binary solvent systems on web structure and CA electrospun fibre morphology. Mag 630x–1260x-electrospinning solvents: (a) acetone : methanol (3 : 2); (b) acetone : methanol (4 : 1); (c) chloroform : methanol (4 : 1); (d) ribbon-like structure in an electrospun CA fibre produced from the Chloroform : methanol (4 : 1) solvent.

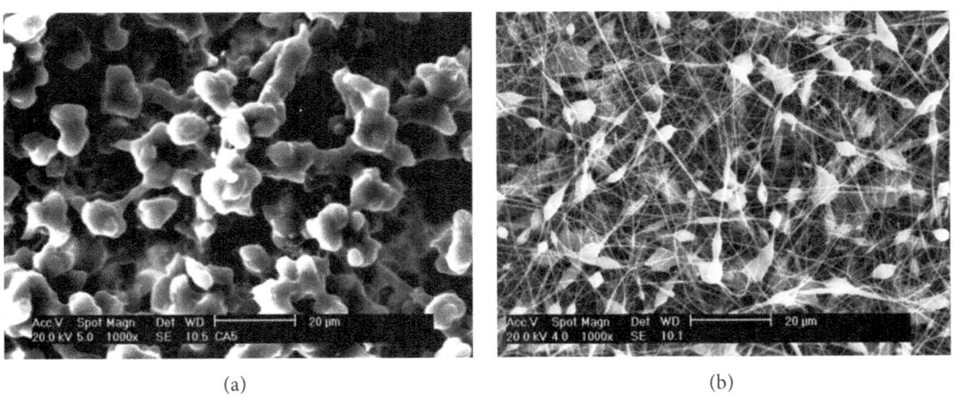

FIGURE 4: SEM micrographs showing the effect of binary solvent systems on resulting CA electrospun fibre morphology and web structure at a polymer concentration of 15% (w/v). Mag 300x: (a) DMAc : acetone (2 : 1) large beads, (b) acetone : DMAc (2 : 1) fibres and small beads.

increased as the CA concentration increased from 15% to 25% (w/v). The range of fibre diameters of beaded-fibres electrospun from a 15% (w/v) CA solution under different electrospinning conditions was around 140 nm–582 nm. At CA concentrations of 20% (w/v) and 25% (w/v), continuous and uniform electrospun fibres were generated with fibre diameters of 192–957 nm (mean = 395 ± 0.5 nm) and 318 nm–1.80 μm (mean = 635 ± 0.5 nm), respectively. The results clearly indicate that electrospun fibre diameter increased by increasing the CA concentration. These results confirm that the most versatile solvent system for electrospinning of CA is a binary system of acetone : DMAc (2 : 1) (Figure 5). Further experiments revealed that optimum conditions for the production of cellulose acetate (CA) in a binary solvent system of acetone : DMAc (2 : 1) can be identified as a CA polymer concentration of 25% (w/v), voltage 25 kV and a flow rate of 0.002 (mL min^{-1}) (Figure 6).

3.1. Effect of Deacetylation of Ultrafine Cellulose Acetate on Fibre Properties. Previously researchers have investigated changes in tensile properties as a result of deacetylation [13]. In this study, the influence of heat treatment (post web

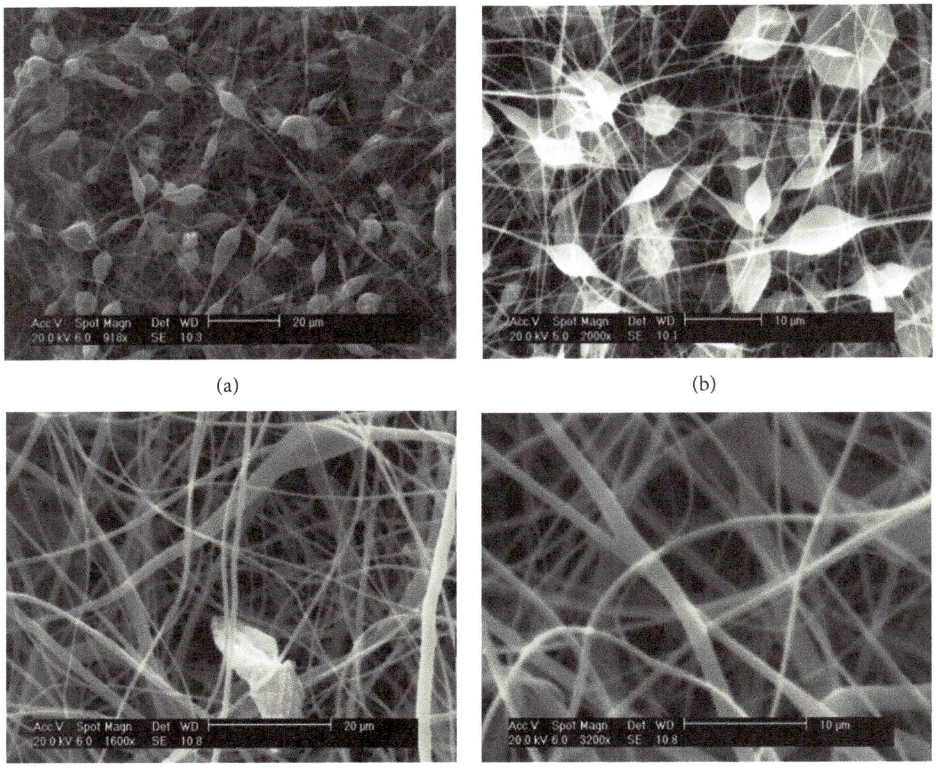

(a) (b)

(c) (d)

FIGURE 5: SEM micrographs of CA webs produced from binary mixed solvent systems: acetone : DMAc (2 : 1) at polymer concentrations: ((a)-(b)) 15% (w/v) mean fibre diameter = 264 ± 0.5 nm:, ((c)-(d)): 20% (w/v) mean fibre diameter = 395 ± 0.5 nm.

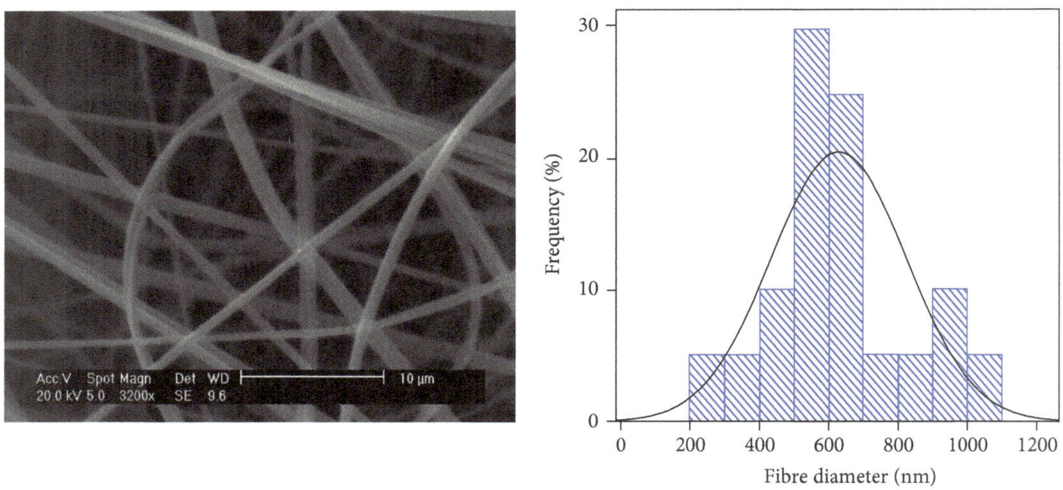

(a) (b)

FIGURE 6: SEM micrograph of CA web produced from the binary solvent system acetone : DMAc at a polymer concentration of 25% (w/v). The operating voltage was 25 kV and the flow rate 0.002 (mL $\min^{-1}$) (b) Fibre diameter distribution obtained using optimum spinning conditions.

formation) on the mechanical properties of the electrospun CA fibre webs (25% w/v in a binary solvent system of acetone : DMAc (2 : 1), voltage 25 kV and a flow rate of 0.002 mL $\min^{-1}$) was explored as a means of maximising their strength.

3.1.1. Heat and Alkaline Treatment of the Cellulose Fibres. Electrospun CA webs were carefully removed from the aluminum foil and then placed into an oven at 208°C for 1 hour. A temperature of 208°C was selected because it is known that the glass transition temperature (T_g) of CA is 198–205°C

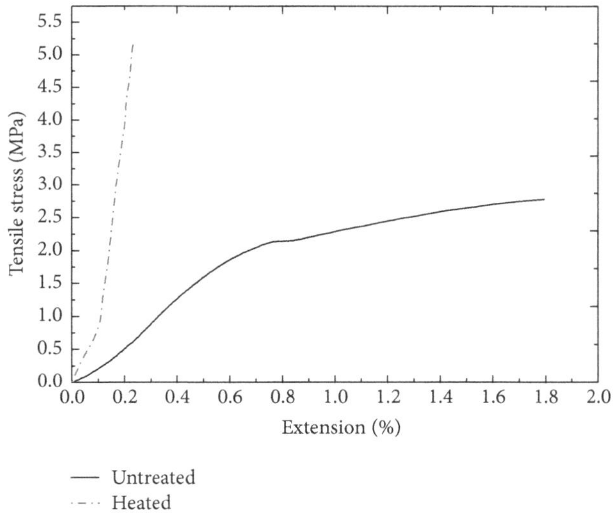

FIGURE 7: Tensile stress-strain curve of the untreated CA fibre web and the heat-treated CA webs. Cross-head speed was 10 mm min^{-1}, web thickness: 150 μm, sample width: 6.35 mm.

and the melting temperature (T_m) is 224–230°C [13]. The heat treated CA webs were then immersed in a 4 g dm^{-3} NaOH solution in an ethanol : water (1 : 4) mixture for 24 hr at room temperature. It has been shown that deacetylation with NaOH is more efficient and complete in ethanol than under aqueous conditions [10, 29]. Acetyl groups in the CA molecules could be completely removed *via* the hydrolysis reaction [30], after deacetylation fibres were rinsed in distilled water and left in desiccators under vacuum until dry.

3.1.2. Tensile Properties. The untreated webs exhibited low tensile strength and high elongation at break compared to the heat treated material. The tensile modulus and tensile strength were improved via the heat treatment process, consistent with previous findings [13]. Figure 7 displays the tensile stress-strain curve of the untreated CA webs compared to the heat-treated CA webs obtained at a cross-head speed of 10 mm min^{-1}. The increase in tensile strength can be attributed to cross-linking between the CA fibres after heat treatment leading to interfacial bonding between fibres at the cross-over points as is evident from Figure 8. As a result, the material's structural integrity and the mechanical strength were dramatically improved [13]. Interestingly, the fibre morphology was not modified by heat treatment and remained intact. All results are in full agreement with previous research reported by Ma and coworkers [13].

An ATR-FTIR spectrum of the untreated cellulose acetate web is shown in Figure 9(a). The broad absorption band seen at 3478 cm^{-1} is attributed to O–H stretching [10, 13, 15]. The band at 1738 cm^{-1} is associated with the overlapped ester carbonyls in CA. This peak is a characteristic peak for CA as other bands may also occur in other cellulosic polymers. Figure 9(b) shows changes in the ATR-FTIR spectra of CA fibres after deacetylation for 24 hr (RC web). As can be observed, the characteristic acetyl band for CA at 1738 cm^{-1} completely disappeared, indicating that the cellulose acetate is converted to regenerated cellulose (Cellulose II). Also, the broad absorption band seen at 3478 cm^{-1} in Figure 9(a) appears to be larger in Figure 9(b). This suggests the presence of more hydroxyl groups after deacetylation. Interestingly, the alkaline treatment did not result in any major morphological changes in the constituent fibres (Figure 10).

4. Conclusion

The aim of this research was to interpret the varying morphology of electrospun cellulose acetate (CA) fibres in single and binary solvent systems based on solubility parameters to identify suitable solvent systems and optimum processing conditions consistent with the production of defect-free CA fibrous webs. In the present study, six solvents, namely, acetone, acetic acid, water, *N,N*-Dimethylacetamide (DMAc), chloroform, and methanol were examined as single and binary solvent systems for the electrospinning of cellulose acetate (CA). The results indicated that fibres could not be continuously produced when acetone, chloroform, and methanol were used as solvents to prepare CA electrospinning solutions because of needle tip blockage as a consequence of their high volatility and low boiling point. Only a few ribbon-like fibres were observed to varying extents in the webs produced from acetone and a relatively large ribbon width was obtained. Similarly, no fibres were obtained from CA solutions comprising acetic acid due to the relatively low dielectric constant of the solvent, which results in unstable electrospinning. The strong hydrogen-bonding interaction between CA and DMAc and also its high boiling point reduces the evaporation rate of DMAc molecules during electrospinning in the short travel distance from the needle tip to the collector.

Based on the Hansen solubility parameters ($D_{(s-p)} <$ 12.40), CA is soluble in acetic acid (2.68), acetone (5.39), DMAc (5.88), chloroform (8.87), and methanol (12.35). Similarly, CA is soluble in solvents when the Hildebrand solubility

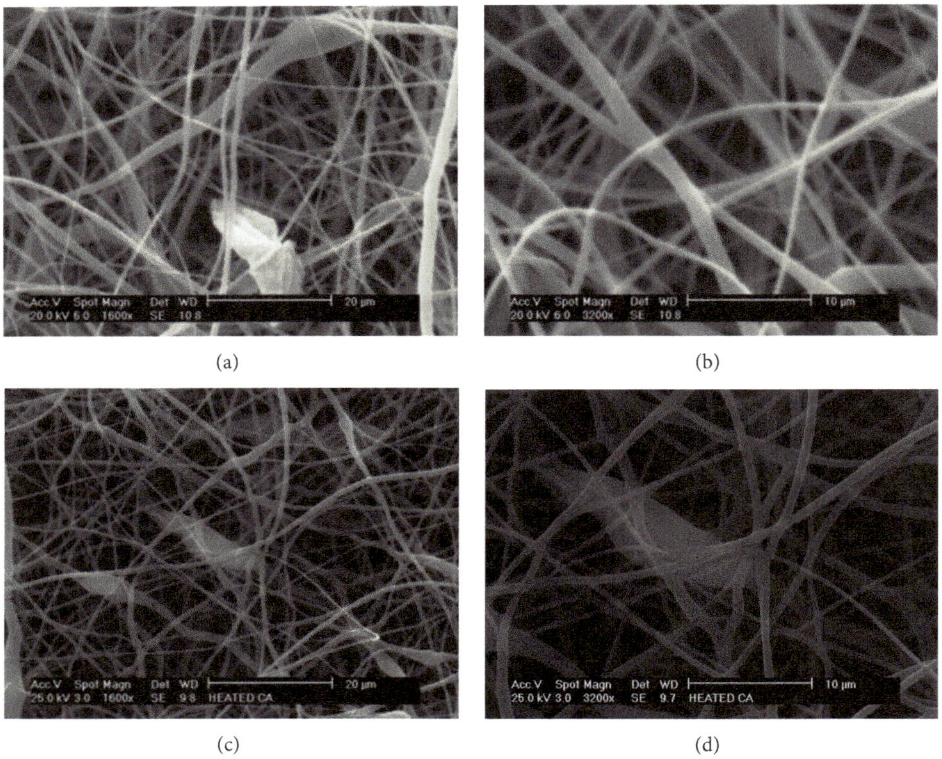

Figure 8: Selected SEM images of electrospun fibres ((a), (b)) cellulose acetate (25% w/v) web; ((c), (d)) heat-treated CA web.

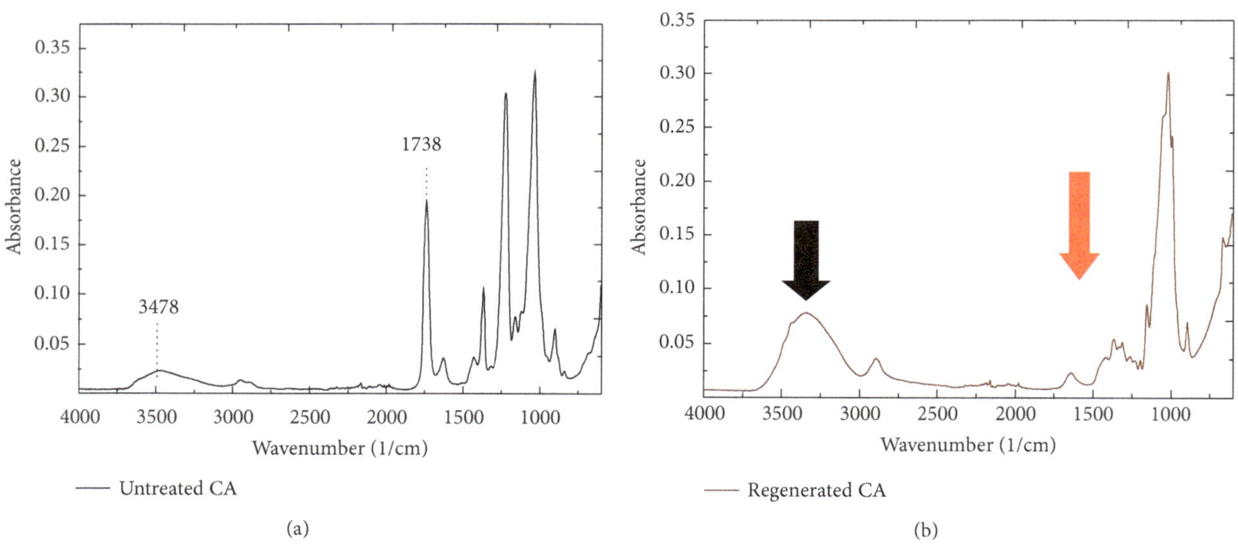

Figure 9: ATR-FTIR spectra of the electrospun CA fibre webs (a) CA web and (b) regenerated Cellulose web.

parameters (δ) lie between 9.5 and 12.5 $(cal/cm^3)^{1/2}$, and therefore, among the six solvents examined, only two solvents meet both solubility criteria : acetone and DMAc with, Hildebrand solubility parameters of 9.77 and 11.1 $(cal/cm^3)^{1/2}$, respectively.

A binary solvent system of acetone : DMAc (2 : 1) was found to be the most suitable solvent system for continuous electrospinning of CA fibres. The preferred conditions for electrospinning of CA in acetone : DMAc (2 : 1) were a polymer concentration of 25% (w/v), voltage of 25 kV, and flow rate of 0.002 (mL min^{-1}). CA fibres exhibited improved tensile strength and modulus after heat and alkaline treatment (deacetylation) of the as-spun web. No substantial fibre morphological degradation occurred during this process.

(a)

(b)

FIGURE 10: SEM images of the surface of the regenerated cellulose fibres at a polymer concentration of 25% (w/v). The operating voltage was 25 kV and the flow rate 0.002 (mL min^{-1}).

References

[1] A. Idris, K. Y. Lee, and H. K. Hing, "Preparation of cellulose acetate dialysis membrane for separation of bovine serum albumin," *Journal of Technology*, vol. 42, pp. 35–46, 2005.

[2] Z.-M. Huang, Y. Z. Zhang, M. Kotaki, and S. Ramakrishna, "A review on polymer nanofibers by electrospinning and their applications in nanocomposites," *Composites Science and Technology*, vol. 63, no. 15, pp. 2223–2253, 2003.

[3] G. Eda, J. Liu, and S. Shivkumar, "Solvent effects on jet evolution during electrospinning of semi-dilute polystyrene solutions," *European Polymer Journal*, vol. 43, no. 4, pp. 1154–1167, 2007.

[4] C. Lu, P. Chen, J. Li, and Y. Zhang, "Computer simulation of electrospinning. Part I. Effect of solvent in electrospinning," *Polymer*, vol. 47, no. 3, pp. 915–921, 2006.

[5] L. Wannatong, A. Sirivat, and P. Supaphol, "Effects of solvents on electrospun polymeric fibers: preliminary study on polystyrene," *Polymer International*, vol. 53, no. 11, pp. 1851–1859, 2004.

[6] T. Uyar and F. Besenbacher, "Electrospinning of uniform polystyrene fibers: the effect of solvent conductivity," *Polymer*, vol. 49, no. 24, pp. 5336–5343, 2008.

[7] T. Jarusuwannapoom, W. Hongrojjanawiwat, S. Jitjaicham et al., "Effect of solvents on electro-spinnability of polystyrene solutions and morphological appearance of resulting electrospun polystyrene fibers," *European Polymer Journal*, vol. 41, no. 3, pp. 409–421, 2005.

[8] Q. Yang, Z. Li, Y. Hong et al., "Influence of solvents on the formation of ultrathin uniform poly(vinyl pyrrolidone) nanofibers with electrospinning," *Journal of Polymer Science B*, vol. 42, no. 20, pp. 3721–3726, 2004.

[9] L. Van der Schueren, B. De Schoenmaker, B. Kalaoglu, and K. De Clerck, "An alternative solvent system for the steady state electrospinning of polycaprolactone," *European Polymer Journal*, vol. 47, no. 6, pp. 1256–1263, 2011.

[10] W. K. Son, J. H. Youk, T. S. Lee, and W. H. Park, "Electrospinning of ultrafine cellulose acetate fibers: studies of a new solvent system and deacetylation of ultrafine cellulose acetate fibers," *Journal of Polymer Science B*, vol. 42, no. 1, pp. 5–11, 2004.

[11] S. O. Han, J. H. Youk, K. D. Min, Y. O. Kang, and W. H. Park, "Electrospinning of cellulose acetate nanofibers using a mixed solvent of acetic acid/water: effects of solvent composition on the fiber diameter," *Materials Letters*, vol. 62, no. 4-5, pp. 759–762, 2008.

[12] S. Tungprapa, T. Puangparn, M. Weerasombut et al., "Electrospun cellulose acetate fibers: effect of solvent system on morphology and fiber diameter," *Cellulose*, vol. 14, no. 6, pp. 563–575, 2007.

[13] Z. Ma, M. Kotaki, and S. Ramakrishna, "Electrospun cellulose nanofiber as affinity membrane," *Journal of Membrane Science*, vol. 265, no. 1-2, pp. 115–123, 2005.

[14] D. Haas, S. Heinrich, and P. Greil, "Solvent control of cellulose acetate nanofibre felt structure produced by electrospinning," *Journal of Materials Science*, vol. 45, no. 5, pp. 1299–1306, 2010.

[15] H. Liu and Y.-L. Hsieh, "Ultrafine fibrous cellulose membranes from electrospinning of cellulose acetate," *Journal of Polymer Science B*, vol. 40, no. 18, pp. 2119–2129, 2002.

[16] J. F. Kadla and R. Korehei, "Effect of hydrophilic and hydrophobic interactions on the rheological behavior and microstructure of a ternary cellulose acetate system," *Biomacromolecules*, vol. 11, no. 4, pp. 1074–1081, 2010.

[17] J. Burke, Ed., *Solubility Parameters: Theory and Application*, vol. 3 of *The Book and Paper Group Annual*, 1984.

[18] C. Hansen, *Hansen Solubility Parameters: A User's Handbook*, CRC Press, Boca Raton, Fla, USA, 2nd edition, 2007.

[19] M. Karimi, W. Albrecht, M. Heuchel et al., "Determination of water/polymer interaction parameter for membrane-forming systems by sorption measurement and a fitting technique," *Journal of Membrane Science*, vol. 265, no. 1-2, pp. 1–12, 2005.

[20] A. Vetere, "A simple modification of the Flory-Huggins theory for polymers in non-polar or slightly polar solvents," *Fluid Phase Equilibria*, vol. 34, no. 1, pp. 21–35, 1987.

[21] B. A. Miller-Chou and J. L. Koenig, "A review of polymer dissolution," *Progress in Polymer Science*, vol. 28, no. 8, pp. 1223–1270, 2003.

[22] N. Olaru and L. Olaru, "Electrospinning of cellulose acetate phthalate from different solvent systems," *Industrial and Engineering Chemistry Research*, vol. 49, no. 4, pp. 1953–1957, 2010.

[23] M. W. Frey, "Electrospinning cellulose and cellulose derivatives," *Polymer Reviews*, vol. 48, no. 2, pp. 378–391, 2008.

[24] S. Koombhongse, W. Liu, and D. H. Reneker, "Flat polymer ribbons and other shapes by electrospinning," *Journal of Polymer Science B*, vol. 39, no. 21, pp. 2598–2606, 2001.

[25] A. Celebioglu and T. Uyar, "Electrospun porous cellulose acetate fibers from volatile solvent mixture," *Materials Letters*, vol. 65, no. 14, pp. 2291–2294, 2011.

[26] C. Appaw, *Rheology and microstructure of cellulose acetate in mixed solvent systems [Ph.D. thesis]*, North Carolina State University, USA, North Carolina, NC, USA, 2004.

[27] G. C. Rutledge, Y. Li, and S. Fridrikh, "Electrostatic spinning and properties of ultrafine fibers," National Textile Center Annual Report, 2000.

[28] O. Suwantong, P. Opanasopit, U. Ruktanonchai, and P. Supaphol, "Electrospun cellulose acetate fiber mats containing curcumin and release characteristic of the herbal substance," *Polymer*, vol. 48, no. 26, pp. 7546–7557, 2007.

[29] P. Goswami, R. S. Blackburn, H. M. El-Dessouky, J. Taylor, and P. White, "Effect of sodium hydroxide pre-treatment on the optical and structural properties of lyocell," *European Polymer Journal*, vol. 45, no. 2, pp. 455–465, 2009.

[30] G. Salihu, P. Goswami, and S. Russell, "Hybrid electrospun nonwovens from chitosan/cellulose acetate," *Cellulose*, vol. 19, no. 3, pp. 739–749, 2012.

[31] M. Dabral, L. F. Francis, and L. E. Scriven, "Drying process paths of ternary polymer solution coating," *AIChE Journal*, vol. 48, no. 1, pp. 25–37, 2002.

18

Effect of Attapulgite Nanorods and Calcium Sulfate Microwhiskers on the Reaction-Induced Phase Separation of Epoxy/PES Blends

Xiaolin Tang, Xinhui Zhong, Guozhu Zhan, Yingfeng Yu, and Hongdong Zhang

State Key Laboratory of Molecular Engineering of Polymers, Department of Macromolecular Science, Fudan University, Shanghai 200433, China

Correspondence should be addressed to Hongdong Zhang; zhanghd@fudan.edu.cn

Academic Editor: Jan-Chan Huang

The influence of two kinds of mesoscale inorganic rod fillers, nanoscale attapulgite and micron-sized $CaSO_4$ whisker, on the reaction-induced phase separation of epoxy/aromatic amine/poly- (ether sulfone) (PES) blends has been investigated by optical microscopy (OM), scanning electron microscopy (SEM), and time resolved light scattering (TRLS). By varying the PES concentration and curing temperature, we found that the incorporation of attapulgite and $CaSO_4$ had dramatic impact on the phase separation process and the final phase morphology of blends. In blends at higher content than critical concentration, the process of phase separation was retarded by the incorporation of nanoscale fillers but accelerated by that of the micron-sized fillers, mainly due to the enhanced viscoelastic effect and the preferential wettable effect, respectively. Meanwhile both mesoscale fillers could change the cocontinuous phase structure of blends with lower PES content than critical concentration into PES-rich dispersed structure due to the surface affinity of fillers to epoxy matrix.

1. Introduction

It has been generally accepted that phase separation plays a significant role in determining the morphology and properties of polymer mixtures. The reaction-induced phase separation in thermosets/thermoplastic blend is a very important physical/chemical process, which has aroused wide interest for many years [1–12].

Recently, it has been found that the incorporation of inorganic filler has dramatic impact on the phase separation behavior and the final phase structure of binary polymer mixture and thermosets/thermoplastic blends [13–18], especially when the fillers are preferentially wettable to one component.

For the ternary mixtures of polymer blends containing glass particles in micron size, Tanaka et al. [19] proved that both the coarsening dynamics and the final morphology were significantly affected by the mobility of glass particles and their selective inclusion of the more wettable phase.

By simulation, Balazs and coworkers pointed out that the addition of hard particles greatly changed both the speed and morphology of phase separation, and the phase separation is arrested in the late stage [20–22]. For the blends with mobile nanoscale rods, their simulation results show that the rods could self-assemble into needle-like and percolating network, and increasing the rod concentration beyond the effective percolation threshold could drive the system to self-assemble in a Lamellar morphology [23, 24]. By a new simulation methodology, Side et al. indicated that the phase pattern of PS-P2VP diblock nanocomposite containing (PSs) functionalized Au particles changes as the nanoparticles density increases, which agreed well with experimental results [25].

For the polymerization-induced phase separation in nanoparticle-monomer-polymer blends, Soulé et al. [26] indicated that a small increase in particle size had an influential effect on extending the immiscibility region of the phase diagram by using the model proposed by Ginzburg [27].

In epoxy/rubber blends containing clay, Balakrishnan et al. [16] found that the clay concentration plays an important role in the clay platelet separation and the dispersion of

TABLE 1: The weight compositions of the samples studied.

Samples	PES-14.3	PES-14.3-A	PES-14.3-Ca	PES-11.4	PES-11.4-A	PES-11.4-Ca
DGEBA (g)	1.00	1.00	1.00	1.00	1.00	1.00
DIM-DDM (g)	0.32	0.32	0.32	0.32	0.32	0.32
PES (g)	0.22	0.22	0.22	0.17	0.17	0.17
ATT (g)	0	0.064	0	0	0.062	0
$CaSO_4$ (g)	0	0	0.064	0	0	0.062

clay aggregates in the epoxy matrix, while the clay platelets were preferentially absorbed into the rubber particles. For the blends of epoxy modified with thermoplastics, Peng et al. [17] had investigated the influence of nanoscale layered silicates on the reaction-induced phase separation behavior, which indicated that the incorporation of silicates had dramatic impact on the phase separation process and the final phase morphology, though the SD mechanism was entirely followed.

While in thermoplastic modified thermoset blends with dynamic asymmetry, the chain mobility difference of the two parts makes the viscoelastic effect play a crucial role in phase separation process [28]. The addition of nanoscale fillers further enhanced this viscoelastic effect from the chain entanglement of slow dynamic part with the filler surface, which significantly hindered the phase structure evolution process and thus results in a refined phase structure with much smaller characteristic length scales [29, 30].

However, it remained unclear that the effect of mesoscale fillers and their size on the phase separation process of systems with moderate curing rate and even secondary phase separation, though we found previously that the incorporation of nanoscale rods changed the final phase morphology due to the nanoscale fillers pinning down the phase separation by distributing along the interfaces between the epoxy-rich and PES-rich phases [31].

This article aimed to investigate the reaction-induced phase separation in Epoxy/PES blends containing different size of mesoscale inorganic rod fillers, ATT (nanoscale), and $CaSO_4$ (microscale) rod, in particular, to examine the influence of incorporation of inorganic rod on the whole phase separation process, which would hopefully give us better understanding of the mechanism of reaction-induced phase separation.

2. Experimental

2.1. Materials and Samples Preparation. The epoxy oligomer DER331, a lower-molecular-weight liquid diglycidyl ether of bisphenol A (DGEBA) with an epoxide equivalent of 182–192, was purchased from Dow Chemical Co. PES ($[\eta]$ (intrinsic viscosity) = 0.36) was supplied by Jilin University (China). The curing agent 4,4′-methylene-bis(2,6-dimethylamiline) (DIM-DDM) was bought from Aldrich Chemical Co. Inc. (Germany). Attapulgite ($Si_8O_{20}Mg_5(Al)(OH)_2(H_2O)_4$, Attagel 50, Engelhard Co., United States) was surface modified by γ-methacryloxypropyltrimethoxysilane to improve its dispersing in epoxy resins. $CaSO_4$ whiskers were provided by Shanghai Fuhua Industrial Co. (China) and used without further treatment.

A certain amount of ATT or $CaSO_4$ rod was first dispersed in acetone under ultrasonication, the epoxy was then added to the suspension, and it was ultrasonicated again. Acetone was vaporized in a circulation oven at room temperature until most of the solvent was removed. After the samples were placed into vacuum oven at 100°C for 4 hrs, PES was added into the mixture of epoxy and inorganic rods and stirred fiercely at 150°C in an oil bath. When the mixture was then cooled to 100°C, stoichiometric amount of DIM-DDM (32 phr to epoxy oligomer) was added, and the mixture was stirred before cooling rapidly to room temperature to avoid further curing. A series of blends were prepared as shown in Table 1, in which the amount of ATT and $CaSO_4$ rod in two hybrid mixtures is the same of 4 wt%, and the content of PES is its weight percent in the resin matrix (without fillers). For example, PES-14.3-A means PES content is 14.3% of PES/epoxy/DIM-DDM and ATT is 4% of total amount of composition.

2.2. Measurements. A PerkinElmer Pyris 1 DSC instrument was used for the study of the curing reaction. The isothermal curing conversion was calculated from residual exotherms observed in scans in the temperature range of 50–350°C, with heating rates of 10°C/min and normalized by the total exotherms for uncured samples. A Tescan TS5163MM scanning electron microscope was employed to examine the morphologies of the fractured surfaces (the samples were first cured at specific temperature for 5 hours, then postcured at 150°C for 5 hours, then fractured in liquid nitrogen, and coated with gold). The phase separation process during isothermal curing was traced *in situ* on a self-made TRLS apparatus with a controllable hot chamber. The change in the light scattering profiles was recorded at appropriate time intervals. The samples for TRLS observation were sandwiched between two glass cover slides and pressed into thin film. In the light scattering experiment, information on the phase morphology evolution could be obtained from the systematic change in the curves of the scattering factor versus the scattering light intensity as a function of the time of phase separation, and the most probable phase domain size was $d_m = 2\pi/q_m$, where q_m is the peak scattering vector. The evolving morphologies of the samples were observed *in situ* with an XDS-1B optical microscope with a hot chamber (Chongqing Optical and Electrical Instrument Co., Ltd., China).

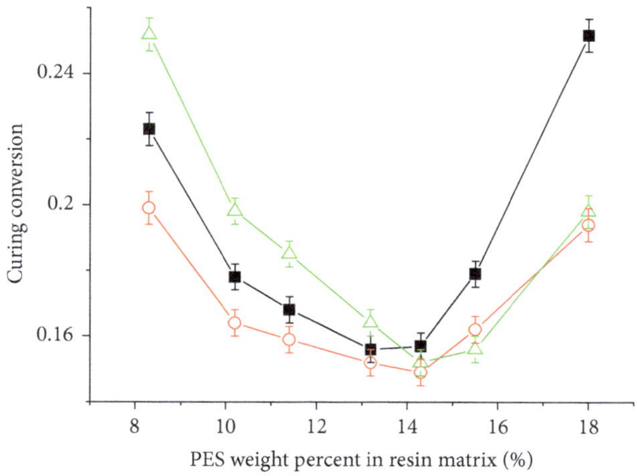

FIGURE 1: The phase diagram of the PES modified epoxy-hardener systems with and without fillers at 120°C.

3. Results and Discussions

3.1. Effect of Mesoscale Fillers on the Phase Diagram. It is well known that the mixture of epoxy precursor and PES exhibits a lower critical solution temperature (LCST) phase behavior [32], including the following: the hardener changes the polarity of the mixture and thus makes the phase decomposition follow a reaction-induced phase separation process, in which the homogenous mixture separates into two phases after the cloud point (with specific curing conversion).

Figure 1 shows the phase diagram of PES/epoxy/DIM-DDM blends without and with mesoscale fillers cured at 120°C. As the epoxy and PES have lower solubility with the increase in temperature, the mixtures begin to phase separate at lower curing conversion when the curing temperature is increased. For the blends cured at 140°C (not shown here), the curing conversions at the initial phase separation are all below 6% of curing conversion at the PES concentration range of 8 wt% to 18 wt%. As the curing conversions are too low to be accurately differentiated of systems with and without fillers at the cloud points, the phase diagram at 140°C was not shown here. In Figure 1, it can be found that the critical concentration of the PES/epoxy/DIM-DDM is located near 13% of PES. The addition of mesoscale fillers changes the phase diagram to some extent. The nanoscale ATT advances the curing conversion of systems a little earlier, while the general shape of the phase diagram keeps similar to that of systems without fillers. This conversion change of phase diagram could be resulted from the nucleation effect of nanoscale fillers, which makes the phase separation be observed a little earlier when the dimension of concentration fluctuation is comparable to the size of fillers.

However, the introduction of micron-sized $CaSO_4$ fillers shifts the critical concentration of PES to about 15% in the resin matrix. As the $CaSO_4$ fillers show better affinity to the epoxy phase, this kind of dilute effect of PES could come from the prior distribution of fillers in epoxy phase [19].

To study the effect of fillers on the phase separation, we therefore choose two compositions of 14.3 wt and 11.4 wt% PES in the resin matrix, which are located at both sides of critical concentration. The selection of blends with near critical composition guarantees the phase separation following spinodal decomposition (SD) mechanism, and thus the whole process could be well characterized by TRLS.

3.2. Double Phase Separation in PES Modified Epoxy without Fillers. The process of morphology evolution in unfilled blends with 14.3 wt% and 11.4 wt% of PES follows the SD mechanism due to the nearly critical concentration. Figure 2 shows both the phase separation process and final morphology of PES-14.3 and PES-11.4 blends.

The morphology evolution of unfilled blend shows the feature of typical viscoelastic phase separation proposed by Tanaka [33]. After incubation time, the epoxy-rich (fast dynamic phase-rich) particles start to appear ((a2) and (e2)) and their size increases with time. Then the PES-rich (slow dynamic phase-rich matrix) particles become network-like with the growth epoxy-rich particles, and the thin parts of the network-like structure are elongated and broken, and finally the epoxy-rich dispersed sea-island phase structure was frozen by gelation or vitrification in the PES-14.3 blend ((a3) and (b)). For the PES-11.4 blend, the cocontinuous phase structure could be observed in the late stage of the phase separation ((e3) and (f)). It is noticeable that the secondary phase separation took place in both primary epoxy- and PES-rich matrix (namely, in fast- and slow-dynamic-rich phase, (a3), (b), (e3), and (f)).

SEM micrographs show the usual phase-in-phase morphology in the PES-14.3 and PES-11.4 blends. The secondary phase separation took place in both primary epoxy-rich and PES-rich phases in these two blends cured at 120 and 140°C. It indicates that the secondary PES-rich particles disperse in the primary epoxy-rich matrix ((c2) and (g2)) and the secondary epoxy-rich particles disperse in the primary silk-like PES-rich domain ((c3) and (g3)). This structure is commonly similar to that described previously [34].

For the PES-11.4 blend, it is could be observed that the co-continuous phase morphology due to the higher concentration of epoxy resin and the lower viscosity (viz., weaker elastic strength of PES-rich domain). In the blend with higher PES concentration (PES-14.3 blend), after the epoxy-rich particles taking place, the increase in size of these particle was restrained by the stronger strength of PES-rich domain. In the above mentioned two blends, the secondary phase separation takes place as a result of the quench going deeper along with the epoxy resin reaction proceeding [1].

3.3. Competition of Enhanced Viscoelastic Effect and Surface Wetting Effect of Mesoscale Fillers on the Phase Separation at Higher PES Content Than Critical Concentration. The phase separation evolution process and final morphology of the two mesoscale fillers added PES-14.3 blends are shown in Figure 3.

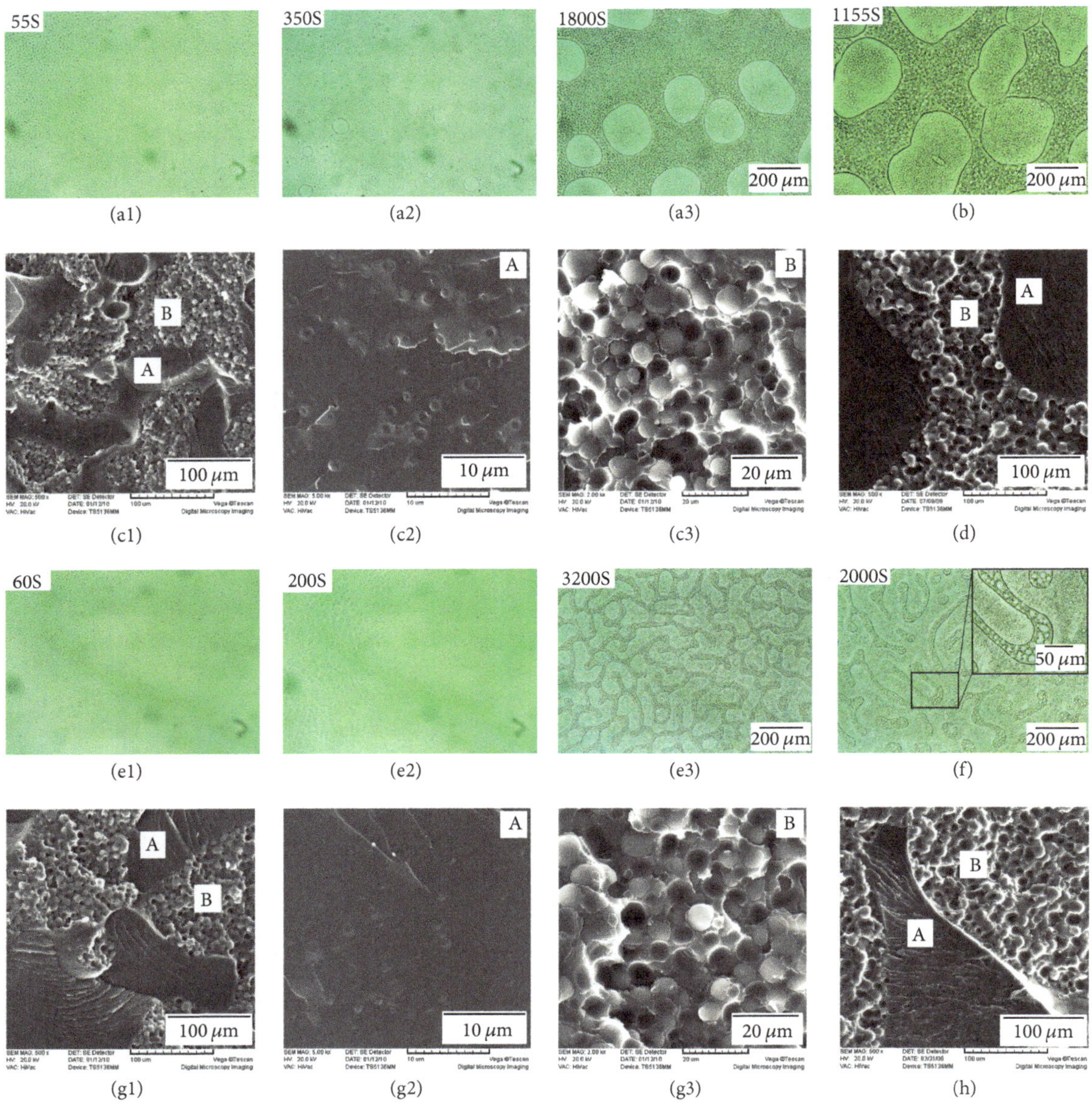

Figure 2: Development of morphologies of the blends followed by OM and SEM micrographs of the fracture surface of the blends. (a1)–(a3) PES-14.3 by OM at 120°C; (b) finial morphology of PES-14.3 by OM at 140°C; (c1)–(c3) PES-14.3 by SEM at 120°C for 5 h; (d) PES-14.3 by SEM at 140°C for 5 h; (e1)–(e3) PES-11.4 by OM at 120°C; (f) finial morphology of PES-11.4 by OM at 140°C; (g1)–(g3) PES-11.4 by SEM at 120°C for 5 h; (h) PES-11.4 by SEM at 140°C for 5 h.

The ATT added systems show similar phase separation behavior as unfilled systems, in which the epoxy-rich phase shows as island structure dispersed in the PES-rich matrix ((a1), (a2), (a3), and (b)). The SEM micrographs ((c), (d1), (d2) is shown and (d3)) show a much refined phase structure of PES-14.3-A system compared with that of PES-14.3 (Figures 2(c1), 2(c2), 2(c3) and 2(d)), in which the interface between secondary PES-rich particle and primary epoxy-rich matrix (Figure 3(d2)), and that between secondary epoxy-rich particle and primary membrane-like PES-rich phase (Figure 3(d3)) is no longer sharp due to the incorporation of the ATT.

The systems with $CaSO_4$ seem to show similar behavior as PES-14.3 at both 120 and 140°C in the early-stage of phase separation. OM micrographs following morphology evolution shows a micro co-continuous structure initially (Figures 3(f1) and 3(f2)), and then small epoxy-rich particles form around the $CaSO_4$ whisker (Figure 3(f3)) due to the $CaSO_4$ whisker's effect of preferential wettable to the epoxy resin. However, SEM results indicate that $CaSO_4$ whiskers are dispersed in both PES-rich and epoxy-rich phase, which implies that there may exist two different effects rather than the preferential affinity to epoxy.

TRLS is employed to investigate the differences between the systems with and without fillers, and the results were shown in Figure 4. The changes in q_m with phase evolution are illustrated for the higher PES content than critical composition cured at 120 and 140°C. Figure 4(a) plots the

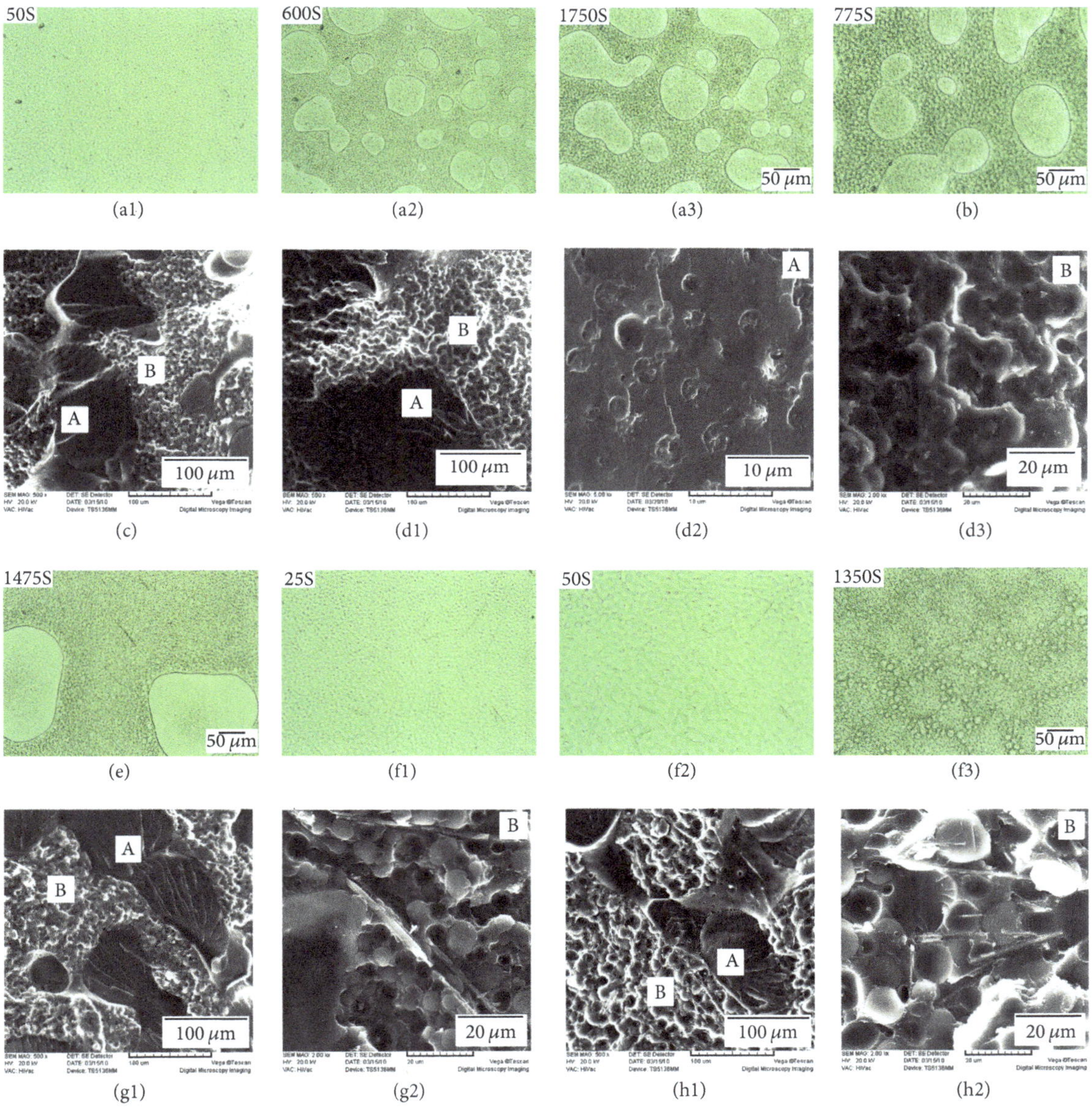

FIGURE 3: Development of morphologies of the PES-14.3 blends with filler followed by OM and SEM micrographs of the fracture surface of these blends with filler. (a1)–(a3) PES-14.3-A by OM at 120°C; (b) finial morphology of PES-14.3-A by OM at 140°C; (c) PES-14.3-A by SEM at 120°C for 5 h; (d1)–(d3) PES-14.3-A by SEM at 140°C for 5 h; (e) finial morphology of PES-14.3-Ca by OM at 120°C; (f1)–(f3) PES-14.3-Ca by OM at 140°C; (g1) and (g2) PES-14.3-Ca by SEM at 120°C for 5 h; (h1) and (h2) PES-14.3-Ca by SEM at 140°C for 5 h.

PES-14.3-Ca cured at 140°C as an example of light scattering vector and intensity changing during phase separation. At the early stage of phase separation, q_m shifted to lower value with a quick increase of light intensity, which indicates the phase structure growth with concentration varying. Then both q_m and I_m dropped quickly to a minimum value, at the same time, a second pick at high vector value grew up. The existence of minimum value in the curve of q_m versus time means that appearance of secondary particles make the average phase domain size decrease.

For PES-14.3 cured at 120 and 140°C, it could be found that the maximal slope of $\log q_m$ versus $\log t$ is 0.71 and 0.89, respectively. Elevated temperature increases the chain mobility and thus makes the phase separation a little quicker than that at lower temperature. The addition of nanoscale filler ATT in PES-14.3-A systems shifts the maximal slope to 0.41 (120°C) and 0.50 (140°C), meaning the phase evolution rate drops significantly. However, with the incorporation of micron-sized $CaSO_4$, the maximal slope increases to 1.00 (120°C) and 1.09 (140°C), these unusually steep slopes mean that the size of the epoxy-rich domain grows rapidly in a short time, which clearly indicate the accelerating effect of this filler on the phase separation process. In other words, the incorporation of fillers show an opposite effect

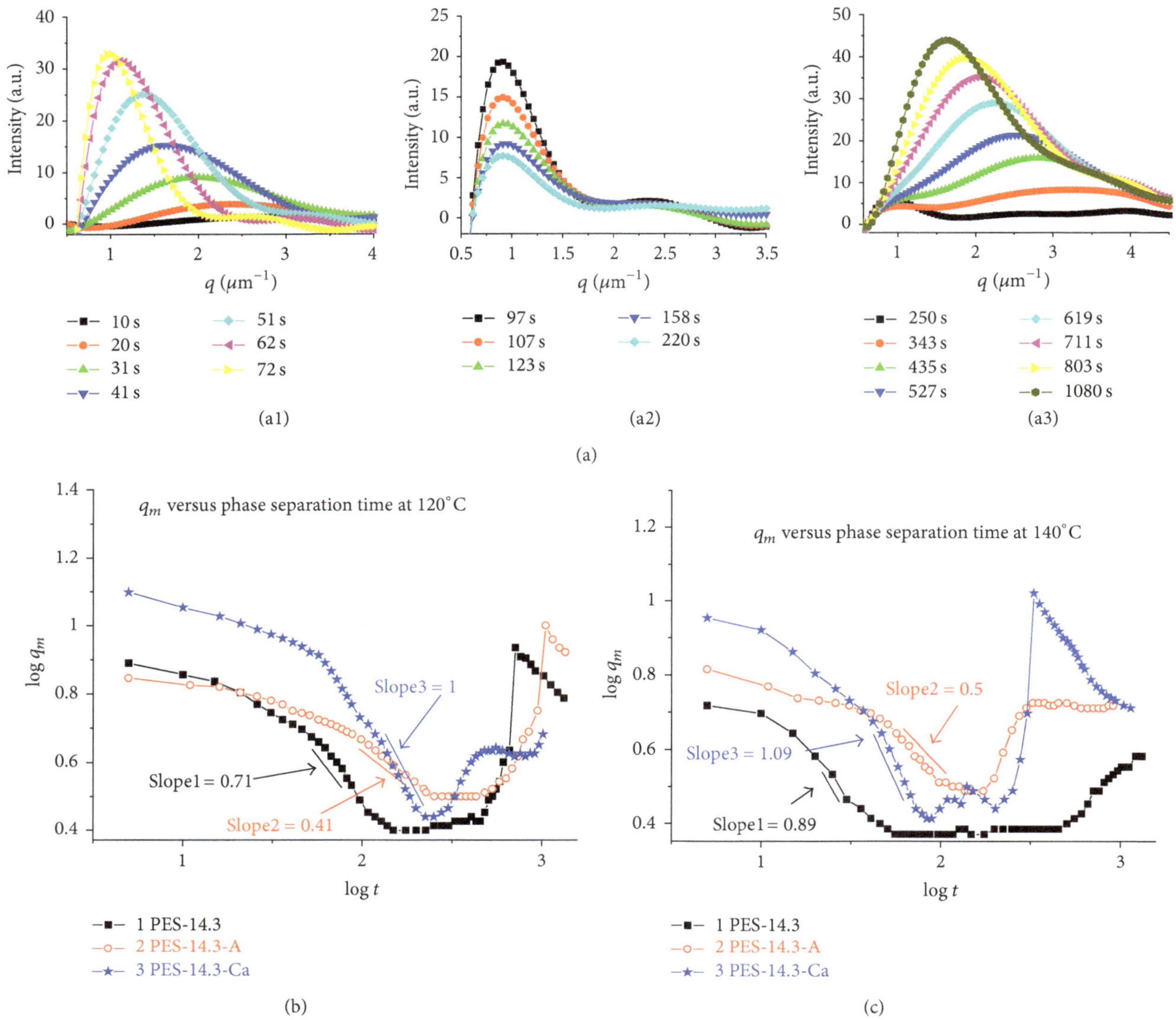

FIGURE 4: The changes of scattering vector, q_m, and light intensity versus time of the PES-14.3 blends with and without fillers. (a) PES-14.3-Ca cured at 140°C, (b) q_m versus time at 120°C, and (c) q_m versus time at 140°C.

on the phase separation: the nanoscale fillers pin down the phase separation while micron-sized fillers accelerate the decomposition process at some stages.

On the basis of the above experimental results and previous works [19, 28–30], a possible scenario for the effect of mesoscale fillers on the phase separation of PES-modified epoxy blend could be as follows: in a dynamic asymmetric system with consider amount of slow dynamic components (PES at higher content than critical concentration), the diffusion of the fast dynamic epoxy-rich phase is prevented by the slow dynamic PES-rich phase, the addition of nanoscale fillers could enhance this process by the chain entanglement of slow dynamic phase with the filler surface due to the nanoscale fillers providing sufficient surface area [29, 30]. Even though the nanoscale fillers have higher affinity to the fast dynamic phase, the elastic deformation of the slow dynamic phase could also form chain entanglement with the filler surface due to the high normal stress from the shear and shrink of the slow-dynamic phase. As a result, the nanoscale filler incorporated systems show a slower phase separation process compared to that of unfilled PES-epoxy-hardener systems, where the slopes of log q_m versus log t of nanoscale filler added systems are only about half values of those of unfilled systems at all the temperatures studied.

While for micron-sized $CaSO_4$ whisker, which has a specific surface area about one tenth of that of ATT, the large filler size provide much smaller amount of entangle points with the slow dynamic phase, and thus this filler show less effect on the viscoelastic behavior of the phase separation process. Moreover, the high affinity of the micron-sized fillers to epoxy-rich phase further facilitate the growth of epoxy-rich along the surface of the rod-fillers, which therefore

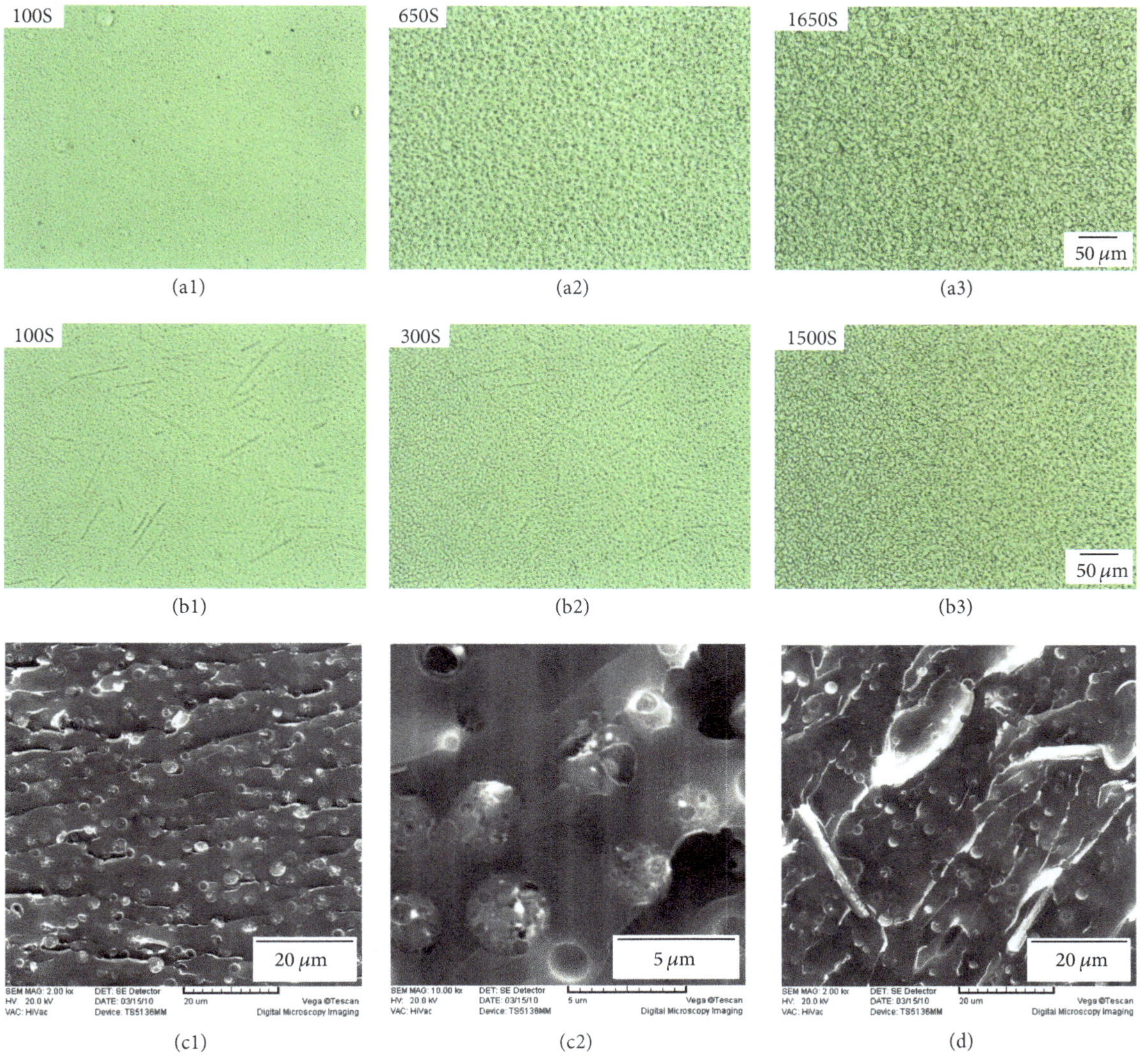

Figure 5: Development of morphologies of the PES-11.4 blend with fillers cured at 120°C followed by OM and SEM micrographs of the fracture surface of the blend with fillers cured at 120°C for 5 h. (a1)–(a3) PES-11.4-A by OM; (b1)–(b3) PES-11.4-Ca by OM; (c1) and (c2) PES-11.4-A by SEM; (d) PES-11.4-Ca by SEM.

demonstrates a quick change in the peak scattering vector q_m in TRLS due to the wetting effect of whiskers and interfacial tension of PES-rich phase [19].

From the above results, it seems that the influence of mesoscale fillers on the phase separation are mainly rely on the competition result of enhanced viscoelastic effect of the slow dynamic phase and surface wetting of the epoxy phase (fast dynamic phase) in our studied systems. Therefore, it can be inferred that surface wetting would be the main factor affecting phase separation in low dynamic asymmetric systems. To verify this point, the lower PES content (than critical concentration) systems are studied due to the diminished viscoelastic effect of these systems from lower volume fraction of slow dynamic phase.

3.4. Dilute Effect of Mesoscale Fillers at Lower PES Content. The phase evolution of mesoscale fillers added PES-11.4 blends (systems of lower content than critical concentration of PES) were followed by OM cured at 120°C (as shown in Figures 5(a1), 5(a2), 5(a3), 5(b1), 5(b2), and 5(b3)). The phase separation of mesoscale filler added systems shows distinct differences from that of unfilled PES-11.4 system (Figures 2(e1), 2(e2), 2(e3) and 2(f)). Although a micro co-continuous phase structure could also be observed at the initial stage of phase separation ((a1) and (b1)), the morphology then changes into PES-rich dispersed structure ((a2) and (b2)) and keeps to the end of the phase separation process ((a3) and (b3)).

SEM micrographs of the cured samples definitely indicate the PES-dispersed phase structure of the two mesoscale filler added systems (Figures 5(c1), 5(c2) and 5(d)). Furthermore, it can be clearly observed that a secondly phase separation still takes place in the PES-rich particles (Figure 5(c2)), while the mesoscale fillers are preferred to disperse in the epoxy matrix (Figure 5(d)) due to their high affinity.

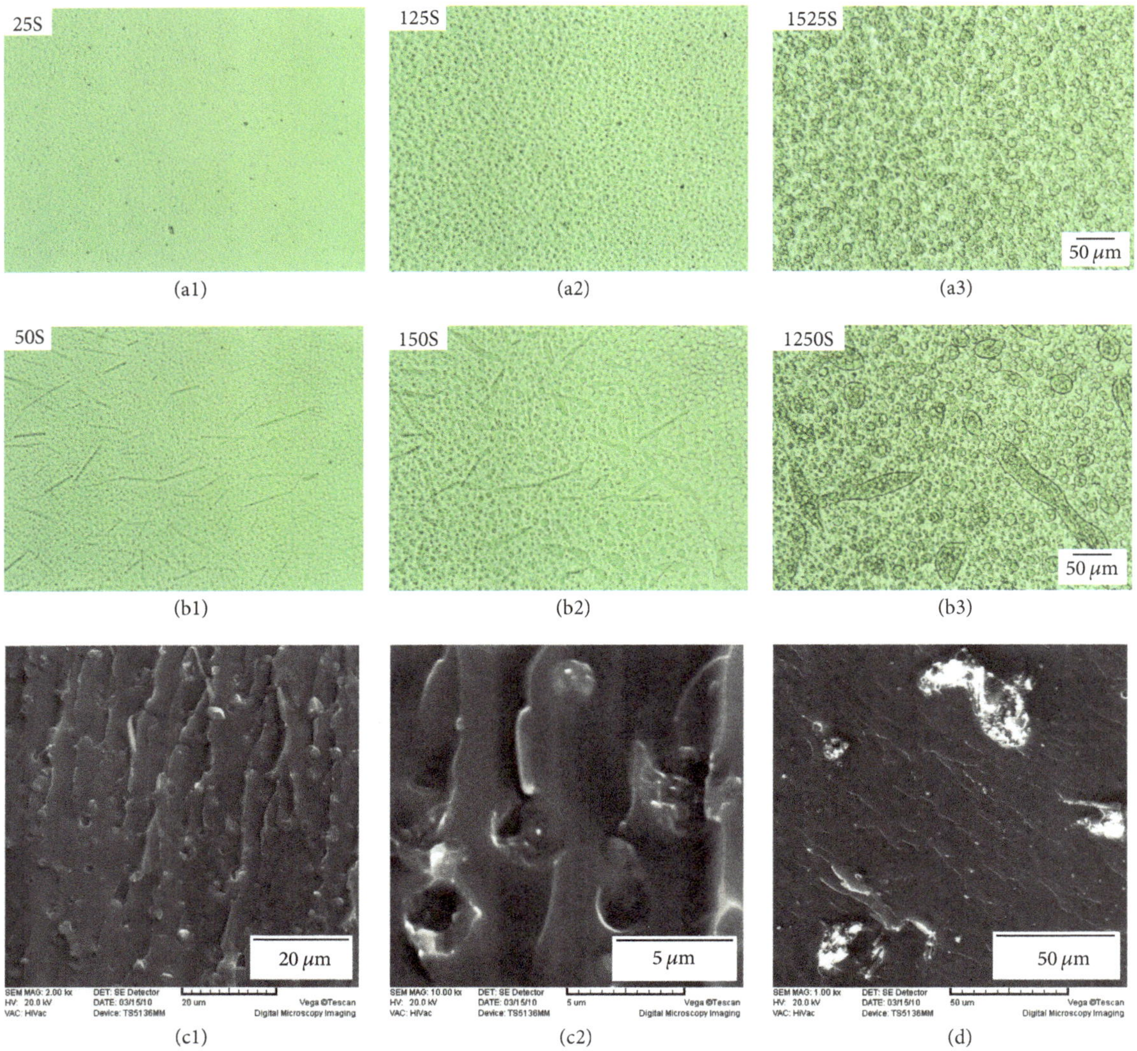

FIGURE 6: Development of morphologies of the PES-11.4 blend with fillers cured at 140°C followed by OM and SEM micrographs of the fracture surface of the blend with fillers cured at 140°C for 5 h. (a1)–(a3) PES-11.4-A by OM; (b1)–(b3) PES-11.4-Ca by OM; (c1) and (c2) PES-11.4-A by SEM; (d) PES-11.4-Ca by SEM.

This kind of morphology change due to mesoscale fillers could also be observed at other curing conditions. As shown in Figure 6, filled systems cured at 140°C also proceed into PES-dispersed phase separation. Furthermore, the SEM micrographs show a PES-dispersed morphology (Figures 6(c1), (c2) and 6(d)) without secondary phase structure (Figure 6(c2)). In Figure 6(b1), 6(b2) and 6(b3), one could also found the aggregation of PES-particles along the whisker axis ((b2) and (b3)), which is deduced by the unmoved $CaSO_4$ whisker blocking the PES-rich particles to reduce the interfacial energy [19].

According to Tanaka's theory, [33] the viscoelastic phase separation takes place most likely near critical concentration at deep quenches, while lower PES content always tend to be shifted to the traditional phase separation mechanisms due to the lower volume fraction of slow dynamic phase having shorter characteristic deformation time.

Even though the PES-11.4 system still follows viscoelastic phase separation, the incorporation of mesoscale fillers changes it into normal spinodal decomposition process due to the dilute effect of fillers and their high affinity to the epoxy-rich phase. The low volume fraction of PES (slow dynamic phase) results a quite short disentangle time of polymer chains, and thus the addition of nano fillers show limited enhancement effect on the viscoelastic phase separation. However, the affinity of both mesoscale fillers to epoxy phase show a dilute effect on PES concentration, furthermore, the domain growth of epoxy-rich along the filler percolation network sharpens the formation of epoxy matrix (or PES particle) structure [19].

TRLS results of PES-11.4 systems cured at 120°C showed quite similar slopes of 0.51, 0.48 and 0.55 for unfilled, nanoscale, and micron-size filled blends, respectively (Figure 7(a)). The addition of nano fillers slightly retards

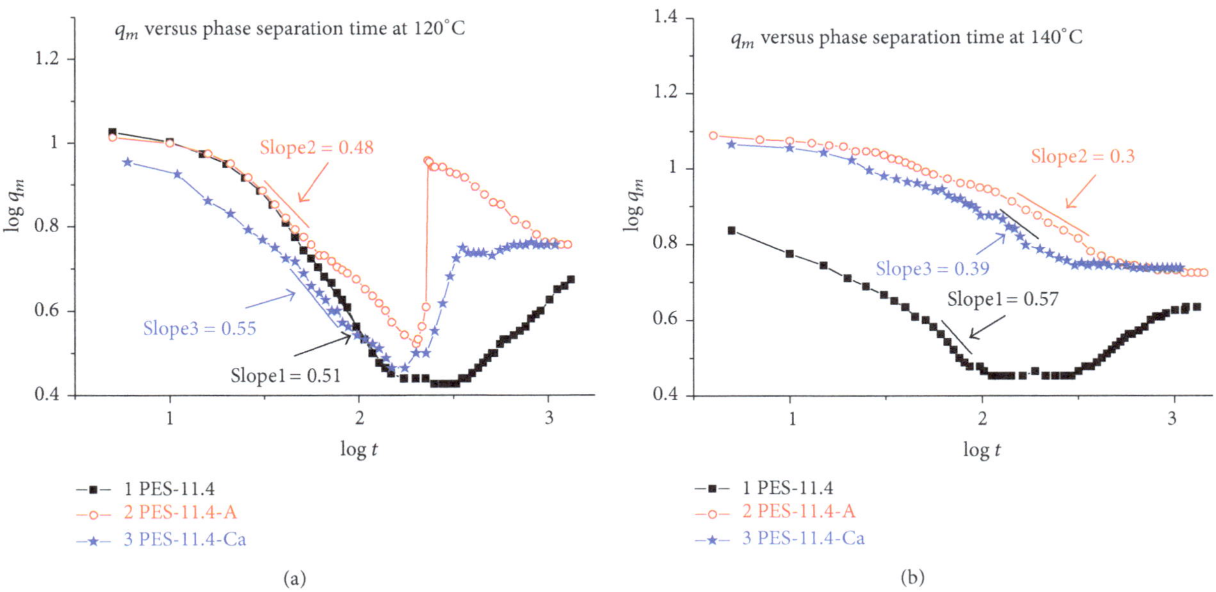

FIGURE 7: The changes of peak scattering vector, q_m, versus time of the blends of 11.4 wt% PES with and without fillers. (a) Cured at 120°C, (b) cured at 140°C.

while that of micron-sized fillers accelerates the phase separation a little due to the above-mentioned reasons. However, both the SEM and TRLS results show that the secondary phase separation disappeared at 120°C at the same time of morphology changes.

As to the effect of curing temperatures on the secondary phase separation, it is discussed before in our previous articles [34]. Hydrodynamic flow and diffusion are the main competition factors which affect the phase separation. If the size growth of the fast-dynamic-rich domain due to the hydrodynamic flow is too fast for the diffusion to keep up with the geometrical growth, secondary phase separation takes place; while no secondary phase separation occurs in the situation of the diffusion balanced geometrical growth systems.

The addition of mesoscale fillers shifts the PES-11.4 systems to normal SD process, and the formation of epoxy-rich matrix along the filler percolation network entraps the PES-rich into particle structure. At lower temperature, the diffusion rate of epoxy and disentangle time of PES chains are not high enough to balance the hydrodynamic flow and growth of epoxy-rich matrix along filler networks, as a result, secondary phase separation takes place among the PES-rich particles. On the contrary, high curing temperature decreases system viscosity and increases the chain mobility sharply, while the percolation network formed by mesoscale fillers changes little at the same time, in this case, the diffusion of epoxy and disentangle of PES chains could follow the geometrical growth of epoxy-rich matrix, therefore no secondary phase separation takes place (Figure 7(b)).

In summery, the TRLS experimental results in Table 2 are quite consistent with the SEM and OM results. Nanoscale fillers enhance the viscoelastic phase separation to some extent, while the surface affinity of fillers also plays a crucial role in the phase separation process. For micron-sized

TABLE 2: Summary of the TRLS experimental results.

Blend	Slope (at 120°C)	Secondary phase separation	Slope (at 140°C)	Secondary phase separation
PES-14.3	0.71	√	0.89	√
PES-14.3-A	0.41	√	0.50	√
PES-14.3-Ca	1.00	√	1.09	√
PES-11.4	0.51	√	0.57	√
PES-11.4-A	0.48	√	0.30	×
PES-11.4-Ca	0.55	√	0.39	×

√: secondary phase separation occurs, ×: does not occur.

fillers, the low surface area provides less entangle with the slow dynamic phase, and thus the surface affinity plays a more conspicuous role in phase separation. The preferential wetting of epoxy and its absorbing on the inorganic rods diminishes the viscoelastic effect and favours the growth of epoxy-rich particle size. The influence of mesoscale fillers on phase separation is a result of the competition between surface affinity and enhanced viscoelastic effect of the fillers. For low dynamic asymmetric systems, the effect of mesoscale fillers results mainly from its surface affinity. While high dynamic asymmetric systems would exhibit complicated phase separation behaviour due to the effects of both surface affinity and enhanced viscoelastic effect of the fillers.

4. Conclusions

The different effects of nanoscale (ATT) and microscale ($CaSO_4$) rod on the reaction-induced phase separation of Epoxy/PES blends originate from the surface affinity of fillers and chain entanglement of polymer chains with filler surfaces. The incorporation of mesoscale rod fillers does not change the spinodal decomposition mechanism of the blends with higher PES content than near-critical concentration, however, nanoscale fillers enhance the viscoelastic phase separation while micron-sized fillers also show dilute effect due to low surface area. For the blends with lower PES content than critical concentration of low dynamic asymmetry, the process of phase separation and final phase morphology are dramatically impacted by the incorporation of mesoscale rod fillers from their surface affinity to the epoxy-rich phase, especially when the filled blends are cured at higher temperature, which comes from the preferential wetting effect of inorganic rod percolation network on the epoxy-rich matrix.

Acknowledgment

This research work was supported by the National Nature Science Foundation of China (Grants 20674014 and 20974027).

References

[1] T. Inoue, "Reaction-induced phase decomposition in polymer blends," *Progress in Polymer Science*, vol. 20, pp. 119–153, 1995.

[2] P. A. Oyanguren, C. C. Riccardi, R. J. J. Williams, and I. Mondragon, "Phase separation induced by a chain polymerization: polysulfone-modified epoxy/anhydride systems," *Journal of Polymer Science B*, vol. 36, no. 8, pp. 1349–1359, 1998.

[3] A. Bonnet, J. P. Pascault, H. Sautereau, J. Rogozinski, and D. Kranbuehl, "Epoxy-diamine thermoset/thermoplastic blends: dielectric properties before, during, and after phase separation," *Macromolecules*, vol. 33, no. 10, pp. 3833–3843, 2000.

[4] K. Mimura, H. Ito, and H. Fujioka, "Improvement of thermal and mechanical properties by control of morphologies in PES-modified epoxy resins," *Polymer*, vol. 41, no. 12, pp. 4451–4459, 2000.

[5] B. Francis, G. Vanden Poel, F. Posada et al., "Cure kinetics and morphology of blends of epoxy resin with poly (ether ether ketone) containing pendant tertiary butyl groups," *Polymer*, vol. 44, no. 13, pp. 3687–3699, 2003.

[6] M. H. Wang, Y. F. Yu, X. G. Wu, and S. J. Li, "Polymerization induced phase separation in poly(ether imide)-modified epoxy resin cured with imidazole," *Polymer*, vol. 45, no. 4, pp. 1253–1259, 2004.

[7] I. A. Zucchi, M. J. Galante, and R. J. J. Williams, "Comparison of morphologies and mechanical properties of crosslinked epoxies modified by polystyrene and poly(methyl methacrylate)) or by the corresponding block copolymer polystyrene-b-poly(methyl methacrylate)," *Polymer*, vol. 46, no. 8, pp. 2603–2609, 2005.

[8] S. Maiez-Tribut, J. P. Pascault, E. R. Soulé, J. Borrajo, and R. J. J. Williams, "Nanostructured epoxies based on the self-assembly of block copolymers: a new miscible block that can be tailored to different epoxy formulations," *Macromolecules*, vol. 40, no. 4, pp. 1268–1273, 2007.

[9] Z. G. Xu and S. X. Zheng, "Reaction-induced microphase separation in epoxy thermosets containing poly(ε-caprolactone)-block-poly(n-butyl acrylate) diblock copolymer," *Macromolecules*, vol. 40, no. 7, pp. 2548–2558, 2007.

[10] M. Bakar, R. Duke, M. Przybylek, and M. Kostrzewa, "Mechanical and thermal properties of epoxy resin modified with polyurethane," *Journal of Reinforced Plastics and Composites*, vol. 28, no. 17, pp. 2107–2118, 2009.

[11] P. Jyotishkumar, C. Özdilek, P. Moldenaers et al., "Dynamics of phase separation in poly(acrylonitrile-butadiene-styrene)-modified epoxy/DDS system: kinetics and viscoelastic effects," *Journal of Physical Chemistry B*, vol. 114, no. 42, pp. 13271–13281, 2010.

[12] Y. Zhang, F. H. Chen, W. C. Shi, Y. R. Liang, and C. C. Han, "Layered structure formation in the reaction-induced phase separation of epoxy/polysulfone blends," *Polymer*, vol. 51, no. 25, pp. 6030–6036, 2010.

[13] J. Fröhlich, R. Thomann, and R. Mülhaupt, "Toughened epoxy hybrid nanocomposites containing both an organophilic layered silicate filler and a compatibilized liquid rubber," *Macromolecules*, vol. 36, no. 19, pp. 7205–7211, 2003.

[14] Y. Ni, S. X. Zheng, and K. M. Nie, "Morphology and thermal properties of inorganic-organic hybrids involving epoxy resin and polyhedral oligomeric silsesquioxanes," *Polymer*, vol. 45, no. 16, pp. 5557–5568, 2004.

[15] M. A. Osman, V. Mittal, M. Morbidelli, and U. W. Suter, "Epoxy-layered silicate nanocomposites and their gas permeation properties," *Macromolecules*, vol. 37, no. 19, pp. 7250–7257, 2004.

[16] S. Balakrishnan, P. R. Start, D. Raghavan, and S. D. Hudson, "The influence of clay and elastomer concentration on the morphology and fracture energy of preformed acrylic rubber dispersed clay filled epoxy nanocomposites," *Polymer*, vol. 46, no. 25, pp. 11255–11262, 2005.

[17] M. Peng, H. b. Li, L. J. Wu, Y. Chen, Q. Zheng, and W. F. Gu, "Organically modified layered-silicates facilitate the formation of interconnected structure in the reaction-induced phase separation of epoxy/thermoplastic hybrid nanocomposite," *Polymer*, vol. 46, no. 18, pp. 7612–7623, 2005.

[18] A. A. Asif, B. John, V. L. Rao, and K. N. Ninan, "Surface morphology, thermomechanical and barrier properties of poly(ether sulfone)-toughened epoxy clay ternary nanocomposites," *Polymer International*, vol. 59, no. 7, pp. 986–997, 2010.

[19] H. Tanaka, A. J. Lovinger, and D. D. Davis, "Pattern evolution caused by dynamic coupling between wetting and phase separation in binary liquid mixture containing glass particles," *Physical Review Letters*, vol. 72, no. 16, pp. 2581–2584, 1994.

[20] A. C. Balazs, V. V. Ginzburg, F. Qiu, G. W. Peng, and D. Jasnow, "Multi-scale model for binary mixtures containing nanoscopic particles," *Journal of Physical Chemistry B*, vol. 104, no. 15, pp. 3411–3422, 2000.

[21] V. V. Ginzburg, F. Qiu, M. Paniconi, G. W. Peng, D. Jasnow, and A. C. Balazs, "Simulation of hard particles in a phase-separating binary mixture," *Physical Review Letters*, vol. 82, no. 20, pp. 4026–4029, 1999.

[22] D. Suppa, O. Kuksenok, A. C. Balazs, and J. M. Yeomans, "Phase separation of a binary fluid in the presence of immobile particles: a lattice Boltzmann approach," *Journal of Chemical Physics*, vol. 116, no. 14, pp. 6305–6310, 2002.

[23] G. W. Peng, F. Qiu, V. V. Ginzburg, D. Jasnow, and A. C. Balazs, "Forming supramolecular networks from nanoscale rods in binary, phase-separating mixtures," *Science*, vol. 288, no. 5472, pp. 1802–1804, 2000.

[24] G. A. Buxton and A. C. Balazs, "Predicting the mechanical and electrical properties of nanocomposites formed from polymer blends and nanorods," *Molecular Simulation*, vol. 30, no. 4, pp. 249–257, 2004.

[25] S. W. Side, B. J. Kim, E. J. Kramer, and G. H. Fredrickson, "Hybrid particle-field simulations of polymer nanocomposites," *Physical Review Letters*, vol. 96, no. 25, Article ID 250601, 4 pages, 2006.

[26] E. R. Soulé, J. Borrajo, and R. J. J. Williams, "Thermodynamic analysis of a polymerization-induced phase separation in nanoparticle-monomer-polymer blends," *Macromolecules*, vol. 40, no. 22, pp. 8082–8086, 2007.

[27] V. V. Ginzburg, "Influence of nanoparticles on miscibility of polymer blends. A simple theory," *Macromolecules*, vol. 38, no. 6, pp. 2362–2367, 2005.

[28] Y. F. Yu, M. H. Wang, W. J. Gan, Q. S. Tao, and S. J. Li, "Polymerization-induced viscoelastic phase separation in polyethersulfone-modified epoxy systems," *The Journal of Physical Chemistry B*, vol. 108, no. 20, pp. 6208–6215, 2004.

[29] X. H. Zhong, Y. Liu, H. H. Su, G. Z. Zhan, Y. F. Yu, and W. J. Gan, "Enhanced viscoelastic effect of mesoscopic fillers in phase separation," *Soft Matter*, vol. 7, no. 7, pp. 3642–3650, 2011.

[30] Y. Liu, X. H. Zhong, G. Z. Zhan, Y. F. Yu, and J. Y. Jin, "Effect of mesoscopic fillers on the polymerization induced viscoelastic phase separation at near- and off-critical compositions," *The Journal of Physical Chemistry B*, vol. 116, no. 12, pp. 3671–3682, 2012.

[31] L. Zhao, G. Z. Zhan, Y. F. Yu, X. L. Tang, and S. J. Li, "Influence of attapulgites on cure-reaction-induced phase separation in epoxy/poly(ether sulfone) blends," *Journal of Applied Polymer Science*, vol. 108, pp. 953–959, 2008.

[32] K. Yamanaka and T. Inoue, "Structure development in epoxy resin modified with poly(ether sulphone)," *Polymer*, vol. 30, no. 4, pp. 662–667, 1989.

[33] H. Tanaka, "Viscoelastic phase separation," *Journal of Physics Condensed Matter*, vol. 12, no. 15, pp. R207–R264, 2000.

[34] X. L. Tang, L. X. Zhang, T. Wang, Y. F. Yu, W. J. Gan, and S. J. Li, "Hydrodynamic effect on secondary phase separation in an epoxy resin modified with polyethersulfone," *Macromolecular Rapid Communications*, vol. 25, pp. 1419–1424, 2004.

19

Are 100% Green Composites and Green Thermoplastics the New Materials for the Future?

Jean Marc Saiter,[1] Larisa Dobircau,[1] and Nathalie Leblanc[1,2]

[1] *AMME International Laboratory, LECAP EA4528, Institut des Matériaux, Faculté des Sciences, Université de Rouen, BP12, 76801 Saint Etienne du Rouvray, France*
[2] *LGMA, Esitpa, 76130 Mont Saint Aignan, France*

Correspondence should be addressed to Jean Marc Saiter, jean-marc.saiter@univ-rouen.fr

Academic Editor: Chantara Thevy Ratnam

A review of the history of the evolution of material science and material technology shows us that one tendency for the future could be the use of agriculture resources. In this work, we review the performances of one of these resources, that is, wheat flour. We show that it is possible to get thermoplastic films with properties quasiequivalent to what is obtained for expensive pure starch. By adding natural fibres, composites are also obtained. These composites exhibit performances which allow their use only for short duration.

1. Introduction

A historical analysis of the evolution and using of materials since the so-called industrial revolution, occurring two centuries ago, shows us that we have to develop new products with the obligation to take into account the end of life and the complete energetic cost associated with their fabrication, use, and afterlife. History tells us that the nineteenth century and the beginning of the twentieth century were the empire of iron and steel or more generally of metallic-based constructions. We have to remember that a train in North America was called iron horse and that the symbol in 1900 of the International Paris Fair was the Eiffel tower. The second important change in materials use occurs after the Second World War which was the use at large scale of plastic materials obtained from petroleum resources. The symbol of these new materials is the nylon used to replace silk in tights.

What will be the future? This is a very difficult question; nevertheless some points can be noted. The first one is linked to the fact that plastic materials have been so largely used that we have too much wastes all over the world. This problem is so drastic today that in many countries the populations want to banish the use of plastic devices like bags even if these plastic bags are often in polyethylene which is a nontoxic polymer. The second problem is in the real composition of the plastic devices, especially those obtained from thermosetting cured resins. As an example, we may report the composition obtained by means of high-energy X-ray fluorescence performed on an unsaturated polyester resin cured with styrene [1]. More than 12 elements of the periodic table have been found and among them strontium, barium, which may pollute environment even if the amounts are very small (of the order of some PPM).

These plastic materials are obtained from the transformation of petroleum. Even if today the main problem is more the distribution of petroleum rather than the stock, everybody knows that the amount of petroleum available will be smaller and smaller leading to two important consequences. The first is the price, which will increase because of its scarcity, and the price of plastic devices will increase as a result. The second problem will be the cost of transportation which will drastically increase, making it difficult to transport manufactured products.

One of the main consequences is the need to develop new materials coming from renewable resources. These new renewable resources can come from four categories: from classical chemistry, by combining biotechnology and synthesis, by using microorganisms, and from agroresources. Figure 1 gives a general scheme of these new possibilities.

In this work we will present a review of results of a series of materials obtained from agroresources. The raw material is obtained from wheat flour. As it is of prime importance that the raw material must not be obtained or produced in detriment of the food production, we propose to use a byproduct of the wheat flour production. The materials obtained are thermoplastic films and different composites prepared by incorporating different natural fibres in the thermoplastic films. These materials are made by an extrusion process.

2. Experimental Section

2.1. Raw Materials. The system we have chosen to study is obtained from a formulation based on a wheatflour byproduct, to which different components were added to obtain a film-forming material by an extrusion process. The initial composition of this system is derived from the work previously done at the Laboratory of Engineering Materials (LGMA) of the Engineering School in Agriculture (Esitpa) and has been the subject of a patent [3]. Wheat flour was supplied by the company Grands Moulins de Paris (France). After a division of dry cereal, flour obtained is separated into two categories. The first is rich in protein and will be used by the food industry. The second has low protein content, about 6-7% by mass. In addition to these proteins, it contains 85–88% starch, and the remaining 3–5% consists mainly of lipids and minerals. This second fraction is often reported as a byproduct of wheat flour production and is the subject of this work. In our formulation, wheat flour is incorporated up to 68.2% by mass. The additives used in this work are the following.

(a) Glycerol provided by the Univar distributor is an agent of both destructive and usually plasticizing properties used in the framework of the transformation and the implementation products containing starch. Its hydroxylated structure confers to this molecule a marked affinity for starch. The glycerol is hygroscopic and has lubricating properties. Its degree of purity is higher than 99%. It is incorporated in height of 12.8% in mass in the formulation of materials.

(b) Sorbitol provided by the company Roquette, of food quality, is incorporated to a total value of 7.2% in mass in our formulation. Sorbitol is also used as a plasticizer agent.

(c) Magnesium stearate provided by the company Riedel de Haën, of laboratory grade, is also called octadecanoic acid of magnesium salt. Its molecular formula is $Mg(C_{18}H_{35}O_2)_2$. It contains very little palmitate (ester of palmitic acid). It has the characteristic to break up in the acids and to be insoluble in water. It serves as lubricant. It is incorporated to a total value of 1.8% in mass in the basic formulation.

(d) Silica (Silicon dioxide) obtained by spraying silicon tetrachloride aerosol in a torch burner. Silica prepared in these conditions is not clustered and thus does not require grinding: it is the silica Aerosil 200 with an average diameter of 2 μm. It facilitates the passage of the formulation within the extruder. The silica comes from the Cabot Company and is referred under name M5 Cabosil. The rate of incorporation of silica in the formulation is of 1% in mass.

(e) Water is an important constituent which destroys native starch granules and plasticizes the formulation, making the passage of the formulation in the processing machine possible. Water, used is distilled water and it is incorporated in height of 9% in mass in the formulation of our materials.

2.2. Materials Preparation. Raw materials were mixed well before extrusion with the help of a turbo mixer (Kaiser, Germany). This turbo mixer functions with superimposed and shifted blades that turn on an axis located at the center of the tank. The number of revolutions used was 750 rpm, during 5 minutes. This mixer has also a system of dual envelope of cooling with water. Our objective was to obtain the final product as films. Experience showed that it was very difficult to extrude the raw materials to films directly from the mixture obtained from the turbo mixer. This problem was solved by a first extrusion allowing us to obtain granules and then in the second stage using these pellets for a second extrusion to get films with excellent reproducible properties [4]. The pellets were obtained by extrusion with a single-screw extrusion machine (Scamex), at a temperature of 120°C, with a screw speed of 60 rpm and using a granulator. For obtaining films, the second extrusion was carried out with the pellets. The same parameters of temperature and speed were used with a spinneret plate at the exit of the extruder. The principle of the die punt is to distribute the molten matter leaving the extrusion machine to make bands of them. The films obtained have 10 cm of width and 0.4 mm thickness.

2.3. Tensile Experiments. Mechanical tests have been performed with a device consisting in a multipurpose machine of mark "Instron 4301" making it possible to carry out tensile tests, compression, and inflection. The sample, which is in the dumbbell shape, is stretched along its main axis at a constant speed until its rupture. Two properties are recorded: the tensile load (F) applied to the specimen and the displacement (Δl) carried out by the pointer of the extensometer. From these elements, two characteristics of material are deduced.

Tensile stress (report of the tensile load (F) per the unit of area of the initial cross section of the sample (S_0)) is noted as σ, and unit is in MPa

$$\sigma = \frac{F}{S_0}. \tag{1}$$

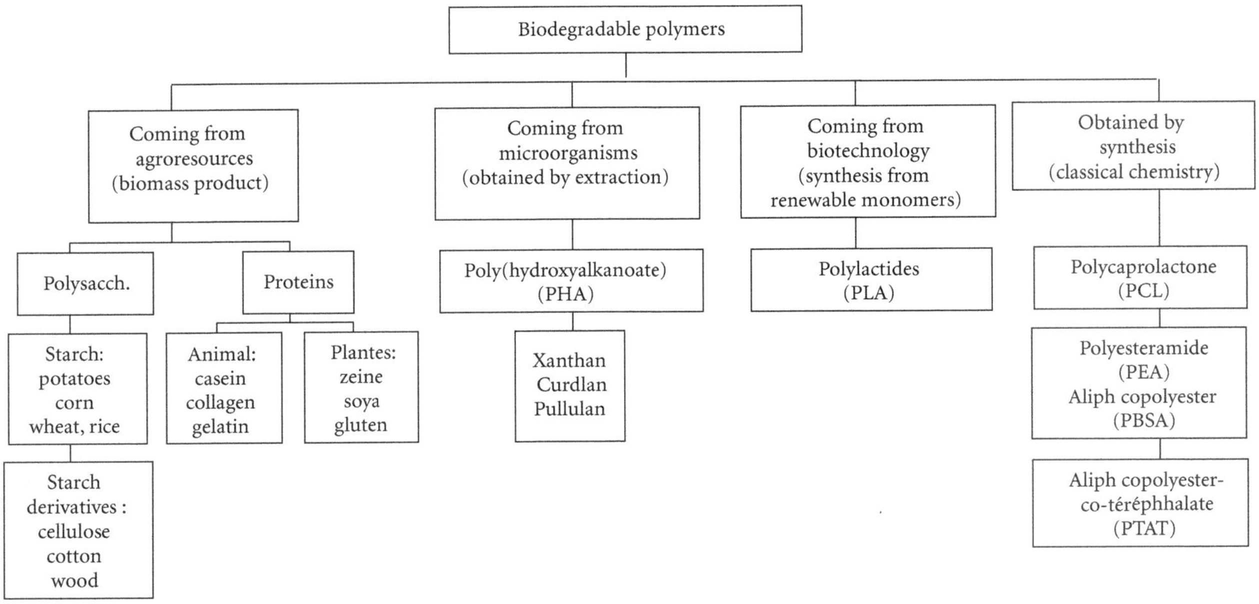

FIGURE 1: General scheme for the possible new renewable resources for developing new materials [2].

Elongation (or deformation) corresponds to the change in the length per the reference length of the sample (l_0). This strain is noted as ε

$$\varepsilon(\%) = \left(\frac{\Delta l}{l_0}\right) \times 100. \quad (2)$$

For each test, a stress-strain graph expresses the behaviour of material where three principal characteristics are found:

(i) breaking strength: tensile stress at the instant of the failure of the sample. It is noted σ_{rupture};

(ii) strain at failure: deformation corresponding to the value of tensile stress to the rupture. It is noted $\varepsilon_{\text{rupture}}$;

(iii) the tensile modulus of elasticity (Young modulus): value of the module provided by the tangent at the origin of stress-strain curve. It is noted E

$$E = \lim_{\varepsilon \to 0} \frac{\Delta\sigma}{\Delta\varepsilon}. \quad (3)$$

The test parameters are the following: Sensor: 1 kN, Length: $L_0 = 100$ mm, Width: $L = 10$ mm, and sSpeed: 2 mm$\cdot$min^{-1}. For our measurements, values of the various mechanical properties obtained for each type of material result from the average carried out on a test specimen of five samples. The uncertainty reported on our results corresponds to the values of the standard deviations obtained.

2.4. Thermogravimetric Analysis. Thermogravimetric studies were performed by thermogravimetric measurements using Netzsch TGA 209 balance (Germany). The measurements were done on 10–15 mg samples between 20 to 800°C at a heating rate of 10°C/min in nitrogen atmosphere at a rate of 20 mL$\cdot$min^{-1}. The calibrations of temperature and mass were made by the manufacturer.

2.5. Dynamical Mechanical Analysis. The dynamic mechanical behaviour of the polymeric material was studied by using TAQ-800 DMA instrument (Thermal Analysis USA). The experiments were performed under tensile mode. The testing temperature ranged from −100°C to 120°C, and the experiments were carried out at frequencies 1, 2, 5, 10, 20, 50, and 100 Hz at a heating rate of 3°C/min. Samples having a dimension of 0.8 ×20 × 0.4 mm were used for the present investigation.

2.6. Scanning Electron Microscopy. Scanning electron microscopy (SEM) (Jeol) was used to analyze the fracture surface morphology. Prior to the SEM imaging, the fracture surface of the sample broken was sputter-coated with about 10nm thick carbon film.

3. Results and Discussion

3.1. SEM. Figure 2 shows the electron microscope pictures, respectively, of wheat flour before extrusion (Figure 2(a2)) and of the film obtained after the second extrusion (Figure 2(b)). As expected the native granule morphology of wheat flour exhibits a roughness surface topology indicating the presence of impurities, but this image is not drastically different of what is expected for a native wheat starch granule shown also on Figure 2(a1). After the second extrusion the film exhibits the same characteristic of a film obtained from purified starch.

FIGURE 2: Electron microscopy performed on wheat starch (a1), wheat flour (a2), wheat flour-based film obtained after two extrusion process (b), and wheat flour-based film with the two phases (c): the small surface is the plasticizer-rich phase, the other part is the starch-rich phase.

From the same analysis, we have also observed the existence of two phases in the film (see Figure 2(c)). It has been demonstrated that these two phases correspond to one mainly composed of plasticizers (plasticizer-rich phase) while the second is mainly composed of starch (starch-rich phase).

3.2. DMA. The existence of these two phases is well identified by means of dynamic mechanical analysis performed at constant frequency and on a large temperature domain. Figure 3 shows the evolution of $\tan\delta$ with temperature and two transitions, one at $\approx -40°$C and a second at $\approx 50°$C, which are well identified for our standard composition. On the same figure we have also reported the effect of the glycerol content. The temperature of each transition is affected by the composition in glycerol. When the amount of glycerol increases, the temperature of maximum $\tan\delta$ decreases for both transitions. Nevertheless, it is clear that this is the starch-rich phase which is more sensitive to the glycerol content.

3.3. Thermogravimetric Analysis. As mentioned previously, to be able to get a film, we have to add plasticizers, and we know also that these polysaccharide-based materials are very sensitive to water molecules. Thermal stability of our materials was analysed by means of thermogravimetry measurements, and, as shown on Figure 4, it appears that a mass loss occurs for temperature close to 100°C.

We have performed many measurements, and this mass loss is always observed, even if pure starch material is used. This mass loss represents 10% of the initial sample mass. We have demonstrated [5] that this mass loss is the result of water vapour and glycerol evaporation. The second mass loss at a high temperature (300°C) is observed for all polysaccharide materials [6] and corresponds to the destruction of the polymeric chains.

3.4. Tensile Experiments. The mechanical performance has been analysed by means of mechanical testing in stretching mode. Figure 5 shows the results obtained with wheat flour-based films and purified starch-based films. On this figure we have superimposed the results of a series of 5 measurements for each material.

The first comment is linked to the excellent reproducibility of the data in spite of the fact that the raw material comes from natural resources. This good reproducibility is in fact obtained because two extrusions are performed. This leads to homogeneous mixtures, including all the initial impurities in the blends. From these measurements, for the wheat flour-based film we have estimated that the value of the strain at break is close to $\varepsilon_{max} = 16\%$ with a value of stress at break close to $\sigma_{max} = 3.2$ MPa. Finally we have estimated a value of modulus at zero strain, $E = 125$ MPa. These values are not drastically different from what is obtained for a purified starch-based film which gives $\varepsilon_{max} = 30\%$; $\sigma_{max} = 3.6$ MPa; $E = 100$ MPa. The wheat flour-based film

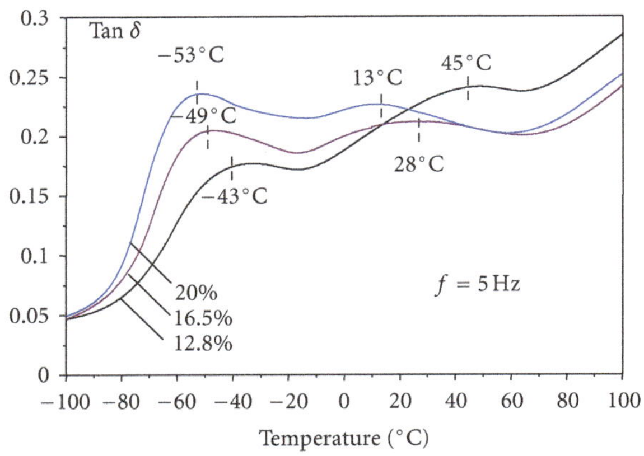

Figure 3: Dynamic mechanical analysis performed for three different contents of plasticizer. The glass transition of the plasticizer-rich phase occurs at low temperature $< -40°C$; the glass transition of the starch-rich phase occurs at high temperature $>20°C$.

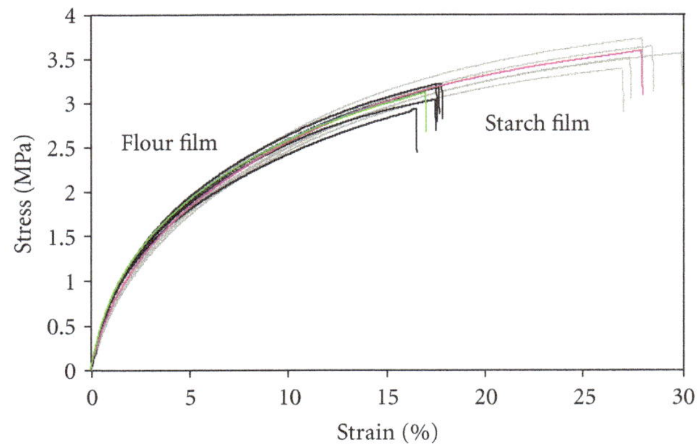

Figure 5: Mechanical test performed on wheat starch- and wheat flour-based films; 5 measurements for each sample are superimposed.

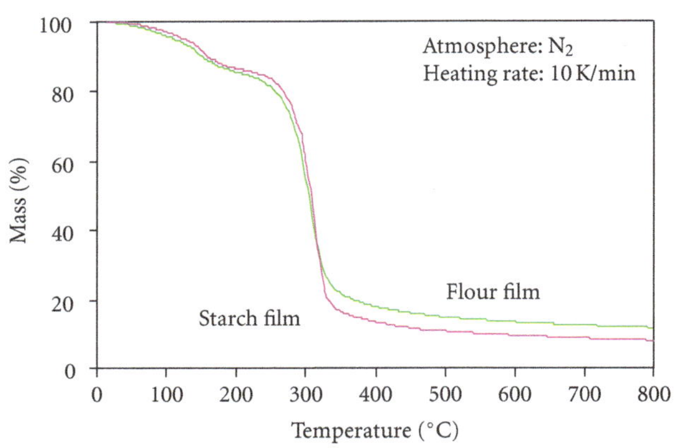

Figure 4: Thermogravimetric curve obtained on films prepared with purified wheat starch and with wheat flour.

mechanical fragility is directly connected to the existence of defects and impurities.

From the same kind of measurement we can analyse the effect of the plasticizer content (Figure 6(a)) and also the effects of other elements as silica used to have a better extrusion performance (Figure 6(b)).

The effect of plasticizer is found to be drastic on the mechanical performance. Indeed by changing the plasticizers composition from 12.8% Glycerol and 9% water to 21.8% glycerol and 0% water we observed a decrease of 16% for ε_{max}, the stress at break σ_{max} decreases by 25%, and the modulus at zero strain E decreases by 35%.

On the other hand, a comparison of the data obtained for films with and without silica (Figure 6(b)) shows that the introduction of silica decreases ε_{max} by 21%, but increases the stress at break σ_{max} by 130% while the modulus E is increased by 250%. If the effects of glycerol are well understood, the effects of silica are more difficult to explain. One possible explanation could be linked to the fact that silica is found mainly in the rich phase in glycerol, and its hygroscopic behaviour traps water molecules around it. So by this way silica plays an indirect role of antiplasticizer.

3.5. 100% Green Composites Analysis. When short fibres are added in the formulation before extrusion, a 100% green composite is obtained if the fibre is flax, cotton, sisal, bamboo, and so on. Up to today, flax and cotton fibres have been incorporated in the wheat flour matrix [7, 8]. Figure 7 shows two scanning electron microscope pictures obtained with a wheat flour-based film containing 20% of flax fibres. Figure 7(a) shows that the fibres are homogeneously dispersed in the matrix and oriented in the direction of the film extrusion. Figure 7(b) is a zoom which allows us to have a better analysis of the interface between the fibre and the matrix. Around the fibre, no cavity is observed and when the composite was broken the fibre and the matrix are together broken, no fibre without surrounding matrix or no hole occurring when the fibres is extracted, are observed.

In other words it appears that the interface between the fibre and the matrix is excellent, and we have a good compatibility between the fibre and the matrix. The same results have been obtained with cotton fibres [9, 10].

3.6. Tensile Experiments. When 20% of flax fibres are added, the mechanical performances of the composite are a decrease of 65% for ε_{max}, the stress at break σ_{max} increases by 178%, and the modulus for zero strain E increases by 270% to reach a value close to 500 MPa. As an example, the mechanical behaviour obtained with 100% green composites made of wheat flour and cotton is displayed in Figure 8. We observe exactly the same behaviour that was observed previously with flax fibre, but the magnitudes of the variations are smaller. Cotton is less efficient than flax.

From this data we may now compare these 100% green materials to other materials that are produced or used today. Table 1 shows the performances of flax fibres in comparison to inorganic fibres.

If the comparison of flax and glass is done, the flax appears to be the best, but if the comparison is performed

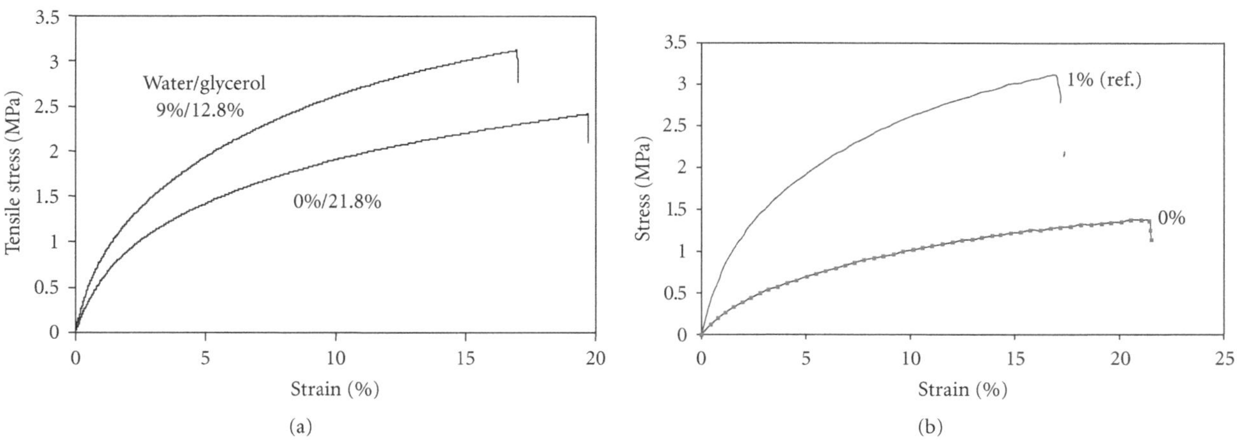

Figure 6: Stress-strain average curves for samples with two ratio water/glycerol contents (a) and two silica contents (b).

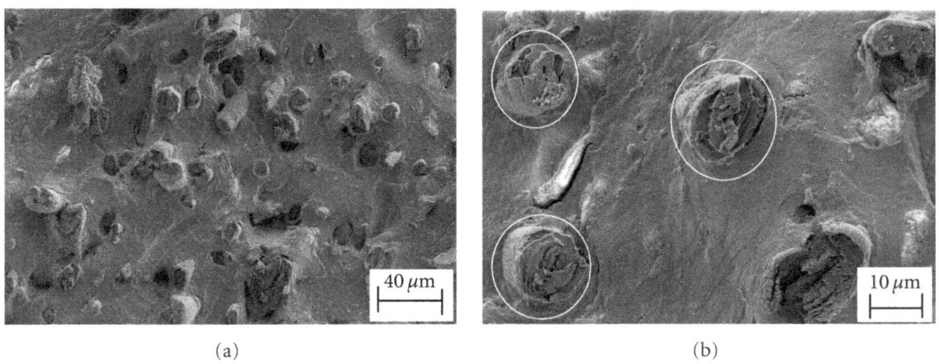

Figure 7: SEM micrographs of a 20% w/w flax fiber composite, at two different magnifications (a) ×300 and (b) ×1000.

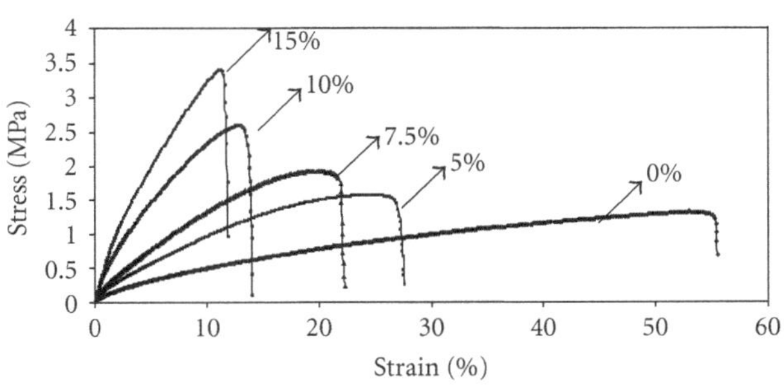

Figure 8: Stress-strain average curves for samples with various cotton contents (0; 5; 7.5; 10; 15% w/w).

with carbon, this is not the case. So, in many published papers, it is claimed that flax fibres are so good that they could be used instead of glass fibre. But that requires also to analyse carefully the interface of the fibre and the matrix, which is often not so good. This is why flax fibres need to be coated or submitted to a surface treatment. On the other hand two other major problems exist. One is the data dispersion of the flax fibre performances. The second is what happens with the 5 to 10% of water molecules which are very difficult to extract.

Table 1: Comparison between flax fibres and inorganic fibres performances.

Nature of fibre	Density	Specific modulus (MNm/kg)
Glass E	2.6	28–30
Carbon	1.7–2	230–600
SiC	2.5	70–80
Flax	1.3–1.5	25–85

Table 2: Average values of physical and mechanical performances of synthetic thermoplastic, thermosetting resin and our natural based thermoplastic.

Matrix	Density	Elastic modulus (Stretching) (MPa)
Synthetic thermoplastic	0.9–1.1	1000–2500
Thermosetting resin (polyester)	1.2	3000–5000
Natural based thermoplastic	1.2–1.4	100

For the matrix, we have reported in Table 2 the average values of the mechanical performances of thermoplastic, thermosetting resin, and our natural-based thermoplastic.

TABLE 3: Average physical and mechanical performances of different composites.

Composite	Density	Elastic modulus (Stretching) (MPa)
Synthetic thermoplastic + 30% glass fibres	1.1–1.6	82–220
Moulding high pressure		4000–6000
Thermo setting Resin (polyester) + 30% glass fibres	1.4–2	80–400
Natural based thermoplastic + 20% flax fibres	1.2–1.4	500

It is clear that wheat flour-based matrix does not exhibit the performance required to make a high-technological product. These results show us that such material must be used for short-term applications and will require drastic evolution and amelioration to be used for long-term applications. For these materials, the study of characteristics at the glass transition has been performed, and it was found that these wheat flour-based thermoplastics present a fragility index value [11] comparable to what it expected for a more conventional thermoplastic materials [12].

Table 3 shows the average performances of different composites. The performances of 100% green composites with regard to the mechanical behaviour under strength are not so different from what is obtained for other synthetic-based composites but remains in the order of low-technological applications.

4. Conclusion

As a conclusion, we may say in regards to the data presented in this paper that it is possible to prepare 100% green thermoplastic with raw materials coming from agricultural resources without being in competition with food production. It is also possible to make 100% green composites by using natural fibres or recycled fibres.

Nevertheless, up to today the problem of durability is not well controlled, and the mechanical performance is not good enough to imagine replacing synthetic thermoplastic or thermosetting-based composites.

New investigations have to be done to increase the water protection. This could be done by adding some natural cross-linkers, or by adding a surface protection using natural oils.

Acknowledgment

The authors would like to thank the “Grand Réseau de Recherche” VATA supported by the Region Haute Normandie, for giving financial support for this study. The authors would like also to thank all the PhD students of AMME-LECAP who participated in this program and who have made a big and interesting work.

References

[1] J. Grenet, S. Marais, M. T. Legras, P. Chevalier, and J. M. Saiter, “DSC and TSDC study of unsaturated polyester resin: influence of the promoter content,” *Journal of Thermal Analysis and Calorimetry*, vol. 61, no. 3, pp. 719–730, 2000.

[2] L. Avérous, “Biodegradable multiphase systems based on plasticized starch: a review,” *Journal of Macromolecular Science: Polymer Reviews*, vol. 44, no. 3, pp. 231–274, 2004.

[3] N. Leblanc and M. Dubois, “Matériaux biodégradables,” Patent FR 01 15451, 2001.

[4] R. Saiah, P. A. Sreekumar, N. Leblanc, M. Castandet, and J. M. Saiter, “Study of wheat-flour-based agropolymers: influence of plasticizers on structure and aging behavior,” *Cereal Chemistry*, vol. 84, no. 3, pp. 276–281, 2007.

[5] R. Saiah, P. A. Sreekumar, N. Leblanc, and J. M. Saiter, “Structure and thermal stability of thermoplastic films based on wheat flour modified by monoglyceride,” *Industrial Crops and Products*, vol. 29, no. 1, pp. 241–247, 2009.

[6] C. Devallencourt, J. M. Saiter, and D. Capitaine, “Characterization of recycled cellulose by dynamic and isothermal thermogravimetry investigations,” *Polymer Engineering and Science*, vol. 39, no. 3, pp. 413–421, 1999.

[7] P. A. Sreekumar, R. Saiah, J. M. Saiter et al., “Effect of chemical treatment on dynamic mechanical properties of sisal fiber-reinforced polyester composites fabricated by resin transfer molding,” *Composite Interfaces*, vol. 15, no. 2-3, pp. 263–279, 2008.

[8] P. A. Sreekumar, R. Saiah, J. M. Saiter et al., “Effect of chemical treatment on dynamic mechanical properties of sisal fiber-reinforced polyester composites fabricated by resin transfer molding,” *Composite Interfaces*, vol. 15, no. 2-3, pp. 263–279, 2008.

[9] R. Saiah, P. A. Sreekumar, P. Gopalakrishnan, N. Leblanc, R. Gattin, and J. M. Saiter, “Fabrication and characterization of 100% green composite: thermoplastic based on wheat flour reinforced by flax fibers,” *Polymer Composites*, vol. 30, no. 11, pp. 1595–1600, 2009.

[10] L. Dobircau, P. A. Sreekumar, R. Saiah et al., “Wheat flour thermoplastic matrix reinforced by waste cotton fibre: agro-green-composites,” *Composites Part A*, vol. 40, no. 4, pp. 329–334, 2009.

[11] J. M. Saiter, L. Dobircau, R. Saiah et al., “Relaxation map of a 100% green thermoplastic film. Glass transition and fragility,” *Physica B: Condensed Matter*, vol. 405, no. 3, pp. 900–905, 2010.

[12] A. Saiter, C. Devallencourt, J. M. Saiter, and J. Grenet, “Thermodynamically ”strong” and kinetically ”fragile” polymeric glass exemplified by melamine formaldehyde resins,” *European Polymer Journal*, vol. 37, no. 6, pp. 1083–1090, 2001.

20

Recognizing Amino Acid Chirality with Surface-Imprinted Polymers Prepared in W/O Emulsions

Min Jae Shin,[1] Young Jae Shin,[2] Seung Won Hwang,[3] and Jae Sup Shin[3]

[1] *Department of Chemical and Biomolecular Engineering, KAIST, Daejeon 305-701, Republic of Korea*
[2] *Macromolecular Science and Engineering, University of Michigan, Ann Arbor, MI 48109, USA*
[3] *Department of Chemistry, Chungbuk National University, Cheongju, Chungbuk 361-763, Republic of Korea*

Correspondence should be addressed to Jae Sup Shin; jsshin@chungbuk.ac.kr

Academic Editor: Jose Ramon Leiza

A molecularly imprinted polymer was prepared by a surface molecular imprinting technique in water-in-oil (W/O) emulsion. In this technique, the solid polymer, which is molecularly imprinted at the internal cavity surface, is prepared by polymerizing W/O emulsions consisting of a water-soluble imprinted molecule, a functional host molecule, an emulsion stabilizer, and a crosslinking agent. Dioleoyl phosphate was used as an emulsion stabilizer, and this compound also acted as a monomer and a host functional group in the imprinted cavity. Divinylbenzene was used as a crosslinker. Tryptophan methyl ester and phenylalanine methyl ester were used as the target template materials. These imprinted polymers exhibited enantiomeric selectivity in absorption experiments, and the maximum separation factor was 1.58. The enantiomeric selectivity with tryptophan methyl ester was higher than that with phenylalanine methyl ester.

1. Introduction

Recently, significant attention has been paid to the development of a molecular imprinting technique that enables polymers to mimic biological receptors. This technique is a very useful approach for the fabrication of a matrix with molecular recognition sites, which are formed by the addition of template molecules during the matrix formation process and the removal of the template molecule after the matrix formation [1–5]. Polymers that were prepared by the molecular imprinting technique have attracted much attention as interesting separation tools, especially for high performance liquid chromatography (HPLC). The imprinting technique is conceptually easy to apply to a wide variety of target molecules. Important applications are optical resolutions of amino acids or amino acid derivatives [6–12], direct enantiomeric separation of drugs [13, 14], and separation of sugar or sugar derivatives [4, 15, 16].

The "surface molecular imprinting technique" was proposed to overcome the inapplicability to water-soluble substances, which are important in the biological or biomedical field [17–20]. In this technique, the solid polymer, which is molecularly imprinted at the internal cavity surface, is prepared by polymerizing water-in-oil (W/O) emulsions consisting of a water-soluble imprint molecule, a functional host molecule, an emulsion stabilizer, and a crosslinking agent. The organic-aqueous interface in W/O emulsions is utilized as the recognition field for a target molecule. The target molecule forms a complex with the functional host molecule, while the orientation of the functional host molecule itself is fixed at the oil-water interface. After polymerization, preparation provides the recognition sites at the inner cavity surface of the imprinted bulk polymer. The complex between the functional host molecule and the target material should not be too hydrophobic or hydrophilic, because otherwise the complex would not be located at the oil-water interface. Thus, a functional host molecule should be amphiphilic, just like a surfactant molecule, in order to yield a high template effect for the target molecule. The crosslinking agent crosslinks the organic phase and stabilizes the water pool or the imprinted water cavity after polymerization. The bulk polymer that is obtained is ground to small particles in order to interact with the target molecules in the solution.

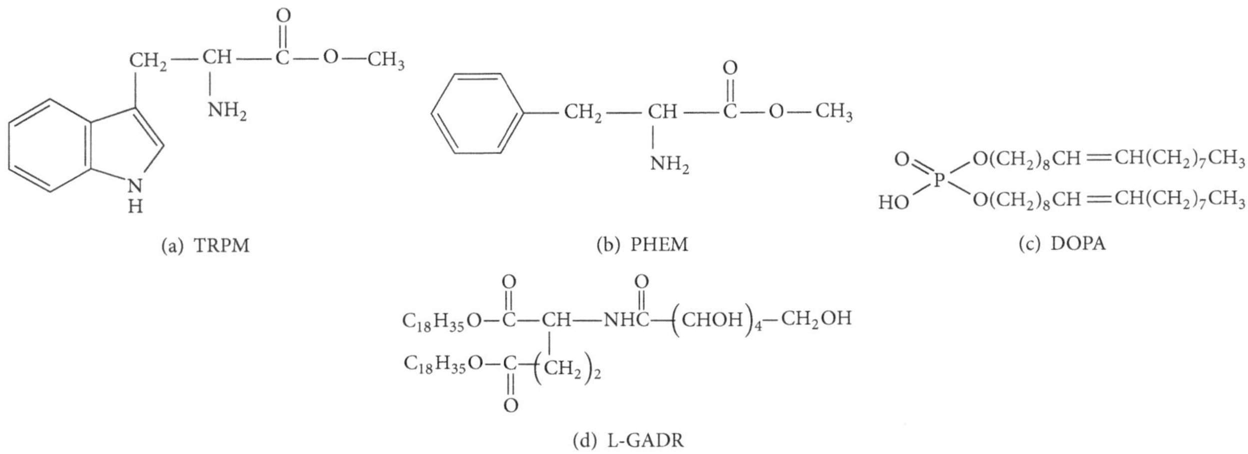

FIGURE 1: Molecular structures of (a) TRPM, (b) PHEM, (c) DOPA, and (d) L-GADR.

In this study, we tried to have a host molecule with a crosslinker participate in the polymerization. So dioleoyl phosphoric acid was selected as a host molecule, and the polymerization was conducted with a large amount of benzoyl peroxide at a high temperature to polymerize both the host molecule and divinylbenzene.

2. Experimental

2.1. Reagents and Instruments. L-Tryptophan methyl ester (L-TRPM), D-tryptophan methyl ester (D-TRPM), divinylbenzene (DVB), L-phenylalanine methyl ester (L-PHEM), D-phenylalanine methyl ester (D-PHEM), benzoyl peroxide (BPO), L-glutamic acid, oleyl alcohol, D-gluconic acid δ-lactone, phosphorus oxychloride, styrene, and xylenes were obtained from Aldrich. Dioleyl phosphoric acid (DOPA) and L-glutamic acid dioleylester ribitol (L-GADR) were synthesized using already reported methods [17, 21–23]. Divinylbenzene was used after treatment with silica gel to remove an inhibitor. Figure 1 shows the chemical structures of TRPM, PHEM, DOPA, and L-GADR.

Infrared spectroscopy (IR) was performed using an FT-IR 680 (Jasco International). Sonication was performed using a Cole-Parmer 4710250W sonicator. Scanning electron microscopy (SEM) was performed using both a Hitachi S-2500C and a Hitachi S-5200V scanning electron microscope.

2.2. Preparation of L- or D-TRPM-Imprinted Polymers. 40.0 g (0.307 mol) of DVB and 1.50 g (2.50 mmol) of DOPA were dissolved in 20 mL of xylenes. Another solution was prepared by dissolving 0.102 g (0.40 mmol) of L- or D-TRPM·HCl in 20 mL of water which contained 0.10 M phosphate buffer solution adjusted to pH 7.0. This buffer solution was added to the xylenes solution. The mixture was sonicated for 5 min to obtain the stable W/O emulsion. Then, 1.60 g (6.61 mmol) of BPO initiator was added to this solution, and then, this solution was sonicated again for 2 min. This mixture was polymerized at 60°C for 30 min, and then, temperature was gradually raised to 140°C, and the reaction was continued at 140°C for 24 h under a flow of nitrogen. The obtained bulk polymer was dried under a vacuum at 50°C for 48 h and ground into particles. The particles were washed with 0.50 M HCl to remove the imprinted L- or D-TRPM and then filtered off. This procedure was repeated several times until the imprinted molecule in the filtrate could not be detected by a UV spectrometer. Finally, the polymer was dried under a vacuum at 50°C for 48 h.

2.3. Preparation of L- or D-PHEM-Imprinted Polymers. These imprinted polymers were prepared by the same method as the preparation for the L- or D-TRPM-imprinted polymers using L- or D-PHEM instead of L- or D-TRPM.

2.4. Adsorption Experiments Using the Imprinted Polymers. The batchwise adsorption experiments of L,D-TRPM and L,D-PHEM were conducted for the L- or D-TRPM-imprinted polymers and L- or D-TRPM-imprinted polymers, respectively. 0.050 g of the imprinted polymer was added to 5.0 mL of aqueous solution containing 0.50 mM of L- or D-TRPM and placed in a sealed 10 mL tube. The pH was adjusted to a desired value between 3.0 and 8.0 with 0.10 M KH_2PO_4 or K_2HPO_4 and 0.10 M HNO_3 or 0.10 M NaOH. The mixture was shaken in a thermostated water bath at 30°C for 24 h. The polymers were then filtered off through a polyethylene membrane. The amount of each amino acid derivative adsorbed to the polymers was calculated from their residual amount in the filtrate. The concentration of amino acid derivatives was analyzed by an HPLC system. The adsorption tests were conducted at least 3 times, and the data was analyzed and compared with average values. The experimental errors were less than 6%.

2.5. Binding Constant of the Substrate for Imprinted Polymer. Binding constants of the substrate for the imprinted polymer were evaluated with the batchwise method. 0.050 g of the imprinted polymer sample was immersed in a sealed 10 mL tube. Then, a 5 mL of the aqueous solution was added, which was buffered with 0.10 M KH_2PO_4 or K_2HPO_4 and 0.10 M

HNO_3 or 0.10 M NaOH containing a substrate that was adjusted to a desired concentration of between 0.050 M and 1.0 M. The mixture was shaken at 30°C for 24 h. The polymers were then filtered off through the polyethylene membrane. The concentration of the substrate in the filtrate was analyzed by means of the HPLC system. The binding constants were calculated by a modified Scatchard equation [24, 25].

3. Results and Discussion

3.1. The Preparation of the Imprinted Polymer. The imprinted polymers were prepared by the surface molecular imprinting technique with W/O emulsions. L- or D-TRPM and L- or D-PHEM were used as the target imprinted molecules, and DOPA was used as both a functional monomer and an emulsion stabilizer. DVB was used as a crosslinker, xylenes were used as a diluent, and BPO was used as an initiator. In order to determine the polymerization conditions, we referred to the result of the radical polymerization for triglyceride oils. Many researchers attempted the radical polymerization of the triglyceride oils with styrene or DVB [26, 27]. Therefore, we used a larger amount of initiator compared to the general polymerization. BPO was used at 2.2 mol % versus the amount of DVB, and this amount is 2.6 times larger than the amount of DOPA. The polymerization was conducted at first at 60°C for 30 min to prevent bumping, and after that the temperature was raised to 140°C gradually, and the polymerization was continued for 24 h under a flow of nitrogen at 140°C. In order to conduct the experiment at 140°C, the xylenes (bp 137–140°C) were used as a diluent.

In order to estimate the exact amount of DOPA which participated in the polymerization, we conducted the preliminary experiment for the polymerization of DOPA with styrene or DVB. 31.9 g (0.307 mol) of styrene and 1.50 g (2.50 mmol) of DOPA were dissolved in 20 mL of xylenes, and the polymerization was conducted with 1.60 g (6.61 mmol) of BPO initiator at 140°C for 24 h under a flow of nitrogen. After finishing the polymerization, the remaining unpolymerized amount of DOPA was estimated using HPLC. The result showed that 56% of DOPA remained in the filtrate. This means that 44% of the DOPA was participated in the polymerization. We also tried another preliminary experiment with DVB. Instead of styrene, 40.0 g (0.307 mol) of DVB was used and the same experiment was conducted, and the remaining amount of DOPA was estimated using HPLC. The results showed that 9% of the DOPA remained in the filtrate. This means that 91% of DOPA stayed in the crosslinked polymer. We did not think that all 91% of the DOPA was polymerized with DVB. The IR spectrum of this polymer showed that some amount of double bonds still remained in the polymer. Because only 44% of the DOPA participated in the polymerization with styrene, we presumed that more than 44% of the DOPA participated in the polymerization with DVB. Therefore, we also presumed that some small part of the DOPAs were possible to anchor onto the highly crosslinked polymer.

After polymerization, the obtained polymer was dried under vacuum, and this polymer was grounded into small particles. After that, the imprinted molecules in the polymer were extracted with 0.50 M HCl solution. The obtained

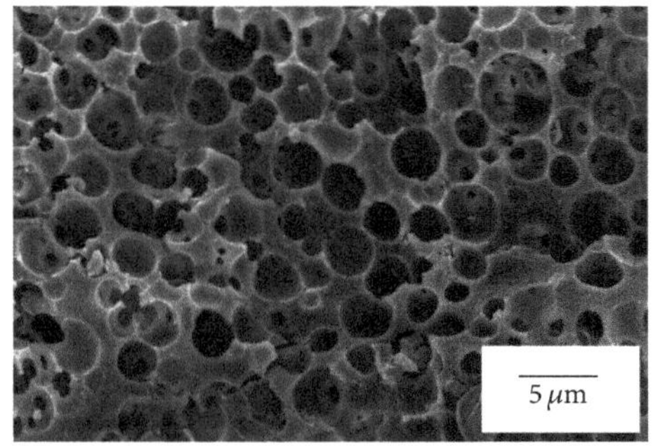

FIGURE 2: SEM photograph of L-TRPM-imprinted polymer.

polymer particles were dried under vacuum. The average size of the particles was about 50 um.

Figure 2 shows the SEM of the particle surface. Figure 2 shows that a lot of round shape rooms were formed by the W/O emulsions. At the surface of these round shape rooms, numerous imprinted cavities were positioned.

3.2. Adsorption Behavior of the Imprinted Polymers. The adsorption experiment was conducted using the L-TRPM-imprinted polymer. The enantiomeric selectivity of this imprinted polymer was estimated using L-TRPM and D-TRPM as substrates. We conducted the adsorption experiment at pH 3.0, pH 5.0, pH 7.0, and pH 8.0. We quantitatively characterized the template effect in the L- and D-imprinted polymers by evaluating the binding constants. The binding constant (K) can be evaluated on the basis of the slope and intercepted by the modified Scatchard plot. The binding constant becomes an indicator to express an adsorption affinity of recognition sites for the target amino acid derivative. To discuss the enantioselectivity quantitatively, we defined the separation factor as follows: $\alpha = K_L/K_D$ or $\alpha = K_D/K_L$.

Table 1 shows the results for the adsorption of L-TRPM and D-TRPM on the L-TRPM-imprinted polymer. The results showed that the binding constant increased as the pH increased. The separation factor was 1.02 at pH 3.0. This means that there was not much difference in the binding constant of the L-TRPM with that of the D-TRPM, but the separation factor was 1.58 at pH 7.0. This means that this imprinted polymer had a high enantiomeric selectivity at pH 7.0. At pH 8.0, the separation factor decreased relative to the data at pH 7.0. This result indicates that the interaction of the functional group in the imprinted cavity with the functional group of the substrate became maximized at pH 7.0.

Using the D-TRPM-imprinted polymer, almost similar results were obtained compared to the data from the L-TRPM-imprinted polymer. Table 2 shows the results for the adsorption of L-TRPM and D-TRPM on the D-TRPM-imprinted polymer. The separation factor was a little lower, but the difference value is within the error range.

The adsorption experiment was conducted using the L-PHEM-imprinted polymer. The enantiomeric selectivity of this imprinted polymer was estimated using L-PHEM and D-PHEM as the substrates. Table 3 shows the results for

Table 1: The adsorption of L-TRPM and D-TRPM on the L-TRPM-imprinted polymer.

pH	Substrate	Binding constant, K (M^{-1})	Separation factor, α
3.0	L-TRPM	1.29×10^2	1.02
	D-TRPM	1.27×10^2	
5.0	L-TRPM	1.02×10^3	1.12
	D-TRPM	0.91×10^3	
7.0	L-TRPM	3.05×10^3	1.58
	D-TRPM	1.93×10^3	
8.0	L-TRPM	3.48×10^3	1.39
	D-TRPM	2.50×10^3	

Table 2: The adsorption of L-TRPM and D-TRPM on the D-TRPM-imprinted polymer.

pH	Substrate	Binding constant, K (M^{-1})	Separation factor, α
3.0	L-TRPM	1.32×10^2	1.04
	D-TRPM	1.37×10^2	
5.0	L-TRPM	0.94×10^3	1.11
	D-TRPM	1.04×10^3	
7.0	L-TRPM	2.02×10^3	1.49
	D-TRPM	3.01×10^3	
8.0	L-TRPM	2.55×10^3	1.32
	D-TRPM	3.36×10^3	

Table 3: The adsorption of L-PHEM and D-PHEM on the L-PHEM-imprinted polymer.

pH	Substrate	Binding constant, K (M^{-1})	Separation factor, α
3.0	L-PHEM	1.48×10^2	1.04
	D-PHEM	1.42×10^2	
5.0	L-PHEM	1.72×10^3	1.06
	D-PHEM	1.62×10^3	
7.0	L-PHEM	3.45×10^3	1.35
	D-PHEM	2.56×10^3	
8.0	L-PHEM	4.17×10^3	1.18
	D-PHEM	3.53×10^3	

the adsorption of L-PHEM and D-PHEM on the L-PHEM-imprinted polymer, and Table 4 shows the results for the adsorption of L-PHEM and D-PHEM on the D-PHEM-imprinted polymer. The results showed that the trend of the data was almost similar with that of the TRPM-imprinted polymer, but the separation factor was much lower in value compared with that of the TRPM-imprinted polymer. The separation factor of the L-TRPM-imprinted polymer was 1.58 at pH 7.0 but the separation factor of the L-PHEM-imprinted polymer was 1.35. We think that the nitrogen in the tryptophan ring of TRPM participates in an important interaction between the functional group in the imprinted cavity and the functional group in the substrate that is needed for increasing enantiomeric selectivity.

3.3. Effect of the Chiral Emulsion Stabilizer. The Goto group introduced the surface molecular imprinting technique. They

Table 4: The adsorption of L-PHEM and D-PHEM on the D-PHEM-imprinted polymer.

pH	Substrate	Binding constant, K (M^{-1})	Separation factor, α
3.0	L-PHEM	1.39×10^2	1.04
	D-PHEM	1.45×10^2	
5.0	L-PHEM	1.52×10^3	1.08
	D-PHEM	1.64×10^3	
7.0	L-PHEM	2.64×10^3	1.33
	D-PHEM	3.51×10^3	
8.0	L-PHEM	3.50×10^3	1.16
	D-PHEM	4.06×10^3	

used L-GADR as an emulsion stabilizer [28]. So, we tried to use this chiral compound in our system. 0.207 g (0.251 mmol) of L-GADR was added during the preparation of the imprinted polymer with L-TRPM as a template molecule. The results showed that the separation factor was 1.60 at pH 7.0. We tried the same experiment 3 times. All of the results ((1) 1.63, (2) 1.58, and (3) 1.59) showed that the separation factor was more than 1.58. This means that L-GADR was effective in achieving enantiomeric selectivity, but compared to the result without L-GADR, the separation factor was bigger by a small amount. We also tried to conduct the experiment using 2 times the amount of L-GADR (0.414 g, 0.502 mmol). The results showed that the separation factor was average 1.61 ((1) 1.58, (2) 1.63, and (3) 1.62). This result value shows a little increase, but the value is within the experimental error. So, we concluded that this chiral compound is not very effective for achieving enantiomeric selectivity.

4. Conclusion

Amino acid-imprinted polymers were prepared by a surface imprinting technique in a W/O emulsion. DOPA was used as the emulsion stabilizer and the material for the host functional group in the imprinted cavity. DVB was used as a crosslinker. TRPM and PHEM were used as the imprinted materials and the substrate materials. The absorption experiments were conducted using these imprinted polymers with enantiomeric substrate materials. These polymers exhibited high enantiomeric selectivity for TRPM and PHEM. The separation factor of the TRPM-imprinted polymer was much higher than that of the PHEM-imprinted polymer.

Conflict of Interests

All of the authors do not have a direct financial relation with Aldrich, Jasco International, Cole-Parmer, and Hitachi.

Acknowledgment

This research was supported by the Basic Science Research Program through the National Research Foundation of Korea (NRF) funded by the Ministry of Education, Science and Technology (Grant no. 600819).

References

[1] R. Gupta and A. Kumar, "Molecular imprinting in sol-gel matrix," *Biotechnology Advances*, vol. 26, no. 6, pp. 533–547, 2008.

[2] A. G. Mayes and K. Mosbach, "Molecularly imprinted polymer beads: suspension polymerization using a liquid perfluorocarbon as the dispersing phase," *Analytical Chemistry*, vol. 68, no. 21, pp. 3769–3774, 1996.

[3] S. W. Lee, D. H. Yang, and T. Kunitake, "Regioselective imprinting of anthracenecarboxylic acids onto TiO_2 gel ultrathin films: an approach to thin film sensor," *Sensors and Actuators B*, vol. 104, no. 1, pp. 35–42, 2005.

[4] D. H. Yang, N. Takahara, S. W. Lee, and T. Kunitake, "Fabrication of glucose-sensitive TiO_2 ultrathin films by molecular imprinting and selective detection of monosaccharides," *Sensors and Actuators B*, vol. 130, no. 1, pp. 379–385, 2008.

[5] D. H. Yang, Y. R. Ham, M. H. Oh et al., "Simple method for the fabrication of 1-hydroxypyrene-imprinted TiO_2 gel nanofilms," *Current Applied Physics*, vol. 9, no. 2, pp. e136–e139, 2009.

[6] B. Sellergren, M. Lepistö, and K. Mosbach, "Highly enantioselective and substrate-selective polymers obtained by molecular imprinting utilizing noncovalent interactions. NMR and chromatographic studies on the nature of recognition," *Journal of the American Chemical Society*, vol. 110, no. 17, pp. 5853–5860, 1988.

[7] L. I. Andersson, D. J. O'Shannessy, and K. Mosbach, "Molecular recognition in synthetic polymers: preparation of chiral stationary phases by molecular imprinting of amino acid amides," *Journal of Chromatography*, vol. 513, pp. 167–179, 1990.

[8] K. J. Shea, D. A. Spivak, and B. Sellergren, "Polymer complements to nucleotide bases. Selective binding of adenine derivatives to imprinted polymers," *Journal of the American Chemical Society*, vol. 115, no. 8, pp. 3368–3369, 1993.

[9] O. Ramström, I. A. Nicholls, and K. Mosbach, "Synthetic peptide receptor mimics: highly stereoselective recognition in non-covalent molecularly imprinted polymers," *Tetrahedron Asymmetry*, vol. 5, no. 4, pp. 649–656, 1994.

[10] M. Kempe, "Antibody-mimicking polymers as chiral stationary phases in HPLC," *Analytical Chemistry*, vol. 68, no. 11, pp. 1948–1953, 1996.

[11] C. Yu and K. Mosbach, "Molecular imprinting utilizing an amide functional group for hydrogen bonding leading to highly efficient polymers," *Journal of Organic Chemistry*, vol. 62, no. 12, pp. 4057–4064, 1997.

[12] S. Vidyasankar, M. Ru, and F. H. Arnold, "Molecularly imprinted ligand-exchange adsorbents for the chiral separation of underivatized amino acids," *Journal of Chromatography A*, vol. 775, no. 1-2, pp. 51–63, 1997.

[13] L. Fischer, R. Müller, B. Ekberg, and K. Mosbach, "Direct enantioseparation of β-adrenergic blockers using a chiral stationary phase prepared by molecular imprinting," *Journal of the American Chemical Society*, vol. 113, no. 24, pp. 9358–9360, 1991.

[14] L. I. Andersson, "Application of molecular imprinting to the development of aqueous buffer and organic solvent based radioligand binding assays for (S)-propranolol," *Analytical Chemistry*, vol. 68, no. 1, pp. 111–117, 1996.

[15] G. Wulff and S. Schauhoff, "Racemic resolution of free sugars with macroporous polymers prepared by molecular imprinting. Selectivity dependence on the arrangement of functional groups versus spatial requirements," *Journal of Organic Chemistry*, vol. 56, no. 1, pp. 395–400, 1991.

[16] G. Wulff and J. Haarer, "Enzyme-analogue built polymers, 29. The preparation of defined chiral cavities for the racemic resolution of free sugars," *Die Makromolekulare Chemie*, vol. 192, no. 6, pp. 1329–1338, 1991.

[17] K. Uezu, H. Nakamura, J. I. Kanno, T. Sugo, M. Goto, and F. Nakashio, "Metal ion-imprinted polymer prepared by the combination of surface template polymerization with postirradiation by γ-rays," *Macromolecules*, vol. 30, no. 13, pp. 3888–3891, 1997.

[18] M. Yoshida, K. Uezu, F. Nakashio, and M. Goto, "Spacer effect of novel bifunctional organophosphorus monomers in metal-imprinted polymers prepared by surface template polymerization," *Journal of Polymer Science A*, vol. 36, no. 15, pp. 2727–2734, 1998.

[19] M. Yoshida, K. Uezu, M. Goto, and S. Furusaki, "Required properties for functional monomers to produce a metal template effect by a surface molecular imprinting technique," *Macromolecules*, vol. 32, no. 4, pp. 1237–1243, 1999.

[20] M. Yoshida, K. Uezu, M. Goto, and S. Furusaki, "Metal ion imprinted microsphere prepared by surface molecular imprinting technique using water-in-oil-in-water emulsions," *Journal of Applied Polymer Science*, vol. 73, no. 7, pp. 1223–1230, 1999.

[21] M. Goto, M. Matsumoto, K. Kondo, and F. Nakashio, "Development of new surfactant for loquid surfactnat membrane process," *Journal of Chemical Engineering of Japan*, vol. 20, no. 2, pp. 157–164, 1987.

[22] K. Uezu, H. Nakamura, M. Goto et al., "Novel metal ion-imprinted resins prepared by surface template polymerization with W/O emulsion," *Journal of Chemical Engineering of Japan*, vol. 27, no. 3, pp. 436–438, 1994.

[23] M. Goto, K. Kondo, and F. Nakashio, "Acceleration effect of anionic surfactants on extraction rate of copper with liquid surfactant membrane containing LIX65N and nonionic surfactant," *Journal of Chemical Engineering of Japan*, vol. 22, no. 1, pp. 79–84, 1989.

[24] B. Sellergren, "Molecular imprinting by noncovalent interactions. Enantioselectivity and binding capacity of polymers prepared under conditions favoring the formation of template complexes," *Die Makromolekulare Chemie*, vol. 190, no. 11, pp. 2703–2711, 1989.

[25] J. Mathew and O. Buchardt, "Molecular imprinting approach for the recognition of adenine in aqueous medium and hydrolysis of adenosine 5′-triphosphate," *Bioconjugate chemistry*, vol. 6, no. 5, pp. 524–528, 1995.

[26] F. S. Güner, Y. Yağci, and A. T. Erciyes, "Polymers from triglyceride oils," *Progress in Polymer Science*, vol. 31, no. 7, pp. 633–670, 2006.

[27] O. S. Kabasakal, "Styrenation of mahaleb and anchovy oils," *Progress in Organic Coatings*, vol. 53, no. 3, pp. 235–238, 2005.

[28] M. Yoshida, K. Uezu, M. Goto, and S. Furusaki, "Surface imprinted polymers recognizing amino acid chirality," *Journal of Applied Polymer Science*, vol. 78, no. 4, pp. 695–703, 2000.

Permissions

The contributors of this book come from diverse backgrounds, making this book a truly international effort. This book will bring forth new frontiers with its revolutionizing research information and detailed analysis of the nascent developments around the world.

We would like to thank all the contributing authors for lending their expertise to make the book truly unique. They have played a crucial role in the development of this book. Without their invaluable contributions this book wouldn't have been possible. They have made vital efforts to compile up to date information on the varied aspects of this subject to make this book a valuable addition to the collection of many professionals and students.

This book was conceptualized with the vision of imparting up-to-date information and advanced data in this field. To ensure the same, a matchless editorial board was set up. Every individual on the board went through rigorous rounds of assessment to prove their worth. After which they invested a large part of their time researching and compiling the most relevant data for our readers. Conferences and sessions were held from time to time between the editorial board and the contributing authors to present the data in the most comprehensible form. The editorial team has worked tirelessly to provide valuable and valid information to help people across the globe.

Every chapter published in this book has been scrutinized by our experts. Their significance has been extensively debated. The topics covered herein carry significant findings which will fuel the growth of the discipline. They may even be implemented as practical applications or may be referred to as a beginning point for another development. Chapters in this book were first published by Hindawi Publishing Corporation; hereby published with permission under the Creative Commons Attribution License or equivalent.

The editorial board has been involved in producing this book since its inception. They have spent rigorous hours researching and exploring the diverse topics which have resulted in the successful publishing of this book. They have passed on their knowledge of decades through this book. To expedite this challenging task, the publisher supported the team at every step. A small team of assistant editors was also appointed to further simplify the editing procedure and attain best results for the readers.

Our editorial team has been hand-picked from every corner of the world. Their multi-ethnicity adds dynamic inputs to the discussions which result in innovative outcomes. These outcomes are then further discussed with the researchers and contributors who give their valuable feedback and opinion regarding the same. The feedback is then collaborated with the researches and they are edited in a comprehensive manner to aid the understanding of the subject.

Apart from the editorial board, the designing team has also invested a significant amount of their time in understanding the subject and creating the most relevant covers. They scrutinized every image to scout for the most suitable representation of the subject and create an appropriate cover for the book.

The publishing team has been involved in this book since its early stages. They were actively engaged in every process, be it collecting the data, connecting with the contributors or procuring relevant information. The team has been an ardent support to the editorial, designing and production team. Their endless efforts to recruit the best for this project, has resulted in the accomplishment of this book. They are a veteran in the field of academics and their pool of knowledge is as vast as their experience in printing. Their expertise and guidance has proved useful at every step. Their uncompromising quality standards have made this book an exceptional effort. Their encouragement from time to time has been an inspiration for everyone.

The publisher and the editorial board hope that this book will prove to be a valuable piece of knowledge for researchers, students, practitioners and scholars across the globe.

List of Contributors

Van Cuong Nguyen and Van Boi Nguyen
Department of Chemical Engineering, Industrial University of Ho Chi Minh City, 12 Nguyen Van Bao Street, Go Vap, Ho Chi Minh City 70000, Vietnam

Ming-Fa Hsieh
Department of Biomedical Engineering, Center for Nanotechnology and Institute of Biomedical Technology, Chung Yuan Christian University, 200 Chung Pei Road, Chungli, 32023 Taoyuan, Taiwan

Anna B. Shipovskaya, Natalia O. Gegel and Sergei L. Shmakov
Institute of Chemistry, Saratov State University, 83 Ulitsa Astrakhanskaya, Saratov 410012, Russia

Sergei Yu. Shchyogolev
Institute of Biochemistry and Physiology of Plants and Microorganisms, 13 Prospect Entuziastov, Saratov 410049, Russia

C. Galán-Marín and C. Rivera-Gómez
Departamento Construcciones Arquitectonicas I, Escuela Tecnica Superior de Arquitectura, University of Seville, Avda. Reina Mercedes, 2, 41012 Seville, Spain

F. Bradley
Department of Architecture, Faculty of Engineering, University of Strathclyde, 16 Richmond Street, Glasgow G11XQ, UK

Perviz Ahmedzade
Transportation Division, Department of Civil Engineering, Faculty of Engineering, Ege University, Izmir, Turkey

Alexander Fainleib and Olga Starostenko
Institute of Macromolecular Chemistry of the National Academy of Sciences of Ukraine (NASU), Kyiv, Ukraine

Taylan Günay
Department of Civil Engineering, Ege University, Izmir, Turkey

Tetyana Kovalinska
Institute of Nuclear Research of the NASU, Kyiv, Ukraine

Gonzalo Martínez-Barrera, Enrique Vigueras Santiago and Susana Hernández López
Laboratorio de Investigacion y Desarrollo de Materiales Avanzados (LIDMA), Facultad de Quımica,
Universidad Autonoma del Estado de Mexico, Km. 12 de la Carretera Toluca-Atlacomulco, 50200 San Cayetano, Mexico

Osman Gencel
Civil Engineering Department, Faculty of Engineering, Bartin University, 74100 Bartin, Turkey

Fernando Ureña-Nuñez
Instituto Nacional de Investigaciones Nucleares, Carretera M´exico-Toluca S/N, 52750 La Marquesa Ocoyoacac, Mexico

Kazuhiro Nakabayashi and Hideharu Mori
Department of Polymer Science and Engineering, Department of Organic Device Engineering, Graduate School of Science and Engineering, Yamagata University, 4-3-16, Jonan, Yonezawa 992-8510, Japan

Keith A. Blackwood and Maria Ann Woodruff
Biomaterials and Tissue Morphology Group, Institute of Health and Biomedical Innovation, Queensland University of Technology, 60 Musk Avenue, Kelvin Grove, QLD 4059, Australia

Tim R. Dargaville
Tissue Repair and Regeneration Program, Institute of Health and Biomedical Innovation, Queensland University of Technology, 60 Musk Avenue, Kelvin Grove, QLD 4059, Australia

Nathalie Bock
Biomaterials and Tissue Morphology Group, Institute of Health and Biomedical Innovation, Queensland University of Technology, 60 Musk Avenue, Kelvin Grove, QLD 4059, Australia
Tissue Repair and Regeneration Program, Institute of Health and Biomedical Innovation, Queensland University of Technology, 60 Musk Avenue, Kelvin Grove, QLD 4059, Australia
Regenerative Medicine Group, Institute of Health and Biomedical Innovation, Queensland University of Technology, 60Musk Avenue, Kelvin Grove, QLD 4059, Australia

Dawei Zhang, Qing Zhang, Xin Gao and Guangzhe Piao
Key Laboratory of Rubber-Plastics of Ministry of Education, Qingdao 266042, China
Shandong Provincial Key Laboratory of Rubber-Plastics, Qingdao 266042, China
School of Polymer Science and Engineering, Qingdao University of Science and Technology, Qingdao 266042, China

Toru Uemukai, Tomoya Hioki and Manabu Ishifune
Department of Applied Chemistry, Faculty of Science and Engineering, Kinki University, 3-4-1 Kowakae, Higashi-Osaka, Osaka 577-8502, Japan

Isaac O. Igwe and Obumneme E. Ezeani
Department of Polymer and Textile Engineering, Federal University of Technology, Owerri, P.M.B 1526, Owerri, Nigeria

Erich K. Odermatt and Rainer Bargon
Aesculap AG, Am Aesculap-Platz, 78532 Tuttlingen, Germany

Lutz Funk
B. Braun Surgical S.A., Carretera de Terrassa 121, 08191 Rubi, Spain

David P.Martin, Said Rizk and Simon F. Williams
Tepha Inc., 99 Hayden Avenue, Lexington, MA 02421, USA

Raúl A. Morales-Luckie, Oscar Olea-Mejia, Alfredo R. Vilchis-Nestor, Gustavo López-Téllez and Víctor Varela-Guerrero
Centro Conjunto de Investigacion en Quımica Sustentable UAEM-UNAM (CCIQS), Facultad de Quımica, Universidad Autonoma del Estado de Mexico, Km. 14.5 de la carretera Toluca-Atlacomulco, 50200 San Cayetano, MEX, Mexico

Víctor Sánchez-Mendieta
Facultad de Quımica, Universidad Autonoma del Estado de Mexico, Paseo Colon y Paseo Tollocan, 50120 Toluca, MEX, Mexico

L. Huerta
Instituto de Investigaciones en Materiales, Universidad Nacional Autonoma de Mexico, Apartado Postal 70-360, 04510 Mexico, DF, Mexico

Jesús Arenas-Alatorre
Instituto de Fısica, Universidad Nacional Autonoma de Mexico, Apartado Postal 20-364, 01000 Mexico, DF, Mexico

Alexandra Mocanu, Edina Rusen and Aurel Diacon
Department of Polymer Science, University Politehnica of Bucharest, 149 Calea Victoriei, 010072 Bucharest, Romania

Gurucharan Chandrasekaran and Rebecca B. Dupaix
Mechanical and Aerospace Engineering Department, The Ohio State University, Columbus, OH 43210, USA

Adriana Gregorova
Institute for Chemistry and Technology of Materials, Graz University of Technology, Stremayrgasse 9, 8010 Graz, Austria

Michal Machovsky
Polymer Centre, Faculty of Technology, Tomas Bata University in Zlin, 76272 Zlin, Czech Republic

Rupert Wimmer
Institute for Wood Technology and Renewable Materials, University of Natural Resources and Life Sciences, Vienna, Konrad Lorenz-Strasse 24, 3430 Tulln, Austria

B. Ghorani, S. J. Russell and P. Goswami
Nonwovens Research Group, Centre for Technical Textiles, University of Leeds, Leeds LS2 9JT, UK

Xiaolin Tang, Xinhui Zhong, Guozhu Zhan, Yingfeng Yu and Hongdong Zhang
State Key Laboratory of Molecular Engineering of Polymers, Department of Macromolecular Science, Fudan University, Shanghai 200433, China

Jean Marc Saiter and Larisa Dobircau
AMME International Laboratory, LECAP EA4528, Institut des Materiaux, Faculte des Sciences, Universite de Rouen, BP12, 76801 Saint Etienne du Rouvray, France

Nathalie Leblanc
AMME International Laboratory, LECAP EA4528, Institut des Materiaux, Faculte des Sciences, Universite de Rouen, BP12, 76801 Saint Etienne du Rouvray, France
LGMA, Esitpa, 76130 Mont Saint Aignan, France

Min Jae Shin
Department of Chemical and Biomolecular Engineering, KAIST, Daejeon 305-701, Republic of Korea

Young Jae Shin
Macromolecular Science and Engineering, University of Michigan, Ann Arbor, MI 48109, USA

Seung Won Hwang and Jae Sup Shin
Department of Chemistry, Chungbuk National University, Cheongju, Chungbuk 361-763, Republic of Korea